D1758910

METHODS IN MOLECULAR BIOLOGY™

Series Editor
John M. Walker
School of Life Sciences
University of Hertfordshire
Hatfield, Hertfordshire, AL10 9AB, UK

For further volumes:
http://www.springer.com/series/7651

Plant Fungal Pathogens

Methods and Protocols

Edited by

Melvin D. Bolton

Northern Crops Science Laboratory, USDA-ARS, Fargo, ND, USA

Bart P.H.J. Thomma

Department of Phytopathology, Wageningen University, Wageningen, The Netherlands

 Humana Press

Editors
Melvin D. Bolton, Ph.D.
Northern Crops Science Laboratory
USDA–ARS
Fargo, ND, USA
melvin.bolton@ars.usda.gov

Bart P.H.J. Thomma, Ph.D.
Department of Phytopathology
Wageningen University
Wageningen, The Netherlands
bart.thomma@wur.nl

ISSN 1064-3745   e-ISSN 1940-6029
ISBN 978-1-61779-500-8   e-ISBN 978-1-61779-501-5
DOI 10.1007/978-1-61779-501-5
Springer New York Dordrecht Heidelberg London

Library of Congress Control Number: 2011943884

Printed on acid-free paper

Humana Press is part of Springer Science+Business Media (www.springer.com)

Preface

Members of the Fungal kingdom are ubiquitous in nature. Over the course of evolution, fungi have adapted to occupy specific niches, from symbiotically inhabiting the flora of the intestinal tract of mammals to saprophytic growth on leaf litter resting on the forest floor. Modern agricultural cropping systems have offered fungal plant pathogens vast amounts of substrate for colonization, resulting in disease development. Although the long-term goal for plant pathologists and geneticists is to breed for genetic resistance to combat fungal pathogens, a glimpse at the history of plant breeding has shown that breeders play an ongoing chess game against plant pathogens without a clear winner so far.

This book brings together over 40 chapters that contribute toward our understanding of the biology and pathology of fungal plant pathogens. Drawing on techniques utilizing model systems such as *Arabidopsis thaliana* as well as agricultural crop plants, this book highlights some of the latest techniques critical for students pursuing molecular plant pathology. However, most protocols can be adapted to any pathosystem or fungal pathogen. Since pathogenicity and disease development may depend on effector proteins secreted by plant pathogens during infection, several chapters are dedicated to various means of identifying effectors. In addition, protocols on various aspects of fungal biology are included, including those utilizing molecular biology, proteomics, and metabolomics. Given the recent deluge of fungal genome sequencing projects, chapters on genome annotation and next-generation sequencing of fungal genomes are included. Finally, protocols for gene knock-outs, fungal transformation, and molecular tools for disease and/or pathogen quantification are included that are critical for revealing the role for a fungal gene of interest in disease development.

Taken together, the protocols and review chapters in this book are timely and provide a current set of techniques that cover a wide-range of methods to study molecular aspects of pathogenesis. We hope that this book will be useful for those new to the field of molecular plant pathology as well as experienced fungal research laboratories.

Fargo, ND, USA   *Melvin D. Bolton, Ph.D.*
Wageningen, The Netherlands   *Bart P.H.J. Thomma, Ph.D.*

Contents

Contributors

AHMED ABD-EL-HALIEM • *Laboratory of Phytopathology, Wageningen University, Wageningen, The Netherlands*

MASOUD ABRINBANA • *Department of Plant Protection, University of Urmia, Urmia, Iran*

ELIZABETH ADAMS • *Delaware Biotechnology Institute, University of Delaware, Newark, DE, USA; Department of Biological Sciences, University of Delaware, Newark, DE, USA*

MOSTAFA AGHAEE • *Seed and Plant Improvement Institute, Karaj, Iran*

MARK ARENTSHORST • *Department Molecular Microbiology and Biotechnology, Institute of Biology Leiden, Leiden University, Leiden, The Netherlands*

ZAHI K. ATALLAH • *Department of Plant Pathology, University of California, Davis, CA, USA*

MELVIN D. BOLTON • *Northern Crop Science Laboratory, USDA-ARS, Fargo, ND, USA*

B.H. BLUHM • *Division of Agriculture, Department of Plant Pathology, University of Arkansas, Fayetteville, AR, USA*

ROEL A.L. BOVENBERG • *DSM Biotechnology Center, DSM Food Specialties B.V., Delft, The Netherlands; Centre for Synthetic Biology, University of Groningen, Groningen, The Netherlands*

GAIL CANNING • *Department of Plant Pathology and Microbiology, Centre for Sustainable Pest and Disease Management, Rothamsted Research, Harpenden, UK*

LIAM CASSIDY • *Research School of Biology, The Australian National University, Canberra, ACT, Australia*

SEAN CROKER • *West Chester University, West Chester, PA, USA*

KIRK CZYMMEK • *Delaware Biotechnology Institute, University of Delaware, Newark, DE, USA; Department of Biological Sciences, University of Delaware, Newark, DE, USA*

JAVIER A. DELGADO • *Department of Plant Pathology, North Dakota State University, Fargo, ND, USA*

ARNOLD J.M. DRIESSEN • *Department of Microbiology, Groningen Biomolecular Sciences and Biotechnology Institute and Zernike Institute for Advanced Materials, University of Groningen, Groningen, The Netherlands*

JULIEN Y. DUTHEIL • *Institut des Sciences de l'Evolution, Université de Montpellier II, Montpellier, France*

JON DUVICK • *Department of Genetics, Development and Cell Biology, Iowa State University, Ames, IA, USA*

RONNIE DE JONGE • *Laboratory of Phytopathology, Wageningen University, Wageningen, The Netherlands*

PIERRE J.G.M. DE WIT • *Laboratory of Phytopathology, Wageningen University, Wageningen, The Netherlands*

QUINN A. EGGERTSON • *Department of Biology, Carleton University, Ottawa, ON, Canada*

DANIELLE EMERSON • *Delaware Biotechnology Institute, University of Delaware, Newark, DE, USA*

LAUREN DU FALL • *Research School of Biology, The Australian National University, Canberra, ACT, Australia*

JUSTIN D. FARIS • *Cereal Crops Research Unit, Northern Crop Science Laboratory, USDA-ARS, Fargo, ND, USA*

METTE FRANDSEN • *Bioneer A/S, Hørsholm, Denmark*

RASMUS JOHN NORMAND FRANDSEN • *Department of Systems Biology, Center for Microbial Biotechnology, Technical University of Denmark, Lyngby, Denmark*

TIMOTHY L. FRIESEN • *Cereal Crops Research Unit, Northern Crop Science Laboratory, USDA-ARS, Fargo, ND, USA; Department of Plant Pathology, North Dakota State University, Fargo, ND, USA*

FLEUR GAWEHNS • *Plant Pathology, Swammerdam Institute for Life Sciences, University of Amsterdam, Amsterdam, The Netherlands*

HENRIETTE GIESE • *Department of Biotechnology, Chemistry and Environmental Engineering, Aalborg University, Ålborg, Denmark*

SCOTT E. GOLD • *Plant Pathology, University of Georgia, Athens, GA, USA*

RUBELLA S. GOSWAMI • *Department of Plant Pathology, North Dakota State University, Fargo, ND, USA*

CHRISTIAN GU • *The Plant Chemetics Laboratory, Chemical Genomics Centre, Max Planck Institute for Plant Breeding Research, Cologne, Germany*

JOEL P.A. GUMMER • *Australian Centre for Necrotrophic Fungal Pathogens, SABC, Murdoch University, Murdoch, Australia and Separation Science Laboratory, Murdoch University, Murdoch, ACT, Australia*

KIM E. HAMMOND-KOSACK • *Department of Plant Pathology and Microbiology, Centre for Sustainable Pest and Disease Management, Rothamsted Research, Harpenden, UK*

ADRIENNE R. HARDHAM • *Plant Science Division, Research School of Biology, College of Medicine, Biology and Environment, The Australian National University, Canberra, ACT, Australia*

S.D. HARRIS • *Department of Plant Pathology and Center for Plant Science Innovation, University of Nebraska, Lincoln, NE, USA*

R.L. HIRSCH • *Division of Agriculture, Department of Plant Pathology, University of Arkansas, Fayetteville, AR, USA*

RAYS H.Y. JIANG • *The Broad Institute of Massachusetts Institute of Technology and Harvard, Cambridge, MA, USA*

MATTHIEU H.A.J. JOOSTEN • *Laboratory of Phytopathology, Wageningen University, Wageningen, The Netherlands; Centre for BioSystems Genomics, Wageningen, The Netherlands*

MÉLANIE JUBAULT • *Department of Plant Pathology and Microbiology, Centre for Sustainable Pest and Disease Management, Rothamsted Research, Harpenden, UK*

A. Justé • *Laboratory for Process Microbial Ecology and Bioinspirational Management, Consortium for Industrial Microbiology and Biotechnology, Department of Microbial and Molecular Systems, K.U. Leuven Association, Lessius Mechelen, Campus De Nayer, Sint-Katelijne-Waver, Belgium; Scientia Terrae Research Institute, Sint-Katelijne-Waver, Belgium*

Shiv D. Kale • *Virginia Bioinformatics Institute, Virginia Polytechnic Institute and State University, Blacksburg, VA, USA*

Seogchan Kang • *Department of Plant Pathology, The Pennsylvania State University, University Park, PA, USA*

Farnusch Kaschani • *The Plant Chemetics Laboratory, Chemical Genomics Centre, Max Planck Institute for Plant Breeding Research, Cologne, Germany*

Gert H.J. Kema • *Plant Research International, Wageningen, The Netherlands*

Hye-seon Kim • *Department of Plant Pathology, The Pennsylvania State University, University Park, PA, USA*

Steven J. Klosterman • *USDA-ARS, Salinas, CA, USA*

Anja Kombrink • *Laboratory of Phytopathology, Wageningen University, Wageningen, The Netherlands*

Andriy Kovalchuk • *Faculty of Agriculture and Forestry, Department of Forest Sciences, Forest Pathology, University of Helsinki, Helsinki, Finland; Department of Microbiology, Groningen Biomolecular Sciences and Biotechnology Institute and Zernike Institute for Advanced Materials, University of Groningen, Groningen, The Netherlands*

Christian Krill • *Australian Centre for Necrotrophic Fungal Pathogens, SABC, Murdoch University, Murdoch, ACT, Australia; Separation Science and Metabolomics Laboratory, Murdoch University, Murdoch, ACT, Australia*

Sang-Jik Lee • *Department of Plant Biology, Cornell University, Ithaca, NY, USA*

Yueqiang Leng • *Department of Plant Pathology, North Dakota State University, Fargo, ND, USA*

C. André Lévesque • *Biodiversity (Mycology), Central Experimental Farm, Agriculture and Agri-Food Canada, Ottawa, ON, Canada; Department of Biology, Carleton University, Ottawa, ON, Canada*

B. Lievens • *Laboratory for Process Microbial Ecology and Bioinspirational Management, Consortium for Industrial Microbiology and Biotechnology, Department of Microbial and Molecular Systems, K.U. Leuven Association, Lessius Mechelen, Campus De Nayer, Sint-Katelijne-Waver, Belgium; Scientia Terrae Research Institute, Sint-Katelijne-Waver, Belgium*

Zhaohui Liu • *Department of Plant Pathology, North Dakota State University, Fargo, ND, USA*

Rohan Lowe • *Department of Plant Pathology and Microbiology, Centre for Sustainable Pest and Disease Management, Rothamsted Research, Harpenden, UK*

Shunwen Lu • *USDA-ARS, Cereal Crops Research Unit, Northern Crop Science Laboratory, Fargo, ND, USA*

Ewa Lukasik • *Plant Pathology, Swammerdam Institute for Life Sciences, University of Amsterdam, Amsterdam, The Netherlands*

Sarrah Ben M'Barek • *Plant Research International, Wageningen, The Netherlands*

LISONG MA • *Plant Pathology, Swammerdam Institute for Life Sciences, University of Amsterdam, Amsterdam, The Netherlands*

SAMUEL G. MARKELL • *Department of Plant Pathology, North Dakota State University, Fargo, ND, USA*

RAHIM MEHRABI • *Seed and Plant Improvement Institute, Karaj, Iran; Laboratory of Phytopathology, Wageningen University, Wageningen, The Netherlands*

STEVEN MEINHARDT • *Department of Plant Pathology, North Dakota State University, Fargo, ND, USA*

VERA MEYER • *Department Molecular Microbiology and Biotechnology, Institute of Biology Leiden, Leiden University, Leiden, The Netherlands; Kluyver Centre for Genomics of Industrial Fermentation, Delft, The Netherlands; Department Applied and Molecular Microbiology, Institute of Biotechnology, Berlin University of Technology, Berlin, Germany*

SHANNON MODLA • *Delaware Biotechnology Institute, University of Delaware, Newark, DE, USA*

JAYMA A. MOORE • *Electron Microscopy Center, North Dakota State University, Fargo, ND, USA*

MARINA NADAL • *Department of Plant Molecular Biology, University of Lausanne, Lausanne, Switzerland*

JEROEN G. NIJLAND • *Department of Microbiology, Groningen Biomolecular Sciences and Biotechnology, Institute and Zernike Institute for Advanced Materials, University of Groningen, Groningen, The Netherlands*

BENJAMIN M. NITSCHE • *Department Molecular Microbiology and Biotechnology, Institute of Biology Leiden, Leiden University, Leiden, The Netherlands*

RICHARD P. OLIVER • *Department of Environment and Agriculture, Australian Centre for Necrotrophic Fungal Pathogens, Curtin University, Perth, ACT, Australia*

SCOTT A. PAYNE • *Electron Microscopy Center, North Dakota State University, Fargo, ND, USA*

ARTHUR F.J. RAM • *Department Molecular Microbiology and Biotechnology, Institute of Biology Leiden, Leiden University, Leiden, The Netherlands; Kluyver Centre for Genomics of Industrial Fermentation, Delft, The Netherlands*

MARTIJN REP • *Plant Pathology, Swammerdam Institute for Life Sciences, University of Amsterdam, Amsterdam, The Netherlands*

J.B. RIDENOUR • *Division of Agriculture, Department of Plant Pathology, University of Arkansas, Fayetteville, AR, USA*

TARA L. RINTOUL • *Biodiversity (Mycology), Central Experimental Farm, Agriculture and Agri-Food Canada, Ottawa, ON, Canada*

W.R. RITTENOUR • *Department of Plant Pathology and Center for Plant Science Innovation, University of Nebraska, Lincoln, NE, USA*

VIVIANA V. RIVERA • *Department of Plant Pathology, North Dakota State University, Fargo, ND, USA*

JOCELYN K.C. ROSE • *Department of Plant Biology, Cornell University, Ithaca, NY, USA*

ABBAS SAIDI • *Department of Biotechnology, College of New Technologies and Energy Engineering, Shahid Beheshti University GC, Tehran, Iran; Energy Engineering, Shahid Beheshti University GC, Tehran, Iran*

PARTHASARATHY SANTHANAM • *Laboratory of Phytopathology, Wageningen University, Wageningen, The Netherlands*

GARY A. SECOR • *Department of Plant Pathology, North Dakota State University, Fargo, ND, USA*

PETER S. SOLOMON • *Research School of Biology, The Australian National University, Canberra, ACT, Australia*

EVA H. STUKENBROCK • *Max Planck Institute for Terrestrial Microbiology, Marburg, Germany*

KRISHNA V. SUBBARAO • *Department of Plant Pathology, University of California, Davis, CA, USA*

MASATOKI TAGA • *Department of Biology, Okayama University, Tsushima-naka, Okayama, Japan*

FRANK L.W. TAKKEN • *Plant Pathology, Swammerdam Institute for Life Sciences, University of Amsterdam, Amsterdam, The Netherlands*

WLADIMIR I.L. TAMELING • *Laboratory of Phytopathology, Wageningen University, Wageningen, The Netherlands*

KAR-CHUN TAN • *Australian Centre for Necrotrophic Fungal Pathogens, Environment and Agriculture, Curtin University, Perth, WA, Australia*

WEI-HUA TANG • *National Key Laboratory of Plant Molecular Genetics, Institute of Plant Physiology and Ecology, Shanghai Institutes for Biological Sciences, Chinese Academy of Sciences, Shanghai, China*

ROBERT D. TRENGOVE • *Separation Science and Metabolomics Laboratory, Murdoch University, Murdoch, WA, Australia*

SUCHETA TRIPATHY • *Virginia Bioinformatics Institute, Virginia Polytechnic Institute and State University, Blacksburg, VA, USA*

BRETT M. TYLER • *Virginia Bioinformatics Institute, Virginia Polytechnic Institute and State University, Blacksburg, VA, USA*

MARTIN URBAN • *Department of Plant Pathology and Microbiology, Centre for Sustainable Pest and Disease Management, Rothamsted Research, Harpenden, UK*

DELPHINE VINCENT • *Research School of Biology, The Australian National University, Canberra, ACT, Australia*

H. CHARLOTTE VAN DER DOES • *Plant Pathology, Swammerdam Institute for Life Sciences, University of Amsterdam, Amsterdam, The Netherlands*

RENIER A.L. VAN DER HOORN • *The Plant Chemetics Laboratory, Chemical Genomics Centre, Max Planck Institute for Plant Breeding Research, Cologne, Germany*

H. PETER VAN ESSE • *Laboratory of Phytopathology, Wageningen University, Wageningen, the Netherlands*

ORMONDE D.C. WATERS • *Department of Environment and Agriculture, Australian Centre for Necrotrophic Fungal Pathogens, Curtin University of Technology, Perth, Australia*

STEFAN S. WEBER • *Department of Microbiology, Groningen Biomolecular Sciences and Biotechnology, Institute and Zernike Institute for Advanced Materials, University of Groningen, Groningen, The Netherlands*

K.A. WILLEMS • *Laboratory for Process Microbial Ecology and Bioinspirational Management, Consortium for Industrial Microbiology and Biotechnology, Department of Microbial and Molecular Systems, K.U. Leuven Association, Lessius Mechelen, Campus De Nayer, Sint-Katelijne-Waver, Belgium; Scientia Terrae Research Institute, Sint-Katelijne-Waver, Belgium*

YAN ZHANG • *National Key Laboratory of Plant Molecular Genetics, Institute of Plant Physiology and Ecology, Shanghai Institutes for Biological Sciences, Chinese Academy of Sciences, Shanghai, China*

SHAOBIN ZHONG • *Department of Plant Pathology, North Dakota State University, Fargo, ND, USA*

Chapter 1

Fungal ABC Transporter Deletion and Localization Analysis

Andriy Kovalchuk, Stefan S. Weber, Jeroen G. Nijland, Roel A.L. Bovenberg, and Arnold J.M. Driessen

Abstract

Fungal cells are highly complex as their metabolism is compartmentalized harboring various types of subcellular organelles that are bordered by one or more membranes. Knowledge about the intracellular localization of transporter proteins is often required for the understanding of their biological function. Among different approaches available, the localization analysis based on the expression of GFP fusions is commonly used as a relatively fast and cost-efficient method that allows visualization of proteins of interest in both live and fixed cells. In addition, inactivation of transporter genes is an important tool to resolve their specific function. Here we provide a detailed protocol for the deletion and localization analysis of ABC transporters in the filamentous fungus *Penicillium chrysogenum*. It includes construction of expression plasmids, their transformation into fungal strains, cultivation of transformants, microscopy analysis, as well as additional protocols on staining of fungal cells with organelle-specific dyes like Hoechst 33342, MitoTracker DeepRed, and FM4-64.

Key words: Fluorescence microscopy, Localization analysis, GFP fusions, Staining, ABC transporter, Transformation

1. Introduction

Fungal transporter proteins have an amazingly wide range of biological functions reaching far beyond the simple nutrient uptake and extrusion of toxic compounds. They are involved, among others, in pheromone export (1, 2), quorum sensing (3), lipid translocation across the membrane (4), and Fe/S cluster protein biogenesis (5–7). A crucial role in the establishment and maintenance of virulence has been demonstrated for a number of transporter proteins of fungal plant pathogens (8–17). Early stages of infection

Melvin D. Bolton and Bart P.H.J. Thomma (eds.), *Plant Fungal Pathogens: Methods and Protocols*,
Methods in Molecular Biology, vol. 835, DOI 10.1007/978-1-61779-501-5_1, © Springer Science+Business Media, LLC 2012

and plant penetration are often impaired in transporter gene mutants. This effect is commonly attributed to the involvement of transporter proteins in the protection against host defense factors such as reactive oxygen species. Another aspect of transporter biology, namely their role in the resistance of fungi against antifungal compounds, traditionally receives a lot of attention owing to the enormous impact of pathogenic species on health care and agriculture. However, there are still a lot of open questions about the function of particular transporters, even in well-characterized fungal species. Many of the performed studies investigated the role of transporters under stress conditions (i.e., upon addition of antifungal drugs or other toxic compounds), yet our knowledge about their cognate substrates and their physiological roles is often insufficiently understood.

ABC transporters are present in virtually every living cell, from archaea and bacteria to higher eukaryotes. They can be easily identified by similarity searches due to the presence of highly conserved motifs: Walker A and Walker B boxes with an ABC signature motif situated between them (18, 19). Numbers of ABC transporters in fungal genomes can range from 19 in *Schizosaccharomyces pombe* to more than 60 in species like *Gibberella zeae* and *Aspergillus oryzae* (20). Known fungal ABC transporters localize to the plasma membrane, vacuolar membrane, or membranes of mitochondria and peroxisomes (21). It is not clear whether ABC transporters reside within endoplasmic reticulum (ER) membranes of fungal cells, as reported localization of *S. pombe* Abc1 to ER-like membranes could be due to mislocalization as a result of overexpression (22).

Knowing the localization of a certain transporter protein is often essential for the understanding of its biological function. One of the commonly used approaches to study protein localization is using fusion of proteins of interest with fluorescent proteins (GFP, YFP, DsRed, and others). This technique is extremely popular as fluorescent proteins do not require laborious staining protocols, they are suitable for both *in vivo* imaging and microscopy of fixed cells, several fluorescent proteins can be expressed simultaneously within a single cell, and the use of fluorescent proteins is time- and cost-effective. At the same time, a drawback of this approach that should be taken into account is the relatively large size of fluorescent protein that might have a negative effect on the recombinant protein stability and/or can cause its mislocalization. Another part of understanding the function of a certain transporter is deletion of the particular protein. Once a transporter is inactivated, detailed analysis in comparison with the progenitor might give an indication of the role the particular transporter plays in the cell.

Here we provide a complete protocol for the construction of knock-outs, GFP fusions, and their localization analysis which we used to characterize ABC transporters of the filamentous fungus *Penicillium chrysogenum*. At the end of the section, protocols for cell labeling with organelle-specific dyes (Hoechst 33342, Mito-Tracker

DeepRed, and FM4-64) are provided. These protocols can be used to confirm the localization of GFP fusion proteins to a certain intracellular compartment.

2. Materials

2.1. Culture Media

1. LB broth: 10 g/L bactotryptone, 5 g/L NaCl, 5 g/L yeast extract.

2. LB kanamycin/ampicillin plates: 10 g/L bactotryptone, 5 g/L NaCl, 5 g/L yeast extract, 20 g/L agar, 50 μg/ml kanamycin or ampicillin (added after autoclaving).

3. YGG medium: mix aseptically using presterilized components: 400 ml KCl-glucose, 100 ml 5× buffered YNB, 10 ml fresh 10% yeast extract, (optional) 1 ml penicillin/streptomycin solution.

4. KCl-glucose: 10 g/L KCl, 20 g/L glucose.

5. 5× buffered YNB: 33.3 g/L YNB, 7.5 g/L citric acid, 30 g/L K_2HPO_4.

6. Penicillin/streptomycin solution: lyophilized penicillin-streptomycin mixture (Roche) dissolved in 20 ml dH_2O.

7. Acetamide medium: 3 g/L NaCl, 10 mg/L $FeSO_4 \cdot 7H_2O$, 0.5 g/L $MgSO_4 \cdot 7H_2O$, 10 g/L glucose, 342 g/L sucrose, 4 ml/L trace elements solution, 20 g/L agar. After autoclaving, add (per 1 l) 10 ml 10% acetamide (filter-sterilized), 10 ml 1.5 M CsCl, and 10 ml 1 M potassium phosphate buffer (pH 6.8).

8. Trace elements solution (add components in the indicated order to prevent precipitation): 31.25 g/L EDTA, 43.75 g/L sodium citrate dihydrate, 24.84 g/L $FeSO_4 \cdot 7H_2O$, 256.4 g/L $MgSO_4 \cdot 7H_2O$, 12.5 mg/L H_3BO_3, 12.5 mg/L $Na_2MoO_4 \cdot 2H_2O$, 625 mg/L $CuSO_4 \cdot 5H_2O$, 2.5 g/L $ZnSO_4 \cdot 7H_2O$, 625 mg/L $CoSO_4 \cdot 7H_2O$, 3.04 g/L $MnSO_4 \cdot H_2O$, 1.6 g/L $CaCl_2 \cdot 2H_2O$. Adjust pH to 6.5 with NaOH. Store at −20°C (long-term) or at 4°C (for 2–3 months).

9. R-agar: 6 ml/L glycerol, 7.5 ml/L beet molasses, 5 g/L yeast extract, 18 g/L NaCl, 50 mg/L $MgSO_4 \cdot 7H_2O$, 60 mg/L KH_2PO_4, 250 mg/L $CaSO_4 \cdot 2H_2O$, 1.6 ml/L $NH_4Fe(SO_4)_2 \cdot 12H_2O$ (1 mg/ml), 10 μl/L $CuSO_4 \cdot 5H_2O$ (10 mg/ml), 20 g/L agar.

2.2. PCR Amplification of Gene of Interest

1. Forward and reverse primers designed according to the Gateway guidelines.

2. Commercially available PCR master mix (we are routinely using Phire Hot Start II PCR Master Mix (Finnzymes) in our lab).

3. Commercially available PCR purification kit (e.g., illustra GFX PCR DNA and gel band purification kit (GE Healthcare)).

2.3. BP Clonase Reactions

1. Gateway BP clonase II enzyme mix (Invitrogen).

2. Gateway pDONR P4-P1R and pDONR P2R-P3 or pDONR221 vector (all Invitrogen). The vectors are amplified in *Escherichia coli* DB3.1 strain since they contain the *ccdB* gene which is toxic to most commonly used *E. coli* laboratory strains.

3. Proteinase K solution (2 mg/ml) (Invitrogen).

2.4. Analysis of Recombinant Plasmids

1. Solution I: 50 mM glucose, 25 mM Tris–HCl (pH 8.0), 10 mM EDTA (pH 8.0), 10 µg/ml RNaseA.

2. Solution II: 0.2 N NaOH, 1% SDS.

3. Solution III: 60 ml 5 M potassium acetate, 11.5 ml glacial acetic acid, 28.5 ml dH₂O.

2.5. LR Clonase Reactions

1. Gateway destination vector pDEST R4-R3 (Invitrogen) for the construction of deletion mutants or modified destination vector pDEST R4-R3-AMDS for GFP fusions (Fig. 1).

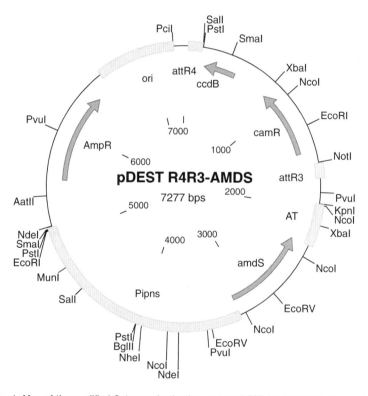

Fig. 1. Map of the modified Gateway destination vector pDEST R4-R3 AMDS. The vector contains the *Aspergillus nidulans* acetamidase gene *amdS* for positive selection of fungal transformants on media with acetamide as a sole nitrogen source. Features: AmpR, β-lactamase gene for the selection in *E. coli*; *ori*, pUC origin of replication; attR4 and attR3, Gateway *att* recombination sites; *camR*, chloramphenicol resistance gene; Pipns, promoter of *P. chrysogenum pcbC* gene; *amdS*, *A. nidulans* acetamidase gene; AT, transcriptional terminator of *P. chrysogenum penDE* gene.

2. Donor vectors pDONR221-AMDS (Fig. 2) for deletion mutants or pDONR-gpdA and pDONR-eGFP-AT (Fig. 3) (for the construction of C-terminal GFP fusions) or pDONR-gpdA-eGFP and pDONR-AT (Fig. 4) (for N-terminal GFP fusions).

3. Gateway LR clonase II Plus enzyme mix (Invitrogen).

4. Proteinase K solution (2 mg/ml) (Invitrogen).

2.6. Transformation of Fungal Protoplasts

1. KC: 60 g/L KCl, 2 g/L citric acid, pH 6.2.

2. Lyticase solution: 25 mg/ml lyticase (SigmaAldrich) in KC buffer.

3. SCT: 1.2 M sorbitol, 50 mM $CaCl_2$, 10 mM Tris–HCl pH 7.5.

4. KC-SCT: mix of KC and SCT buffer in ratio 1:1.

5. ATA solution: 0.4 M aurintricarboxylic acid, ammonium salt (SigmaAldrich) in SCT buffer.

6. 20% PEG-4000: 20% PEG-4000 in SCT buffer.

7. 60% PEG-4000: 60% PEG-4000 in SCT buffer.

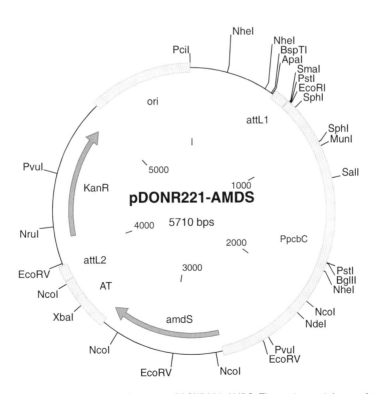

Fig. 2. Map of the modified Gateway destination vector PDONR221-AMDS. The vector contains an *Aspergillus nidulans* acetamidase gene *amdS* for the positive selection of fungal transformants on media with acetamide as a sole nitrogen source. Features: KanR, kanamycin resistance gene for the selection in *E. coli*; pUC origin of replication; *att*L1 and *att*L2, Gateway *att* recombination sites; P*pcbC*; *pcbC* promoter region; *amdS A. nidulans* acetamidase gene; AT, transcriptional terminator of the *P. chrysogenum pen*DE gene.

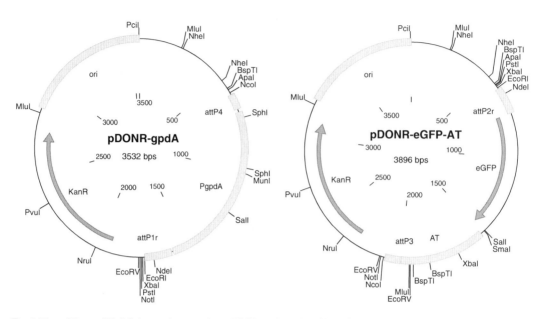

Fig. 3. Map of the modified Gateway donor vectors pDONR-gpdA and pDONR-eGFP-AT, which can be used for construction of C-terminal GFP fusions for genes of interest. Features of the vectors: KanR, kanamycin resistance gene for the selection in *E. coli*; *ori*, pUC origin of replication; attP1r, attP2r, attP3, and attP4, Gateway *att* recombination sites; P*gpdA*, promoter of *A. nidulans gpdA* gene; eGFP, gene encoding enhanced green fluorescent protein; AT, transcriptional terminator of *P. chrysogenum penDE* gene.

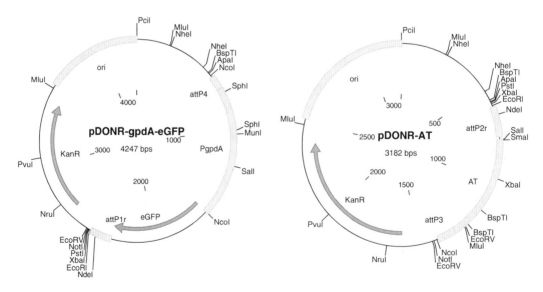

Fig. 4. Map of the modified Gateway donor vectors pDONR-gpdA-eGFP and pDONR-AT, which can be used for construction of N-terminal GFP fusions for genes of interest. Features of the vectors: KanR, kanamycin resistance gene for the selection in *E. coli*; *ori*, pUC origin of replication; attP1, attP2r, attP3, and attP4, Gateway *att* recombination sites; *gpdA*, promoter of *A. nidulans gpdA* gene; eGFP, gene encoding-enhanced green fluorescent protein; AT, transcriptional terminator of *P. chrysogenum penDE* gene.

2.7. P. chrysogenum Organelle-Specific Staining

1. FM4-64 stock solution: 1 mg/ml FM4-64 (Invitrogen) in dH$_2$O.

2. Mitotracker Deep Red stock solution: 1 mM Mitotracker Deep Red FM (Invitrogen) in DMSO.

3. 37% formaldehyde solution.

4. PBS: 8 g/L NaCl, 0.2 g/L KCl, 1.44 g/L Na$_2$HPO$_4$, 0.24 g/L KH$_2$PO$_4$, pH 7.4.

5. Hoechst 33342 stock solution: 1 mg/ml Hoechst 33342 in dH$_2$O.

3. Methods

Gene knock-outs are performed in a *P. chrysogenum* Δ*ku*70 strain (23). This strain is impaired in non-homologous end-joining, which usually results in a targeting efficiency of >50%. It is possible to use these protocols for strains without this deletion, but the targeting efficiency will decrease to 0.1–0.5%.

During the transformation process, the genome of multipenicillin cluster-containing *P. chrysogenum* is relatively unstable and side effects might occur (24). To avoid problems, it is recommended to construct and analyze multiple transformed strains.

3.1. Design of Gateway Primers for Deletion Cassettes (see Note 1)

1. For the construction of inactivation mutants, at least 1,000 bp of both flanking regions of the gene of interest should be amplified (see Note 2). Therefore, two sets of primers should be constructed.

2. The forward primer of the first set should include the *att*B4 site and the reverse primer should contain the *att*B1 sequence. First set forward: 5′-GGGGACAACTTTGTATAGAAAAGT-TG-(template-specific sequence)-3′ and reverse: 5′-GGGGAC-T-GCTTTTTTGTACAAACTTG-(template-specific sequence)-3′.

3. The forward primer of the second set should include the *att*B2 site and the reverse primer should include *att*B3 sequence. Second set forward: 5′-GGGGACAGCTTTCTTGTACAAAGTGG-(template-specific sequence)-3′ and reverse: 5′-GGGGACAA-CTTTGTATAATAAAGTTG-(template-specific sequence)-3′.

4. The template-specific sequence of the primers should be 18–25 nucleotides long.

3.2. PCR Amplification of Region of Interest

1. PCRs are performed in total volume of 50 μl. When using commercially available PCR master mixes, combine in a PCR tube: 20 μl PCR-grade water, 25 μl of PCR master mix, 2 μl forward primer (10 μM), 2 μl reverse primer (10 μM), and 1 μl of fungal genomic DNA (20–50 ng).

2. Amplification with Phire Hot Start II PCR Master Mix (Finnzymes) is performed under following conditions: initial denaturation at 98°C for 1 min, 30 cycles of 15 sec at 98°C, 5 sec at 55–60°C, and 20 sec per 1 kb of the template at 72°C, followed by the final extension of 1 min at 72°C (see Note 3).

3. Load 2 μl of PCR reaction on 0.8% agarose gel to confirm the amplification of the product of interest.

4. Purify the amplification products with commercially available PCR purification kits. Elute the PCR products with 30 μl of PCR-grade water (see Note 4).

3.3. BP Clonase Reaction with pDONR P4-P1R and pDONR P2R-P3

1. Inactivation cassettes consist of three parts: two flanking regions and an *amd*S gene for the selection using acetamide. In this step, we create two different plasmids both containing one of the flanking regions.

2. Add to a 1.5 ml microcentrifuge tube 0.5 μl of pDONR P4-P1R vector (30–40 ng), and to another 1.5 ml microcentrifuge tube 0.5 μl of pDONR P2R-P3 vector (30–40 ng). Add to both tubes 1–4 μl of the desired PCR product (50–150 ng) and PCR-grade water to 4.5 μl (see Note 5).

3. Add 0.5 μl of BP clonase II enzyme mix, gently mix the reaction by pipetting and spin down briefly.

4. Incubate at 25°C for 1–3 h (see Note 6).

5. Inactivate the BP clonase by incubation with 1 μl of proteinase K at 37°C.

6. Use the BP reaction to transform competent *E. coli* DH5α cells.

3.4. Analysis of Recombinant Plasmids (see Note 7)

1. For the analysis, select 12–24 colonies appearing after transformation of *E. coli* DH5α with the BP reaction and grow them overnight in 2 ml of LB broth with kanamycin.

2. Collect the cells by spinning down at 13,000 rpm for 1 min in a microcentrifuge.

3. Resuspend them in 100 μl of Solution I.

4. Add 200 μl of Solution II and gently mix by inverting the tube.

5. Add 150 μl of Solution III and mix gently.

6. Centrifuge the tubes at 13,000 rpm for 10 min in a microcentrifuge.

7. Transfer the supernatant into fresh 1.5 ml tube.

8. Add 600 μl of isopropanol.

9. Spin down the tubes at 13,000 rpm for 10 min in a microcentrifuge.

10. Wash the DNA pellet once with 200 μl of 70% ethanol.

11. Dissolve DNA in 40 μl of PCR-grade water.

12. Use 1–1.5 µl of plasmid DNA for the digestion with appropriate enzyme.

13. Perform the digestion at 37°C for 1 h.

14. Mix the digestion reaction with 3 µl of loading buffer and load it on 0.8% agarose gel.

15. Select the clones carrying the right insert for further experiments.

3.5. LR Clonase Reaction with the Destination Vector pDEST R4-R3

1. Mix in a 1.5-ml microcentrifuge tube 60 ng of each of the plasmids pDEST R4-R3, pDONR P4-P1R carrying the flanking region of interest, pDONR221-AMDS, and pDONR P2R-P3 carrying the flanking region of interest (see Note 7). Add PCR-grade water to 4.5 µl, if necessary.

2. Add 0.5 µl of LR clonase II Plus, mix gently by pipetting, and spin down briefly.

3. Incubate at 25°C for 1–3 h or overnight (see Note 6).

4. Inactivate LR clonase by adding 1 µl of proteinase K and incubating at 37°C for 15 min.

5. Use the whole LR reaction to transform chemically competent *E. coli* DH5α cells.

6. Plate the cells on LB plates with ampicillin.

3.6. Analysis of Recombinant Clones

1. Select 12–24 clones appearing after the transformation of *E. coli* DH5α with LR reaction and analyze them as described in Protocol 3.4.

2. Amplify clones carrying correct constructs and scale-up the isolation protocol to generate at least 4–5 µg of plasmid DNA.

3. At this point, the plasmid for the construction of a knock-out is ready. Continue from here with the preparation for fungal transformation at Subheading 3.10.

3.7. Gateway Primers for Generation of GFP Fusions (see Note 1)

1. Primers are designed to amplify the complete gene of interest. Forward primer should include the *attB1* sequence (underlined below), and it is recommended to include Kozak sequence if you want to express your protein as C-terminal GFP fusion (see Note 8). If you wish to fuse your PCR product in frame with N-terminal tag, the primer must include two additional nucleotides to maintain the proper reading frame (shown as N). These two additional nucleotides cannot be AA, AG, or GA because these combinations will create a stop codon. The forward primer: 5′-GGGG<u>ACAAGTTTGTACAAAAAAGCAGGC-T</u>NN-(template specific sequence)-3′.

2. Reverse primer should include the *attB2* sequence (underlined below). Additionally, a stop codon should be added if the protein of interest will be expressed as N-terminal GFP fusion. Importantly, no stop codon should be placed between the gene

of interest and the GFP gene. In the last case, an additional nucleotide (shown as N) should be added to maintain the proper reading frame. The reverse primer. 5'-GGGGA<u>CCACTTTGTA-</u><u>CAAGAAAGCTGGGT</u>N-(template specific sequence)-3'.

3. Template-specific sequence in primers should be 18–25 nucleotides long.

4. For the amplification of the gene of interest, please refer to Subheading 3.2.

3.8. BP Clonase Reaction with the Donor Vector pDONR221

1. Mix in a 1.5 ml microcentrifuge tube 0.5 μl of pDONR221 vector (30–40 ng), 1–4 μl of PCR product (50–150 ng), and PCR-grade water to 4.5 μl (see Note 5).

2. Add 0.5 μl of BP clonase II enzyme mix, gently mix the reaction by pipetting, and spin down briefly.

3. Incubate at 25°C for 1–3 h (see Note 6).

4. Inactivate the BP clonase by incubation with 1 μl of proteinase K at 37°C.

5. Use the BP reaction to transform competent *E. coli* DH5α cells.

6. Plate the cells on LB plates with kanamycin.

7. The analysis of recombinant clones can be done the same way as described in Subheading 3.4.

3.9. LR Clonase Reaction with the Destination Vector pDEST R4-R3-AMDS

1. Mix in a 1.5-ml microcentrifuge tube 60 ng of each of the plasmids pDEST R4-R3-AMDS, pDONR-gpdA, pDONR-GFP-AT, and pDONR221 carrying the gene of interest in case you want to create C-terminal GFP fusion or pDEST R4-R3-AMDS, pDONR-gpdA-GFP, pDONR-AT, and pDONR221 carrying the gene of interest if you want to make N-terminal fusion (see Note 9). Add PCR-grade water to 4.5 μl if necessary.

2. Add 0.5 μl of LR clonase II Plus, mix gently by pipetting, and spin down briefly.

3. Incubate at 25°C for 1–3 h or overnight (see Note 6).

4. Analysis of these strains can be done as described in Subheading 3.6.

3.10. Preparation of Plasmid DNA for Fungal Transformation

1. Mix in 1.5 ml microcentrifuge tube 5 μg of plasmid DNA (ca. 40 μl of plasmid sample), 5 μl of appropriate buffer, and PCR-grade water to 47 μl (see Note 10).

2. Add 3 μl of appropriate enzyme, vortex the mix, and spin it down briefly. Incubate at 37°C for 2–4 h or overnight.

3. Purify the DNA by one of the commercially available DNA purification kits. Elute it with or dissolve it in ≤30 μl of PCR-grade water.

3.11. Transformation of Fungal Protoplasts

1. Inoculate *P. chrysogenum* conidia immobilized on rice (as described in Subheading 3.12) in 25 ml of YGG medium and incubate with shaking (200 rpm) at 25°C for 24 h.

2. Transfer 10 ml of culture into 90 ml of fresh YGG medium. Grow overnight at 25°C.

3. Collect the mycelium by centrifugation at $4,000\times g$ for 5 min.

4. Wash the pellet once with 50 ml of KC solution.

5. Gently resuspend the pellet in 1.5 volumes (relative to the volume of the pellet) of KC solution.

6. Add 1 ml of lyticase solution to every 9 ml of mycelium suspension.

7. Incubate with slow shaking at 25°C for 1–2 h. Follow the protoplast formation under the microscope.

8. Add KC solution to 50 ml and spin down at $4,000\times g$ for 5 min.

9. Wash the protoplasts once with 50 ml of KC-SCT solution.

10. Wash the protoplasts once with 50 ml of SCT buffer.

11. Resuspend the protoplasts in SCT buffer to the final concentration of approximately $1\cdot10^8$ protoplasts/ml (usually, we add 2–4 ml of SCT).

12. Mix in a 1.5 ml microcentrifuge tube 200 µl of protoplast suspension with 15 µl ATA solution and add approximately 5 µg of linearized plasmid DNA in a volume not exceeding 30 µl.

13. Add 100 µl 20% PEG-4000 solution and incubate on ice for 30 min.

14. Add 1.5 ml 60% PEG-4000 solution and incubate at 25°C for 15 min without agitation.

15. Add 5 ml SCT buffer and spin down at $500\times g$ for 5 min.

16. Gently resuspend the pellet in 800 µl of SCT buffer and distribute the suspension over 5–6 selective plates with acetamide medium.

17. Incubate the plates at 25°C for 5–7 days until the colonies of transformants appear.

3.12. Selection and Maintenance of Fungal Transformants (see Note 11)

1. Transfer colonies that appear after protoplast transformation to fresh acetamide plates and incubate them at 25°C for 5–7 days.

2. Discard clones that failed to grow on acetamide. Transfer several putative transformants to R-agar plates and incubate them at 25°C for 5–6 days before they start to produce conidia.

3. Streak a small amount of conidia on selective acetamide plate to obtain single colonies. Incubate the plates containing single-spore-derived colonies at 25°C for 4–5 days.

4. Select a few colonies to be transferred to R-agar plates. Try to break a colony with a sterile tooth-pick in a number of small

pieces (25–30) and transfer all of them to a single plate. Incubate the plates at 25°C for 3–4 days to the stage when the conidia just start to appear.

5. Resuspend all colonies growing on the plate in 5–6 ml of sterile tap water and transfer the suspension to a batch of autoclaved rice grains (25 ml). Incubate the rice with occasional shaking at 25°C for 8–10 days until it becomes greenish. Freeze-dry the rice. Conidia immobilized on rice in this way can be stored at room temperature for several months.

6. Correct integration of a construct can be determined via Southern analysis (25).

3.13. Microscopy Analysis of Transformants Expressing GFP Fusions

1. Inoculate several rice grains with immobilized conidia in 10 ml YGG and incubate them overnight at 25°C with agitation (200 rpm) (see Note 12).

2. The next day, dilute the culture tenfold with fresh YGG to the final volume of 15 ml.

3. Grow at 25°C with agitation for 16–18 h (see Note 13).

4. Directly analyze the sample under fluorescent microscope using appropriate filter set (Fig. 5).

3.14. Staining of P. chrysogenum Vacuolar Membranes with FM4-64 Dye

1. Take 0.5 ml of *P. chrysogenum* culture grown in YGG for 16–36 h.

2. Briefly spin down the mycelium (10,000 rpm, 30 sec) in a microcentrifuge, resuspend in fresh YGG medium containing 2 μg/ml FM4-64, and incubate at 25°C for 30 min.

3. Briefly spin down the mycelium (10,000 rpm, 30 sec) in a microcentrifuge and wash it once with fresh YGG.

4. Resuspend the mycelium in 1 ml of fresh YGG and incubate it at 25°C for 90 min.

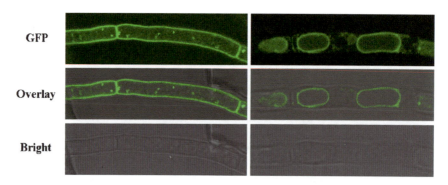

GFP

Overlay

Bright

Fig. 5. Examples of intracellular localization of ABC transporters in *P. chrysogenum*. The two transporters shown localize to plasma membrane (*left*) and vacuolar membrane (*right*).

5. Analyze the sample under fluorescent microscope using an appropriate filter set. FM4-64 accumulates in vacuolar membranes, which typically can be seen as red-colored ring-like structures (Fig. 6). The accumulation occurs via endocytosis, so the staining can only be performed on live material. If the internalization of the dye was not complete, some residual staining can be observed in plasma membrane. In this case, the incubation time at Step 4 should be increased.

3.15. Staining of P. chrysogenum Mitochondria with Mitotracker Deep Red Dye

1. Take 0.5 ml of *P. chrysogenum* culture grown in YGG for 16–36 h.

2. Add Mitotracker Deep Red to the final concentration of 100 nM, mix the suspension gently, and incubate the mycelium at 25°C for 15–30 min.

3. Briefly spin down the mycelium (10,000 rpm, 30 sec) in a microcentrifuge and wash once with fresh YGG.

4. Resuspend mycelium in 1 ml of fresh YGG and analyze sample under fluorescent microscope using appropriate filter set. Mitochondria appear as numerous small red dots, sometimes forming kind of reticulum (Fig. 6).

3.16. Staining of P. chrysogenum Nuclei with Hoechst 33342 Dye

1. Take 0.5 ml of *P. chrysogenum* culture grown in YGG for 16–36 h.

2. Briefly spin down mycelium (10,000 rpm, 1 min) in a microcentrifuge and fix in 1 ml of 2% formaldehyde in PBS (see Note 14).

3. Add Hoechst 33342 to the final concentration of 1 μg/ml and incubate the mycelium at room temperature for 15 min.

4. Wash the mycelium twice with PBS.

5. Analyze the sample under fluorescent microscope using appropriate filter set. Nuclei can be seen as large blue-colored round structures, more or less evenly distributed throughout the mycelium (Fig. 6).

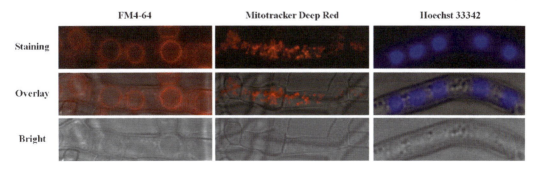

Fig. 6. Examples of staining of *P. chrysogenum* cells with fluorescent organelle-specific dyes. From left to right: FM4-64 (vacuolar membrane staining), Mitotracker Deep Red (mitochondria-specific dye), and Hoechst 33342 (nuclear staining).

4. Notes

1. In our knock-out and localization analysis of ABC transporters of *P. chrysogenum*, we made use of Gateway cloning system (Invitrogen) (26). It allows us to prepare a large number of constructs in a straightforward and time-efficient manner, so that it is especially recommended for large-scale projects.

2. The flanking regions should include (preferably endogenous) restriction sites to be able to create a linear inactivation cassette from the final plasmid. If natural restriction sites cannot be used, then suitable restriction sites should be included in the primer between the *att* sites and the actual primer sequence.

3. We included here an amplification protocol for the Phire Hot Start II PCR Master Mix (Finnzymes) routinely used in our lab. If you prefer other commercially available PCR enzymes, please refer to the manufacturer's recommendations.

4. It is important to purify PCR products prior to BP reaction as primers containing *att* sites can recombine with pDONR221 vector resulting in high background of colonies carrying plasmids without the correct insert.

5. High concentrations of DNA in BP reaction can inhibit BP recombinase.

6. Performing the BP and LR reactions overnight can increase number of colonies after transformations by several folds.

7. Alternatively, commercially available plasmid isolation kits can be used at this step.

8. It is recommended to construct and express both N-terminal and C-terminal GFP fusions of the transporters of interest, as certain combinations might be unstable and prone to mislocalization.

9. Equimolar amounts of all four plasmids are required for the efficient recombination.

10. To increase the efficiency of fungal transformation, plasmid DNA has to be linearized prior to the transformation. The choice of the enzyme largely depends on the sequence of the gene of interest, but two of the options that can be considered are DraI and AatII. Relatively large amounts of DNA are required for the fungal transformation, in our lab we are generally using 4–5 μg of DNA per transformation.

11. As fungal protoplasts often carry more than one nucleus, it is necessary to perform additional round of selection to obtain isogenic clones.

12. *P. chrysogenum* transformants often differ significantly in the expression level of GFP-fused proteins. This difference is likely caused by the integration of expression cassette into different

loci in the fungal genome. Thus, it is recommended to check several (up to ten) independent clones to find the one with the optimal expression level.

13. We repeatedly noticed that young cultures of *P. chrysogenum* give better results in localization analysis than older ones. Green fluorescent signal is often accumulated within vacuoles in cells grown for more than 24 h, indicating partial degradation of GFP fusions.

14. While some published protocols describe the use of live yeast and fungal cells for Hoechst 33342 staining, we could obtain satisfactory results only with cells fixed with formaldehyde.

Acknowledgments

This project was financially supported by the Netherlands Ministry of Economic Affairs and the B-Basic partner organizations (http://www.b-basic.nl) through B-Basic, a public–private NWO-ACTS programme (ACTS = Advanced Chemical Technologies for Sustainability) and by the Kluyver Centre for Genomics of Industrial Fermentation, which is part of the Netherlands Genomics Initiative/Netherlands Organization for Scientific Research.

References

1. Kuchler, K., Sterne, R. E., and Thorner, J. (1989) Saccharomyces cerevisiae STE6 gene product: a novel pathway for protein export in eukaryotic cells, *EMBO J. 8*, 3973–3984.

2. McGrath, J. P. and Varshavsky, A. (1989) The yeast STE6 gene encodes a homologue of the mammalian multidrug resistance P-glycoprotein, *Nature 340*, 400–404.

3. Hlavacek, O., Kucerova, H., Harant, K., Palkova, Z., and Vachova, L. (2009) Putative role for ABC multidrug exporters in yeast quorum sensing, *FEBS Lett. 583*, 1107–1113.

4. Smriti, Krishnamurthy, S., Dixit, B. L., Gupta, C. M., Milewski, S., and Prasad, R. (2002) ABC transporters Cdr1p, Cdr2p and Cdr3p of a human pathogen Candida albicans are general phospholipid translocators, *Yeast 19*, 303–318.

5. Kispal, G., Csere, P., Guiard, B., and Lill, R. (1997) The ABC transporter Atm1p is required for mitochondrial iron homeostasis, *FEBS Lett. 418*, 346–350.

6. Kispal, G., Csere, P., Prohl, C., and Lill, R. (1999) The mitochondrial proteins Atm1p and Nfs1p are essential for biogenesis of cytosolic Fe/S proteins, *EMBO J. 18*, 3981–3989.

7. Leighton, J. and Schatz, G. (1995) An ABC transporter in the mitochondrial inner membrane is required for normal growth of yeast, *EMBO J. 14*, 188–195.

8. Gupta, A. and Chattoo, B. B. (2008) Functional analysis of a novel ABC transporter ABC4 from Magnaporthe grisea, *FEMS Microbiol. Lett. 278*, 22–28.

9. Sun, C. B., Suresh, A., Deng, Y. Z., and Naqvi, N. I. (2006) A multidrug resistance transporter in Magnaporthe is required for host penetration and for survival during oxidative stress, *Plant Cell 18*, 3686–3705.

10. Urban, M., Bhargava, T., and Hamer, J. E. (1999) An ATP-driven efflux pump is a novel pathogenicity factor in rice blast disease, *EMBO J. 18*, 512–521.

11. Barhoom, S., Kupiec, M., Zhao, X., Xu, J. R., and Sharon, A. (2008) Functional characterization of CgCTR2, a putative vacuole copper transporter that is involved in germination and pathogenicity in Colletotrichum gloeosporioides, *Eukaryot. Cell 7*, 1098–1108.

12. Callahan, T. M., Rose, M. S., Meade, M. J., Ehrenshaft, M., and Upchurch, R. G. (1999)

CFP, the putative cercosporin transporter of Cercospora kikuchii, is required for wild type cercosporin production, resistance, and virulence on soybean, *Mol. Plant Microbe Interact.* *12*, 901–910.

13. Chague, V., Maor, R., and Sharon, A. (2009) CgOpt1, a putative oligopeptide transporter from Colletotrichum gloeosporioides that is involved in responses to auxin and pathogenicity, *BMC. Microbiol. 9*, 173.

14. Fleissner, A., Sopalla, C., and Weltring, K. M. (2002) An ATP-binding cassette multidrug-resistance transporter is necessary for tolerance of Gibberella pulicaris to phytoalexins and virulence on potato tubers, *Mol. Plant Microbe Interact. 15*, 102–108.

15. Stefanato, F. L., bou-Mansour, E., Buchala, A., Kretschmer, M., Mosbach, A., Hahn, M., Bochet, C. G., Metraux, J. P., and Schoonbeek, H. J. (2009) The ABC transporter BcatrB from Botrytis cinerea exports camalexin and is a virulence factor on Arabidopsis thaliana, *Plant J. 58*, 499–510.

16. Stergiopoulos, I., Zwiers, L. H., and De Waard, M. A. (2003) The ABC transporter MgAtr4 is a virulence factor of Mycosphaerella graminicola that affects colonization of substomatal cavities in wheat leaves, *Mol. Plant Microbe Interact. 16*, 689–698.

17. Wahl, R., Wippel, K., Goos, S., Kamper, J., and Sauer, N. (2010) A novel high-affinity sucrose transporter is required for virulence of the plant pathogen Ustilago maydis, *PLoS. Biol. 8*, e1000303.

18. Walker, J. E., Saraste, M., Runswick, M. J., and Gay, N. J. (1982) Distantly related sequences in the alpha- and beta-subunits of ATP synthase, myosin, kinases and other ATP-requiring enzymes and a common nucleotide binding fold, *EMBO J. 1*, 945–951.

19. Bairoch, A. (1992) PROSITE: a dictionary of sites and patterns in proteins, *Nucleic Acids Res. 20 Suppl*, 2013–2018.

20. Kovalchuk, A. and Driessen, A. J. (2010) Phylogenetic analysis of fungal ABC transporters, *BMC. Genomics 11*, 177.

21. Jungwirth, H. and Kuchler, K. (2006) Yeast ABC transporters-- a tale of sex, stress, drugs and aging, *FEBS Lett. 580*, 1131–1138.

22. Iwaki, T., Giga-Hama, Y., and Takegawa, K. (2006) A survey of all 11 ABC transporters in fission yeast: two novel ABC transporters are required for red pigment accumulation in a Schizosaccharomyces pombe adenine biosynthetic mutant, *Microbiology 152*, 2309–2321.

23. Snoek, I. S., van der Krogt, Z. A., Touw, H., Kerkman, R., Pronk, J. T., Bovenberg, R. A., van den Berg, M. A., and Daran, J. M. (2009) Construction of an hdfA Penicillium chrysogenum strain impaired in non-homologous end-joining and analysis of its potential for functional analysis studies, *Fungal. Genet. Biol. 46*, 418–426.

24. Nijland, J. G., Ebbendorf, B., Woszczysnka, M., Boer, R., Bovenberg, R. A. L., and Driessen, A. J. M. A nonlinear biosynthetic gene cluster dose effect on penicillin production by *Penicillium chrysogenum*.

25. Sambrook, J., Fritsch, E. F., and Maniatis, T. (1989) *Molecular cloning: a laboratory manual, second ed.* Cold Spring Harbor Laboratory Press.

26. Walhout, A. J., Temple, G. F., Brasch, M. A., Hartley, J. L., Lorson, M. A., van den, H. S., and Vidal, M. (2000) GATEWAY recombinational cloning: application to the cloning of large numbers of open reading frames or ORFeomes, *Methods Enzymol. 328*, 575–592.

Chapter 2

Targeted Gene Replacement in Fungal Pathogens via *Agrobacterium tumefaciens-* Mediated Transformation

Rasmus John Normand Frandsen, Mette Frandsen, and Henriette Giese

Abstract

Genome sequence data on fungal pathogens provide the opportunity to carry out a reverse genetics approach to uncover gene function. Efficient methods for targeted genome modifications such as knock-out and *in locus* over-expression are in high demand. Here we describe two efficient single-step cloning strategies for construction of vectors for *Agrobacterium tumefaciens*-mediated transformation (ATMT). Targeted genome modifications require integration by a homologous double crossover event, which is achieved by placing target sequences on either side of a selection marker gene in the vector. Protocols are given for two single-step vector construction techniques. The In-Fusion cloning technique is independent of compatible restriction enzyme sites in the vector and the fragment to be cloned. The method can be directly applied to any vector of choice and it is possible to carry out four fragment cloning without the need for subcloning. The cloning efficiency is not always as high as desired, but it still presents an efficient alternative to restriction enzyme and ligase-based cloning systems. The USER technology offers a higher four fragment cloning efficiency than In-Fusion, but depends on specific structures in the binary vector. The available fungal binary vectors adapted for the USER system are described and protocols are provided for vector design and construction. A general protocol for verification of the resulting gene replacement events in the recipient fungal cells is also given. The cloning systems described above are relevant for all transformation vector constructs, but here we describe their application for ATMT compatible binary vectors. Protocols are provided for ATMT exemplified by *Fusarium graminearum*. For large-scale reverse genetic projects, the USER technology is recommended combined with ATMT.

Key words: *Agrobacterium tumefaciens*-mediated transformation, ATMT, AMT, Vector construction, USER cloning, In-fusion cloning, Fungal pathogens, Gene replacement, Fungi

1. Introduction

1.1. Agrobacterium tumefaciens-Mediated Transformation

The natural ability of the plant pathogenic bacterium *A. tumefaciens* to transfer part of its DNA to plants has for the last 30 years been utilized extensively to manipulate plant genomes. Initially the technique was restricted to the natural dicot host plants (1), but has been developed to also work in monocots (2). The transferred

Melvin D. Bolton and Bart P.H.J. Thomma (eds.), *Plant Fungal Pathogens: Methods and Protocols*,
Methods in Molecular Biology, vol. 835, DOI 10.1007/978-1-61779-501-5_2, © Springer Science+Business Media, LLC 2012

DNA (T-DNA) resides on a >100 kb large tumour-inducing plasmid (Ti) and is demarcated by two imperfect 25 bp long direct repeats known as left and right border (LB and RB). The LB and RB sequences constitute recognition sites for a nicking endonuclease (VirD2) that, in cooperation with a helicase (VirD1), cuts out a single-stranded T-DNA element, which is subsequently transferred to the host cell (reviewed by (3)). The T-DNA includes a set of oncogenes and opine biosynthesis encoding genes, which once expressed in the recipient plant cell result in uncontrolled proliferation and production of opines for bacterial catabolism (4).

The adaptation of the *A. tumefaciens* virulence system for genetic transformation experiments has resulted in a binary vector system where the natural Ti plasmid has been divided into two plasmids: A large plasmid (100 kb) with the genes encoding the T-DNA transfer machinery and a small shuttle plasmid (5–10 kb) with a disarmed T-DNA region. The opine biosynthetic machinery and oncogenes have been removed from the T-DNA to obtain symptomless transformants and give room for larger fragments of heterogeneous DNA to be transferred (5). The small size of the T-DNA containing shuttle vector allows for easy modification of the T-DNA region via standard molecular techniques in *Escherichia coli*. The shuttle vector can subsequently be transferred to an *A. tumefaciens* strain that contains the modified Ti plasmid.

The ability of *A. tumefaciens* to transform fungi was first shown in *Saccharomyces cerevisiae* (6). However, it was only after the discovery that *A. tumefaciens* mediated transformation (ATMT) could be applied to both ascomycetes and basidiomycetes without the need to produce protoplasts (7) that the fungal research community realized the potential of this new tool. Subsequent studies showed that the transformation frequency for the majority of fungal species is higher compared to that obtained by other techniques, for example, biolistics, electroporation, lithium-acetate, or $CaCl_2$/PEG. An additional advantage is that ATMT typically results in stable integration of a single T-DNA molecule, making it ideal for random insertional mutagenesis experiments (7, 8). Since 1995, more than 125 fungal species have successfully been transformed via ATMT (own recording), and the technique has become the standard transformation method for many fungal species.

Following the discoveries made by de Groot, ATMT has been used extensively for generating libraries of random mutants for linking mutant phenotypes to specific genes (forward genetics) (9–12). However, the increased availability of fungal genome sequences now allows for the application of reverse genetics (gene to function), an area where ATMT also has proven its value.

In addition to the requirement of sequence data, the success of reverse genetics depends on methods for performing targeted genome modifications. Fungi are generally amenable to targeted integration of heterologous DNA through homologous recombination (HR). The frequency of integration via HR varies greatly

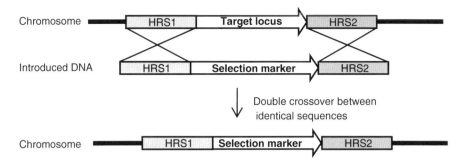

Fig. 1. Integration via homologous recombination by double crossover between identical sequences in the genome and the introduced DNA fragment, denoted as HRS1 and HRS2. Here resulting in a replacement of the target locus with a selection marker gene.

between species and even between strains (13, 14). Targeted integration via HR is dependent on a double crossover event between the sequences bordering the genomic target locus and identical sequences adjacent to a selection marker gene in the introduced DNA fragment (see Fig. 1). The two sequences in the vector will hereafter be referred to as homologous recombination sequences (HRSs).

The majority of reported targeted genome modifications have relied on custom-designed cloning strategies for construction of vectors with T-DNA harbouring a selection marker gene surrounded by the two HRSs. Only a handful of systems that can be generally applied have emerged, such as (1) *in vitro* transposon mutagenesis where the selection marker gene is randomly introduced via a T-DNA bearing plasmid containing a segment of the targeted gene (15, 16), (2) sequential restriction enzyme and ligase-dependent cloning (17), (3) sequential Xi-cloning into a T-DNA barring vector (18), (4) Fusion-PCR (19), and (5) split-marker-based ATMT (20). These systems all require multiple cloning steps and offer little or no control over the placement of the HRSs, or have limitations with regard to the placement of HRSs due to the requirement of unique restriction enzyme sites during the construction process.

In the following section, we describe two generally applicable single-step vector construction techniques that allow efficient construction of binary vectors for targeted integration. Both techniques offer complete freedom with respect to the sequences that are used as HRSs.

1.2. Construction of Binary Vectors for Targeted Genome Modifications

1.2.1. In-Fusion Cloning Technique

The In-Fusion cloning technique is dependent on the 3′–5′ single-strand exonuclease activity of poxvirus DNA polymerase (commercially sold as In-Fusion enzyme) (21) for generating 5′ compatible overhangs on a PCR-amplified fragment and the recipient vector (22). Fusion of the DNA fragments is catalyzed *in vivo* in the recipient bacterial strain by the natural DNA repair system. The recipient vector is digested with a single unique cutting restriction enzyme generating sticky or blunt ends. To introduce compatible ends for cloning, 10–15 bp long 5′ overhangs are included in the primers used for amplification of the fragment to be cloned. The added 5′ overhangs should be identical to the sequences adjacent to the chosen restriction enzyme recognition site in the vector. Incubation of the digested vector and PCR amplicon with the In-Fusion enzyme results in the formation of 10–15 bp long compatible 5′ overhangs on the free DNA ends. The length of the 5′ overhang increases with incubation time, but when compatible overhangs have been generated, the sticky ends will anneal and delay further 3′–5′ digestion. Following inactivation of the In-Fusion enzyme, the resulting chimeric DNA fragments are stable enough to survive transformation into *E. coli* where they are covalently joined and replicated.

We have previously used this technique in a two-step vector construction strategy, by sequential cloning of the HRSs into unique restriction sites found on either side of the hygromycin resistance gene (*hph*) in the binary pAg1-H3 vector (data not published). However, (38) and others (23) have found that the In-Fusion cloning reaction can be used for the simultaneous fusion of multiple DNA fragments in a single cloning reaction, suggesting that efficient construction of vectors via a single cloning step is achievable. Here, the pAg1-H3 vector is digested with the two unique cutting blunt-end restriction enzymes *Sma*I and *Swa*I, resulting in an *hph* fragment and a vector backbone fragment. The two HRSs are then amplified by PCR with primers containing 15 bp long 5′ overhangs identical to the sequences surrounding the *Sma*I and *Swa*I restriction enzyme sites (see Fig. 2).

The single-step In-Fusion construction strategy presents several advantages over standard multistep construction methods. The most important is that it is independent of unique restriction enzyme sites in the HRSs, permitting any desired HRS to be used for vector construction. This results in a procedure with complete control over which modifications are introduced into the fungal genome. The technique allows for easy automation of primer design as the 5′ overhangs that are added to the four primers are identical for all vector constructs (insert-independent design). The method only relies on a single cloning step, making it ideal for large-scale projects aimed at the construction of genome-wide single gene knockout libraries. In addition, the technology is compatible with all existing binary vectors that have unique restriction enzyme sites on either side of the selection marker, or which have sites with a similar placement for a double-cutting enzyme. The only disadvantage is the relative low frequency of correct assembly of the four fragments (approximately 10–20% correct transformants) and the subsequent need to screen a larger number of transformants. However, the time and materials saved by the reduced number of construction steps easily make up for the required screening work.

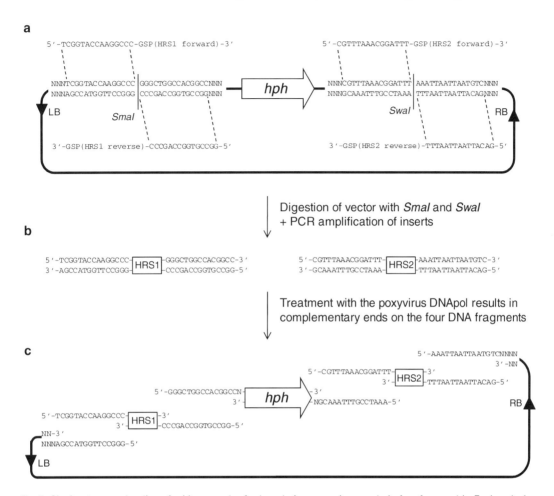

Fig. 2. Single-step construction of a binary vector for targeted gene replacement via four fragment In-Fusion cloning. (a) Sequences adjacent to the *Smal* and *Swal* restriction site in pAg1-H3 are included as 5′ overhangs in the primers designed for amplification of the two HRSs. (b) PCR amplifications of the fragments to be cloned result in double-stranded DNA with ends that are identical to the ends of the vector following *Smal/Swal* digestion. (c) Single-stranded 3′–5′ digestion of the inserts and vector fragments by the In-Fusion enzyme results in the formation of complementary 5′ sticky ends that allows for directional annealing of the two insert with the two vector fragments. (Gene-Specific primer (GSP) sequence).

1.2.2. USER Cloning Technique

The Uracil Specific Excision Reagent cloning technique, USER cloning, also allows multi-fragment cloning in specially designed vectors (24, 25). The assembly efficiency is much higher than observed for In-Fusion cloning. The USER system relies on the combined action of Uracil DNA glycosylase and DNA glycosylase-lyase Endo VII to identify and excite 2-deoxyuridines from the ends of PCR-amplified fragments. The introduction of single-stranded breaks at predetermined positions in the double-stranded DNA ends results in 3′ sticky-ends through strand denaturation. The 2-deoxyuridines are placed in the PCR primers as a 5′ extension to fit with overhangs that can be generated in vectors with two USER Cloning Sites (UCSs) (see Fig. 3a). The UCSs consist

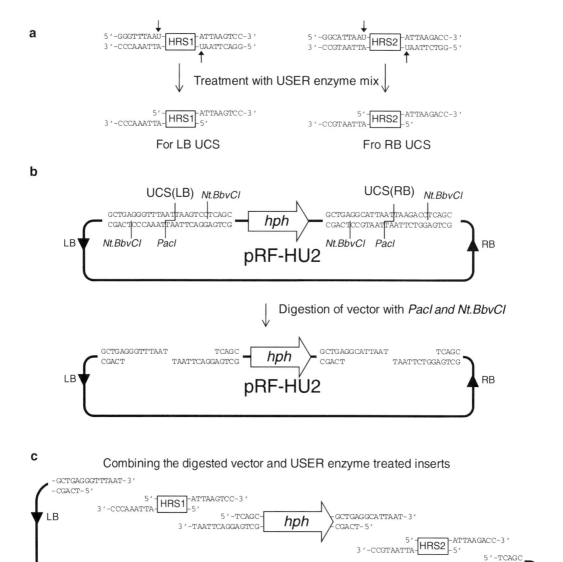

Fig. 3. Single-step construction of a binary vector for targeted gene replacement via four fragment USER cloning. (a) Generation of sticky ends on inserts, which have been amplified with 2-deoxyuridine containing primers (marked with arrows). (b) Digestion of the two USER cloning sites (UCSs) surrounding the *hph* selection marker gene with *PacI* and *Nt.BbvCI*. Note that the two UCSs differ in four positions and that the two halves of each UCS also differ to ensure directional assembly of the fragments. (c) Mixing of the two inserts (HRSs) with the digested vector allows for directional annealing due to the unique complementary sticky ends.

of recognition sites for a standard restriction enzyme and two sites for a nicking enzyme (see Fig. 3b). The combined action of these endonucleases allows for the introduction of a double-stranded break and two single-stranded breaks, which combined results in the generation of 3′ overhangs that can be used for cloning (see Fig. 3c).

We have shown that vectors containing two unique UCSs can be used for directional assembly of four DNA fragments in a single cloning reaction (see Fig. 3), allowing for a single-step construction of vectors for targeted gene replacement (24). Digestion of the two UCSs results in the generation of four unique 9 bp long overhangs that allow for directional annealing of the two PCR-amplified inserts. The resulting chimeric molecule is stable enough to survive transformation into *E. coli* where the four fragments are covalently joined and replicated. Compared to In-Fusion, the USER system offers higher assembly efficiency (approximately 85% correct) reducing the need for extensive screening to identify correctly assembled vectors.

The USER cloning technique has the same advantages as the In-Fusion system with respect to the experimental design and the placement of the HRSs. It differs because it is only compatible with vectors that contain UCSs, but the superior performance compared to the more versatile In-Fusion system makes USER cloning a more attractive technique for large-scale gene replacement projects.

Four binary vectors with UCSs and the hygromycin B selection marker gene are available for use in ascomycetes (24). New selection marker cassettes can easily be introduced into these vectors, and the series have been extended to include vectors with geneticin/G418 (*nptII* gene) and glufosinate ammonium (*bar* gene) resistance markers (data not published). Existing binary vectors can easily be converted into USER compatible vectors by introducing UCSs. We recommend that the UCSs described in (24) are used in new USER constructions to ensure compatibility with existing systems?

1.2.3. Available Vectors for USER Cloning

The vector series consist of four vectors, each designed for specific research purposes. They all contain the same selection marker cassette: The hygromycin B resistance gene from *E. coli* controlled by the *Aspergillus nidulans* trpC promoter and terminator originating from pANT-hyg(R) (26).

1. Random integration into the genome

The pRF-HU and pRF-HUE vectors contain a single UCS allowing for efficient cloning of a single PCR amplicon into the T-DNA region of the vector (see Fig. 4). The pRF-HU vector can be used for random mutagenesis, heterologous expression, and complementation experiments, where the introduced gene should be controlled by its native promoter and expressed from a random position in the genome. The pRF-HUE vector contains a constitutive *A. nidulans gpdA* promoter next to the UCS (see Fig. 4). The vector can be used for constitutive expression of heterologous/reporter genes or complementation experiments requiring constitutive expression.

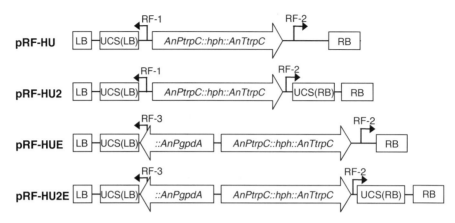

Fig. 4. T-DNA region of the USER vectors with placement and direction of standard test primers used for screening and validation of the constructs in *E. coli* and correct integration in the fungal transformants.

2. Targeted integration into the genome

The pRF-HU2 and pRF-HU2E vectors contain two UCSs to accommodate cloning of two PCR amplicons into different positions in the T-DNA, as shown in Fig. 3. The pRF-HU2 vector is intended for targeted gene replacement, while pRF-HU2E allows for *in locus* over-expression by inserting the constitutive *AnPgpdA* promoter in front of the coding sequence to be over expressed (see Fig. 4). Both vectors can also be used for more complex tasks such as the targeted introduction of expressional and translational reporter gene constructs by utilizing the USER Fusion technique described by Geu-Flores and co-workers (27).

1.3. Screening and Validation of Fungal Transformants

Introduction of a targeted gene replacement cassette into a fungus by ATMT can result in different recombination events due to competing DNA repair mechanisms or protection systems that are active in the recipient cell. The reported fates of T-DNA in *S. cerevisiae* (28) include:

1. Desired integration by HR via double crossover (dcHR) between linear T-DNA and the genomic target.

2. Degradation by exonuclease.

3. Formation of circular double-stranded T-DNA and autosomal replication.

4. Random integration into the genome via non-homologous end joining (NHEJ).

5. Integration by HR via a single crossover reaction between upstream or downstream HRSs in circular T-DNA and the target locus (scHR).

6. Or combinations of the above.

Complete degradation of the T-DNA or extensive truncation of the T-DNA ends results in the loss of the selection marker gene and the recipient cell will not survive the imposed selection regime. The formation of circular T-DNA molecules typically leads to unstable replication due to a lack of an autosomal replication sequence in the T-DNA and will eventually result in loss of the selection marker. However, integration by NHEJ, scHR, or dcHR all result in stable replication of the selection marker gene and thereby viable transformants (see Note 1).

To discriminate between the latter and the correct homologous integration event, a diagnostic PCR procedure based on four primer pairs was designed. The first PCR reaction checks the quality of the gDNA isolated from the transformant and targets the selection marker gene (hyg588U/L primers). The second reaction tests the presence/absence of the targeted locus in the cells (*gene*-T1/T2). Finally, it is very important to show that the correct homologous crossover on both sides of the target locus has taken place, which is addressed by the third and fourth PCR reaction. These reactions use the primers *gene*-T3 and *gene*-T4, which are designed to anneal in the genome outside the used HRSs and are used in combination with primers located in the introduced selection marker gene (RF-1 and RF-2) (see Fig. 5).

The diagnostic PCR reactions allow recognition of the T-DNA integration events and identification of transformants with the desired gene replacement (see Fig. 5). However, it cannot determine if correct gene replacement has occurred in combination with an additional random ectopic integration event. It is therefore imperative that the transformants are further analysed by Southern blot analysis or via a sexual backcrossing (if possible) to show that the strain contains only a single copy of the T-DNA and selection marker gene.

For the majority of fungal species, it is essential to produce single spore cultures or homokaryotic mycelium to ensure that the transformant strains contain only nuclei of the same genotype. Single spore colonies are typically obtained by plating a spore dilution series onto selective plates and then isolating the resulting colonies as they appear. Hyphal tip propagation and sonic disruption techniques have been used for preparing homokaryotic mycelium for fungal species that do not readily produce spores (29). For fungal species with a low gene replacement frequency (<20%), it is most cost-efficient to perform a preliminary screening (*gene*-T1/T2 primers) to identify likely correct transformants and develop homokaryotic strains from these followed by verification using the diagnostic primers.

In our experience, it is not necessary to produce single spore cultures of *F. graminearum* transformants obtained by the described ATMT protocol. A gene replacement analyses of a mycelium pigment biosynthetic pathway in *F. graminearum* (30) showed that all transformants had a stable and uniform colour phenotype (recessive character), proving that only a single genotype was present. This was confirmed by diagnostic PCR and Southern blot analysis.

a

	Type	hyg	T1/T2	T3	T4
1	Wild type / non transformed	-	+	-	-
2	dcHR resulting in replacement	+	-	+	+
3	Degradation of T-DNA	+/-	+	-	-
4	Autosomal replication	+	+	-	-
5	Ectopic integration (NHEJ)	+	+	-	-
6	scHR with upstream HRS	+	+/-	+	-
7	scHR with downstream HRS	+	+/-	-	+
8	Type 2 + Type 5	+	-	+	+

b Scenario no. 2

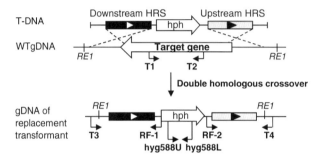

c Scenario no.5

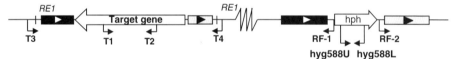

d Scenario no.6

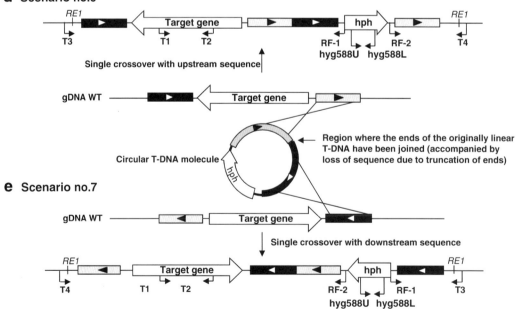

2. Materials

2.1. Equipment

1. Heating block or water bath.
2. 25, 28, and 37°C incubators with and without orbital shaker.
3. PCR thermocycler.
4. Erlenmeyer flasks: 50 and 300 mL, 1 L.
5. Tabletop centrifuge for Eppendorf tubes.
6. Centrifuge with a capacity to process up to 400 mL solution at $4,000 \times g$ and 4–25°C.
7. Electroporation cuvette with a 0.2-mm electrode gap.
8. Electroporation apparatus such as Bio-Rad Gene Pulser II or similar.
9. Black filter paper such as AGF 220 85 mm or similar (see Note 2).
10. Sterile toothpicks, Drigalski spatula, 1.5-mL centrifuge tubes, 15-mL centrifuge tubes, and 50-mL centrifuge tubes.
11. Plastic Petri dishes 5.5 and 9-cm and a 9-cm glass Petri dish.
12. Bottle top and syringe 0.2 μm filters for sterilization of solutions.
13. Miracloth (EMD Chemicals).
14. Light microscope and haemocytometer.
15. Spectrophotometer and agarose gel electrophoresis system for analysis of DNA.

2.2. Solutions and Media

1. MilliQ water or distilled water (sterile).
2. Glycerol 10% v/v, 100 mL: 10 mL 100% glycerol, 90 mL MilliQ water.
3. Glycerol 20% v/v, 100 mL: 20 mL 100% glycerol, 80 mL MilliQ water.
4. Glucose 20% w/v, 1 L: Dissolve 200 g glucose (CAS: 14431-43-7) in 1,000 mL MilliQ water and filter-sterilize.
5. TE buffer 10:1: 10 mM Tris–HCl (CAS: 77-86-1) and 1 mM EDTA (CAS: 60-00-4), pH 8.

Fig. 5. (**a**) Expected outcome of a PCR screening for the various fates of the T-DNA using the four primer pairs described in the main text. (**b**) Integration via dcHR between the HRSs in the linear T-DNA and the genomic target. (**c**) Integration of the T-DNA via NHEJ into a random position in the genome, the strain retains the target locus (positive for T1/T2). (**d**) Integration via scHR between the upstream HRS in a circular T-DNA and the upstream HRS of the genomic target. The resulting transformant is positive for the RF-2/gene-T4 reaction but negative for the RF-1/gene-T3 reaction. (**e**) Integration via scHR between the downstream HRS in the circular T-DNA and the downstream HRS in the genome (see Note that the genomic sequence is inverted compared to the other situation in the figure). The resulting transformants will be positive for the RF-1/gene-T3 reaction but negative for the RF-2/gene-T4 reaction.

6. L-aspargine 50 mM, 1 L: Dissolve 6.6 g L-aspargine (CAS: 70-47-3) in 1,000 mL MilliQ water and filter-sterilize.

7. $MgSO_4$ 210 mM, 500 mL: Dissolve 23.99 g $MgSO_4 \cdot 6H_2O$ (CAS: 30-18-1) in 500 mL MilliQ water and autoclave.

8. KH_2PO_4 1.12 M+KCl 0.7 M, 500 mL: Dissolve 79.21 g KH_2PO_4 (CAS: 7778-77-0) and 26.09 g KCl (CAS: 7447-40-7) in 500 mL MilliQ water, adjust pH to 6 and autoclave.

9. Acetosyringone 10 mM, 10 mL: Dissolve 19.62 mg acetosyringone (CAS: 2478-38-8) in 10 mL MilliQ-water. Stir for 1 h. Adjust the pH to 8 with 5 M KOH. Filter-sterilize and store at −20°C.

10. MES 1 M, 100 mL: Dissolve 19.52 g MES (CAS: 145224-94-8) in 80 mL MilliQ water, adjust pH to 5.3 with 5 M KOH and then bring the volume to 100 mL. Filter-sterilize and store at −20°C.

11. $CaCl_2$ 0.1 M, 50 mL: Dissolve 5.55 g $CaCl_2$ (CAS: 10043-52-4) in 500 mL MilliQ, autoclave and store at 4°C.

12. 200× Trace solution for Vogels salts, 1 L: Dissolve 5 g citric acid ($C_6H_8O_7 \cdot H_2O$) (CAS: 5949-29-1), 5 g $ZnSO_4 \cdot 7H_2O$ (CAS: 7446-20-0), 1 g $Fe(NH_4)_2(SO_4)_2 \cdot 6H_2O$ (CAS: 7783-85-9), 0.25 g $CuSO_4 \cdot 5H_2O$ (CAS: 7758-99-8), 0.05 g $MgSO_4 \cdot H_2O$ (CAS: 30-18-1), 0.05 g H_3BO_3 (CAS: 10043-35-3), 0.05 g $Na_2MoO_4 \cdot 2H_2O$ (CAS: 10102-40-6) in 1,000 mL MilliQ water and sterilize by filtration.

13. 50× Vogels salts, 1 L: Dissolve 125 g Na_3-citrate $\cdot 2H_2O$ (CAS: 6132-04-3), 250 g KH_2PO_4 (CAS: 7778-77-0), 9.3 g $MgSO_4 \cdot 6H_2O$ (CAS: 30-18-1), 5 g $CaCl_2 \cdot 2H_2O$ (CAS: 10035-04-8), and 5 mL 200× Trace solution in 1,000 mL MilliQ water and sterilize by filtration.

14. 1,000× Trace elements for DFM medium, 1 L: Dissolve 40 mg $Na_2B_4O_7 \cdot 10H_2O$ (CAS: 1303-96-4), 400 mg $CuSO_4 \cdot 5H_2O$ (CAS: 7758-99-8), 1.2 g $FeSO_4 \cdot 7H_2O$ (CAS: 7782-63-0), 700 mg $MnSO_4 \cdot H_2O$ (CAS: 10034-96-5), 800 mg $Na_2MoO_4 \cdot 2H_2O$ (CAS: 10102-40-6), 10 g $ZnSO_4 \cdot 7H_2O$ (CAS: 7446-20-0) in 1,000 mL MilliQ water and sterilize by filtration (the iron will precipitate as an insoluble oxide if autoclaved).

15. 2.5× Salt solution, 1 L: Dissolve the following salts one at a time to avoid formation of insoluble complexes: 3.625 g KH_2PO_4 (CAS: 7778-77-0), 5.125 g K_2HPO_4 (CAS: 88-57-1), 0.375 g NaCl (CAS: 7647-14-5), 1.160 g $MgSO_4 \cdot 6H_2O$ (CAS: 30-18-1), 0.165 g $CaCl_2 \cdot 2H_2O$ (CAS: 10035-04-8), 0.0062 g $FeSO_4 \cdot 7H_2O$ (CAS: 7782-63-0), 1.250 g $(NH4)_2SO_4$ (CAS: 7783-20-2) in 1,000 mL MilliQ water, and sterilize by filtration.

16. Solid and liquid LB medium, 1 L: Mix 5 g Bacto-tryptone, 2.5 g Bacto-yeast extract, 5 g NaCl with 950 mL MilliQ water.

Adjust pH to 7.4 for *E. coli* and 7.7–7.8 for *A. tumefaciens* media. Add MilliQ water to 1,000 mL. For solid medium, add 10 g agar before adjusting the pH. Autoclave solutions.

17. Liquid SOC medium, 500 mL: Mix 10 g Bacto-tryptone, 2.75 g Bacto-yeast extract, 0.3 g NaCl, 0.37 g KCl in 480 mL MilliQ water. Autoclave and cool to 50°C. Add 0.95 g $MgCl_2$, 10 mL 20% glucose. Adjust pH to 7.0. Add MilliQ water to 500 mL.

18. RA medium, 1 L: Mix 50 g succinic acid ($C_4H_4O_4Na_2 \cdot 6H_2O$) (CAS: 6106-21-4), 12.1 g $NaNO_3$ (CAS: 7631-99-4), and 20 mL 50× Vogels salts (–N, –C) with 950 mL MilliQ water and autoclave. Before use, add 50 mL sterile 20% glucose.

19. Water agar for IMAS plates, final volume 300 mL: Mix 6 g Bacto agar and 146 mL MilliQ water in a 300 mL bottle. Autoclave and re-melt in microwave oven before adding the remaining components of IMAS media (see next point).

20. IMAS-medium solid, 300 mL: Mix 120.0 mL 2.5× Salt solution (pre-heated to 60°C), 7 mL 20% glucose (w/v), 7.5 mL 20% glycerol (v/v) with melted water agar (final agar concentration 2%). Cool the solution to 55°C before adding: 12.0 mL MES-solution (1 M) (see Note 3), 6.0 mL acetosyringone (10 mM).

21. Water agar for DFM plates, final volume 500 mL: Mix 10 g Bacto agar and 358 mL MilliQ water in a 500 mL bottle. Autoclave and re-melt in microwave oven before adding the remaining components of DFM media (see next point).

22. Defined *Fusarium* Medium (DFM), 500 mL: Mix 31.25 mL 20% glucose, 100.0 mL 50 mM L-aspargine (pre-heated to 60°C), 5.0 mL 210 mM $MgSO_4$, 5.0 mL 1.12 M $KH_2PO_4 + 0.7$ M KCL (pH 6), and 0.5 mL 1,000× Trace elements with melted water agar (final agar concentration 2%). Cool the solution to 55°C before adding the required antibiotics.

2.3. Antibiotic Stocks (Hazardous and Toxic Compounds. Work in Fume Hood and Use Appropriate Gloves)

All stock solutions are sterilized by filtration and stored at –20°C in 1–2 mL aliquots, except HygB, which is stored at 4°C.

1. Kanamycin (Kan) stock: Dissolve 10 mg kanamycin sulphate (CAS: 25389-94-0) in 1 mL MilliQ water and use a work concentration of 50 µg/mL in the final medium.

2. Cefoxitin (Cef) stock: Dissolve 50 mg cefoxitin sodium (CAS: 33564-30-6) in 1 mL MilliQ water and use a work concentration of 300 µg/mL in the final medium.

3. Rifampicin (Rif) stock: Dissolve 50 mg rifampicin (CAS: 13292-46-1) in 1 mL DMSO and a use a work concentration of 10 µg/mL in the final medium. The compound is light-sensitive,

and to prevent degradation, keep the stock solution wrapped in tinfoil; media and plates should be stored in the dark.

4. Hygromycin B (HygB) stock: Dissolve 100 mg hygromycin B (CAS: 31282-04-9) in 1 mL MilliQ water and use a work concentration between 100 and 150 μg/mL for *F. graminearum* in the final medium. We normally buy this as a 100 mg/mL solution from InVivoGen.

2.4. Enzymes and Molecular Biological Kits

1. Taq DNA polymerase and PCR reagents for screening of transformants.

2. Proof-reading DNA polymerase:

PfuX7 DNA polymerase or PfuTurbo Cx hotstart DNA polymerase (Stratagene) (see Note 4).

3. In-Fusion dried-down PCR cloning kit (Clontech).

4. USER cloning enzyme mix (New England Biolabs).

5. Suitable restriction enzymes and buffers: *SmaI* and *SwaI* for In-Fusion cloning into pAg-H3 or *PacI* and *Nt.BbvCI* for USER cloning.

6. PCR purification kit (GFX PCR and gel purification kit or similar).

7. Plasmid purification kit (Qiagen Miniprep kit or similar).

2.5. Organisms and Cells

1. *F. graminearum* pH-1 (*Gibberella zeae*).

2. Homemade or commercial chemical competent *E. coli* cells (DH5a, JM109 or similar) for USER cloning.

3. Fusion-Blue chemical competent cells (>10^8 cfu/μL) (Clontech).

4. *A. tumefaciens* LBA4404.

2.6. Primers

1. Primers with appropriate 5′ overhangs for amplifying the two HRSs (for instruction on the primer design for In-Fusion cloning, see Subheading 3.1.1, and USER cloning, see Subheading 3.2.1).

2. Standard primers for screening and verification of fungal transformants.
RF-1: 5′-AAATTTTGTGCTCACCGCCTGGAC
RF-2: 5′-TCTCCTTGCATGCACCATTCCTTG
RF-3: 5′-TTGCGTCAGTCCAACATTTGTTGCCA
Hyg588U: 5′-AGCTGCGCCGATGGTTTCTACAA
Hyg588L: 5′-GCGCGTCTGCTGCTCCATACAA

In addition to these standard primers, you will also need to design four test primers (T1–T4) for each targeted genome modification (see Subheading 1.3). The target *gene*-T1/T2 primer pair should amplify part of the replaced region in knockout experiments, while for *in locus* over-expression experiments they should be designed to amplify the junction between the endogenous promoter and the coding sequence. The target *gene*-T3 and *gene*-T4 primers are used in combination with the RF-1 and RF-2 standard primers, respectively. Design the primers so that they anneal outside the used HRS and amplify part of the genome+HRS+part of selection marker gene.

3. Methods

3.1. In-Fusion Cloning Strategy (Can Be Used with All Vectors)

This vector construction strategy is applicable for all vectors with two unique restriction sites, one located on either side of the fungal selection marker gene. We have mainly used the pAg1-H3 vector (17), depending on the two unique blunt-end cutting restriction enzymes *SwaI* and *SmaI*. The protocol below is optimized for pAg1-H3, but can easily be applied to other plasmids with minor modifications.

3.1.1. Primer Design

1. Design primers for amplification of an upstream and a downstream HRS using standard primer design criteria (or a suitable software). The required length of the HRSs is species-specific; for *F. graminearum*, we normally use 1–2 kb HRSs.

2. Add 5′ overhangs to the primers identical to the sequences surrounding the two restriction enzyme sites you want to use for insertion of the HRS into the binary vector.

   In the case of the pAg1-H3 vector combined with the *SwaI* and *SmaI* restriction enzymes, add the following overhangs to the gene-specific primers (GSP):

   Upstream primer forward:

       5′- TCGGTACCAAGGCCC-(GSP sequence).

   Upstream primer reverse:

       5′- GGCCGTGGCCAGCCC-(GSP sequence).

   Downstream primer forward:

       5′- CGTTTAAACGGATTT-(GSP sequence).

   Downstream primer reverse:

       5′- GACATTAATTAATTT-(GSP sequence).

   If you are using a different vector/enzyme combination, please consult the "In-Fusion™ PCR cloning Kit User Manual" or use the online primer design tool offered by Clontech (http://bioinfo.clontech.com/infusion/convertPcrPrimers-Init.do) to determine which overhangs you should add.

3. Order the primers (see Note 5).

4. Amplify the two HRS via PCR (We use a two-step PCR program starting with five cycles at a low annealing temperature based on the calculated temperature for the gene-specific part of the primer followed by 25 cycles with an annealing temperature of 60°C).

3.1.2. Vector Preparation

1. Prepare 10 μg of the pAg1-H3 vector following a protocol of your choice.

2. Digest the vector with *Sma*I:

*Sma*I	2 μL = 40 units
pAg1-H3	187 μL = 10 μg
BSA	24 μL
NEBuffer 4	24 μL
Total	240 μL

3. Incubate the reaction overnight at 25°C.

4. Purify the digested vector using two GFX spin columns, or similar, by loading 120 μL onto each. Elute in 50 μL EB buffer (see Note 6).

5. Digest the vector with *Swa*I:

*Swa*I	4 μL = 40 units
pAg1-H3 (*Sma*I)	75 μL = approximately 7 μg
BSA	10 μL
NEBuffer 3	10 μL
Total	100 μL

6. Incubate the reaction overnight at 25°C.

7. Check the digestion by gel electrophoresis (load 4 μL digestion mix). You should see two bands (2,565 and 3,851 bp). The digestion results in a 1 to 1 molar ratio of the two fragments.

8. Purify the digested vector using a single GFX-column and elute in 50 μL EB buffer.

9. Determine the concentration using a spectrophotometer.

3.1.3. In-Fusion Cloning (Dried-Down Kit)

1. Mix the inserts and vector fragments in a 2:2:1:1 molar ratio, respectively, with the vector fragments amounting to approximately 200 ng.

2. Add MilliQ water to a total volume of 10 μL.

3. Add the 10 μL of vector+inserts to a tube with dried-down In-Fusion cloning mix.

4. Mix by pipetting up and down to ensure that the pellet with the dried-down enzyme mix has completely dissolved.

5. Incubate the reaction at 40°C for 30 min.

6. Add 40 μL 10:1 TE buffer, bringing the volume to 50 μL.

7. Use the solution directly for transformation or store at –20°C for later use.

8. Transformation: Use 2.5 μL of the diluted In-Fusion reaction to transform 50 μL of chemical competent *E. coli* cells with a transformation efficiency of >10^8 cfu/μg (such as Fusion-Blue cells from Clontech). Use a protocol of your choice.

   After the transformation, pellet the cells by centrifugation and remove most of the supernatant, leaving 70 μL. Resuspend the cells by pipetting or vortexing. Plate the cells on a LB+Kan$_{50}$ agar plate and incubate overnight at 37°C.

9. Screen the resulting transformants by colony PCR using the RF-1 and RF-2 primers in combination with the primers you used for amplifying the inserts.

10. Set up liquid cultures, incubate and prepare the plasmid (Qiagen Miniprep or similar).

11. Verify the obtained transformants via restriction enzyme digestion and sequence the HRS using the RF-1 and RF-2 primers (or RF-3 and RF-3 for –HU2E vectors).

3.2. USER Friendly Cloning Strategy (Requires Vectors with Two UCSs)

3.2.1. Primer Design

1. Design primers for amplification of an upstream and a downstream HRS using standard primer design criteria (or a suitable software). The required length of the HRSs is species-specific; for *F. graminearum*, we normally use 1–2 kb HRSs. Add the following 5′ overhangs to the gene-specific primer (GSP):

For pRF-HU and pRF-HUE:

   LB UCS primer forward: 5′- GGACTTAAU-(GSP sequence).

   LB UCS primer reverse: 5′- GGGTTTAAU-(GSP sequence).

For pRF-HU2 and pRF-HU2E:

   RB UCS primer forward: 5′- GGTCTTAAU-(GSP sequence).

   RB UCS primer reverse: 5′- GGCATTAAU-(GSP sequence).

   LB UCS primer forward: 5′- GGACTTAAU-(GSP sequence).

   LB UCS primer reverse: 5′- GGGTTTAAU-(GSP sequence).

If you are planning to perform both targeted gene replacement and *in locus* over-expression of a gene, you can reuse the HRS located in the promoter region for both constructs (24).

The pRF-HUE and pRF-HU2E vectors contain the *A. nidulans gpdA* promoter next to the LB UCS allowing for constitutive over-expression (see Fig. 3). For in-frame fusion with the first codon of your gene, design the "LB UCS forward" primer to start with the G of the start codon (ATG). The (AT) of the start codon will be added when you add the 5′ overhangs for USER cloning. The LB UCS forward and LB UCS reverse primers should amplify the coding sequence. In case of *in locus* over-expression (pRF-HU2E), it is not necessary to amplify the entire CDS, just enough to ensure integration via crossover. However, for expression from a random locus in the genome (pRF-HUE), the entire coding sequence and terminator regions should be amplified (see Note 7).

3.2.2. Vector Preparation

1. Digest 10 µg of pRF-HU2 with 70 units *PacI* (7 µL) overnight at 37°C in a total volume of 300 µL with NEBuffer 4+BSA.

2. The next day, add an additional 20 units (2 µL) of *PacI* and 40 units (4 µL) *Nt.BbvCI*, and incubate for 1 h at 37°C.

3. Verify the linearization of the vector by gel electrophoresis (2–3 µL).

4. Purify the linearized vector using two GFX spin columns and elute each column with 50 µL EB buffer.

5. Determine the DNA concentration using a spectrophotometer. We normally obtain a DNA concentration of approximately 50 ng/µL and a volume of 90 µL, sufficient for 20 cloning reactions (see Note 8). Divide the app. 90 µL into four aliquots and store them at −20°C until use; this will prevent degradation of the DNA by repeated freezing and thawing.

6. Optional but recommended: When preparing larger batches of the vectors for USER friendly cloning, it is a good idea to check the quality by performing a test transformation reaction without inserts. This will allow you to establish the background caused by undigested plasmid.

3.2.3. Preparing Chemical Competent E. coli Cells

The following steps are performed with sterile materials under sterile conditions.

1. Inoculate 5 mL LB medium with *E. coli* and incubate overnight at 37°C with 150 rpm (see Note 9).

2. Pre-heat 100 mL LB medium and inoculate with 2 mL of the overnight culture.

3. Incubate at 37°C with 150 rpm until OD_{600} reaches 0.5.

4. Cool the culture on ice for 15 min.

5. Pellet the cells by centrifugation ($4,000 \times g$) for 10 min at 4°C.

6. Discard the supernatant and resuspend the pellet in 50 mL ice-cold 0.1 M $CaCl_2$.

7. Incubate on ice for 30 min.

8. Pellet the cells by centrifugation and discard the supernatant.

9. Resuspend the pellet in 2.5 mL ice-cold 0.1 M $CaCl_2$.

10. Transfer 100 µL aliquots to pre-cooled 1.5 mL tubes and store at −80°C until use.

1. Amplify the two inserts with PfuTurbo C_x Hotstart or PfuX7 DNA polymerase following the manufacturer's recommendations in a reaction volume of 15 μL per reaction.

2. Check the success of the PCR reactions by agarose gel electrophoresis of 5 μL PCR per reaction. Note that it is not necessary to purify the PCR amplicon before USER Friendly cloning.

3. Mix the following in a 0.2 mL PCR tube (see Note 10).

PCR product 1	5 μL (app. 100 ng)
PCR product 2	5 μL (app. 100 ng)
Linearized vector	4 μL (200 ng)
USER enzyme mix	1 μL
Total volume	15 μL

4. Incubate at 37°C for 20 min followed by 25°C for 20 min (we use a PCR cycler for this).

5. Transformation (see Note 11): Use the 15 μL of the USER cloning reaction for transformation of 50 μL of competent *E. coli* cells (>10^6 cfu/mg). Following a heat shock transformation protocol of your choice (see Note 12). Following the transformation process, pellet the cells by centrifugation ($2,500 \times g$ for 60 s) and remove most of the supernatant, leaving 70 μL. Resuspend the cells by pipetting or vortexing. Plate the resuspended cells on a LB+Kan$_{50}$ agar plate and incubate overnight at 37°C.

6. Screen an appropriate number of the resulting colonies (typically 2–10) by colony PCR using the insert-specific primers, used for amplification of the two inserts in step 1 or in combination with appropriate RF-1, RF-2, or RF-3 primers (two reactions pr colony) (see Note 13).

7. Set up liquid cultures, incubate, and prepare the plasmid (Qiagen Miniprep or similar).

8. Verify the obtained transformants via restriction enzyme digestion and sequence the HRS using the RF-1 and RF-2 primers (or RF-2 and RF-3 for –HU2E vectors).

3.3. Validation of the Constructed Vector for Targeted Integration via ATMT

It is important that the cloned HRSs do not contain any PCR-induced mutations. Genes in many fungi are separated by very short intergenic regions and HRSs will in many cases overlap with surrounding genes or their regulatory elements. The HRSs in plasmids, constructed by In-Fusion or USER cloning, should therefore be sequenced using the RF-1 and RF-2 primers.

3.4. Introducing the Binary Vector into A. tumefaciens

3.4.1. Preparing Electro Competent A. tumefaciens Cells

The following steps are performed with sterile material under sterile conditions.

1. Inoculate 5 mL LB+Rif$_{10}$ medium with *A. tumefaciens* and incubate overnight at 28°C with 100 rpm (see Note 9).

2. Pre-heat 200 mL LB+Rif$_{10}$ medium and inoculate each flask with 150 μL of the overnight culture.

3. Incubate at 28°C with 100 rpm until OD$_{600}$ reaches 0.5.

4. Cool the culture on ice for 15 min.

5. Pellet the cells by centrifugation ($4,000 \times g$) for 10 min at 4°C.

6. Discard the supernatant and resuspend the pellet in 50 mL ice-cold MilliQ water and pellet again.

7. Discard the supernatant and resuspend the pellet in 10 mL ice-cold 10% glycerol and incubate for 1 h on ice.

8. Pellet the cells, discard the supernatant, and resuspend the cells in 3 mL ice-cold 10% glycerol (keep the cells on ice).

9. Transfer 100 μL aliquots to pre-cooled 1.5 mL tubes and store at –80°C until use.

3.4.2. Electroporation of A. tumefaciens

1. Place an electroporation cuvette on ice.

2. Thaw the electro competent *A. tumefaciens* cells on ice.

3. Using a pipette, place 1 μL of plasmid DNA on the internal side of the cuvette.

4. Use 50 μL of competent cells to gently flush the DNA containing droplet to the bottom of the cuvette. Gently thump the cuvette on a table a couple of times to remove air bubbles. Wipe down the exterior sides of the cuvette with a paper towel (to prevent short circuits).

5. Place the cuvette in the electroporation apparatus (Voltage = 2.50 kV, Capacitance = 25 μF, Resistance = 200 W) and shock. The accumulator should discharge the stored energy in approximately 5 ms (see Note 14).

6. Remove the cuvette from the apparatus and add 450 μL SOC medium.

7. Pour the cells into a sterile 1.5 mL Eppendorf tube.

8. Incubate the cells at 28°C for 90 min with 350 rpm.

9. Plate the cells onto two LB+Kan$_{50}$+Rif$_{10}$ plates (1/10 and 9/10 of the volume).

10. Incubate the plates at 28°C for 2–3 days.

11. Select a single colony and verify that it contains the vector (either by PCR or restriction enzyme digestion).

Day 1:

Inoculate 2×1 L baffled Erlenmeyer flasks, containing 250 mL RA medium, with ten agar-plugs (5×5 mm) from a fresh plate containing the *F. graminearum* wild type. Incubate at 20°C for 3 days with 150 rpm horizontal shake.

Day 4:

1. Filter the cultures through a sterile Miracloth into sterile centrifuge tubes.

2. Wash the Erlenmeyer flasks with 250 mL MilliQ water and filter it through the Miracloth into new sterile centrifuge tubes.

3. Pellet the spores by centrifugation ($14,000 \times g$) at 4°C for 40 min.

4. Discard the supernatant.

5. Resuspend the pellet with 10 mL of MilliQ water and transfer the solution to sterile centrifuge tubes and pellet the spores by centrifugation as above.

6. Discard the supernatant.

7. Resuspend the pellet in 5 mL MilliQ water.

8. Determine the spore concentration using a light microscope and a haemocytometer.

9. Pellet the spores by centrifugation as above but at 8°C for 30 min.

10. Resuspend the spores with 10% glycerol to obtain a spore concentration of 1×10^8 per mL.

11. Store the spores as 1 mL aliquots at −80°C for up to 1 year.

3.5. A. tumefaciens-Mediated Transformation of the Target Fungus

The following protocol is optimized for transformation of *F. graminearum* and typically results in 1,500–2,000 transformants per 1×10^7 conidia with a targeted gene replacement frequency of 70% (see Note 15). For fungal species where no protocol has been developed, please consult the review by Michielse and co-workers (31), which provides a run-through of the various biotic and abiotic factors that can affect the transformation frequency of ATMT and the T-DNA copy number.

Day 1:

Inoculate 10 mL of LB+Kan$_{50}$+Rif$_{10}$ medium in a 50 mL Erlenmeyer flask with a single *A. tumefaciens* (pRF-HU2::Δgene) colony and incubate for 1–2 days at 28°C while shaking at 100 rpm (see Note 9).

Day 3:

• Place eight black AGF220 80 mm filters in a glass Petri dish and autoclave them.

• Prepare 50 mL liquid IMAS medium.

- Inoculate 10 mL IMAS+Kan$_{50}$ in a 50 mL Erlenmeyer flasks with 100 µL *A. tumefaciens* pRF-HU2::Δgene and incubate at 28°C while shaking (100 rpm) until OD$_{600}$ reaches 0.5–0.7 (typically the next day) (see Note 16).

Day 4:

1. Cast 8 IMAS plates (diameter 9 cm). Label plates: "IMAS—Δgene #1 to #8."

2. Place sterile filters onto six of the eight IMAS plates (see Note 17).

3. When the *A. tumefaciens* cells have reached an OD$_{600}$ of 0.5–0.7, dilute the *F. graminearum* spores with liquid IMAS medium to a final spore concentrations of 2×10^6 spores/mL.

4. Mix the *A. tumefaciens* culture in a 1:1 (v:v) ratio with the *F. graminearum* spores (resulting in 1×10^6 spores/mL).

5. Spread 200 µL of the bacterial/fungal mixture onto each of the sterile filters (equalling 2×10^5 spores/plate) using a sterile Drigalski spatula. Keep mixing the bacterium/fungal suspension before pipetting the 200 µL to obtain an even concentration of cells. Make the following control plates: (1) Sterile filters, (2) 100 µL of *A. tumefaciens* strain (without filter), and (3) *F. graminearum* spores (2×10^5 spores, without filter).

6. Incubate the eight plates for 2–3 day at 28°C in darkness (see Note 18).

Day 8:

1. Check the control plates: the filter control should be clean, while the bacterial and fungal control plates should contain bacterial and fungal colonies, respectively.

2. Cast 5 DFM+HygB$_{150}$+Cef$_{300}$ plates (first set of selection plates = S1) (see Note 19).

3. Peel off the filters under sterile conditions and transfer them onto the S1 plates using sterile tweezers. Discard the IMAS plates.

4. Incubate the five plates for 3–5 days at 25°C.

Day 12:

1. For some fungal species, colonies will be visible at this stage allowing you to skip the second selection round and proceed to the tasks outlined for day 15.

2. Cast 5 DFM+HygB$_{150}$ plates (second set of selection plates = S2).

3. Under sterile conditions, transfer the filters onto the DFM+HygB$_{150}$ plates.

4. Incubate the five S2 plates at 25°C for 3–5 days, depending on the growth rate of the transformants.

Day 15:

1. Cast an appropriate number of 5.5 cm DFM+HygB$_{100}$ plates for isolation of the transformants (see Note 20).

2. Gently peel off the filters from the S2 plates with sterile tweezers and discard the filters. Transformants that have grown into the medium are transferred to the small DFM+HygB$_{100}$ plates using a sterile toothpick, simply stab the toothpick into the agar at the edge of the colony, and then repeat the action at the centre of the isolation plate (see Note 21).

3. Incubate the plates for 2–7 days, until the colonies reach the edge of the plates.

3.6. Screening for and Validation of Gene Replacement

The identification of transformants that are the result of gene replacement is easily achieved by diagnostic PCR, relying on four primer pairs. The process consists of four steps: (1) Purification of gDNA, (2) testing that the gDNA is of PCR quality, (3) testing for loss of the target locus, and (4) testing for crossover events at the two ends of the construct.

3.6.1. Fast Preparation of gDNA (Based on (32)) for Screening of Transformants (see Note 22)

1. Transfer approximately 2 mg of mycelium (use sterile pipette tip or toothpick) to 50 μL of 10:1 TE buffer in a 1.5 mL tube.

2. Cook the sample for 1-2 min in a microwave oven at maximum power.

3. Let the samples cool for 2 min at room temperature (25°C).

4. Spin the cells for 5 min at 10,000 rpm in a tabletop centrifuge.

5. Transfer the supernatant to a new tube (see Note 23), dilute 100-fold with MilliQ water, and store at –20°C until use.

6. For a 15 μL PCR reaction, use 1 μL of the diluted supernatant.

3.6.2. PCR Screening of Transformants

1. Test the quality of the purified gDNA using the primers that amplify part of the selection marker gene. We use the Hyg588U/L primers which amplify a 588 bp fragment of the *hph* gene in the pAg1-H3-derived vectors (see Note 24).

2. Test the transformants that were positive in step 1 for loss of the targeted locus using the *gene*-T1/T2 primers. Transformants where the T-DNA has replaced the targeted locus will give negative PCR results, due to the loss of annealing sites for the used primers, while transformants still carrying a wild-type locus will give positive results.

3. Test transformants that were positive in step 1 and negative in step 2 using the *gene*-T3/RF-1 primers which amplify the left border (for the pRF-HU2E vector use RF-3).

4. Test transformants that gave a positive result in step 3 with the *gene*-T4/RF-2 primers which amplify the right border.

The gene-specific T3 and T4 diagnostic primers should be located outside the HRS, to ensure locus specific amplification (see Subheading 1.3).

3.6.3. Southern Blotting   The final confirmation of the genetic modification should be performed via Southern blot analysis to identify transformants that have ectopic integrated copies (see Note 25).

4. Notes

1. If you are working with a fungal species where the NHEJ pathway is dominant, increased gene targeting efficiency can be obtained by mutating genes that are involved in the NHEJ such as Ku70 or Lig4. However, it is important to note that such mutations are likely to affect fitness and virulence. The NHEJ affecting mutation should be removed from the genome, by a backcross if possible, prior to phenotypical characterization (33).

2. Why black filter paper? We have opted to use black filters for the co-transformation, as it makes it easier to identify transformants. However, standard filter paper, nitrocellulose, cellulose, and other types of filters/membranes have also been used with equal success.

3. The thawed MES stock contains a white precipitate. Heat the solution to room temperature and vortex the solution until the precipitate has completely dissolved.

4. These DNA polymerases are currently the only available proof-reading DNA polymerases compatible with 2-deoxyuridin containing primers. The PfuX7 is a non-commercial DNA polymerase expressed in *E. coli* that is easily and inexpensively purified via His/Ni affinity chromatography (34).

5. Though the primers are rather long, there is no need for ordering high-quality primers, as it is our experience that the In-Fusion cloning reaction selects against fragments with errors in the primer sequence, probably due to a lower annealing affinity with the vector backbone compared to primers without errors.

6. This step is included as the buffers for the two used enzymes are not compatible. For other combinations of enzymes, it will be possible to digest with both enzymes in a single reaction tube or to use a single enzyme that cuts twice in the vector at appropriate sites.

7. The terminator region, including the 3′ untranslated region and the transcriptional terminator, is difficult to predict with

existing gene prediction software. To ensure that we have a functional polyA signal and a transcriptional terminator, we normally include 500 bp downstream of the stop codon. When ordering the primers, note that the uracil by some suppliers is called 2-deoxyuridin.

8. If the vector is completely degraded during either of the restriction enzyme digestion steps, it is most likely due to a contamination with exonucleases. Repeat digestion with new chemicals and enzymes.

9. *A. tumefaciens* sometimes forms aggregates due to filamentous growth. It is our experience that cultures that show signs of filamentous growth retain their ability to transform the target fungus. However, this growth pattern makes it difficult to control the number of cells that are added to the co-incubation plates and makes it difficult to obtain an even spread of cells over the entire plate—therefore we try to avoid using such cultures. We have found that using a LB medium with a start pH of 7.7–7.9 and rotation at 100 rpm eliminates filamentous growth.

10. If the PCR reactions yield a specific product, we normally do not measure or adjust the DNA concentration of the added PCR products, but simply add 1–10 μL of the unpurified PCR reaction. However, if there is a need for gel-based purification of the PCR product due to unspecific products, it is important to add PfuTurbo PCR buffer or similar to the USER cloning reactions. If it is the first time you are performing USER cloning in your laboratory, it is recommended to try with variable amounts of the PCR inserts (1, 5, and 10 μL) to find the amount that yield the highest number of colonies.

11. Electrocompetent cells cannot be used as the electrical shock will cause the hybridized DNA fragments to disassociate.

12. We have found that the most common cause for low transformation efficiency is that pipetting of the chemical competent *E. coli* cells are not done carefully enough. Therefore, use a pipette with a long piston travelling length and pipette slowly.

13. If you consistently obtain a high background (high number of false positive) in the colony PCR screening using the amplicon-specific primers, it is most likely due to PCR amplicons found on the surface of the LB medium (leftovers from the cloning reaction). This problem can be solved either by using the insert-specific primers in combination with RF-1, RF-2, or RF-3 primers or by re-streaking the transformants onto new plates before performing the screening.

14. If the salt concentration in the DNA sample is too high, the electric charge will be discharged too fast resulting in a small explosion, which typically can be heard as a short loud "CLICK"

or "BANG." Repeat the procedure with 0.5 μL of your plasmid Miniprep, and new cells as the competent cells are killed by the "explosion."

15. The high gene replacement frequency observed in *F. graminearum* with ATMT is likely due to a dominant HR system similar to that found in *S. cerevisiae*. This notion is lent additional support by the observation that random mutagenesis experiments in *F. graminearum* via ATMT only result in few transformants (13). A similar situation has been reported for *F. fujikuroi* (35).

16. Several publications report that prolonged pre-induction of the *A. tumefaciens* cells with acetosyringone leads to higher transformation efficiency, but also a higher percentage of transformants with multiple T-DNA inserts (36, 37). For *F. graminearum*, we have not experienced this problem.

17. Make sure that no air pockets exist between the medium and filters; this is done by adding 50–400 μL of water (depending on the moisture of the plates) onto the centre of the filters and spreading it with a sterile Drigalski spatula.

18. The duration of the co-cultivation period and the incubation temperature have a large effect on the transformation efficiency, and if you are optimizing ATMT, it is worth focusing on these parameters.

19. When setting up an ATMT protocol for a fungus that has not been transformed before, it is important to test if the antibiotic you are planning to use is functional against the fungus and to establish the concentration you should use during the selection phase. Note that the sensitivity often varies between different growth media.

20. We normally use 5.5-cm petri dishes to reduce the amount of required medium. It is not recommendable to grow multiple transformants on the same isolation plate, as you risk cross-contamination and the formation of heterokaryons due to anastomosis. If you experience problems with intergrowing transformants due to rapid growth on the co-culturing plates or selection plates, try to reduce incubation temperature or reduce the used amount of spores.

21. It is our experience that the transfer of as few cells as possible gives the purest cultures and there is no need for cutting out agar blocks or similar. If you are using a toothpick as suggested in the protocol, make sure that these do not contain antimicrobial compounds.

22. This DNA extraction method yields low-quality genomic DNA that works for the PCR screening of the transformants, but not other techniques such as southern blotting or more complex PCR reactions.

23. Sometimes it can be an advantage to store the gDNA samples in PCR tubes in strip format as it allows use of multichannel pipettes during the subsequent PCR steps—However, be careful with the lids as it is easy to cross-contaminate the samples by mixing up the lids.

24. If you do not obtain any PCR products, try to dilute the gDNA solution 50 or 100 times and increase the number of PCR cycles. This will typically dilute any PCR inhibiting factors in the solution sufficiently to allow for a good PCR result.

25. Southern blot analysis—we have chosen to use the introduced selection marker gene as the standard probe and restriction enzymes cutting outside the homologous recombination flanks. The advantage of this is that the same probe can be used for all constructs. The use of a probe targeting the selection marker gene rather than the sequences from the targeted locus still allows you to identify transformants with multiple T-DNA insertions and errors in the size of the locus. This probe does not yield bands with wild-type gDNA; however, it is recommended to include wild-type gDNA for trouble shooting purposes, as described below.

    It is important to use a restriction enzyme, or a set of restriction enzymes, that cuts outside the region that was targeted for homologous recombination as this ensures that ectopic and targeted integration can be differentiated.

    If the size of the band observed in the Southern blot analysis does not match the theoretical expected size, it can be due to an error in the genome sequence that results in the "appearance" or "disappearance" of recognition sites for the used restriction enzyme(s). To test this hypothesis: strip the blot for probe and re-probe it with a probe against the target locus. This should result in a single band in lanes loaded with wild-type gDNA and allow you to check that the fragment in the wild type is as predicted based on the available genome sequence.

References

1. Matzke, A.J. and Chilton, M.D. (1981) Site-specific insertion of genes into T-DNA of the *Agrobacterium* tumor-inducing plasmid: an approach to genetic engineering of higher plant cells. *Journal of Molecular and Applied Genetics* 1(1), 39–49.

2. Chan, M.T., Chang, H.H., Ho, S.L., Tong, W.F. and Yu, S.M. (1993) *Agrobacterium*-mediated production of transgenic rice plants expressing a chimeric alpha-amylase promoter/beta-glucuronidase gene. *Plant Molecular Biology* 3, 491–506.

3. Lee, L.-Y. and Gelvin, S.B. (2008) T-DNA Binary Vectors and Systems. *Plant Physiology* 146, 325–332.

4. Gelvin, S.B. (2003) *Agrobacterium*-Mediated Plant Transformation: the Biology behind the "Gene-Jockeying" Tool. *Microbiology and Molecular Biology Reviews* 67(1), 16–37.

5. Pasternak, J.J., Gruber, M.Y., Thompson, J.E. and Glick, B.R. (1983) Development of DNA-mediated transformation systems for plants. *Biotechnology Advances* 1(1), 1–15.

6. Bundock, P., Dendulkras, A., Beijersbergen, A. and Hooykaas, P.J. (1995) Trans-kingdom T-DNA transfer from *Agrobacterium tumefaciens* to *Saccharomyces cerevisiae*. *EMBO Journal* 14(13), 3206–3214.

7. de Groot, M.J., Bundock, P., Hooykaas, P.J. and Beijersbergen, A.G. (1998) *Agrobacterium tumefaciens*-mediated transformation of filamentous fungi. *Nature Biotechnology* 16(9), 839–842.

8. Meyer, V., Mueller, D., Strowig, T. and Stahl, U. (2003) Comparison of different transformation methods for *Aspergillus giganteus*. *Current Genetics* 43(5), 371–377.

9. Walton, F.J., Idnurm, A. and Heitman, J. (2005) Novel gene functions required for melanization of the human pathogen *Cryptococcus neoformans*. *Molecular Microbiology* 57(5), 1381–1396.

10. Betts, M.F., Tucker, S.L., Galadima, N., Meng, Y., Patel, G., Li, L., Donofrio, N., Floyd, A., Nolin, S., Brown, D., Mandel, M.A., Mitchell, T.K., Xu, J.R., Dean, R.A., Farman, M.L. and Orbach, M.J. (2007) Development of a high throughput transformation system for insertional mutagenesis in *Magnaporthe oryzae*. *Fungal Genetics and Biology* 44(10), 1035–1049.

11. Blaise, F., Remy, E., Meyer, M., Zhou, L.G., Narcy, J.P., Roux, J., Balesdent, M.H., Rouxel, T. (2007) A critical assessment of *Agrobacterium tumefaciens*-mediated transformation as a tool for pathogenicity gene discovery in the phytopathogenic fungus *Leptosphaeria maculans*. *Fungal Genetics and Biology* 44(2), 123–138.

12. Jeon, J., Park, S.Y., Chi, M.H., Choi, J., Park, J., Rho, H.S., Kim, S., Goh, J., Yoo, S., Choi, J., Park, J.Y., Yi, M., Yang, S., Kwon, M.J., Han, S.S., Kim, B.R., Khang, C.H., Park, B., Lim, S.E., Jung, K., Kong, S., Karunakaran, M., Oh, H.S., Kim, H., Kim, S., Park, J., Kang, S., Choi, W.B., Kang, S. and Lee, Y.H. (2007) Genome-wide functional analysis of pathogenicity genes in the rice blast fungus. *Nature Genetics* 39(4), 561–565.

13. Malz, S., Grell, M.N., Thrane, C., Maier, F.J., Rosager, P., Felk, A., Albertsen, K.S., Salomon, S., Bohn, L., Schäfer, W. and Giese, H. (2005) Identification of a gene cluster responsible for the biosynthesis of aurofusarin in the *Fusarium graminearum* species complex. *Fungal Genetics and Biology* 42(5), 420–433.

14. Cardoza, R.E., Vizcaino, J.A., Hermosa, M.R., Monte, E. and Gutiérrez, S. (2006) A comparison of the phenotypic and genetic stability of recombinant *Trichoderma spp.* generated by protoplast- and *Agrobacterium*-mediated transformation. *The Journal of Microbiology* 44(4), 383–395.

15. Zwiers, L.H. and De Waard, M.A. (2001) Efficient *Agrobacterium tumefaciens*-mediated gene disruption in the phytopathogen *Mycosphaerella graminicola*. *Current Genetics* 39(5–6), 388–393.

16. Gardiner, D.M. and Howlett, B.J. (2004) Negative selection using thymidine kinase increases the efficiency of recovery of transformants with *targeted gene*s in the filamentous fungus *Leptosphaeria maculans*. *Current Genetics* 45(4), 249–255.

17. Zhang, A., Lu, P., Dahl-Roshak, A.M., Paress, P.S., Kennedy, S., Tkacz, J.S. and An, Z. (2003) Efficient disruption of a polyketide synthase gene (pks1) required for melanin synthesis through *Agrobacterium*-mediated transformation of *Glarea lozoyensis*. *Molecular Genetics and Genomics* 268(5), 645–655.

18. Frandsen, R.J., Albertsen, K.S., Stougaard, P., Sørensen, J.L., Nielsen, K.F., Olsson, S. and Giese, H. (2010) Methylenetetrahydrofolate reductase activity is involved in the plasma membrane redox system required for pigment biosynthesis in filamentous fungi. *Eukaryotic Cell* 9(8), 1225–1235.

19. Yi, M., Chi, M.H., Khang, C.H., Park, S.Y., Kang, S., Valent, B. and Lee, Y.H. (2009) The ER Chaperone LHS1 Is Involved in Asexual Development and Rice Infection by the Blast Fungus *Magnaporthe oryzae*. *The Plant Cell* 21(2), 681–695.

20. Wang, Y., DiGuistini, S., Wang, T.C., Bohlmann, J. and Breuil, C. (2010) *Agrobacterium*-mediated gene disruption using split-marker in *Grosmannia clavigera*, a mountain pine beetle associated pathogen. *Current Genetics* 56(3), 297–307.

21. Hamilton, M.D., Nuara, A.A., Gammon, D.B., Buller, R.M. and Evans, D.H. (2007) Duplex strand joining reactions catalyzed by vaccinia virus DNA polymerase. *Nucleic Acids Research* 35(1), 143–151.

22. Clontech Laboratories (2008) In-Fusion™ PCR Cloning Kit User Manual PT3650-1

23. Zhu, B., Cai, G., Hall, E.O. and Freeman, G.J. (2007) In-fusion assembly: seamless engineering of multidomain fusion proteins, modular vectors, and mutations. *Biotechniques* 43(3), 354–359.

24. Frandsen, R.J., Andersson, J.A., Kristensen, M.B. and Giese, H. (2008) Efficient four fragment cloning for the construction of vectors for targeted gene replacement in filamentous fungi. *BMC Molecular Biology* 9:70.

25. New England Biolabs (2008) Instruction manual: USER™ Friendly Cloning Kit - A Novel Tool for Cloning PCR Products by Uracil Excision. Version 1.3.

26. Fulton, T.R., Ibrahim, N., Losada, M.C., Grzegorski, D. and Tkacz, J.S. (1999) A melanin polyketide synthase (PKS) gene from *Nodulisporium* sp. that shows homology to the pks1 gene of *Colletotrichum lagenarium*. *Molecular Genetics and Genomics* 262, 714–720.

27. Geu-Flores, F., Nour-Eldin, H.H., Nielsen, M.T. and Halkier, B.A. (2007) USER fusion: a rapid and efficient method for simultaneous fusion and cloning of multiple PCR products. *Nucleic Acids Research* 35(7), e55.

28. van Attikum, H. and Hooykaas, P.J.J. (2003) Genetic requirements for the targeted integration of *Agrobacterium* T-DNA in *Saccharomyces cerevisiae*. *Nucleic Acids Research* 31(3), 826–832.

29. Bashi, Z.D., Khachatourians, G. and Hegedus, D.D. (2010) Isolation of fungal homokaryotic lines from heterokaryotic transformants by sonic disruption of mycelia. *Biotechniques* 48(1), 41–46.

30. Frandsen, R.J., Nielsen, N.J., Maolanon, N., Sørensen, J.C., Olsson, S., Nielsen, J. and Giese, H. (2006) The biosynthetic pathway for aurofusarin in *Fusarium graminearum* reveals a close link between the naphthoquinones and naphthopyrones. *Molecular Microbiology* 61(4), 1069–1080.

31. Michielse, C.B., Hooykaas, P.J., van den Hondel, C.A. and Ram, A.F. (2005) *Agrobacterium*-mediated transformation as a tool for functional genomics in fungi. *Current Genetics* 48(1), 1–17.

32. Tendulkar, S.R., Gupta, A. and Chattoo, B.B. (2003) A Simple protocol for isolation of fungal DNA. *Biotechnology Letters* 25, 1941–1944.

33. Nielsen, J.B., Nielsen, M.L. and Mortensen, U.H. (2007) Transient disruption of non-homologous end-joining facilitates targeted genome manipulations in the filamentous fungus *Aspergillus nidulans*. *Fungal Genetics and Biology* 45(3), 165–170.

34. Nørholm, M.H. (2010) A mutant Pfu DNA polymerase designed for advanced uracil-excision DNA engineering. *BMC Biotechnology* 10, 21.

35. Fernández-Martín, R., Cerdá-Olmedo, E. and Avalos, J. (2000) Homologous recombination and allele replacement in transformants of *Fusarium fujikuroi*. *Molecular Genetics and Genomics* 263(5), 838–845.

36. Abuodeh, R.O., Orbach, M.J., Mandel, M.A., Das, A. and Galgiani, J.N. (2000) Genetic transformation of *Coccidioides immitis* facilitated by *Agrobacterium tumefaciens*. *Journal of Infectious Diseases* 181(6), 2106–2110.

37. Mullins, E.D., Chen, X., Romaine, P., Raina, R., Geiser, D.M. and Kang, S. (2001) *Agrobacterium*-Mediated Transformation of *Fusarium oxysporum*: An Efficient Tool for Insertional Mutagenesis and Gene Transfer. *Phytopathology* 91(2), 173–180.

38. Frandsen, R.J., Schütt, C., Lund, B.W., Staerk, D., Nielsen, J., Olsson, S. and Giese, H. (2011) Two novel classes of enzymes are required for the biosynthesis of aurofusarin in *Fusarium graminearum*. *Journal of Biological Chemistry* 286(12), 10419–10428.

Chapter 3

Activity-Based Protein Profiling of Infected Plants

Farnusch Kaschani, Christian Gu, and Renier A.L. van der Hoorn

Abstract

Activity-based protein profiling (ABPP) is a powerful analytical method to detect and compare the activity of proteins in proteomes. This is achieved using specific activity-based probes that are often derived from inhibitors and are linked to reporter groups like rhodamine or biotin for fluorescence detection and/or affinity purification, respectively. The probes react with the active site residue of proteins and become covalently and irreversibly attached, facilitating the separation, detection and identification of the labelled proteins. In this protocol we describe all the steps required for labelling, purification and identification of labelled proteins from gels and show how activities in two proteomes can be compared. The identification of serine hydrolases from Arabidopsis plants infected with *Botrytis cinerea* using the trifunctional probe TriFP is used as an example.

Key words: Activity-based protein profiling, In-gel digest, ABPP, Serine hydrolases, Fluorophosphonate, *Botrytis cinerea*, *Arabidopsis thaliana*

1. Introduction

Activity-based protein profiling (ABPP) is a diagnostic tool developed to record and study the changes in activity of proteins in proteomes (1–3). Recently, we introduced this approach in plant science (4). The key to successful ABPP is the probe (Fig. 1a). This is often a small molecule inhibitor of the enzyme class to be investigated which is chemically modified with a reporter group (e.g. rhodamine for fluorescent detection and/or biotin for affinity purification). Trifunctional FP (TriFP, Fig. 1b) is an example of an activity-based probe for serine hydrolases. The probe will react with the serine in the active site of serine hydrolases by forming an irreversible, covalent bond (Fig. 2a). This covalent bond facilitates the purification and detection of the labelled proteins under denaturing conditions (Fig. 2b). The ABPP methodology allows us to compare changes in the activity of proteins after different treatments (comparative ABPP, Fig. 2b).

Melvin D. Bolton and Bart P.H.J. Thomma (eds.), *Plant Fungal Pathogens: Methods and Protocols*,
Methods in Molecular Biology, vol. 835, DOI 10.1007/978-1-61779-501-5_3, © Springer Science+Business Media, LLC 2012

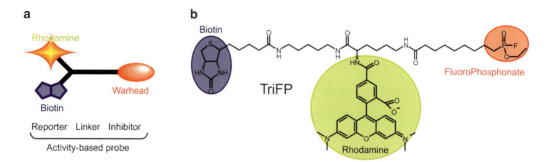

Fig. 1. Trifunctional activity-based probes. (**a**) Components of a trifunctional probe. The trifunctional probes carry a fluorescent group (e.g. rhodamine) for detection, a biotin group for affinity purification and a warhead that confers selective, covalent and irreversible labelling of the catalytic residue of the tageted protein class. (**b**) TriFP, an example of a trifunctional probe. TriFP contains biotin, rhodamine and a fluorophosphonate (FP) group that reacts with the catalytic serine of serine hydrolases.

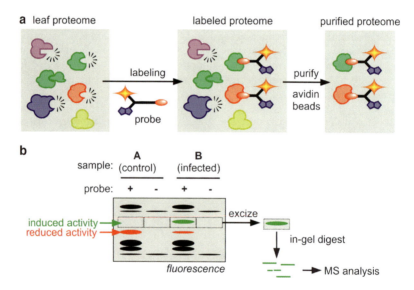

Fig. 2. ABPP principle and procedure. (**a**) General principle of ABPP. Proteomes are labelled with a trifunctional probe in an activity-dependent manner. Labelled proteins are purified, detected and identified. (**b**) Procedure of ABPP analysis. Probe-labelled proteins are purified and separated by gel electrophoresis and detected by fluorescent scanning. Fluorescent signals are excised and proteins digested in-gel with trypsin. Tryptic peptides are eluted and analysed by mass spectrometry to identify the labelled protein. Comparison with the no-probe control ("–" lane) confirms that the identified protein is not a background contamination. Comparison between probe-labelled samples will display proteins with differential activities.

The impact of ABPP has been particularly large in the field of plant-pathogen interactions (5–8). Comparative studies where changes in the activity of e.g. cysteine proteases (5, 6), serine hydrolases (7) or the proteasome (8) were displayed with the help of specific probes, revealed exciting new functions for these proteins.

The success of ABPP experiments relies to a large extent on the identification of the targeted proteins. The following protocol contains all the information to perform a comparative ABPP experiment on 1D protein gels, followed by affinity purification and preparation of samples for identification by mass spectrometry (MS). As an example we explain the procedure for serine hydrolase profiling of *Arabidopsis thaliana* infected with the necrotrophic pathogen *Botrytis cinerea* using TriFP (7, 9, 10).

2. Materials

Prepare all solutions immediately before use unless otherwise stated. Always wear gloves to reduce keratin contamination. Use chemicals and water suitable for mass spectrometry when preparing solutions.

2.1. Components for the ABPP Reaction

1. Bovine serum albumin (BSA) protein standard (2 mg/mL).

2. Amber glass threaded vials (4 mL) with screw cap.

3. Phosphate buffered saline (10× PBS): 10.6 mM KH_2PO_4, 1,552 mM NaCl, 30 mM Na_2HPO_4, pH 7.4. In a 1-L glass beaker mix 1.44 g KH_2PO_4, 90.70 g NaCl and 4.26 g Na_2HPO_4. Add 800 mL water and stir the initial suspension with the help of a magnetic stirrer until all the salts are completely dissolved. Then adjust the pH (if necessary) to pH 7.4 by adding hydrochloric acid. Transfer the solution to a graduated cylinder and adjust the volume to 1 L with water. Sterile filter the buffer using a 0.22-μM filter unit. The solution can be stored at RT indefinitely.

4. Activity-based probe: for large scale TriFP labelling experiments the probe is diluted to a concentration of 1 mM with dimethyl sulfoxide (DMSO) and the stock stored at –20°C. Handle the stocks always at room temperature. This reduces the likelihood of water getting into the stock by condensation.

5. Phosphate buffered saline (1× PBS): 1.1 mM KH_2PO_4, 155 mM NaCl, 3 mM Na_2HPO_4, pH 7.4. Transfer 100 mL 10× PBS to a 1-L measuring cylinder and add water up to the 1 L mark. Transfer the solution to a 1-L bottle and mix well by inverting the bottle several times. The solution can be stored at RT indefinitely.

2.2. Affinity Purification Components

1. Prepacked desalting column.

2. 0.22-μM filter unit attached to a syringe.

3. Sodiumdodecylsulfate (10% SDS): 350 mM SDS. Dissolve 10 g SDS in 80 mL sterile water (see Note 1). Stir gently until all the SDS is dissolved and the solution is clear. Then adjust volume to 100 mL. The solution can be stored at RT indefinitely.

4. Avidin beads: Sigma-Aldrich, St. Louis, MO, USA (see Note 2).

5. Sodiumdodecylsulfate (1% SDS): 35 mM SDS. Transfer 20 mL 10% SDS solution to a measuring cylinder and add water up to the 200 mL mark. Store this solution at room temperature.

2.3. Elution and Separation of Proteins

1. Tris (tris(hydroxymethyl)aminomethane) (1 M Tris pH 6.8): dissolve 60.57 g Tris in 400 mL sterile water. Adjust the pH to 6.8 with concentrated HCl. Then transfer the solution to a 500 mL measuring cylinder and add water up to the 500 mL mark. Transfer the solution to a 500-mL bottle and autoclave it. The solution can be stored at RT indefinitely.

2. SDS-PAGE gel-loading buffer (4× GLB): 280 mM SDS, 400 mM Tris, 40% glycerol, 1.4 M β-mercaptoethanol, 0.6 mM Bromophenol Blue, pH 6.8. Add the following components to a 50-mL Falcon tube: 2 g SDS, 10 mL 1 M Tris (pH 6.8) and 10 mL glycerol (see Note 3). Tumble this solution until all the components are dissolved. Then add 2.5 mL 14.2 M β-mercaptoethanol (see Note 4) and bring the volume up to 25 mL with water. Finally, add a pinch (~10 mg) of Bromophenol blue (see Note 5). Make 1 mL aliquots and store these at −20°C.

3. Tris (tris(hydroxymethyl)aminomethane) (1.5 M Tris pH 8.8): dissolve 90,85 g Tris in 400 mL sterile water. Adjust the pH to 8.8 with hydrochloric acid (HCl). Then transfer the solution to a 500-mL measuring cylinder and add water up to the 500 mL mark. Transfer the solution to a 500-mL bottle and autoclave it. The solution can be stored at RT indefinitely.

4. Ammoniumpersulfate (10% APS): 400 mM APS. Add 1 g APS $((NH_4)_2S_2O_8)$ to a 15-mL tube and add sterile water up to 10 mL. Mix well until the chemical is completely dissolved. The solution can be used for at least 1 week when stored at 4°C.

5. Ready-made Acrylamide-Solution (30% Acrylamide/Bis): 29% Acrylamide, 1% Bisacrylamide (see Note 6).

6. Resolving gel (12%): Mix the following components in a clean glass beaker on ice: 33 mL water, 40 mL 30% Acrylamide/Bis solution, 25 mL 1.5 M Tris (pH 8.8), 1 mL 10% SDS, 1 mL 10% APS, 40 μL TEMED. Pour the gel between the glass plates and overlay the surface with 2-Butanol. You have to be quick. Once the TEMED has been added the gel starts to polymerize within minutes when left at room temperature. Once the resolving gel is polymerized (~20–60 min), pour the 2-Butanol off and rinse the gel surface with water. Dry the gel surface by tipping (not sliding!) the gel with Whatman paper.

7. Stacking gel (6%): Mix the following components in a glass beaker on ice: 19.6 mL water, 6 mL 30% Acrylamide/Bis solution, 3.8 mL 1.0 M Tris (pH 6.8), 300 μL 10% SDS, 300 μL 10% APS, 30 μL TEMED. Pour the stacking gel solution onto the resolving gel until it reaches the brim of the gel-cassette and add the comb. Let the gel rest for 1 h at RT. Then wrap the gel-cassette in saran wrap and store it over night at 4°C. This is essential to complete the polymerization.

8. SDS Running Buffer (10×): 248 mM Tris, 2 M glycin, 35 mM SDS. Mix 30 g Tris, 144 g glycin and 10 g SDS in a 1-L glass beaker with 800 mL water. Stir gently until all the components are dissolved. Then transfer the solution to a 1-L measuring cylinder and adjust the volume to 1 L with water. Store the solution at RT in a 1-L bottle. It is not necessary to sterilise this solution.

2.4. In-Gel-Digestion (IGD)-Components

1. Disposable steel blades.

2. Ammonium bicarbonate solution (100 mM): Dissolve 395 mg ammonium bicarbonate $((NH_4)CO_3H)$ in 50 mL MS-water.

3. TCEP solution (10 mM): Dissolve 28.7 mg Tris(2-carboxyethyl) phosphine (TCEP) in 10 mL MS-water. Prepare immediately before use.

4. Iodoacetamide (IAA, 55 mM): Dissolve 40.7 mg IAA in 4 mL MS-water. Keep in darknes; IAA is sensitive to light. Prepare immediately before use and store at a dark place.

5. Acetonitrile (ACN).

6. ACN and 100 mM ammonium bicarbonate (50:50 solution): mix 25 mL 100 mM ammonium bicarbonate with 25 mL ACN.

7. Ammonium bicarbonate (25 mM): mix 1 mL 100 mM ammonium bicarbonate with 3 mL water.

8. Trypsin (10 ng/μL): Dissolve 20 μg lyophilised trypsin in 200 μL of trypsin resuspension buffer (provided by manufacturer, usually 50 mM acetic acid). Leave on the bench for 15 min to activate trypsin. Then dilute with 25 mM ammonium bicarbonate solution to a final concentration of 10 ng/μL (1:10 dilution).

9. Formic acid (FA, 5%): 1.33 mM FA. Mix 50 μL 100% FA (HCOOH) in 950 μL MS-water. Formic acid is volatile and corrosive. Handle always in a chemical hood.

10. Formic acid (FA, 0.1%): 0.27 mM FA. Mix 1 μL FA with 999μL water.

3. Methods

3.1. Sample Preparation

1. Collect the material of interest from non-infected (sample A) and infected plants (sample B).Collect enough material corresponding to at least 6 mg protein for each sample.

2. Grind the tissues in a mortar. Add enough water (see Note 7) to the sample to obtain a homogenous solution. Transfer the extract to a centrifugation tube and clear it by centrifugation (13 k, 5 min, RT). Transfer the supernatant to a fresh tube.

3. Determine the protein concentration (see Note 8).

4. Prepare 2 mg/mL stock solutions of each proteome in 1× PBS (see Note 9). Bring the equivalent of 10 mg protein up to a volume of 4.5 mL with water and add 500 μL 10× PBS. Vortex for 5 s.

3.2. Labelling of Plant Extracts with Activity-Based Probes

1. Label four 4-mL amber glass vials (or equivalent, see Note 10) with A+, A–, B+and B–.

2. Transfer 1 mL of the 2 mg/mL protein solutions to the different vials. The extract from non-infected plants is added to vials labelled A– and A+, and the extract from infected samples is added to vials labelled B– and B+(Fig. 3.2b).

3. Then add 5 μL of the 1 mM probe to those vials labelled with "+" and 5 μL DMSO to vials labelled with "–" (see Note 11). Vortex for 5 s.

4. Incubate at room temperature in the dark for 1–2 h (see Notes 10 and 12).

5. Meanwhile, equilibrate a prepacked desalting column with 1× PBS. Snap off the bottom lip of the column and discard the buffer solution. Equilibrate the columns by passing twice 15 mL PBS through the column by gravity.

3.3. Large Scale Affinity Purification of Labelled Proteins

1. After 1 h incubation dilute the reaction mix with 1× PBS to a final volume of 2.5 mL and mix well. The reaction mix should be clear (see Note 13).

2. Apply the diluted reaction mix onto the 1× PBS equilibrated prepacked desalting column and allow the solution to enter by gravity (see Note 14). Discard the flow-through.

3. Place a 15-mL tube containing 183 μL 10% SDS solution (see Note 15) under the desalting column, add 3.5 mL 1× PBS to the desalting column and collect the flow-through in the SDS-containing tube. Mix well by inverting the tube several times. Then incubate the tube for 5 min at 90°C in a water bath (see Note 16).

4. Immediately afterwards place the falcon tube on ice for 1 min and invert the tube a couple of times. Do not leave your samples on ice for too long otherwise SDS and proteins may precipitate. Then add 5 mL 1× PBS (see Note 17).

5. Equilibrate the avidin beads by adding the equivalent of a 100 μL aliquot (50 μL bed volume) of the avidin beads (see Note 18) to a 1.5-mL Eppendorf tube and wash the beads twice with 1 mL 1× PBS (see Note 19). Re-suspend each aliquot in 100 μL 1× PBS.

6. Add one aliquot of the washed avidin bead suspension to the diluted protein solution and incubate for 1 h at RT by gently inverting the tube (see Note 20).

7. Collect the beads by centrifugation ($400 \times g$, 3 min, swinging out rotor; see Note 21) and remove the supernatant with a 10-mL disposable pipette (see Note 22).

8. Add 10 mL 1% SDS to the beads and re-suspend them by inverting the tube several times. Incubate at RT for 10 min by gently inverting the tube (see Note 20).

9. Collect the beads by centrifugation ($400 \times g$, 3 min, swinging out rotor; see Note 21) and remove the supernatant with a 10-mL disposable pipette (see Note 22).

10. Repeat the washes with 1% SDS (see steps 8 and 9) five more times (see Note 23).

11. Transfer the beads to a new 1.5-mL tube. Collect the beads by centrifugation (1 min, $16,000 \times g$) and remove the supernatant carefully.

3.4. Separation of Biotinylated Proteins

1. Add 30 µL of the 4× GLB to the tube containing the avidin beads with the captured proteins. Mix well.

2. Incubate this mix for 10 min at 90°C. Agitate every 2 min by ticking the bottom of the tube.

3. Place the tubes on ice for 10 s and centrifuge briefly (10 s, $16,000 \times g$) to spin down the water that has condensed in the lid. Vortex briefly again and centrifuge (5 min, $16,000 \times g$).

4. Assemble the protein gel device using previously prepared gels. Fill the reservoirs with 1× SDS running buffer and make sure the apparatus is not leaking.

5. Load the pre-stained fluorescent marker in the first and the last lane (see Note 24). Then load 35–50 µL of the eluted proteins. Leave one lane empty between samples to prevent cross-contaminations. Fill the empty lanes with 2× GLB. Separate the proteins by SDS-PAGE (see Note 25).

3.5. In-Gel-Digestion Protocol

1. After electrophoresis carefully transfer the gel to a plastic box containing water suitable for MS. Wash the gel with several changes of water for at least 1 h.

2. Next, place the gel on a fluorescent scanner and detect TriFP-targets in-gel using the appropriate settings (see Note 26).

3. Printout the scan and put an overhead transparency sheet on top of your printout. Place the gel on the transparency and superimpose it with the gel. Use the marker bands as guide. Excise the regions where you see fluorescent signals on the printout with a new disposable steel blade. Excise the corresponding region from the no-probe control ("–" samples) for

background control. Cut the gel slices into at least eight pieces and transfer them to a properly labelled, low protein binding Eppendorf tube (or equivalent). Indicate on the printout which gel fragment was removed and in which tube this fragment was transferred. Do not squash the gel pieces. After the bands of interest have been excised, re-scan to verify that the right part of the gel has been removed.

4. Wash the gel pieces twice with 500 μL water for 15 min while vigorously shaking (see Note 27). Briefly centrifuge (10 s, 16,000×g) and discard the supernatant.

5. Wash the gel pieces twice for 15 min with 500 μL 100 mM ammonium bicarbonate (see Note 28). Briefly centrifuge (10 s, 16,000×g) and discard the supernatant.

6. Add 200 μL 10 mM TCEP or enough to completely cover the gel slices. Incubate at 62°C for 30 min (see Note 29). Briefly centrifuge (10 s, 16,000×g) and discard the supernatant.

7. Add immediately 200 μL 55 mM IAA (see Note 30). The gel slice has to be completely covered by the IAA solution. Tumble gently in the dark for 30 min at RT. Briefly centrifuge (10 s, 16,000×g) discard the supernatant.

8. Wash the gel slices three times15 min with 500 μL 50:50 ACN: 100 mM ammonium bicarbonate (see Note 31). Briefly centrifuge (10 s, 16,000×g) and discard the supernatant.

9. Add 50 μL 100% ACN to dry gel slices (they become completely white). Remove ACN and dry the samples in a vacuum concentrator (see Note 32).

10. Add 20 μL of the 10 ng/μL trypsin solution to the gel slice and incubate for 10 min at RT (see Note 33). Then completely cover the slices with 25 mM ammonium bicarbonate and seal the tubes with parafilm. Incubate over night under constant shaking at 37°C (see Note 34).

11. After 16 h, briefly centrifuge (10 s, 16,000×g) and transfer the supernatant to a new Eppendorf tube (low protein binding or equivalent). Do NOT discard this solution as this is the major peptide fraction!

12. Add sufficient 5% formic acid to the gel slices to cover them (~100–200 μL) and incubate at RT for 15 min (see Note 35).

13. Combine this supernatant with the fraction obtained from the overnight digestion (see step 11).

14. Add sufficient ACN to cover the gel slices and incubate for 15 min at RT while vigorously shaking. Combine this supernatant with the other supernatants. Repeat this step three times until the gel slices have become opaque. (e.g. use 100 μL ACN, then 70 μL and finally 50 μL).

15. Reduce the volume of the combined supernatants in a vacuum concentrator (~3–5 h at 30°C) to a final volume of 10 µL (see Note 36).

16. Use this peptide concentrate for MS analysis (see Note 37).

4. Notes

1. When handling SDS powder always wear a dust mask, gloves and protective clothing.

2. The protocol has been thoroughly tested with these beads. Beads from other manufacturers may also work. The avidin or streptavidin on the beads should be able to bind biotinylated proteins at 0.2% SDS and the interaction with the biotinylated proteins should not be affected during washes with 1% SDS and 6 M Urea.

3. Glycerol is a viscous liquid that can be hard to handle. To dispense this liquid, always calculate the respective weight (using the density of glycerol), place your tube on a balance and directly pour the required amount in. In this case add ~12.5 g of glycerol directly to your Falcon tube.

4. β-mercaptoethanol is a strong reducing agent. It is toxic (11) and smells bad. Wear protective clothing and gloves when handling this chemical. Always work in chemical hood.

5. Very little of Bromophenol Blue is required. It is added as a pH indicator. As long as the pH is higher than 4.6 it will be violet-blue. If the pH drops to a lower pH the colour of the indicator will change to yellow (12).

6. To reduce the health hazard of acrylamide we use a commercially available ready-made solution, which is always stored at 4°C. Acrylamide is a neurotoxin (13) and causes cancer in rats (14). The powder tends to form dust clouds. When you opt to make your own 30% Acrylamide/Bis always work in a chemical hood and wear a dust mask, gloves, and protective clothing.

7. You can also use a buffer like 1× PBS or 1× TBS to extract your proteins as long as it will be compatible with your following ABPP reaction.

8. We use the RCDC kit from Bio-Rad for measuring the protein concentration. The changes in absorbance are measured at 750 nm which reduces interference of chlorophyll at this wavelenght. Expect protein concentrations for leaf-extracts to be around 2–5 mg/mL.

9. PBS is the standard buffer for labelling reactions with FP-derived probes. Other probes may require a different buffer. Adjust this step accordingly.

10. The fluorescent group rhodamine on TriFP is light sensitive. Therefore, samples should be kept in the dark as much as possible. The protection from light can also be achieved by wrapping the sample tubes in aluminium foil.

11. It is very important to include a no-probe control. It will allow us later to identify proteins that bind non-specifically to the affinity purification matrix. This negative control increases confidence for identified targets.

12. If you plan to profile the activity of a different enzyme class, keep in mind that the probe concentration, reaction times and buffer components have to be optimised individually for each probe and for each plant extract.

13. The reaction mix should be clear. In case a precipitate has formed remove it by passing the reaction mix through a 0.22-µM filter unit, attached to a syringe.

14. The desalting step removes non-reacted, excess probe and other small molecules. This will increase the yield for targeted, biotinylated proteins in the following affinity purification steps.

15. SDS is added to unfold proteins (15). Activity-based probes react with the active site residue of an enzyme. Since these are often buried deep inside the protein, affinity tags like biotin on TriFP may not be accessible to affinity purification matrices. The SDS treatment and the subsequent incubation at 90°C will permanently unfold the proteins and therefore significantly increase the yield of affinity purification. At the same time this SDS treatment will also abrogate the activity of proteases and other degradative enzymes which might interfere with efficient affinity purification.

16. If you do not have a traditional water bath at hand: place 1 L water in a 2-L glass beaker and bring it to the boil in a microwave oven. Let the water cool down for 10 min. Then place the closed tubes upright into the hot water (~90°C) for 5 min. Invert the tubes from time to time. Wear heat protective gloves when you handle the hot tubes.

17. The sample is diluted at this step with 1× PBS to reduce the SDS concentration from ~0.5 to 0.2%. The reduced SDS concentration is compatible with the subsequent affinity purification with avidin beads.

18. We use avidin beads from Sigma (A9207). The beads are delivered as 1 or 5 mL aliquots in 50% glycerol. Please make 100 µL aliquots and store at −20°C to reduce freeze–thaw cycles. With a cut-off yellow tip, transfer 100 µL of the slurry to a new Eppendorf tube. Because the beads tend to settle down quickly, it is necessary to vortex before pipetting. The aliquots are stored at −20°C. One aliquot is sufficient for one affinity purification.

19. Add 1 mL PBS to the beads and mix them well. Then spin the beads down (5 min, 800–1,000 ×g) and gently remove 900 µL of the supernatant. Do not attempt to remove all the supernatant. The closer you get to the beads the more beads you will remove. This will reduce your capture efficiency. Repeat this step once more.

20. During incubation and washes it is important to constantly agitate the beads. Otherwise they will settle down quickly. It does not matter how this is achieved (horizontal, by rolling or overhead) but it should be a gentle process.

21. It is important to pellet the beads at low g-values. If the centrifugal force is too high the beads will stick together with precipitating proteins, thereby compromising the quality of the purification. For optimal settings consult the manufacturer's instructions.

22. Remove only 90% of the supernatant (e.g. of 10 mL supernatant only remove 9 mL). Do not attempt to remove all the supernatant. The closer you get to the beads the more beads you will remove, which will reduce the yield of your purification.

23. If you know that you are dealing with sticky contaminants that are not removed by SDS alone you may consider washing the beads twice with 10 ml 6 M Urea (Dissolve 36 g Urea in 80 mL water. Then adjust the volume with water to 100 mL). Wash first twice with 10 mL 1% SDS then twice with 10 mL 6 M Urea and finally twice with 10 mL 1% SDS.

24. Fluorescent protein markers are expensive and not essential. Many of the regular protein ladders that contain red marker bands are fluorescent under the settings used to detect rhodamine-labelled proteins.

25. We have successfully used both homemade gels and commercial pre-cast gels for the separation and identification of probe-labelled targets. When making your own gels please keep in mind that keratin is a frequent contamination that will have a drastic effect on the success of the downstream MS identification. You can minimise the risk of keratin contamination by always using fresh solutions. Also use MS pure ingredients (particularly MS-water) whenever possible. Avoid wearing clothes that contain wool. The choice in length of the gel and acrylamide percentage depends on the number of targets you expect and how close they migrate on the gel. The more targets you expect the longer the gel should be. For TriFP target identification we usually run proteins for 350 V-h on a 10 or 12% 20×18 cm gel.

26. We detect fluorescent proteins in protein gels using the Typhoon 8600 (GE Healthcare, Munich, Germany). Any other scanner or camera system that can excite at 532 nm (green laser) and

that has a TAMRA filter (580 nm BP30) for detection of rhodamine can be used. These systems can, however, differ significantly in sensitivity. For other fluorescent tags the settings might be different. Please check before proceeding.

27. It is necessary to remove all the excess SDS from the gel. Small traces of residual SDS may inhibit the following alkylation reaction (16) and may interfere with the separation of peptides later.

28. The following alkylation step requires a basic pH (16). This is achieved by washing the gel slices with the ammonium bicarbonate solution.

29. TCEP will reduce disulfide-bonds (R^1–S–S–$R^2 \rightarrow R^1$–SH + HS–R^2) which is essential for the subsequent alkylation of the cysteines (17).

30. The IAA treatment is necessary to irreversibly alkylate the thiol-group of cysteins (R–SH + I–CH_2–C(O)–$NH_2 \rightarrow$ R–S–CH_2–C(O)–NH_2 + HI) which are prone to oxidation (18). Through this chemical modification all thiol groups (–SH) are uniformly modified with a 57 Da carbamidomethyl-rest, which facilitates the automated detection of cysteine-containing peptides by MS.

31. With every washing step the gel slices become more opaque/white. This is because acetonitrile extracts water from the gel slices. What remains is solid polyacrylamide.

32. It is essential to use a vacuum centrifuge. Simply using a desiccator attached to a pump will dry your samples nicely but retardation of boiling may cause your gel pieces to jump out of their Eppendorf tubes. This may lead to a loss of your samples and cause contaminations.

33. The polyacrylamide flakes will start to swell once the trypsin mix is added. This way they absorb trypsin.

34. It is important to seal the tubes with parafilm so they will not open accidentally. A nice trick is to put the tubes in 200-mL plastic beaker and stuff the beaker with paper towels so the tubes cannot move. This beaker is then placed in a 37°C shaker for bacteria and left there shaking over night.

35. The formic acid treatment inactivates trypsin (19) and protonates all the peptides (the peptides will become positively charged).

36. Do not dry samples completely as this causes loss of peptides. If samples have dried accidentally then add 10 μL 0.1% formic acid solution to bring the peptides back in solution.

37. The preparation of the samples for the subsequent MS analysis is dependent on the mass spectrometer you use in your lab. We usually load the samples directly on a 10 cm 100-μM equilibrated C_{18}-column (7), which is directly attached to a LTQ or Velos (Thermo Fisher Scientific Inc., Waltham, MA, USA). Please check with your MS department for details.

Acknowledgements

This work was supported by the Max Planck Society and the Deutsche Forschungsgemeinschaft projects HO3983/3-1,2 and HO3983/4-1

References

1. Kolodziejek, I., and Van der Hoorn, R. A. L. Mining the active proteome in plant science and biotechnology. *Curr Opin Biotechnol* 21, 225–33.

2. Simon, G. M., and Cravatt, B. F. Activity-based proteomics of enzyme superfamilies: serine hydrolases as a case study. *J Biol Chem* 285, 11051–5.

3. Puri, A. W., and Bogyo, M. (2009) Using small molecules to dissect mechanisms of microbial pathogenesis. *ACS Chem Biol* 4, 603–16.

4. Van der Hoorn, R. A. L., Leeuwenburgh, M. A., Bogyo, M., Joosten, M. H. A., and Peck, S. C. (2004) Activity profiling of papain-like cysteine proteases in plants. *Plant Physiol* 135, 1170–8.

5. Song, J., Win, J., Tian, M., Schornack, S., Kaschani, F., Ilyas, M., Van der Hoorn, R. A. L., and Kamoun, S. (2009) Apoplastic effectors secreted by two unrelated eukaryotic plant pathogens target the tomato defense protease Rcr3. *Proc Natl Acad Sci USA* 106, 1654–9.

6. Shabab, M., Shindo, T., Gu, C., Kaschani, F., Pansuriya, T., Chintha, R., Harzen, A., Colby, T., Kamoun, S., and Van der Hoorn, R. A. L. (2008) Fungal effector protein AVR2 targets diversifying defense-related cys proteases of tomato. *Plant Cell* 20, 1169–83.

7. Kaschani, F., Gu, C., Niessen, S., Hoover, H., Cravatt, B. F., and Van der Hoorn, R. A. L. (2009) Diversity of serine hydrolase activities of unchallenged and botrytis-infected Arabidopsis thaliana. *Mol Cell Proteomics* 8, 1082–93.

8. Gu, C., Kolodziejek, I., Misas-Villamil, J., Shindo, T., Colby, T., Verdoes, M., Richau, K. H., Schmidt, J., Overkleeft, H. S., and Van der Hoorn, R. A. L. (2010) Proteasome activity profiling: a simple, robust and versatile method revealing subunit-selective inhibitors and cytoplasmic, defense-induced proteasome activities. *Plant J* 62, 160–70.

9. Kidd, D., Liu, Y., and Cravatt, B. F. (2001) Profiling serine hydrolase activities in complex proteomes. *Biochemistry* 40, 4005–15.

10. Liu, Y., Patricelli, M. P., and Cravatt, B. F. (1999) Activity-based protein profiling: the serine hydrolases. *Proc Natl Acad Sci USA* 96, 14694–9.

11. White, K., Bruckner, J. V., and Guess, W. L. (1973) Toxicological studies of 2-mercaptoethanol. *J Pharm Sci* 62, 237–41.

12. Zikolov, P., and Budevsky, O. (1973) Acid-base equilibria in ethylene glycol--I: definition of pH and determination of pk-values of acid-base indicators. *Talanta* 20, 487–93.

13. Calleman, C. J., Wu, Y., He, F., Tian, G., Bergmark, E., Zhang, S., Deng, H., Wang, Y., Crofton, K. M., Fennell, T., and et al. (1994) Relationships between biomarkers of exposure and neurological effects in a group of workers exposed to acrylamide. *Toxicol Appl Pharmacol* 126, 361–71.

14. Friedman, M. A., Dulak, L. H., and Stedham, M. A. (1995) A lifetime oncogenicity study in rats with acrylamide. *Fundam Appl Toxicol* 27, 95–105.

15. Gavina, J. M. A., and Britz-McKibbin, P. (2007) Protein Unfolding and Conformational Studies by Capillary Electrophoresis. *Curr. Anal. Chem.* 3, 17–31.

16. Galvani, M., Hamdan, M., Herbert, B., and Righetti, P. G. (2001) Alkylation kinetics of proteins in preparation for two-dimensional maps: a matrix assisted laser desorption/ionization-mass spectrometry investigation. *Electrophoresis* 22, 2058–65.

17. Andrews, P. C., and Dixon, J. E. (1987) A procedure for in situ alkylation of cystine residues on glass fiber prior to protein microsequence analysis. *Anal Biochem* 161, 524–8.

18. Herbert, B., Galvani, M., Hamdan, M., Olivieri, E., MacCarthy, J., Pedersen, S., and Righetti, P. G. (2001) Reduction and alkylation of proteins in preparation of two-dimensional map analysis: why, when, and how? *Electrophoresis* 22, 2046–57.

19. Smillie, L. B., and Neurath, H. (1959) Reversible inactivation of trypsin by anhydrous formic acid. *J Biol Chem* 234, 355–9.

Chapter 4

The Use of Agroinfiltration for Transient Expression of Plant Resistance and Fungal Effector Proteins in *Nicotiana benthamiana* Leaves

Lisong Ma, Ewa Lukasik, Fleur Gawehns, and Frank L.W. Takken

Abstract

Agroinfiltration is a versatile, rapid and simple technique that is widely used for transient gene expression in plants. In this chapter we focus on its use in molecular plant pathology, and especially for the expression of plant resistance (*R*) and fungal avirulence (*Avr*) (effector) genes in leaves of *Nicotiana benthamiana*. Co-expression of an *R* gene with the corresponding *Avr* gene triggers host-defence responses that often culminate in a hypersensitive response (HR). This HR is visible as a necrotic sector in the infiltrated leaf area. Staining of the infiltrated leaves with trypan blue allows visual scoring of the HR. Furthermore, fusion of a fluorescent tag to the recombinant protein facilitates determination of its sub-cellular localization by confocal microscopy. The matching gene pair *I-2* and *Avr2*, respectively from tomato and the fungal root-pathogen *Fusarium oxysporum* f. sp. *lycopersici*, is presented as a typical example.

Key words: Agroinfiltration, *Nicotiana benthamiana*, Resistance and effector protein expression, Hypersensitive response, Trypan blue staining, Protein Localization

1. Introduction

Agrobacterium-mediated transient expression (agroinfiltration) is based on infiltration of *Agrobacterium tumefaciens* cultures into intact plant leaves. The bacterium subsequently transfers a DNA segment, called transfer-DNA or T-DNA, into the plant cells. In nature this T-DNA is part of the bacterial tumour-inducing (Ti) plasmid that, besides the T-DNA, carries the genes required for its transfer. The T-DNA carries effector genes that allow the pathogen to cause crown-gall disease on species in the Rosaceae family. In disarmed laboratory strains the effector genes are deleted and the two essential parts of the T-DNA, its left- and right border, are

Melvin D. Bolton and Bart P.H.J. Thomma (eds.), *Plant Fungal Pathogens: Methods and Protocols*,
Methods in Molecular Biology, vol. 835, DOI 10.1007/978-1-61779-501-5_4, © Springer Science+Business Media, LLC 2012

placed on a separate plasmid. Genes placed between these borders will be transferred to the plant. Since the genes required for T-DNA transfer now reside on a different plasmid, this system is called the "binary vector" system. The binary vectors containing the T-DNA can carry inserts of up to over 100 kbp (1). T-DNA transferred to a plant cell will relocate to the nucleus, where its genes can be transcribed and expressed (2). The majority of the plant cells in the infiltrated region express the transgene and the expression typically reaches its highest level 2–3 days after infiltration. At later stages the expression is quenched by RNA silencing (3). In addition to the regular binary vectors, variants have been developed for specific purposes. One example is the pEAQ series that contain deconstructed viral genes that enhance recombinant protein production by producing a highly stable mRNA that is very efficiently transcribed (4). Other types of binary vectors carry viral genomes, such as that of Potato Virus X or Tobacco Mosaic Virus (5). Use of these binary viral vectors, referred to as agroinfection, result in the production of an autonomous replicating virus that can spread systemically through a plant. A limitation of agroinfection is the relative small size of the insert tolerated (<2 kb) in the viral genome and the formation of a recombinant plant-pathogenic virus that requires elevated containment measures, which hampers broad application (6). We here focus on the use of regular binaries. For details on application of the alternative binary vectors, please refer to Kanneganti et al. (7).

Unarguably, the main advantage of agroinfiltration over stable transformation procedure is its speed. In addition, compared to other transient expression systems (protoplast transformation (8), particle bombardment (9), and microinjection (10)), it has the advantages that it is inexpensive—as it relies on simple technology—and can be exploited in intact plants of which a relative large leaf sector is transformed (2). Agroinfiltration has been exploited in many types of experiments, for instance to study gene function (11, 12), protein production (13), host–pathogen interaction (14, 15), protein–protein interaction (16) and protein localization (17). *Agrobacterium* is applied either using vacuum infiltration or by syringe infiltration. Whereas vacuum infiltration has the advantage that whole leaves, and even entire plants, can be infiltrated at once, syringe infiltration is easier to perform and is more frequently utilised for protein production at a small scale (18). The main advantage of syringe infiltration is that different genes, either alone or in combination, can be expressed together in a single leaf. Syringe agroinfiltration has been applied successfully in a variety of plant species, including *Nicotiana* spp., tomato, lettuce, Arabidopsis, flax, pea, grapevine, pepper and rose (11, 12, 15, 19–22). Among them, *Nicotiana benthamiana* and *Nicotiana tabacum* (tobacco) are preferably used. Especially *N. benthamiana*, as the plant has a short life cycle, carries relatively large and easily infiltratable leaves that produce recombinant proteins at high levels and the leaf does not show necrosis upon infiltration with most *Agrobacterium* strains (18). These advantages made it a favourite model to study plant–pathogen interactions (23).

A major challenge in plant pathology is the functional analysis of resistance (*R*) and avirulence (*Avr*) genes. Agroinfiltration in *N. benthamiana* plants has been highly instrumental to tackle this challenge. In our laboratory agroinfiltration has been used for a variety of purposes. Co-expression of the tomato resistance gene *I-2* with candidate *Avr2* genes was used to identify the avirulent allele, as only this candidate triggered the hypersensitive response (24). Moreover, agroinfiltration allowed us to demonstrate physical interactions of I-2 with plant proteins such as (a) Heat-Shock Protein 90, (b) its co-chaperone Protein Phosphatase 5 and (c) the small Heat-Shock Protein RSI2 (25, 26). Also in structure/function analyses this method has shown its merits, as it allowed the rapid testing of a series of *I-2* mutants for either loss-of-function or constitutive activity (27). The ability to co-express constructs encoding different R protein sub-domains allowed us to analyse their ability to trans-complement and to examine physical interactions between the domains (28, 29). Finally, expression of fluorescent-tagged AVR2 allowed determination of the sub-cellular localization of the recombinant protein in plant cells, providing clues for the endogenous location of protein activity (unpublished data).

The procedures in this chapter describe a generic method for transient production of R and AVR proteins in *N. benthamiana* leaves. The method is exemplified by expression of the tomato *R* gene *I-2* and the corresponding *Avr2* gene from the root-invading pathogenic fungus *Fusarium oxysporum* f. sp. *lycopersici*. Also detailed protocols for scoring HR and the use of the vital stain (trypan blue) to visualise cell death are provided. The last section describes two methods that can be used to detect the recombinant proteins produced: western blot analysis and confocal microscopy using a GFP-labelled AVR2 protein.

2. Materials

2.1. Seeds and Plant Growth Materials

1. Seeds of *N. benthamiana*.
2. 10 cm square plastic pots.
3. Standard germination soil (substrate No. 1) and potting soil (No. 3) for plant growth.
4. Edding 400 permanent marker, color 001 black.

Table 1
List of plasmid constructs

Constructs	Aims
CTAPi (30)	Empty vector; negative control for HR
CTAPi-I-2 (29)	I-2, triggers HR when combined with AVR2
CTAPi-ΔspAvr2 (24)	c.p. localised AVR2, triggers HR together with I-2
pGWB451-ΔspAvr2-GFP (31)	c.p. localised AVR2 labelled with GFP
pB451-GFP	c.p. localised GFP
pGWB454-Avr2-RFP	Apoplastic localised AVR2 labelled with RFP
pBin19-ZmHVR-YFP[a]	Plasmamembrane marker labelled with YFP

c.p. cytoplasmic
[a]Select on 50 mg/L kanamycin, others are 100 mg/L spectinomycine

2.2. Binary Plasmids and Agrobacterium Strains

1. Plasmid constructs that are used as examples in this chapter are listed in Table 1 (see Note 1).

2. *Agrobacterium tumefaciens* strain GV3101 is used for all infiltrations (see Note 2).

2.3. Media, Buffers and Solutions

1. Luria-Bertani mannitol (LBman) medium: 10 g/L Bacto-tryptone, 5 g/L yeast extract, 2.5 g/L NaCl, 10 g/L mannitol, pH 7.0, autoclave at 120°C.

2. Antibiotics: 10 mg/mL rifampicin, 100 mg/mL spectinomycin and 100 mg/mL kanamycin, filter sterilise the latter two.

3. 0.2 M acetosyringone: dissolve 0.784 g in 20 mL dimethyl sulfoxide (DMSO), filter sterilise.

4. 1 M MES (2-[N-morpholino]ethanesulfonic acid): dissolve 42.64 g MES in 200 mL H_2O and adjust pH to 5.6 with KOH, filter sterilise.

5. LBmani: 100 mL LBman supplemented with 10 μL 0.2 M acetosyringone and 1 mL 1 M MES.

6. MMAi (made fresh): 5 g/L MS salts (Murashige and Skoog medium without gamborg B5 vitamins, Duchefa, the Netherlands), 20 g/L sucrose, 10 mM MES and 200 μM acetosyringone, non-sterile.

7. Protein extraction buffer: 25 mM Tris–HCl pH 8.0, 1 mM EDTA, 150 mM NaCl, 5 mM dl-Dithiothreitol (DTT), 0.1% NP40, 2% Poly-(Vinyl-Poly-Pyrolidone) (PVPP), 10% glycerol and 1× Roche Complete protease inhibitor cocktail (Roche Diagnostics, Germany).

8. Primary antibodies: PAP antibody (Sigma, USA) to detect TAP tag fused to AVR2 and primary antibody from rabbit against I-2 (29).

9. Secondary antibody: goat anti-rabbit conjugated with horseradish peroxidase (Rockland, USA).

10. Trypan blue staining solution: 10 mL lactic acid (DL Sigma L-1250), 10 mL phenol (Buffer-saturated, Invitrogen, USA), 10 mL glycerol, 10 mL H_2O and 10 mg trypan blue (Sigma-Aldrich, USA).

11. Trypan blue destaining solution: dissolve 250 g chloral hydrate (Sigma-Aldrich, USA) into 100 mL H_2O.

12. 800 mM mannitol: dissolve 14.61 g mannitol into 100 mL H_2O.

2.4. Laboratory Materials and Equipments

1. Spectrophotometer (HITACHI, Japan).

2. Capped 50 mL polypropylene tubes.

3. ROTINA 46 R Centrifuge (Depex, the Netherlands).

4. New Brunswick Incubator (EDISON, NJ, USA).

5. 1- or 2-mL syringes and one syringe needle (BD Plastipak, Madrid, Spain).

6. Liquid nitrogen, mortar and pestle.

7. Miracloth™ (EMD Chemicals).

8. Eppendorf 5415 R Centrifuge (Eppendorf, Germany).

9. Perfection 1200 Scanner (EPSON, USA).

10. PVDF membrane (Millipore, Germany).

11. LSM 510 confocal microscopy (Zeiss, Germany).

3. Methods

3.1. Nicotiana benthamiana Plant Growth

1. Germinate 15 seeds in 10 cm square plastic pots containing standard germination soil at 25°C with 70% humidity and 15 h light and 9 h dark photoperiod with 74 μmol/m²s light intensity.

2. Transfer the seedlings 8 days after sowing to 10 cm plastic pots containing potting soil (one seedling per pot) and grow the plants at 22°C with 60% humidity and 12 h light and 12 h dark photoperiod with 159 μmol/m²s light intensity (see Note 3).

3. Three weeks after transplanting, the plant has the optimal developmental stage to be used for agroinfiltration (4–5 week old). At this stage the plant has at least five fully developed true leaves and no visible flower buds (Fig. 1) (see Note 4).

Fig. 1. Optimal growth stage of *N. benthamiana* used for agroinfiltration. Counting from the top of the plant, leaves indicated with number 2, 3 and 4 are best used for agroinfiltration.

3.2. A. tumefaciens Culturing and Preparation of the Bacterial Suspension

1. Inoculate 4 mL LBman starter cultures containing 25 mg/L rifampicin and 100 mg/L spectinomycin or 50 mg/L kanamycin in a capped 25-mL glass culture tube with the *A. tumefaciens* strain GV3101 harbouring the desired plasmid (Table 1) directly from the glycerol stock. Incubate the starter cultures at 28°C under agitation (220 rpm) for 24 h to allow growth (see Note 5).

2. Inoculate 10 µL starter cultures into 20 mL LBmani containing 25 mg/L rifampicin and 100 mg/L spectinomycin or 50 mg/L kanamycin in 50-mL polypropylene tube with screw cap and grow approximately 18 h at 28°C with shaking (220 rpm) (see Note 6).

3. Measure the OD_{600} of the overnight cultures with a spectrophotometer, this should be in the range of 0.6–2.0; an OD_{600} of 0.8 is optimal.

4. Pellet the cells by centrifugation for 15 min at $2,800 \times g$ using ROTINA 46 R (see Note 7) and gently resuspend the cells in MMAi medium to the required final OD_{600} value (see Note 8) and incubate minimally for 1 h at room temperature.

3.3. Leaf Selection, Agroinfiltration and Plant Incubation

1. One day prior to infiltration, the 4–5 weeks old *N. benthamiana* plants have to be transferred to a greenhouse compartment (20°C) with low light (see Note 9).

2. Use the second, third and fourth leaf for protein expression (Fig. 1). The third and fourth leaves are best used for visualisation of HR and determination of the *in planta* localization of R or AVR proteins (see Note 10).

3. Use a syringe needle to make a small scratch at the lower side of the leaf. Fill a 1-mL needleless syringe with the *Agrobacterium*

Fig. 2. (a) *Agrobacterium* suspension is infiltrated into a *N. benthamiana* leaf. (b) To visualise specific HR triggered by co-expression of an *R* and *Avr* gene pair they are infiltrated such that both genes are expressed in the overlapping region, which is marked by a *black* marker.

suspension, and place the syringe tip on the scratch and apply slight counter-pressure to the upper side of the leaf with your gloved finger (Fig. 2a).

4. Push the piston slowly down to force entry of the *Agrobacterium* suspension into the leaf. Filling of the apoplastic spaces with the bacterial culture is visible by the formation of a dark-green sector (Fig. 2a) (see Note 11). To infiltrate an entire leaf (for protein expression) multiple infiltration points might be necessary.

5. For visualisation of HR, infiltrate the *Agrobacterium* suspension carrying either the *R* gene or *Avr* gene. Dry the leaf surface with a tissue and mark the edge of the infiltrated circle with a black marker. Wait until the infiltration zone is dry (2–3 h) and infiltrate the *Agrobacterium* suspension containing the other gene to form a second overlapping circle, also mark the second circle (Fig. 2b) (see Note 12).

6. Keep the plants at 20°C and shield them from direct light, for instance by placing them under a table that is surrounded with black plastic.

7. For protein localization studies (see Subheading 3.5.2), collect the leaves at approximately 36 h post infiltration.

8. For protein expression studies (see Subheading 3.5.1), collect the leaves at approximately 48 h after infiltration and either use them fresh, or transfer them to 50 mL polypropylene tubes and snap-freeze in liquid nitrogen. The frozen leaves can than be stored at –80°C (see Note 13).

9. For HR staining studies (see Subheading 3.4), collect the leaves at approximately 60 h after infiltration.

10. HR is normally apparent 3 days after infiltration (see Note 14).

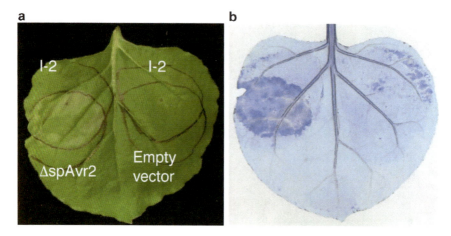

Fig. 3. Typical example of an *N. benthamiana* leaf showing HR. (**a**) HR is induced in the overlapping region that expresses both *Avr2* and *I-2*. (**b**) Cell death representing HR is visualised with trypan blue staining. There is no HR induction in the negative control (*empty vector*) or in the regions expressing the genes alone.

3.4. Trypan Blue Staining to Visualise Cell Death During Development of the Hypersensitive Response

1. Collect the leaves for trypan blue staining 60 h after agroinfiltration. At this stage HR just becomes visible to the naked eye (Fig. 3a) (see Note 15).

2. Mix 50 mL staining solution with 50 mL 96% ethanol (ratio 1:1) in a 1-L glass beaker.

3. Cover the beaker with a glass plate to avoid evaporation of the ethanol and heat the mixture Bain-marie (by placing the beaker in a boiling water bath) until boiling.

4. Place one leaf in the 100 mL solution.

5. Boil the leaf for approximately 5 min until its green colour has vanished completely (see Note 16).

6. Transfer maximum two leaves to a 50-mL polypropylene tube and add 5 mL destaining solution, rotate the tubes overnight at room temperature (see Note 17).

7. Transfer the leaf when it is fully destained to a petri dish and scan the leaf using a Perfection 1200 scanner (Fig. 3b) (best results are obtained with light from the top of the scanner).

3.5. Detection of Resistance and Effector Protein Expression

3.5.1. Protein Extraction for Western Blot Analysis

1. Crush the frozen leaves in a 50-mL polypropylene tube with a pestle and weigh the tube.

2. Place the leaf pieces into a mortar that is pre-cooled with liquid nitrogen and grind the leaves in liquid nitrogen to a fine powder.

3. Transfer the powder to a second mortar pre-cooled with ice and add 2 mL ice-cold protein extraction buffer per 1 g of leaf powder (see Note 18).

4. Keep the mortar on ice and transfer the slurry using a pipet to a 2-mL Eppendorf vial (see Note 19).

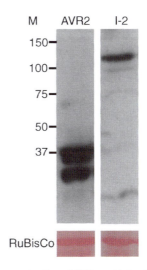

Fig. 4. Western blot showing production of AVR2 and I-2 extracted from agroinfiltrated *N. benthamiana* leaves. The Ponceau S staining of RuBisCo (*lower panel*) is used to confirm equal loading and efficient protein transfer during blotting. The protein size ladder is indicated on the *left*.

5. Centrifuge for 20 min at 13,000 rpm at 4°C in an Eppendorf centrifuge. Pour supernatant through four layers of Miracloth into an Eppendorf tube (see Note 20).

6. The protein extract can now be used to analyse R and AVR protein expression by western blot analysis (western blot) (see Note 21). Typically 30 µL protein extract is used for sodium dodecyl sulphate polyacrylamide gel electrophoresis (SDS-PAGE) gel. After running the gel, the size-separated proteins are transferred to PVDF membrane and detection is done according to (28). An example showing the production of AVR2 and I-2 is provided in Fig. 4.

3.5.2. In planta
Localization
of Recombinant Proteins

1. Harvest the infiltrated leaves 36 h after infiltration. This example describes the use of (a) cytoplasmically localised AVR2 labelled with GFP or (b) apoplastically localised AVR2 labelled with RFP.

2. Cut the leaf into 3×3 mm pieces and place a piece with its lower side facing up on a microscope slide.

3. Mount the leaf piece with H_2O and cover it with a cover slide.

4. Place a drop of immersion oil on the objective of a LSM 510 confocal microscope and place the sample under the objective.

5. Excite the GFP at 488 nm and capture emission with a 505–530 nm pass filter. The images are typically scanned eight times (Fig. 5a).

6. To detect apoplastic localisation of Avr2-RFP, treat the small leaf pieces co-infiltrated with Avr2-RFP and a plasma membrane marker labelled with YFP, with 800 mM mannitol for

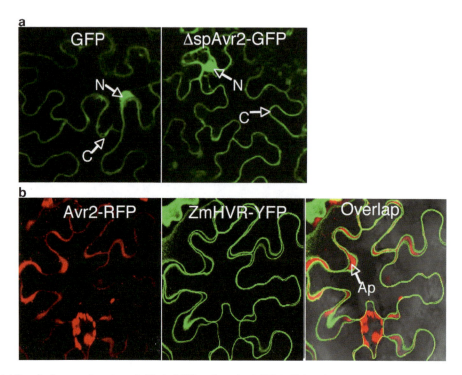

Fig. 5. (a) Transient expression of agroinfiltrated GFP and ΔspAvr2-GFP in *N. benthamiana* leaves. GFP and ΔspAvr2-GFP are localised in the nucleus and cytoplasm. Position of nucleus and cytoplasm are indicated with N and C, respectively. (b) Co-transient expression of Avr2-RFP and the plasma membrane marker ZmHVR-YFP in the epidermal cells of *N. benthamiana* after plasmolysis. Avr2-RFP is clearly presented in the apoplastic space (Ap), which is enlarged due to plasmolysis.

30 min to induce plasmolysis. Subsequently, mount the pieces in 30% glycerol on the slide (see Note 22).

7. For the localization of Avr2-RFP, excite the RFP at 543 nm and the YFP at 514 nm. Capture YFP with a 505–530-nm filter and RFP with 565–615-nm filter. The images are scanned eight times (Fig. 5b).

4. Notes

1. Other binary vectors can also be used for agroinfiltration (32). However, we typically obtain highest expression levels using pPZP200 derived vectors (30, 31).

2. The GV3101 strain is routinely used in our laboratory. Other *Agrobacterium* strains including LBA4404, C58C1, EHA105 and AGL0 have also been used successfully by us and others (18, 33).

3. Various growth conditions have to be tested to identify optimal conditions for protein accumulation and HR. High humidity

and temperature during the growth of the plants are detrimental to both processes.

4. During winter, recombinant protein yield and HR development are often compromised. Although the underlying mechanism is not known we suspect this is due to altered physiology of the plants grown during this season.

5. Transformation of *Agrobacterium* is performed as described by Takken et al. (34). Transformants need to be stored at −80°C in 30% glycerol.

6. When inoculating the starter cultures after 16:00, the OD_{600} should range between 0.6 and 2.0 after overnight growth.

7. When the cultures have reached the desired OD_{600}, it is no longer necessary to work under sterile conditions.

8. The optimal OD_{600} used for infiltration depends on the gene-of-interest. We recommend testing different ODs, but typically for expression of *R* or *Avr* genes we use an OD_{600} of 1.0. To visualise HR, the OD_{600} for the *Agrobacterium* culture carrying the *R* gene lays between 0.2 and 0.5 and for the ones carrying the *Avr* gene at 0.5. An OD_{600} that is too high for the *Agrobacterium* containing the *R* gene can trigger AVR-independent HR.

9. Placing the plants under the table to shield them from direct sun and artificial light gives more consistent results.

10. Although the younger (second) leaves normally shows highest protein accumulation, the third and fourth leaves can also be used for protein expression. For HR visualisation, the second leaf often shows aspecific HR development, which is possibly due to overexpression of the *R* gene. Therefore, it is best to use the third and fourth leaves for these purposes.

11. Wear glasses and gloves when doing syringe infiltrations.

12. The overlapping method is used to have internal controls on the leaf in which only one of the two genes is expressed. Try to keep the scratches small to confer minimal damage to the leaves.

13. When using autoactive R proteins, which trigger HR in the absence of an AVR protein, it is necessary to collect leaves already at 24 h after infiltration, or earlier when the infiltrated sector starts to collapse (29).

14. Onset and timing of HR differs for each *R* and *AVR* gene combination. For *I-2* and *Avr2* the first symptoms are visible two days after infiltration and the overlapping region at the lower side of leaf becomes "shiny". HR developing after day five is considered to be aspecific.

15. For trypan blue staining the leaves have to be harvested before the HR has fully developed. Leaves can be stored overnight at 4°C for staining next day. If the HR has already fully developed

staining of the necrotic sector is often poor. Furthermore, a late staining increases background for the sectors only expressing the *R* gene.

16. Fresh ethanol can be added to the staining solution during boiling. The solution should be replaced after boiling five to six leaves.

17. Replacing the destaining solution before overnight incubation reduces background as do longer incubation times (24 h).

18. Since PVPP is not soluble, it is necessary to suspend it in the protein extraction buffer before usage and to use wide-bore pipet tips.

19. The slurry can be transferred to a 2-mL Eppendorf tube using a regular pipet-tip. For efficient protein extraction, the leaves need to be ground quickly and thoroughly while keeping them cold.

20. Instead of Miracloth to clear the protein lysate also a sephadex G-25 column pre-equilibrated with extraction buffer can be used. The latter method results in less background on western blot.

21. Freshly prepared protein extracts give the best results on western blot.

22. The epidermal cells of *N. benthamiana* have big vacuoles making it difficult to determine the exact localization of proteins residing in either cytoplasm, plasma membrane or apoplast. Plasmolysis can facilitate determination of their localization. However, plasmolysis is inefficient in *N. benthamiana* leaves, but can be increased by incubating an intact leaf in plasmolysis solution (700 mM sucrose) for 5–6 h before analysing it by confocal microscopy (35).

Acknowledgements

We thank Anna Pietraszewska (CAM, University of Amsterdam) for the assistance using the confocal microscope. We are grateful to Ludek Tikovsky, Harold Lemereis and Thijs Hendrix for plant care.

References

1. Hamilton, C. M. (1997) A binary-BAC system for plant transformation with high-molecular-weight DNA, *Gene* 200, 107–116.

2. Kapila, J., De Rycke, R., Van Montagu, M., and Angenon, G. (1997) An Agrobacterium-mediated transient gene expression system for intact leaves, *Plant Sci.* 122, 101–108.

3. Voinnet, O., Rivas, S., Mestre, P., and Baulcombe, D. (2003) An enhanced transient expression system in plants based on suppression of gene silencing by the p19 protein of tomato bushy stunt virus, *The Plant J.* 33, 949–956.

4. Sainsbury, F., Thuenemann, E. C., and Lomonossoff, G. P. (2009) pEAQ: versatile expression vectors for easy and quick transient expression of heterologous proteins in plants, *Plant Biotech. J.* 7, 682–693.

5. Scholthof, H. B., Scholthof, K.-B. G., and Jackson, A. O. (1996) Plant virus gene vectors for transient expression of foreign proteins in plants, *Annu. Rev. Phytopathol.* 34, 299–323.

6. Fischer, R., Liao, Y. C., Hoffmann, K., Schillberg, S., and Emans, N. (2005) Molecular farming of recombinant antibodies in plants, *Biol. Chem.* 380, 825–839.

7. Kanneganti, T.-D., Huitema, E., and Kamoun, S. (2007) In planta expression of Oomycete and fungal genes, in *Plant-Pathogen Interactions* (Ronald, P. C., Ed.), pp 35–43, Humana Press.

8. Sheen, J. (2001) Signal Transduction in Maize and Arabidopsis mesophyll protoplasts, *Plant Physiol.* 127, 1466–1475.

9. Schweizer, P., Christoffel, A., and Dudler, R. (1999) Transient expression of members of the germin-like gene family in epidermal cells of wheat confers disease resistance, *The Plant J.* 20, 541–552.

10. Bilang, R., Zhang, S., Leduc, N., Iglesias, V. A., Gisel, A., Simmonds, J., Potrykus, I., and Sautter, C. (1993) Transient gene expression in vegetative shoot apical meristems of wheat after ballistic microtargeting, *The Plant J.* 4, 735–744.

11. Wroblewski, T., Tomczak, A., and Michelmore, R. (2005) Optimization of Agrobacterium-mediated transient assays of gene expression in lettuce, tomato and Arabidopsis, *Plant Biotech. J.* 3, 259–273.

12. Zottini, M., Barizza, E., Costa, A., Formentin, E., Ruberti, C., Carimi, F., and Lo Schiavo, F. (2008) Agroinfiltration of grapevine leaves for fast transient assays of gene expression and for long-term production of stable transformed cells, *Plant Cell Rep.* 27, 845–853.

13. Vaquero, C., Sack, M., Chandler, J., Jürgen, D., Schuster, F., Monecke, M., Schillberg, S., and Fischer, R. (1999) Transient expression of a tumor-specific single-chain fragment and a chimeric antibody in tobacco leaves, *Proc. Natl. Acad. Sci. USA* 96, 11128–11133.

14. Tang, X., Frederick, R. D., Zhou, J., Halterman, D. A., Jia, Y., and Martin, G. B. (1996) Initiation of plant disease resistance by physical interaction of AvrPto and Pto kinase, *Science* 274, 2060–2063.

15. Van den Ackerveken, G., Marois, E., and Bonas, U. (1996) Recognition of the Bacterial avirulence protein AvrBs3 occurs inside the host plant cell, *Cell* 87, 1307–1316.

16. Ihara-Ohori, Y., Nagano, M., Muto, S., Uchimiya, H., and Kawai-Yamada, M. (2007) Cell death suppressor Arabidopsis Bax inhibitor-1 is associated with calmodulin binding and ion homeostasis, *Plant Physiol.* 143, 650–660.

17. Bhat, R., Lahaye, T., and Panstruga, R. (2006) The visible touch: in planta visualization of protein-protein interactions by fluorophore-based methods, *Plant Methods* 2, 12.

18. D'Aoust, M.-A., Lavoie, P.-O., Belles-Isles, J., Bechtold, N., Michèle, M., and Louis-P., V. (2008) Transient expression of antibodies in plants using syringe Agroinfiltration, in *Recombinant Proteins From Plants: Methods and Protocols* (Loïc, F., and Véronique, G., Eds.), pp 41–50, Humana Press.

19. Tai, T. H., Dahlbeck, D., Clark, E. T., Gajiwala, P., Pasion, R., Whalen, M. C., Stall, R. E., and Staskawicz, B. J. (1999) Expression of the Bs2 pepper gene confers resistance to bacterial spot disease in tomato, *Proc. Natl. Acad. Sci. USA* 96, 14153–14158.

20. Van der Hoorn, R. A. L., Laurent, F., Roth, R., and De Wit, P. J. G. M. (2000) Agroinfiltration as a versatile tool that facilitates comparative analyses of Avr9/Cf-9-induced and Avr4/Cf-4-induced necrosis, *Mol. Plant-Microbe Interact.* 13, 439–446.

21. Yasmin, A., and Debener, T. Transient gene expression in rose petals via *Agrobacterium* infiltration, *Plant Cell, Tissue and Organ Culture* 102, 245–250.

22. Abramovitch, R. B., Janjusevic, R., Stebbins, C. E., and Martin, G. B. (2006) Type III effector AvrPtoB requires intrinsic E3 ubiquitin ligase activity to suppress plant cell death and immunity, *Proc. Natl. Acad. Sci. USA* 103, 2851–2856.

23. Goodin, M. M., Zaitlin, D., Naidu, R. A., and Lommel, S. A. (2008) *Nicotiana benthamiana*: its history and future as a model for plant-pathogen interactions, *Mol. Plant-Microbe Interact.* 21, 1015–1026.

24. Houterman, P. M., Ma, L., Van Ooijen, G., De Vroomen, M. J., Cornelissen, B. J. C., Takken, F. L. W., and Rep, M. (2009) The effector protein Avr2 of the xylem-colonizing fungus *Fusarium oxysporum* activates the tomato resistance protein I-2 intracellularly, *The Plant J.* 58, 970–978.

25. De La Fuente Bentem van, S., Vossen, J. H., de Vries, K. J., van Wees, S., Tameling, W. I. L., Dekker, H. L., de Koster, C. G., Haring, M. A., Takken, F. L. W., and Cornelissen, B. J. C. (2005) Heat shock protein 90 and its co-chaperone phosphatase 5 interact with distinct regions of the tomato I-2 disease resistance protein, *The Plant J.* 43, 284–298.

26. Van Ooijen, G., Lukasik, E., Van Den Burg, H. A., Vossen, J. H., Cornelissen, B. J. C., and Takken, F. L. W. The small heat shock protein 20 RSI2 interacts with and is required for stability and function of tomato resistance protein I-2, *The Plant J.* 63, 563–572.

27. Tameling, W. I. L., Vossen, J. H., Albrecht, M., Lengauer, T., Berden, J. A., Haring, M. A., Cornelissen, B. J. C., and Takken, F. L. W. (2006) Mutations in the NB-ARC domain of I-2 that impair ATP hydrolysis cause autoactivation, *Plant Physiol.* 140, 1233–1245.

28. Van Ooijen, G., Mayr, G., Albrecht, M., Cornelissen, B. J. C., and Takken, F. L. W. (2008) Transcomplementation, but not physical association of the CC-NB-ARC and LRR domains of tomato R protein Mi-1.2 is altered by mutations in the ARC2 subdomain, *Molecular plant* 1, 401–410.

29. Van Ooijen, G., Mayr, G., Kasiem, M. M. A., Albrecht, M., Cornelissen, B. J. C., and Takken, F. L. W. (2008) Structure-function analysis of the NB-ARC domain of plant disease resistance proteins, *J. Exp. Biol.* 59, 1383–1397.

30. Rohila, J. S., Chen, M., Cerny, R., and Fromm, M. E. (2004) Improved tandem affinity purification tag and methods for isolation of protein heterocomplexes from plants, *The Plant J.* 38, 172–181.

31. Nakagawa, T., Suzuki, T., Murata, S., Nakamura, S., Hino, T., Maeo, K., Tabata, R., Kawai, T., Tanaka, K., Niwa, Y., Watanabe, Y., Nakamura, K., Kimura, T., and Ishiguro, S. (2007) Improved gateway binary vectors: high-performance vectors for creation of fusion constructs in transgenic analysis of plants, *Bioscience, Biotechnology, and Biochemistry* 71, 2095–2100.

32. Hellens, R., Allan, A., Friel, E., Bolitho, K., Grafton, K., Templeton, M., Karunairetnam, S., Gleave, A., and Laing, W. (2005) Transient expression vectors for functional genomics, quantification of promoter activity and RNA silencing in plants, *Plant Methods* 1, 13.

33. Pruss, G. J., Nester, E. W., and Vance, V. (2008) Infiltration with *Agrobacterium tumefaciens* induces host defense and development-dependent responses in the infiltrated zone, *Mol. Plant-Microbe Interact.* 21, 1528–1538.

34. Takken, F. L. W., Luderer, R., Gabriëls, S. H., Westerink, N., Lu, R., De Wit, P. J. G. M., and Joosten, M. H. A. J. (2000) A functional cloning strategy, based on a binary PVX-expression vector, to isolate HR-inducing cDNAs of plant pathogens, *The Plant J.* 24, 275–283.

35. Lim, H.-S., Bragg, J. N., Ganesan, U., Ruzin, S., Schichnes, D., Lee, M. Y., Vaira, A. M., Ryu, K. H., Hammond, J., and Jackson, A. O. (2009) Subcellular localization of the Barley Stripe Mosaic Virus triple gene block proteins, *J. Virol.* 83, 9432–9448.

Chapter 5

Proteomic Techniques for Plant–Fungal Interactions

Delphine Vincent, Kar-Chun Tan, Liam Cassidy, Peter S. Solomon, and Richard P. Oliver

Abstract

Proteomics is a key technique that is helping elucidate many complex biological processes. The analysis of plant–pathogen interactions using proteomics is complicated by the presence of the proteomes of two species, but is benefiting from the developing maturity and power of these techniques. More and more pathogen genomes are being sequenced, so fungal proteomics is reaching its full potential and remains the chosen technology to unravel the molecular pathways of pathogenicity and resistance. In this chapter, we suggest proteomic strategies that have proved successful on various plant-interacting fungal species. Several protein extraction methods are described. For adequate quantitative analyses of protein abundances, we recommend either separation using two-dimensional gel electrophoresis or labelling with isobaric tags followed by two-dimensional HPLC separation. Proteins of interest are then identified using mass spectrometry. Identified proteins can assist in refining genome annotations, otherwise known as proteogenomics.

Key words: PLANT fungal pathogen, Proteomics, Protein extraction, IEF, SDS-PAGE, 2-DE, iTRAQ, MS, Genome annotation, Proteogenomic

Abbreviations

2-DE	Two-dimensional electrophoresis
2-ME	2-Mercaptoethanol
APS	Ammonium persulfate
CAs	Carrier Ampholites (IPG buffers from Bio-Rad for instance)
CHAPS	3-[3-(Cholamidopropyl)dimethylammonio]-1-propanesulfonate
DMSO	Dimethyl sulfoxide
DTT	Dithiothreitol
EDTA Na$_2$	Ethylenediaminetetraacetic acid disodium salt
IEF	Isoelectric focusing
IPG	Immobilised pH gradient
iTRAQ	Isobaric tag for relative and absolute quantitation
LC	Liquid chromatography

Melvin D. Bolton and Bart P.H.J. Thomma (eds.), *Plant Fungal Pathogens: Methods and Protocols*,
Methods in Molecular Biology, vol. 835, DOI 10.1007/978-1-61779-501-5_5, © Springer Science+Business Media, LLC 2012

MALDI-ToF/ToF	Matrix assisted laser desorption ionisation- time of flight/time of flight
MeOH	Methanol
MMTS	Methyl methanethiosulfonate
PMSF	Phenylmethanesulfonide fluoride
PVP	Polyvinylpyrrolidone
SCX	Strong cation exchange
SDS	Sodium dodecyl sulphate
SPE	Solid phase extraction
TCA	Trichloroacetic acid
TCEP	Tris-(2-carboxyethyl)-phosphine-HCl
TCEP	Tris(2-carboxyethyl)phosphine
TEAB	Triethylammonium bicarbonate
TEMED	Tetramethylethylenediamine

1. Introduction

1.1. Proteomics for Unravelling Plant–Fungal Interactions

Proteomics is the term coined to encompass studies that aim to identify and quantify *all* the proteins within a given sample. Such studies can identify the fungal or plant proteins that control the outcome of an encounter between a pathogen and a potential host. A second use of proteomics—to assist the annotation of the genomes—has been termed proteogenomics and is described below.

The completion of genome sequencing projects of various fungal pathogens has made proteomic analysis feasible. As most fungal genomes are relatively small, their acquisition is becoming a routine activity. In contrast, host genomes are vastly larger and only a few model species genome sequences are currently available. As plant pathology is the study of two (or more) organisms, the full potential of proteomics will only become apparent when host genomes are also acquired.

Initial proteomic studies of fungal pathogens dealt with axenically-grown samples. Both intracellular soluble proteins (1, 2) and the secreted proteome have been studied. Secreted proteins play a key role in the interaction between a fungus and the plant host (3). Furthermore, many interactions are controlled by secreted effectors, many of which are proteins (4–6).

The analysis of infected plant material is a significantly more complex challenge. Whether we are dealing with obligate pathogens such as *Blumeria graminis* (7) or specific infection structures such as *Magnaporthe oryzae* appressoria, the main limitations are the small quantity of material and the complexity caused by mixing two species' proteomes. Nonetheless, these problems are being overcome by modern proteomic strategies (8). Table 1 lists selected proteomic and secretome studies published on fungal plant pathogens. This growing list illustrates the range of approaches and provides sources for many variations in techniques not covered fully below.

Table 1
Selected proteomic and secretome studies of fungal plant pathogens.

References	Fungal species	Material and collection method	Protein extraction	Protein precipitation	Protein solubilization	Protein separation	Peptide separation	Protein identification	Proteins with hit (including unknown function)	Identified proteins (known function only) (b)
(7)	*Blumeria graminis* f. sp. *Hordei* Powdery mildew	Sporulating hyphae collected from leaves	1-3/TCA/acetone, 4/CHCl₃/MeOH	Cold 10%TCA/0.07% 2-ME/acetone or MeOH	7 M urea, 2 M thiourea, 2% CHAPS, 20 mM DTT	1-DE shotgun	nLC (C18 column)	nESI-MS/MS (HCT or Orbitrap)	N.S.	441, 775, 47, N.D.
(23)	*Ustilago maydis* (haploid strains). Corn smut	In vitro cell sedimentation by centrifugation	Tris/EtOH	no	7 M urea, 2 M thiourea, 2% CHAPS, 20 mM DTT, 0.5% CAs, 0.1% BPB	2-DE	No	MALDI-TOF-MS + LC-MS/MS	N.S.	250
(16)	*Stagonospora nodorum* (SN15 wild type and gna1 strain). Tan spot	In vitro 1/lyophilisation 2/cheese-cloth +0.2 μm Millipore filtration	1/Tris buffer, 2/acetone	2/Cold acetone	1/10 mM Tris pH 7.5, 2/0.5 M TEAB pH 8.5	LC shotgun	2D-LC (SXC-RP/nLC)	MALDI-TOF/TOF-MS	8,717	N.S.
(12)	*Stagonospora nodorum* (SN15 wild type and gna1 strain). Tan spot	In vitro cheese-cloth filtration + lyophilisation	Tris-Cl	No	Tris-Cl pH 7.5	No	2-D LC (SCX/C18)	MALDI-TOF/TOF-MS	1,336	N.S.
(24)	*Fusarium graminearum* (strain PH-1). Head blight.	1/In vitro, 2/in planta 1/Miracloth filtration, 2/centrifugation/desalting/lyophilisation	TCA/EtOH	11% TCA	50 mM ammonium bicarbonate	1-DE shotgun	LC	ESI-q-TOF-MS/MS	N.S.	1/228, 2/120

(continued)

Table 1
(continued)

References	Fungal species	Material and collection method	Protein extraction	Protein precipitation	Protein solubilization	Protein separation	Peptide separation	Protein identification	Proteins with hit (including unknown function)	Identified proteins (known function only) (b)
(25)	*Fusarium graminearum.* Head blight.	In vitro centrifugation + concentration (Vivacell MWCO 10 kD)	Phosphate buffer	No	8 M urea, 50 mM DTT, 2% CHAPS, 0.2% CAs, 0.002% BPB	2-DE	nLC	ESI-q-TOF-MS/MS	88	88
(26)	*Botrytis cinerea* (strain BO5.10). Gray mold	In vitro (membrane overlaid on strawberry/tomato/ Arabidopsis pulp media)	SDS sample buffer	No	2× Laemmli buffer	1-DE shotgun	nLC (C18 column)	ESI-q-TOF-MS/MS	89	76
(27)	*Botrytis cinerea* (strain BO5.10). Gray mold	In vitro (pectin medium) paper filtration (No. 41 Whatman) + lyophilisation	Laemmli sample buffer	No	2X Laemmli buffer	1-DE shotgun	nLC (C18 column)	ESI-q-TOF-MS/MS	N.S.	126
(13)	*Leptosphaeria maculans* (isolate v23.1.3). Blackleg	In vitro (Fries medium)	1/TCA/ acetone, 2/ filtration/ dialysis/ prefractionation/ lyopholisation/ resuspension	1/cold 10% TCA/ 0.07% 2-ME/ acetone	1-2/7 M urea, 2 M thiourea, 2% CHAPS, 1% DTT, 0.5% proteimase inhibitor cocktail, 0.5% CAs	2-DE	nLC	ESI-q-TOF-MS/MS	1/76, 2/52	1/75, 2/47
(28)	*Sclerotinia sclerotiorum.*	In vitro (minimal salt medium)	1/TCA/ acetone, 2/ filtration/ dialysis/ lyopholisation/ resuspension	1/cold 10%TCA/ 0.07% 2-ME/ acetone	1-2/2 mM TBP buffer	2-DE	LC	ESI-q-TOF-MS/MS	1/95, 2/18	1/58, 2/14

Research in our laboratories has used a series of interlocking protocols to achieve the desired result. The first step is the extraction of the proteins from either mycelial or supernatant fractions (see Subheading 1.2 below). The extracted proteins can then be separated by two-dimensional gel electrophoresis (2-DE) (see Subheading 1.3) or by gel-free liquid chromatography (LC) (see Subheading 1.4) methods. Finally the gel spots containing the protein of interest or peaks are analysed by mass spectrometry (MS). This can be applied to multiple (2, 4 or 8) samples to identify peptides differentially present in different samples as in isobaric tag for relative and absolute quantitation (iTRAQ) (9). We lastly described the large scale analysis of peptides used to validate gene models—proteogenomics (see Subheading 1.5).

1.2. Protein Extraction Methods

Many different methods have been published to extract proteins from mycelial samples. The two methods we describe are optimised for both protein yield and purity. The first method precipitates proteins in trichloroacetic acid (TCA)/acetone followed by several washes before final resuspension of the pelleted proteins (10). The second is a procedure based on phase-partition during which proteins are first recovered in a phenol phase settling above a dense sucrose/aqueous phase, then precipitated by ammonium acetate followed by various washes prior to final resuspension (11). Although fully compatible with 2-DE and isoelectric focusing (IEF) in particular, these extraction methods are not suitable for iTRAQ labelling. In this case, proteins must then be extracted using a Tris/acetone procedure (12).

Methods to recover secreted proteins are much more diverse. We detail a protocol that should prove appropriate for most culture filtrates by maximising contaminant removal while minimising protein loss. Such procedures involve sequential steps of filtration, dialysis, and lyophilisation prior to final resuspension. Fungal secretomes often include a few extremely abundant proteins which obscure rarer molecules. Vincent et al. (13) have demonstrated a significant gain in secreted protein recovery and resolution when an initial pre-fractionation step using liquid-phase IEF was included. In this technique, liquid-phase IEF is performed prior to resolution via SDS-polyacrylamide gel electrophoresis (SDS-PAGE).

1.3. Analysis of Fungal Proteomes Using 2-DE

Having obtained a protein sample, the next step is to determine the complexity of the mixture. Two-dimensional gel electrophoresis (2-DE) is a powerful and mature technology for the separation and analysis of complex protein samples. Solubilised protein mixtures are separated firstly by isoelectric point (pI) and secondly by molecular weight (MW). The final gel is stained and the individual spots are available for excision, proteolytic cleavage and MS identification. Whilst 2-DE is gradually being superseded by multidimensional LC methods (see Subheadings 3 and 4), it retains several key advantages that will ensure its use for years to come. Firstly, the result is a powerful visual representation of the similarities and differences in the sample that can be used to uncover novel features of the pathogen (14, 15). Secondly, the equipment needed is relatively inexpensive. Thirdly, less training is required than for LC techniques.

The identification of proteins within a proteome is normally performed by identification of tryptic peptides. Following gel-based separation, the resolved proteins are digested by trypsin, resulting in a large number of peptides which are much more amenable to chromatographic separation and detection on a mass analyser.

1.4. Analysis of Fungal Proteomes Using Gel-Free Proteomics

Proteomic analysis of complex samples requires a high degree of sample separation before they are subjected to analysis by MS. Unlike gel-based proteomic techniques (in which the intact proteins are separated) gel-free proteomics relies on the separation of enzymatically-digested peptides. Two-dimensional high performance liquid chromatography (2D HPLC) is well suited for this role and provides enough resolving power to allow for the identification of a significant percentage of proteins within the sample. The injected liquid sample is first eluted using a strong cation exchange (SCX) column which separates peptides based on their charge, and then further separated using a reverse phase (RP) C18 column which resolves peptides according to their hydrophobicity.

The separated peptides can then be analysed by a mass spectrometer such as a matrix-assisted laser desorption/ionisation time-of-flight (MALDI-TOF) analyser with a high degree of mass accuracy. Peptide mass fingerprinting compares the experimentally obtained masses against those of the predicted tryptic peptides. This can identify the peptides and link them to a predicted gene. An extension of this utilises a second linked mass-analyser (in this case a second TOF analyser) and measures the masses of fragmentation products of tryptic peptides which are then compared against a database of predicted fragmentation products (peptide MS/MS identification). This greatly aids the certainty of assignment of peptides to proteins and subsequently the validation of genes (16).

A further refinement quantitatively compares 2, 4 or 8 samples and detects peptides that differ in abundance within the mixture. Each sample is digested and then labelled with one or up to eight mass-tags. These isobaric tags are identical in mass but fragment to give unique reporter ions. The relative frequency of the reporter ions is a measure of the relative abundance of the peptide in the original mixture. Here, we describe the iTRAQ labelling of protein samples, via the iTRAQ 8-plex system, for 2D HPLC with subsequent analysis on a MALDI TOF/TOF mass spectrometer.

1.5. Proteogenomics   The power of LC-based MS techniques to separate and analyse very large numbers of individual peptides can be optimised to identify as many peptides as possible and thereby identify protein encoding regions of genomes. This technique has become known as proteogenomics and rivals EST analysis as a method for gene validation (16). The technique is a variation of the LC-MS based methodology.

2. Materials

2.1. Labware   We list the equipment needed for all protocols and in some cases nominate manufacturers and items that have worked in our labs. Doubtless other suppliers and items would be suitable.

1. Porcelain mortar with pour lip and pestle.

2. Plastic tubes with screw closure allowing for a volume at least 5 times to that of the fungal sample. Tubes must resist several centrifugation steps at high speed and subfreezing temperatures, as well as acidic solutions and organic solvents. Nalgene® centrifuge tubes Oak Ridge style (Sigma T1418, nominal capacity 50 mL) made of polypropylene copolymer offering approximately 40 mL capacity are adequate.

3. Bench centrifuge allowing high speed ($12,000 \times g$) and low temperature ($-10°C$) with a fixed angle rotor.

4. Eppendorf tubes (1.5 or 2 mL) for storage of the protein extracts.

5. Glass or metal spatula.

6. Scalpel blade.

7. Dialysis membrane (MW cut-off to be adapted to the need). We recommend Spectra/Por 3 Regenerated Cellulose (RC) dialysis membranes with a 3,500 Da cut-off (Spectrum Europe, Breda, The Netherlands).

8. Freeze-drier, SpeedVac or vacuum device to dry pellets.

9. Spectrophotometer: We use a Lambda 25 UV/VIS spectrophotometer.

10. Sonicator: We use a probe tip Misonix Sonicator XL2015.

11. Gel scanner: ProExpress scanner (Perkin Elmer).

12. Protein spot densitometry analysis: ProGenesis Workstation 2005 software (Nonlinear Dynamics, United Kingdom).

13. Mass spectrometry peptide analysis: We use an Agilent 1100 series capillary LC system coupled to a QSTAR Pulsar i LC-MS/MS system (Applied Biosystems, MA, USA) or XCT Ultra IonTrap mass spectrometer (Agilent, Santa Clara, USA) with the ion spray set at positive ion mode.

14. Orbital rocker.

15. Protein identification search engine: We use the Mascot software (Matrix Science, Boston, USA).

16. Immobilised pH gradient (IPG) strips: The method described from here on focuses on the use of 7 cm linear pH 5–8 and 3–10 ReadyStrip IPG strips (Bio-Rad, Hercules, USA) but can be adapted to accommodate larger IPG strips.

17. Disposable plastic trays that can accommodate the length of IPG strips used.

18. PROTEAN IEF focusing trays (Bio-Rad).

19. PROTEAN IEF Cell (Bio-Rad).

20. Polyacrylamide gel casting and electrophoresis unit: We manually cast 12% (w/v) SDS polyacrylamide gels using the mini PROTEAN III system (Bio-Rad).

21. Mini PROTEAN III system compatible glass plates with 1 mm spacers.

22. Electrophoresis powerpack that can be set to maintain constant amperage. We used a Bio-Rad Model 3000xi unit.

23. Polyvinyllidene chloride plastic film.

24. Equilibration tray (Bio-Rad).

25. Whatman paper cut to 0.3×0.5 cm.

26. Starta-X 33um polymeric reverse phase column.

27. Polysulfoethyl 4.6×100 mm 5 μm 300A column (The Nest Group, MA, USA).

28. Probot micro fraction collector (LC Packings, CA, USA).

29. Ultimate 3000 nano HPLC system (LC Packings-Dionex, CA, USA).

30. 4800 MALDI-TOF/TOF® mass spectrometer (Applied Biosystems).

31. 384-well Opti-TOF® plates (Applied Biosystems).

2.2. Solutions for TCA/Acetone Method

Solutions were prepared as described in Damerval et al. (10) with modifications.

1. Precipitation solution (P): 10% (w/v) TCA, 0.07% (v/v) 2-mercaptoethanol (2-ME) in acetone stored at –20°C until use. P should exceed (at least 5 times) the volume of starting material.

2. Washing solution (W): prepare a 0.07% (v/v) 2-ME in acetone solution and store it at –20°C until required. Since a minimum of three washes should be performed, W should be at least 3 times the volume of P.

3. Resuspension solution (R): mix the following chemicals: 7 M urea, 2 M thiourea, 4% (w/v) 3-(3-(cholamidopropyl)

dimethyla-mmonio]-1-propanesulfonate (CHAPS), 1% (w/v) dithiothreitol (DTT), 1% (v/v) 2-ME, 1 mM Tris-(2-carboxyethyl)-phosphine–HCl (TCEP), 0.5% (v/v) carrier ampholytes, 0.5% (v/v) proteinase inhibitor cocktail. The molarity of this solution is close to urea saturation, therefore milliQ (MQ) H_2O amount should be kept to a minimum while chemicals are dissolving. Aliquots of R can be stored for months at –20°C (see Note 1).

4. Alternative solubilisation solution: Multiple surfactant solution (MSS) (17–19): 40 mM ultrapure Tris, 2% (w/v) CHAPS, 2% (w/v) sulfobetaine 3–10, 5 M electrophoresis grade urea, 2 M thiourea, 2 mM tributylphosphine, 0.2% (v/v) carrier ampholytes and 0.002% (w/v) bromophenol blue (Bio-Rad). The type of carrier ampholytes used is dependent on the pH range of the immobilised pH gradient (IPG) strip. For instance, we used Bio-Lyte 3-10 (Bio-Rad) for immobilised pH gradient (IPG) strips with a linear resolving range of pH 3–10 and 5–8 (see Note 2).

5. Precipitated proteins can be resuspended 100–500 μL R or MSS solution in a 1.5-mL Eppendorf tube by vortexing (see Note 3).

6. Non-soluble materials can be removed from the sample by centrifugation at $20,000 \times g$ for 5 min at room temperature. Transfer the supernatant containing soluble proteins to a new tube and determine the protein concentration.

2.3. Solutions for Phenol/Ammonium Acetate Method

Solutions were prepared as described in Hurkman and Tanaka (11) with modifications.

1. Homogenization buffer (H): mix the following chemicals: 0.7 M sucrose, 0.1 M KCl, 50 mM EDTA, 1 mM phenylmethanesulfonide fluoride (PMSF; 1 M PMSF stock solution dissolved in DMSO), 2% (v/v) 2-ME, 0.5% (w/v) deoxycholic acid, 0.5 M Tris-Cl pH 8.0. Aliquots of H can be stored for at –20°C. Prior to use, thaw H overnight at 4°C.

2. Phenol solution: a water-saturated phenol pH 8.0 must be employed. If necessary, its pH can be adjusted using 10 mM Tris-Cl pH 10.0. This can be purchased from Sigma (P4557). Store at 4°C.

3. Precipitation solution (P): prepare a 0.1 M ammonium acetate/methanol solution and store at –20°C. P should exceed (at least 5 times) the equivalent volume of starting material.

4. Washing solution 1 (W1): store 100% methanol at –20°C until use. W1 should be at least twice the volumes of P (see Subheading 3.1).

5. Washing solution 2 (W2): store 100% acetone at –20°C until use. W2 should be at least twice the volumes of P (see Subheading 3.1).

6. Resuspension solution (R): see step 3 in Subheading 3.1.

2.4. Solutions for Second Dimension Protein Separation: SDS-PAGE

1. SDS-PAGE reagents for the Laemmli denaturing discontinuous gel electrophoresis method (20). All reagents must be purchased as electrophoresis grade where possible.

2. SDS electrophoresis buffer (1): 3 g/L ultrapure Tris, 1.0 g/L electrophoresis grade SDS, 14.4 g/L glycine.

3. 4× (1.5 M) Tris-Cl, pH 8.8.

4. 10% (w/v) SDS solution.

5. Acrylamide/bisacrylamide solution. We used 40% (w/v) acrylamide/bisacrylamide (37.5:1) from Bio-Rad.

6. 10% (w/v) ammonium persulfate (APS). Make solution prior to use.

7. Tetramethylethylenediamine (TEMED).

8. Overlay agarose: 0.5% (w/v) agarose in SDS electrophoresis buffer. A trace of bromophenol blue dye can be added to the molten agarose as an electrophoresis dye.

9. Isobutanol.

10. Equilibration buffer: 6 M electrophoresis grade urea, 0.375 M ultrapure Tris, 20% (w/v) glycerol, 2–4% (w/v) electrophoresis grade SDS and 2% (w/v) DTT or 2.5% (w/v) iodoacetamide (see Subheading 3.7).

11. Protein MW standard: We used Precision Plus protein standard (Bio-Rad).

12. Gel fixing solution: 40% (v/v) methanol and 10% (v/v) glacial acetic in MQ water.

13. Colloidal Coomassie G250 stain (9): 0.1% (w/v) Coomassie G250 (Bio-Rad), 10% (w/v) ammonium acetate and 2% (w/v) H_3PO_4.

14. Gel destaining solution: 0.5% (w/v) ortho-phosphoric acid in ultrapure MQ H_2O.

2.5. Solutions for Protein Digestion

1. HPLC-grade acetonitrile.

2. HPLC-grade formic acid.

3. Destain solution: 50% (v/v) acetonitrile, 10 mM NH_4HCO_3.

4. Trypsin solution: 12.5 µg/mL sequencing grade trypsin (Roche) in 25 mM NH_4HCO_3.

2.6. Solutions for iTRAQ Labelling and 2DHPLC

1. Triethylammonium bicarbonate (TEAB): 500 mM TEAB in MQ H_2O.

2. Sodium dodecyl sulphate (SDS): 2% (w/v) SDS in MQ H_2O.

3. Tris(2-carboxyethyl)phosphine (TCEP): 50 mM TCEP in MQ H_2O.

4. Methyl methanethiosulfonate (MMTS): 200 mM MMTS, pH 9.4.

5. iTRAQ® Reagent-8Plex Multiplex Kit.(AB Sciex, Mt Waverley, VIC, Australia).

6. HPLC-grade isopropanol.

7. HPLC-grade methanol.

8. HPLC-grade acetonitrile.

9. SCX buffer A: 10 mM KH_2PO_4, 10% (v/v) acetonitrile in MQ H_2O at pH 3.0.

10. SCX buffer B: 10 mM KH_2PO_4, 1 M KCl, 10% (v/v) acetonitrile in MQ H_2O at pH 3.0.

11. RP buffer A: 2% (v/v) acetonitrile in MQ H_2O with 0.1% (v/v) formic acid.

12. RP buffer B: 100% (v/v) acetonitrile with 0.1% (v/v) formic acid.

13. Loading buffer: 2% (v/v) acetonitrile, 0.05% (v/v) trifluoroacetic acid in MQ H_2O.

14. Matrix solution: 5 mg/mL α-cyano-4-hydroxycinnamic acid, 10 mM ammonium citrate, 80% (v/v) acetonitrile, 0.1% (v/v) trifluoroacetic acid, 0.1 μL/mL Calibration mixture 1 (AB Sciex, Mt Waverley, VIC, Australia).

2.7. Solutions for Solid Phase Extraction

1. HPLC-grade acetonitrile.

2. HPLC-grade methanol.

3. Elution solution: 80% (v/v) acetonitrile, 0.1% (v/v) formic acid in MQ H_2O.

3. Methods

3.1. Extraction of Mycelial Proteins Using a TCA/Acetone Method

1. Mycelium samples are ground in liquid N_2 using a pre-chilled mortar and pestle to pulverise the tissue into a fine frozen powder. Keep sample cold at all time.

2. Add 5 volumes of cold P to the frozen powdered sample in the mortar. Mix well using the pestle until homogenization and transfer the solution into a pre-chilled plastic tube. Incubate at –20°C for at least 2 h (possibly overnight to increase protein yield) with regular tube inversions. Centrifuge tube to pellet proteins and insoluble material for 30 min at 12,000×g and –10°C. Discard supernatant.

3. Fill up tube with cold W and vortex to dissolve pellet. Incubate at –20°C for at least 2 h with regular tube inversions. Centrifuge tube to pellet proteins and insoluble material for 30 min at 12,000×g and –10°C. Discard supernatant.

4. Repeat step 2 twice, for a total of three washes. These washes remove TCA. The pH of the successive discarded supernatants

can be checked using pH paper to confirm TCA removal (pH should increase from 2 to 6). Additional washes can be performed if necessary (see Note 4). Dry the pellet to remove all traces of solvent and resuspend in 1 volume of R by agitation. Vortex for at least 1 min to completely resuspend the pellet. A water-bath sonicator may be used to finalise protein solubilisation but the water temperature should not exceed 37°C as urea will then generate protein carbamylation.

5. Centrifuge the tube for 5 min at $12,000 \times g$ and room temperature. Transfer supernatant in Eppendorf tube and store at –20 or –80°C until use.

3.2. Extraction of Mycelial Proteins Using a Phenol/ Ammonium Acetate Method

1. Add two volumes of chilled H to the frozen powdered sample in the mortar. Mix well using the pestle until homogenization and transfer the solution into a pre-chilled plastic tube. Vortex briefly and incubate for 30 min at 4°C (see Note 5).

2. Add the same volume of chilled water-saturated phenol pH 8.0, vortex to homogenise, and incubate for 30 min at 4°C with regular mixing.

3. Centrifuge tube for 30 min at $12,000 \times g$ and 4°C to achieve phase separation. Pipette upper phenol phase without disturbing the interface and transfer it into a fresh plastic tube.

4. Re-extract the phenol phase by adding the same volume of chilled H. Vortex to homogenise and incubate for 30 min at 4°C.

5. Repeat step 3.

6. Add 5 volumes of cold P and homogenise by tube inversion and vortexing. Incubate for at least 2 h (possibly overnight) at –20°C with regular tube inversion. Centrifuge tube to pellet proteins and insoluble material for 30 min at $12,000 \times g$ and –10°C. Discard supernatant.

7. Fill up tube with cold W1 and homogenise by tube inversion and vortexing. Incubate for at least 1 h at –20°C with regular tube inversion. Centrifuge tube for 30 min at $12,000 \times g$ and –10°C. Discard supernatant.

8. Repeat step 7.

9. Fill up the tube with cold W2 and homogenise by tube inversion and vortexing. Incubate for at least 1 h at –20°C with regular tube inversion. Centrifuge the tube for 30 min at $12,000 \times g$ and –10°C. Discard the supernatant.

10. Repeat step 9.

11. Dry the pellet to remove all traces of solvent, resuspend in 1 volume of R, centrifuge briefly and transfer the supernatant in a fresh tube.

3.3. Extraction of Fungal Secreted Proteins from Culture Filtrates

1. Filter the culture filtrate using 0.2-µm syringe filters with 25 mm membranes (Nalgene, NY, USA) to remove any cellular debris.

2. Dialyse the filtrates using Spectra/Por 3 Regenerated Cellulose (RC) dialysis membranes with a 3,500 Da cut-off (Spectrum Europe, Breda, Netherlands). To ensure maximal removal of growth medium component, dialyse 50 mL of secretome sample against 5 L MQ H_2O at 4°C with constant stirring for 12 h, and repeat this step with fresh cold MQ H_2O.

3. Transfer the dialysed filtrates into plastic tubes kept on ice, and add 1 mM PMSF and 0.001% (w/v) polyvinylpyrrolidone to quench serine protease activities and chelate polyphenols, respectively. Mix until complete homogenization and store at –80°C until use.

4. Lyophilise the filtrate until complete removal of water to concentrate sample (see Note 6).

5. Dry the pellet to remove all traces of solvent, resuspend the pellet in 1 volume of R (see Subheading 3.1), centrifuge briefly and transfer the supernatant in a fresh tube.

3.4. Extraction of Mycelial Proteins Prior to iTRAQ Labelling

To ensure compatibility with iTRAQ tags, proteins must be extracted via acetone precipitation (12). Protein concentration of the samples can be determined using methods described in Subheading 3.5. 500–700 µg of protein is required for the labelling process.

1. Add 4.5 mL 10 mM Tris–HCl pH 7.6 to ground frozen mycelia and homogenise.

2. Centrifuge (15 min 12,000×g at 4°C) and recover supernatant.

3. Repeat step 2.

4. Perform a protein concentration assay (see Subheading 3.5) on an aliquot.

5. Add 6 volumes of cold (–20°C) acetone to a volume of sample equivalent to 500–700 µg.

6. Transfer the tubes to a freezer and allow protein to precipitate overnight.

7. Following acetone precipitation and the formation of a flocculent, pulse spin the Eppendorf tubes at 2,000×g for 5 s at 20°C, carefully decant the acetone.

8. Repeat the pulse centrifugation (2,000×g, 5 s at 20°C) and remove the remaining acetone with a 10-µL pipette. This should be completed as quickly as possible for each sample to avoid drying of the protein pellet. It is recommended that the samples be processed individually until they are resuspended in TEAB.

9. Add 100 µL of TEAB to the sample tube and vortex heavily to break the pellet.

10. Add 5 μL of 2% (w/v) SDS to the sample tube and vortex to mix. Samples are now ready for alkylation, reduction and trypsin digestion (see Note 7).

3.5. Protein Concentration Assay

Protein concentration can be determined using the Lowry-based colorimetric *RC DC* protein assay (Bio-Rad) according to the kit instruction as it can effectively remove most R and MSS reagents that interfere with the assay (see Subheading 3.1). Alternatively, 2-D Quant kit (GE Healthcare) can be used for the quantification of diluted protein samples.

Protein-induced colorimetric development can be measured spectrophotometrically at 750 nm (*RC DC* assay), whereas 2-D Quant is assayed at 480 nm (see Note 8).

3.6. First Dimension Protein Separation: IEF

The methodology is carried out using Bio-Rad ReadyStrip IPG strips and associated products, but IPG strips from other companies may work equally well.

1. Apply 125–160 μL MSS buffer containing 200–300 μg proteins for IPG strip rehydration.

2. IPG strip rehydration is carried out using the "active" rehydration method. To perform this, the protein sample in MSS is applied to length of the IEF focusing tray for 7 cm strips. Following this, the IPG strip is carefully laid gel-side onto the focusing tray with the acidic side overlapping the anode and the basic side overlapping the cathode.

3. Cover the tray and leave for 60 min.

4. Overlay and cover the strip with mineral oil.

5. Secure the focusing tray in the PROTEAN IEF Cell and apply a 50 V current. Active rehydration is performed for 16 h at a cell temperature of 20°C.

6. After active rehydration, carefully remove the mineral oil without moving the IPG strip in the focusing tray.

7. Pre-wet pieces of Whatman paper with ultrapure MQ H_2O and carefully place it at the contact points between the IPG strip and the focusing tray electrode.

8. Overlay and cover the strip with mineral oil.

9. For IEF using the Bio-Rad IEF Cell, intracellular proteins are focused at 250 V for 15 min and 14,000 V-h using rapid ramping. For IEF of extracellular protein samples, 12,000 V-h is used. Current is maintained at 50 μA per IPG strip (see Notes 9 and 10).

3.7. Second Dimension Separation: SDS-PAGE and Gel Staining

1. We use SDS-polyacrylamide gels for the Laemmli denaturing discontinuous gel electrophoresis method.

2. Prepare gel electrophoresis apparatus for casting. The Bio-Rad mini PROTEAN III system is used with glass plate spacers to cast gels to the thickness of 1 mm.

3. Prepare 12% (w/v) polyacrylamide solution for casting resolving gels. The solution consists of 7.5 mL 4× Tris-Cl pH 8.8, 100 µL 10% (w/v) APS, 10 µL TEMED and 9 mL 40% (w/v) acrylamide/bisacrylamide. Make solution to a volume of 30 mL with MQ H$_2$O. This is sufficient to cast two gels.

4. Pour the polyacrylamide solution between the glass plates and up to 1 cm below the short plate.

5. Cover the top of the gel with isobutanol to ensure a flat surface. The volume required should be sufficient to just cover the top of the polyacrylamide solution between the glass plates.

6. After 60 min or when the gel is set, the isobutanol layer can be removed by washing with distilled water. Store the gel in the casting apparatus at 4°C wrapped in polyvinyllidene chloride plastic film until required.

7. Prior to second dimension protein separation, it is necessary to equilibrate the IPG strip in SDS and to reduced/alkylate proteins. Mineral oil must firstly be drained from the IPG strip prior to this procedure. Place IPG strip onto the equilibration tray (see Note 11).

8. Cover the IPG strip in equilibration buffer supplemented with DTT. Equilibrate the strip by gentle inversion using a rocking platform for 20 min at room temperature.

9. Replace the current equilibration buffer with iodoacetamide-supplemented equilibration buffer and equilibrate for 20 min at room temperature.

10. Following this, the equilibration buffer is removed by draining and immersing the strip in SDS electrophoresis buffer.

11. For second dimension separation via SDS-PAGE, the strip is placed onto the flat surface of the resolving polyacrylamide gel with protein MW standard loaded onto a piece of Whatman paper. Orientate the IPG strip so that the backing plastic adheres to the tall glass plate.

12. Seal the IPG strip and protein marker-saturated Whatman paper on the resolving gel with overlay agarose. Avoid introducing air bubbles where possible. Allow 10 min for the agarose to set prior to electrophoresis.

13. Electrophoresis is carried at 15 mA per gel for PROTEAN III until the bromophenol blue dye-front reaches the end of the gel (see Note 12).

14. Remove the gel from the glass plates, place in a glass container and rinse briefly in MQ H$_2$O.

15. Immerse gel in fixing solution. The volume required should be sufficient to just cover the entire gel. Place the container on a rocking platform and fix it for 2 h under gentle agitation.

16. Replace the gel fixing solution with 80% (v/v) colloidal Coomassie G250 staining solution and 20% (v/v) methanol in sufficient volume to cover the entire gel.

17. Protein spots can be visualised after 2 h of staining but it is recommended that the gel be stained overnight.

18. Remove the staining solution and destain the gel in several washes of the destaining solution to attain a sufficiently clear background for visualisation.

19. The gel can be stored at 4°C in destaining solution.

3.8. Gel Scanning and Image Analysis

1. Protein gels can be scanned using the ProExpress scanner.

2. Scanning parameters for the ProFinder software (Perkin Elmer) are set for a resolution of 100 μ, using bottom illumination and an emission wavelength spectrum of 530 nm.

3. Image acquisition is carried out using an exposure time of 3 s after a flatfield has been established.

4. Images are saved as bitmap files.

5. Protein spot detection and analysis is performed by eye in conjunction with the ProGenesis Workstation 2005 software.

6. Software parameters as follows: background subtraction is set at the proprietary "ProGenesis background" option, spots are manually matched and volume normalisation is performed using the "total spot volume" option. Matching protein spots from "treatment A" and "treatment B" are considered differentially abundant if the normalised densitometry values are significantly different according to an unpaired t-test ($p < 0.05$) with an average fold difference of ≥ 2.0.

7. For comparative purposes, proteomes derived from two or more treatments require the use of biological replicates (a minimum of three replicates is desirable). Prior to protein spot comparisons between proteomes derived from multiple treatments, biological replicates must be merged into an average gel for each treatment.

3.9. In-Gel Trypsin Digestion of Proteins for MS Analysis

1. Protein spots of interest derived from 2-DE can be excised as gel pieces ($1 \times 1 \times 1$ mm) with a clean scalpel blade and placed in standard 1.5-mL Eppendorf tubes. In-gel protein digestion is performed as described by Shevchenko et al. (10) with some modifications.

2. Add 50 μL of destain solution to the gel piece and shake on an orbital shaker for 45 min.

3. Discard the destain solution and repeat step 1.

4. Dry the gel piece in a 50°C oven for 30 min.

5. Add 15 μL of trypsin solution to the gel piece overnight to facilitate protein digestion.

6. Incubate reaction overnight at 37°C.

7. Add 15 µL acetonitrile and shake for 15 min on an orbital shaker.

8. Retain supernatant and add 15 µL 50% (v/v) acetonitrile and 5% (v/v) formic acid.

9. Repeat steps 6 and 7.

10. Dry sample in SpeedVac.

11. Peptides are resuspended in 16 µL 5% (v/v) acetonitrile, 0.1% (v/v) formic acid prior to LC-MS/MS analysis. Detailed mass spectrometry protocols are described in Taylor et al. (11) and Lee et al. (12).

12. The Mascot software uses public, proprietary or in-house protein sequence databases to compare peptide mass spectra to predicted spectra based on amino acid sequences in the databases. Mascot is set to utilise error tolerances of ±1.2 for MS and ±0.6 for MS/MS, "max missed cleavages" set to 1, and the "oxidation (M)" variable modification and peptide charge set at 2+ and 3+. Results are filtered using "standard scoring," "max number of hits" set to 20, "significance threshold" at a p value of 0.05, and "ion score cutoff" at 15.

3.10. Trypsin Digestion and Preparation of Proteins for iTRAQ Labelling

1. Add 10 µL of TCEP to the sample, vortex for 30 s then pulse spin (2,000×g, 5 s at 20°C).

2. Incubate at 60°C for 1 h.

3. Add 5 µL of MMTS, vortex to mix and pulse spin (2,000×g, 5 s at 20°C) to bring the sample to the bottom of the tube.

4. Incubate for 10 min at 20°C.

5. Centrifuge samples (10,000×g, 10 min at 20°C) to remove any insoluble material.

6. Transfer the supernatant into clean Eppendorf tubes.

7. Quantify the protein concentration of the supernatant to determine the amount of sample to be used for labelling (see Subheading 3.5).

8. Following the protein concentration assay, add 35 µg of protein from each sample to a fresh tube.

9. Reconstitute 2×25 µg vials of trypsin with 2×25 µL of MQ H_2O to produce a 1 µg/µL solution.

10. Add 3.5 µL of trypsin solution to each sample tube and incubate overnight at 37°C.

3.11. iTRAQ Labelling of Peptides

1. Following overnight trypsin digestion, dry the digested proteins in a SpeedVac.

2. Resuspend the digested samples in 30 µL TEAB buffer pH 8.5.

3. Resuspend the eight individual iTRAQ reagents in 70 μL each of isopropyl alcohol.

4. Note which iTRAQ label (113–121) will be paired to each protein sample and add the contents of one individual iTRAQ reagent tube to one individual sample tube.

5. Vortex to mix, then pulse spin on a bench top microcentrifuge.

6. Incubate the tubes at room temperature for 2 h.

7. Add 900 μL of MQ H_2O to each tube to quench the reaction.

8. Vortex to mix and pulse spin.

9. Combine the contents of all eight tubes into one clean 10-mL tube, ensuring that all the liquid from each tube has been transferred.

10. Clean up the combined iTRAQ-labelled sample using solid phase extraction (SPE).

3.12. SPE of iTRAQ-Labelled Peptides

1. Condition the Starta-X 33 μm polymeric reverse phase (RP) column by adding 2×1 mL of 100% methanol, then 2×1 mL of 100% acetonitrile. Discard the flow through.

2. Wash the column with 2×1 mL of MQ H_2O. Discard the flow through.

3. Load the sample onto the column discarding the flow through.

4. Wash the sample tube with 2×1 mL of MQ H_2O and load this onto the column (this ensures the entire sample is transferred from the tube). Discard the flow through.

5. Elute the column by adding 500 μL of elution solution and collect the flow through in an Eppendorf tube.

6. Dry the sample down on a SpeedVac.

3.13. Strong Cation Exchange (SCX) of iTRAQ-Labelled Peptides

1. Resuspend the labelled peptides in 80 μL of SCX buffer A (see Subheading 3.5).

2. Place 40×1.5 mL glass tubes into the fraction collection rack and set the auto-sampler to collect fractions in 1 min intervals.

3. Load the sample and separate on a polysulfoethyl 4.6×100 mm 5 μm 300A column with an increasing gradient of SCX buffer B, collecting fractions as they are eluted off the column. Record the absorbance at 214 nm.

4. After collection of the fractions and analysis of the absorbance spectrum, group the fractions to give eight pooled fractions of relatively equal peak area.

5. Repeat the SPE procedure (see Subheading 3.12) on the eight pooled fractions to generate eight dry samples.

3.14. Reverse Phase (RP) Nano-LC with Probot Micro Fraction Collection of iTRAQ-Labelled Peptides

1. Resuspend the eight samples in 20 µL of loading buffer, vortex the samples for 1 min to mix, then pulse spin.

2. Dilute 2 µL of the resuspended sample in 9 µL of loading buffer.

3. Initiate the Probot micro fraction collector and set the flow rate of the matrix to 0.3 µL/min.

4. Align the Opti-TOF 384-well plates and determine that the plate spotter is spotting correctly by running a test plate to ensure matrix is present, spot size is consistent and spot placement is correct.

5. Inject the sample and separate the peptides on a C18 PepMap 100, 3 µm column with an increasing gradient of RP buffer B at a flow rate of 0.3 µL/min using the Ultimate 3000 nano HPLC system.

6. The eluent is mixed with the matrix solution following RP separation and spotted onto the Opti-TOF 384-well plate over a 90 min collection period resulting in 180 spots per sample for mass spectrometry analysis.

3.15. MALDI TOF/TOF MS/MS Analysis

1. Analyse the labelled peptides on a 4800 MALDI-TOF/TOF™ analyzer operating in reflector positive mode.

2. Acquire MS data over a mass range of 800–4,000 m/z with 400 shots accumulated for each spectrum.

3. Perform a job-wide interpretation method selecting the 20 most abundant precursor ions above the signal/noise ratio of 20 for each spectrum for MS/MS acquisition.

4. Acquire MS/MS spectra for the 20 selected precursor ions with 4,000 shots per selected ion with a mass range of 60 to the precursor ion –20 m/z.

3.16. Proteogenomics

Mass spectral data from individual biological replicates are combined and analysed using the Mascot sequence matching software (Matrix Science, Boston, USA). The search parameters are: Enzyme: Trypsin; Max missed cleavages: 1; Fixed modifications: iTRAQ-4plex (K), iTRAQ4plex(N-term), Methylthio(C); Variable modifications: Oxidation(M); Peptide tolerance: ±0.6 Da; MS/MS tolerance: ±0.6 Da.

The MOWSE algorithm (MudPIT scoring) of Mascot is used to score the significance of peptide/protein matches with $p < 0.05$ for each protein identification. Protein datasets can be from the presumed genes, from a between-stop codon 6-frame translation of the genome assembly, from 6-frame translated, CAP3-generated (21) contigs of un-assembled reads, and 6-frame translated singleton un-assembled reads which did not assemble into contigs via CAP3. All 6-frame open reading frames (ORFs) can be subjected to a ten amino acid minimum length threshold.

The false discovery rate can be calculated by using a database of randomised sequences from the 6-frame translated assembly protein datasets generated as Mascot decoy databases (22) as detailed in the Mascot help section.

4. Notes

1. The resuspension solutions R and MSS are compatible with 2-DE (and IPG-IEF in particular) downstream separation technique. R or MSS can be replaced by Laemmli sample buffer if only SDS-PAGE is to be attempted (20).

2. R and MSS solutions can be prepared in bulk and stored frozen as single-use aliquots. The solutions must be made at room temperature. Care must be taken to avoid exposing the solutions to temperatures 37°C and above as urea may decompose.

3. Sonication may be used to assist in protein resolubilisation. An output of 95 W and a 25%/s pulsar duty cycle is used for sonication. The sample is placed on ice during sonication to prevent excessive heat accumulation.

4. As the acidity decreases, insoluble material becomes harder to pellet and the latter may be dislodged upon discarding the supernatant.

5. All solutions used here are described in Subheading 2.3.

6. Note that a prefractionation step such as liquid-phase IEF can be performed at this stage or after step 3. For details see ref. (13).

7. Please note that the SDS concentration of the digestion mixture must be less than 0.059% in order for effective trypsin digestion of the proteins to occur. If the concentration is outside of this range, dilute the SDS by addition of MQ H_2O.

8. It is advisable to construct a protein standard curve using bovine serum albumin solubilised in R or MSS solutions (see Subheading 3.5) each time the assay is performed.

9. For protein samples that regularly result in horizontal streaking of protein spots in 2-D gels, we found that increasing the V-h by 2,000 during IEF reduces the problem.

10. While the IPG strips are undergoing rehydration, it is recommended that SDS polyacrylamide gels be cast and allowed to set overnight prior to SDS-PAGE (see Subheading 3.7).

11. Avoid touching the polyacrylamide layer of the IPG strip.

12. The large PROTEAN II Xi electrophoresis system can be used for second dimension electrophoresis of longer IPG strips (e.g. 18 cm ReadyStrip). Gel electrophoresis is carried at 25 mA per gel.

If multiple gels were simultaneously electrophoresed, it is recommended that a refrigerated recirculator unit be used to prevent overheating. We use a Bio-Rad refrigerated recirculator model 4860 set at 21°C.

References

1. Tan, K. C., Simon V. S, I., Robert D, T., Richard P, O., and Peter S, S. (2009) Assessing the impact of transcriptomics, proteomics and metabolomics on fungal phytopathology, *Mol. Plant Pathol.* **10**, 703–709.

2. Mehta, A., Brasileiro, A. C. M., Souza, D. S. L., Romano, E., Campos, M. A., Grossi-De-Sa, M. F., Silva, M. S., Franco, O. L., Fragoso, R. R., Bevitori, R., and Rocha, T. L. (2008) Plant-pathogen interactions: What is proteomics telling us?, *FEBS J.* **275**, 3731–3746.

3. Ellis, J. G., Dodds, P. N., and Lawrence, G. J. (2007) The role of secreted proteins in diseases of plants caused by rust, powdery mildew and smut fungi, *Curr. Opin. Microbiol.* **10**, 326–331.

4. Oliver, R. P., and Solomon, P. S. (2010) New developments in pathogenicity and virulence of necrotrophs., *Curr Opin Plant Biol* **13**, 415–419.

5. Liu, Z., Faris, J. D., Oliver, R. P., Tan, K. C., Solomon, P. S., McDonald, M. C., McDonald, B. A., Nunez, A., Lu, S., Rasmussen, J. B., and Friesen, T. L. (2009) SnTox3 acts in effector triggered susceptibility to induce disease on wheat carrying the Snn3 gene, *PLoS Path.* **5**. e10005816.

6. Friesen, T. L., Faris, J. D., Solomon, P. S., and Oliver, R. P. (2008) Host-specific toxins: Effectors of necrotrophic pathogenicity, *Cell. Microbiol.* **10**, 1421–1428.

7. Bindschedler, L. V., Burgis, T. A., Mills, D. J. S., Ho, J. T. C., Cramer, R., and Spanu, P. D. (2009) In planta proteomics and proteogenomics of the biotrophic Barley fungal pathogen Blumeria graminis f. sp. hordei, *Mol. Cell. Proteomics* **8**, 2368–2381.

8. Kim, S. T., Yu, S., Kim, S. G., Kim, H. J., Kang, S. Y., Hwang, D. H., Jang, Y. S., and Kang, K. Y. (2004) Proteome analysis of rice blast fungus (Magnaporthe grisea) proteome during appressorium formation, *Proteomics* **4**, 3579–3587.

9. Ross, P. L., Huang, Y. N., Marchese, J. N., Williamson, B., Parker, K., Hattan, S., Khainovski, N., Pillai, S., Dey, S., Daniels, S., Purkayastha, S., Juhasz, P., Martin, S., Bartlet-Jones, M., He, F., Jacobson, A., and Pappin, D. J. (2004) Multiplexed protein quantitation in Saccharomyces cerevisiae using amine-reactive isobaric tagging reagents, *Mol. Cell. Proteomics* **3**, 1154–1169.

10. Damerval, C., de Vienne, D., Zivy, M., and Thiellement, H. (1986) Technical improvements in two-dimensional electrophoresis increase the level of genetic variation detected in wheat-seedling proteins. *Electrophoresis* **7**, 52–54.

11. Hurkman, W. J., and Tanaka, C. K. (1986) Solubilization of Plant Membrane Proteins for Analysis by Two-Dimensional Gel Electrophoresis. 1986, **81**, 802-806. *Plant Physiol* **81**, 802–806.

12. Casey, T., Solomon, P. S., Bringans, S., Tan, K. C., Oliver, R. P., and Lipscombe, R. (2010) Quantitative proteomic analysis of G-protein signalling in Stagonospora nodorum using isobaric tags for relative and absolute quantification, *Proteomics* **10**, 38–47.

13. Vincent, D., Balesdent, M. H., Gibon, J., Claverol, S., Lapaillerie, D., Lomenech, A. M., Blaise, F. O., Rouxel, T., Martin, F., Bonneu, M., Amselem, J., Dominguez, V., Howlett, B. J., Wincker, P., Joets, J., Lebrun, M. H., and Plomion, C. (2009) Hunting down fungal secretomes using liquid-phase IEF prior to high resolution 2-DE, *Electrophoresis* **30**, 4118–4136.

14. Tan, K. C., Heazlewood, J. L., Millar, A. H., Thomson, G., Oliver, R. P., and Solomon, P. S. (2008) A signaling-regulated, short-chain dehydrogenase of *Stagonospora nodorum* regulates asexual development, *Eukaryot. Cell* **7**, 1916–1929.

15. Tan, K.-C., Heazlewood, J. L., Millar, A. H., Oliver, R. P., and Solomon, P. S. (2009) Proteomic identification of extracellular proteins regulated by the *Gna1* Gα subunit in *Stagonospora nodorum*, *Mycol. Res.* **113**, 523–531.

16. Bringans, S., Hane, J. K., Casey, T., Tan, K. C., Lipscombe, R., Solomon, P. S., and Oliver, R. P. (2009) Deep proteogenomics; high throughput gene validation by multidimensional liquid chromatography and mass spectrometry of proteins from the fungal wheat pathogen Stagonospora nodorum, *BMC Bioinformatics* **10**, 301.

17. Rabilloud, T. (1998) Use of thiourea to increase the solubility of membrane proteins in two-dimensional electrophoresis, *Electrophoresis 19*, 758–760.

18. Herbert, B. (1999) Advances in protein solubilisation for two-dimensional electrophoresis, *Electrophoresis 20*, 660–663.

19. Herbert, B. R., Molloy, M. P., Gooley, A. A., Walsh, B. J., Bryson, W. G., and Williams, K. L. (1998) Improved protein solubility in two-dimensional electrophoresis using tributyl phosphine as reducing agent, *Electrophoresis 19*, 845–851.

20. Laemmli, U. K. (1970) Cleavage of structural proteins during assembly of head of bacterio-phage-T4, *Nature 227*, 680–685.

21. Huang, X., and Madan, A. (1999) CAP3: A DNA sequence assembly program, *Genome Res 9*, 868–877.

22. Elias JE, Haas W, Faherty BK, and SP:, G. (2005) Comparative evaluation of mass spectrometry platforms used in large-scale proteomics investigations, *Nat Methods 2*, 667–675.

23. Bohmer, M., Colby, T., Bohmer, C., Brautigam, A., Schmidt, J., and Bolker, M. (2007) Proteomic analysis of dimorphic transition in the phytopathogenic fungus *Ustilago maydis*, *Proteomics 7*, 675–685.

24. Paper, J. M., Scott-Craig, J. S., Adhikari, N. D., Cuomo, C. A., and Walton, J. D. (2007) Comparative proteomics of extracellular proteins *in vitro* and *in planta* from the pathogenic fungus *Fusarium graminearum*, *Proteomics 7*, 3171–3183.

25. Phalip, V., Delalande, F., Carapito, C., Goubet, F., Hatsch, D., Leize-Wagner, E., Dupree, P., Dorsselaer, A. V., and Jeltsch, J. M. (2005) Diversity of the exoproteome of Fusarium graminearum grown on plant cell wall, *Curr. Genet. 48*, 366–379.

26. Shah, P., Gutierrez-Sanchez, G., Orlando, R., and Bergmann, C. (2009) A proteomic study of pectin-degrading enzymes secreted by Botrytis cinerea grown in liquid culture, *Proteomics 9*, 3126–3135.

27. Shah, P., Atwood, J. A., Orlando, R., Mubarek, H. E., Podila, G. K., and Davis, M. R. (2009) Comparative proteomic analysis of botrytis cinerea secretome, *Journal of Proteome Research 8*, 1123–1130.

28. Yajima, W., and Kav, N. N. (2006) The proteome of the phytopathogenic fungus *Sclerotinia sclerotiorum*, *Proteomics 6*, 5995–6007.

Chapter 6

Identification of HR-Inducing cDNAs from Plant Pathogens via a Gateway®-Compatible Binary Potato Virus X-Expression Vector

H. Peter van Esse

Abstract

Identification of pathogen effectors that elicit a hypersensitive response (HR) in resistant plant hosts is essential to study disease resistance. In this method, it is described how to generate a cDNA library, how to transfer the library into a binary PVX-expression vector, and finally how to set up a high-throughput screen for HR-inducing cDNAs from plant pathogens.

Key words: Effector, cDNA library, Resistance, Effector-triggered immunity, Potato virus X, PVX, High throughput

1. Introduction

Plants are under constant threat of microbes that aim to parasitize their host. As a consequence, plants have evolved receptors that recognize pathogen-associated molecular patterns (PAMPs) of potential pathogens (1, 2). Upon recognition of a potential pathogen, these receptors initiate plant defense responses which lead to PAMP-triggered immunity (PTI) (3). Pathogens, in turn, have evolved effector molecules that contribute to disease establishment by subverting the host immune system (3–6). Subsequently, plants have evolved novel receptors which are able to detect (the activities of) these effectors and activate effector-triggered immunity (ETI), which effectively makes the effector an avirulence determinant (3). This last type of immunity is governed by the classical gene-for-gene interaction in which a plant *R* gene product conveys resistance via direct or indirect recognition of a cognate *Avr* gene product from the pathogen (7). ETI initiates various defense responses that are stronger than typical

Melvin D. Bolton and Bart P.H.J. Thomma (eds.), *Plant Fungal Pathogens: Methods and Protocols*,
Methods in Molecular Biology, vol. 835, DOI 10.1007/978-1-61779-501-5_6, © Springer Science+Business Media, LLC 2012

PTI responses and often include a hypersensitive response (HR) in which plant cells near the infection site die rapidly (3, 8–10).

To unravel ETI responses, it is essential that, in addition to the *R* gene, the corresponding pathogen effectors are identified. Furthermore, the effectors that are recognized by *R* gene products are often potent virulence factors (4–6, 11–14) and therefore of great interest to study pathogenicity mechanisms of plant pathogens.

In the method described here, we use a Gateway®-compatible binary Potato virus X (PVX) expression vector that enables high-throughput screening of individual cDNAs present in pathogen cDNA libraries to identify cDNAs that code for HR-inducing effectors that induce HR. The use of Gateway® technology implies that construction of the cDNA library is different from previously described methods that use a binary PVX-expression vector (15). Although binary PVX-mediated cDNA screens have so far been successfully performed on tomato and potato (15, 16), they may be functional for all plants that can be infected by PVX which includes species from the families *Aizoaceae, Amaranthaceae, Chenopodiaceae, Compositae, Cucurbitaceae, Labiatae, Leguminosae, Pedaliaceae, Portulacaceae, Solanaceae,* and *Umbelliferae* (17).

The first step in construction of the binary PVX cDNA library involves construction of a multifunctional Gateway® cDNA library. This library is subsequently transferred by a single recombination reaction such that individual cDNAs are inserted behind the duplicated coat protein promoter of a modified PVX backbone that is present in a binary vector and driven by the Cauliflower Mosaic Virus (CaMV) 35S promoter (15). This construct can be transferred to plant cells via *Agrobacterium tumefaciens*-mediated transformation, after which the CaMV 35S promoter will drive production of infectious recombinant PVX transcripts. Subsequent translation of these transcripts will result in the production of fully functional, recombinant virus particles that will infect the surrounding tissues. The recombinant virus will also produce the protein that is encoded by the cDNA inserted behind the duplicated coat promoter (18, 19). When HR-associated recognition of a produced protein by the plant takes place, a necrotic and/or chlorotic lesion will develop around the inoculation site which indicates that the binary vector inoculated at that site contains a cDNA that codes for an HR-inducing protein.

2. Material

Prepare all solutions with autoclaved, deionized water (18.2 MΩ/cm). Be careful when working with mRNA to keep your working environment clean and free from RNases.

2.1. mRNA Isolation

1. Trizol® reagent (Invitrogen) (see Note 1).
2. Liquid nitrogen (see Note 2).

3. 7.5 M ammonium acetate (CH_3COONH_4).

4. Phenol:chloroform:isoamylalcohol 24:24:1 (see Note 1).

5. Chloroform (see Note 1).

6. Oligotex ® direct mRNA mini kit (Qiagen).

2.2. Gateway® Donor Library

1. Cloneminer™ II cDNA library construction kit (Invitrogen).

2. pDONR™/Zeo (Invitrogen).

3. Zeocin™ (Invitrogen) (see Note 3).

4. Sterile 50% glycerol.

5. Cryogenic 2 mL tubes.

6. Low salt LB with zeocin™: 10 g/L tryptone, 5 g/L NaCl, 5 g/L yeast extract, pH 7.5 with NaOH. Allow the media to cool to 55°C before adding the zeocin™ to a 50 µg/mL final concentration. For plates, add 15 g/L agar before autoclaving. Keep all media that contains zeocin™ in the dark at 4°C. Plates with this antibiotic are stable for 2 weeks.

7. M13 forward primer: 5′-TCGCGTTAACGCTAGCATGGATC TC, M13 reverse primer 5′-CAGGAAACAGCTATGAC.

8. Electro-competent *Escherichia coli* DH5α (efficiency higher than 10^9 cfu/µg).

9. Qiagen plasmid midi kit (Qiagen).

10. Binary PVX-expression vector (courtesy of Adrian Valli, Centro Nacional de Biotecnología, Madrid, Spain).

11. Gateway® LR Clonase® II enzyme mix (Invitrogen).

2.3. Gateway® PVX Library

1. Competent *Agrobacterium tumefaciens* strain GV3101 or mog800 (efficiency >10^6 cfu/µg).

2. PVX forward primer: 5′-CAATCACAGTGTTGGCTTGC.

3. LB mannitol with kanamycin and rifampicin: 10 g/L tryptone, 10 g/L NaCl, 5 g/L yeast extract, pH 7.5 with NaOH. Allow the media to cool to 55°C before adding the kanamycin to a 50 µg/mL final concentration and the rifampicin to a 25 µg/mL final concentration. For plates, add 15 g/L agar before autoclaving.

4. Resistant and susceptible host plant cultivars.

3. Methods

3.1. mRNA Isolation for Generation of the Gateway® Donor cDNA Library

1. Obtain 10 g of in vitro-grown material from the pathogen of choice (see Note 4). (recommended) Material can be freeze-dried prior to extraction; this avoids dilution of the Trizol® reagent and reduces the amount of Trizol® needed in subsequent steps.

2. Flash-freeze the sample by transfer to liquid nitrogen.

3. Grind the sample with a precooled mortar and pestle such that the sample does not thaw.

4. Add the appropriate amount of Trizol® (per 50 mg of freeze-dried tissue, add 1 mL of Trizol® reagent).

5. Homogenize the sample with the Trizol® reagent using a mortar and pestle.

6. Aliquot the homogenized sample into eppendorff tubes in portions of 1 mL

7. Add 0.2 mL of chloroform per 1 mL of Trizol® reagent.

8. Shake sample(s) vigorously and centrifuge at $12,000 \times g$ for 5 min. After centrifugation, the homogenized mixture will be divided in a lower phenol phase and an upper aqueous phase.

9. Transfer the upper aqueous phase to a clean RNase-free tube and add one volume of phenol:chloroform:isoamyl alcohol (24:24:1). Shake vigorously and centrifuge at $12,000 \times g$ for 5 min (see Note 5).

10. Transfer the upper aqueous phase to a clean RNase-free tube and add one volume of chloroform. Shake vigorously and centrifuge at $12,000 \times g$ for 5 min. Repeat this step one more time.

11. Transfer the upper aqueous phase to a clean RNase-free tube and precipitate the RNA by adding two volumes of isopopanol and 1/10 volume of 7.5 M ammonium acetate.

12. Wash pellet twice with 70% ethanol and dissolve in 600 μL MQ.

13. Determine concentration and quality of the total RNA (tRNA) by resolving 5 μL of the sample on a 2% agarose gel and determining OD 260/280 and OD 260/230 ratios, which both should be between 1.8 and 2.0 before continuing. The total amount of tRNA should be no lower than 4 mg before continuing.

14. Divide the tRNA into aliquots of 1 mg and follow the Oligotex mRNA Spin column protocol according to manufacturer's specifications. Add RNase-free water to the tRNA sample so that a final volume of 500 μL is reached. Subsequently, add 500 μL buffer OBB and 55 μL Oligotex suspension (included in kit) to each aliquot. Use the small spin columns.

15. Determine the quality of the obtained mRNA by resolving 5 μL of the sample on a 2% agarose gel and determining OD 260/280 and OD 260/230 ratios, which both should be between 1.8 and 2.0 before continuing. The total amount of mRNA should be no lower than 1 μg before continuing.

3.2. Generation of the Gateway® Donor cDNA Library and PVX cDNA Library

1. For this method, at least 1 µg of high-quality mRNA is needed.

2. Generate the cDNA library according to manufacturer's specifications until the BP® recombination reaction.

3. Perform the BP® recombination according to manufacturer's specifications; however, instead of using the pDONR™ 222, use pDONR™/Zeo vector for construction of the library and also incubate the BP® reaction for 18 h.

4. Use 1 µL of the BP® recombination reaction to transform an aliquot of electro-competent *E. coli* DH5α and plate several dilutions to determine which concentration should be used in the final transformation (approximately 1,500–2,000 colonies per plate). Also, retain one plate with fewer colonies to check library quality. Store the remainder of the reaction at –20°C.

5. Perform a colony PCR on 50 independent colonies with the PVX forward and reverse primer to determine average insert length of the library. If the average insert size is less than 1 kb, it is recommended to isolate new mRNA and to repeat the previous steps (see Note 6).

6. Sequence at least 30 clones of the resulting library with the PVX forward primer to ascertain the library quality (see Note 7).

7. After quality control, use the rest of the BP reaction to transform electro-competent *E. coli* DH5α (3×2 µL).

8. The next day, harvest all colonies by adding ~3 mL of ice-cold LB medium to the plates and suspending the bacteria using a Drigalski spatula (see Note 8). Transfer the LB medium to two 50-mL tubes. Before centrifugation of the harvested bacteria, generate glycerol stocks by transferring 6×1 mL into cryogenic tubes and adding 500 µL of 50% glycerol. This will enable replating of the library at a later date. Spin down the rest of the bacteria at $4,000 \times g$ and decant the supernatant.

9. Spin down the remaining bacteria in the 50 mL tubes and isolate plasmid DNA from the pellets using the Qiagen plasmid midi kit according to manufacturers' specifications.

10. Perform an LR recombination according to manufacturer's specifications. Use the plasmid mixture isolated in step 8 as donor/entry vector and the PVX Gateway® vector as expression/destination vector.

11. Use 1 µL of the LR recombination reaction to transform an aliquot of electro-competent *E. coli* DH5α and plate several dilutions to determine which concentration should be used in the final transformation (approximately 1,500–2,000 colonies per plate).

12. After determination of the optimal dilution, use the remainder of the LR reaction to transform electro-competent *E. coli* DH5α (3×2 µL).

13. The next day, harvest all colonies by adding 5 mL of ice-cold LB medium to the plates and suspending the bacteria using a Drigalski spatula. Before centrifugation of the harvested bacteria, generate glycerol stocks as described above which will enable replating at a later date. Spin down the rest of the bacteria at 4,000×*g* and decant the supernatant.

14. Isolate plasmid DNA from the pellets using the Qiagen plasmid midi kit according to manufacturer's specifications.

3.3. Screen for HR-Inducing cDNAs with the PVX cDNA Library

1. Use a several dilutions of the plasmid DNA isolated in Subheading 3.2, step 14 to transform electro-competent *A. tumefaciens* to determine which concentration should be used in the final transformation (colonies should be well separated to allow easy picking later on).

2. Sequence at least 30 *A. tumefaciens* clones with the PVX forward primer to check library fidelity. For this, it is necessary to isolate plasmid DNA from these 30 clones and perform PCR, or retransform to *E. coli* and isolate plasmid DNA from the resulting transformants which can be used for sequencing. Of the sequenced clones, more than 65% of the inserts should contain full-length cDNAs before continuing.

3. If the library contains enough full-length cDNAs, transform enough electro-competent *A. tumefaciens* cells with the plasmid DNA isolated in Subheading 3.2, step 13, to obtain 9,600 colonies.

4. Transfer these colonies to 96-well culture plates that contain 100 µL LB mannitol with kanamycin and rifampicin. Grow for 48 h at 28°C, add 50 µL 50% (v/v) glycerol, and store at −80°C until further use.

5. For screening of plants, transfer the library from the glycerol stock with a 96-well pin stamp to 96-well plates that contain 100 µL LB mannitol with kanamycin and rifampicin and 15 g/L agar and grow for 60 h at 28°C (see Note 9). The resulting colonies can now be inoculated with stamps or toothpicks onto a suitable host plant (see Fig. 1). Next to the resistant cultivar include a susceptible cultivar as a control to identify cDNAs that induce necrosis in absence of the *R* gene (see Note 10). When using a stamp, be sure to puncture the leaf to increase transformation efficiency of the *A. tumefaciens* clones. To reduce the number of false positive colonies, it is recommended to perform the screen in duplicate. Positive colonies should be rescreened 3–4 additional times in a smaller scale assay.

6. Scoring can be done at 10 and 14 days after inoculation.

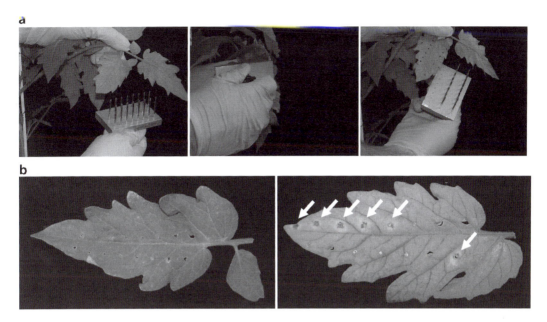

Fig. 1. (**a**) Inoculation of *Agrobacterium tumefaciens* colonies that carry binary PVX plasmids with pathogen cDNA from a 96-well plate using a stamp with 16 pins. (**b**) Result of inoculation after 14 days on a susceptible (*left*) and a resistant (*right*) tomato cultivar. A spreading chlorosis and necrosis can be observed in the resistant tomato cultivar at those inoculation sites where expression of the pathogen cDNA via Potato Virus X results in the production of proteins that trigger an HR (indicated with *white arrows*).

4. Notes

1. Trizol®, phenol, and chloroform are all highly toxic, volatile substances. Use the proper protection and work in a fume hood.

2. Liquid nitrogen is a cryogenic fluid with a boiling temperature of −196°C, take appropriate safety precautions and wear safety glasses when handling this liquid.

3. Zeocin™ is light-sensitive, so always store it in the dark. Since zeocin™ is highly toxic, the manufacturer recommends wearing gloves and safety glasses when handling this antibiotic.

4. We get the best quality mRNA when grinding the material in the presence of liquid nitrogen. This way, the material stays in optimal condition when adding the Trizol® reagent. The Trizol® reagent will thaw together with the powdered material.

5. Although not in the manufacturer's specifications, it is my experience that the additional phenol:chloroform:isoamyl and chloroform extractions are necessary to ensure isolation of high quality and a high yield of mRNA.

6. Different organisms such as fungi, oomycetes, and nematodes might have different average mRNA length; therefore, the

>1 kb insert length might not be a good reference point for all organisms.

7. Although product length of colony PCR will provide an indication for quality, we recommend sequencing at least 30 clones to get an impression of the library quality. An important indicator for the initial quality of the mRNA that was used to generate the library is the ratio of full-length transcripts compared to nonfull-length transcript. For example, when only 50% of the transcripts are full length, either double the amount of colonies has to be screened, or the mRNA isolation and subsequent steps have to be optimized. A small loss in fidelity of the cDNA library is possible since there is some bias towards smaller plasmids during the sequential transformation steps.

8. When harvesting the *E. coli* colonies, it is important to keep them cool so that no division occurs during the harvesting process, which might lead to a bias toward smaller cDNA inserts in the library.

9. Although it is also possible to use liquid medium for this step, we obtain the best results when the colonies have grown on solid media, likely because higher bacterial titers are reached. Furthermore, handling is also more practical when using solid media.

10. Many cDNAs such as those that code for heat-shock proteins and ubiquitin ligases will be able to induce nonspecific HR-like symptoms when overexpressed via the PVX-expression system. Therefore, it is important that a control cultivar is included that does not provide resistance against the pathogen. Those cDNAs that trigger HRs in both cultivars are nonspecific and can therefore be disregarded for further screening.

Acknowledgments

H. P. V. E. is supported by a Veni grant of the Research Council for Earth and Life sciences (ALW) of the Netherlands Organization for Scientific Research (NWO). The author thanks Adrian Valli for kindly providing a Gateway®-compatible PVX-vector.

References

1. Felix, G., Regenass, M., and Boller, T. (1993) Specific perception of subnanomolar concentrations of chitin fragments by tomato cells. Induction of extracellular alkalinization, changes in protein phosphorylation, and establishment of a refractory state. *Plant J.* 4, 307–316.

2. Felix, G., Duran, J.D., Volko, S., Boller, T. (1999) Plants have a sensitive perception system for the most conserved domain of bacterial flagellin. *Plant J.* 18, 265–276.

3. Jones, J.D.G., and Dangl, J.L. (2006) The plant immune system. *Nature* 444, 323–329.

4. van Esse, H.P., Bolton, M.D., Stergiopoulos, I., de Wit, P.J.G.M., Thomma, B.P.H.J. (2007) The chitin–binding *Cladosporium fulvum* effector protein Avr4 is a virulence factor. *Mol. Plant–Microbe interact.* 20, 1092–1101.

5. van Esse, H.P., van 't Klooster, J.W., Bolton, M.D., Yadeta, K.A., van Baarlen, P., Boeren, S., Vervoort, J., de Wit, P.J.G.M., Thomma, B.P.H.J. 2008. The *Cladosporium fulvum* virulence protein Avr2 inhibits host proteases required for basal defense. *Plant Cell* 20, 1948–1963.

6. de Jonge R, van Esse HP, Kombrink A, Shinya T, Desaki Y, Bours R, van der Krol S, Shibuya N, Joosten MHAJ, Thomma BPHJ. (2010) Conserved fungal LysM effector Ecp6 prevents chitin-triggered immunity in plants. *Science* 20, 953–955.

7. Flor, H.H. (1942) Inheritance of pathogenicity in *Melampsora lini*. *Phytopathology* 32, 653–669.

8. Greenberg, J. T., and Yao, N. (2004) The role and regulation of programmed cell death in plant–pathogen interactions. *Cell. Microbiol.* 6, 201–211.

9. Shen, Q., Saijo, Y., Mauch, S., Biskup, C., Bieri, S., Keller, B., Seki, H., Ülker, B., Somssich, I.E., and Schulze-Lefert, P. (2007) Nuclear activity of MLA immune receptors links isolate–specific and basal disease–resistance responses. *Science* 315, 1098–1103.

10. Tao, Y. Xie, Z., Chena, W., Glazebrook, J., Changa, H., Han, B., Zhua, T., Zou, G., and Katagiri, F. (2003) Quantitative nature of *Arabidopsis* responses during compatible and incompatible interactions with the bacterial pathogen *Pseudomonas syringae*. *Plant Cell* 15, 317–330.

11. Kim, M.G., da Cunha, L., McFall, A.J., Belkhadir, Y., DebRoy, S., Dangl, J.L., Mackey, D. (2005) Two *Pseudomonas syringae* type III effectors inhibit RIN4-regulated basal defense in Arabidopsis. *Cell* 121, 749–759.

12. Bos, J.I., Kanneganti, T.D., Young, C., Cakir, C., Huitema, E., Win, J., Armstrong, M.R., Birch, P.R., and S. Kamoun. (2006) The C-terminal half of *Phytophthora infestans* RXLR effector AVR3a is sufficient to trigger R3a-mediated hypersensitivity and suppress INF1-induced cell death in *Nicotiana benthamiana*. *Plant J.* 48, 165–176.

13. Bolton MD, van Esse HP, Vossen JH, de Jonge R, Stergiopoulos I, Stulemeijer IJ, van den Berg GC, Borrás-Hidalgo O, Dekker HL, de Koster CG, de Wit PJGM, Joosten MHAJ, Thomma BPHJ. (2008) The novel *Cladosporium fulvum* lysin motif effector Ecp6 is a virulence factor with orthologues in other fungal species. *Mol. Microbiol.* 119–136.

14. Houterman, P.M., Cornelissen, B.J.C., and Rep, M. (2008) Suppression of plant resistance gene-based immunity by a fungal effector, PLoS Pathog. 4, e1000061.

15. Takken, F.L.W., Luderer, R., Gabriels, S.H., Westerink, N., Lu, R., de Wit, P.J.G.M., and Joosten, M.H.A.J. (2000) A functional cloning strategy, based on a binary PVX–expression vector, to isolate HR–inducing cDNAs of plant pathogens. *Plant J.* 24, 275–283.

16. Torto, T.A., Li, S., Styer, A., Huitema, E., Testa, A., Gow, N.A.R., van West, P., and Kamoun, S. (2003) EST mining and functional expression assays identify extracellular effector proteins from the plant pathogen *Phytophthora*. *Genome Res.* 13, 1675–1685.

17. Moreira, A., Jones, R.A.C., and Fribourg, C.E. (1980) Properties of a resistance-breaking strain of potato virus X. *Ann. Appl. Biol.* 95, 93–103.

18. Chapman, S., Kavanagh, T., Baulcombe, D. (1992) Potato virus X as a vector for gene expression in plants. *Plant J.* 2, 549–557.

19. Jones L, Hamilton AJ, Voinnet O, Thomas CL, Maule AJ, Baulcombe DC. (1999) RNA-DNA interactions and DNA methylation in post-transcriptional gene silencing. *Plant Cell* 11, 2291–2301.

Chapter 7

Freeze-Fracture of Infected Plant Leaves in Ethanol for Scanning Electron Microscopic Study of Fungal Pathogens

Jayma A. Moore and Scott A. Payne

Abstract

Fungi often are found within plant tissues where they cannot be visualized with the scanning electron microscope (SEM). We present a simple way to reveal cell interiors while avoiding many common causes of artifact. Freeze-fracture of leaf tissue using liquid nitrogen during the 100% ethanol step of the dehydration process just before critical point drying is useful in exposing intracellular fungi to the SEM.

Key words: Fungus, Leaf, Cross-section, Liquid nitrogen, Scanning electron microscopy, Cryofracture

1. Introduction

Because of its high resolution and unparalleled depth of field, the scanning electron microscope (SEM) is a versatile and valuable tool for examining plant surfaces. Morphological detail is greater than with light microscopy, images are easily interpretable, and sample preparation is relatively simple (1). However, fungi may not just colonize plant surfaces, but also invade internal tissues where they are hidden from view of the SEM. The described technique provides a quick, simple, and inexpensive way to visualize the inner as well as the outer surfaces of fungal-infected leaves.

Leaf tissue is fixed routinely in glutaraldehyde and dehydrated for critical point drying, which preserves fine structural detail quite well (1). As the plant material is being dehydrated, however, we incorporate the additional step of freezing the tissue in 100% ethanol and fracturing it while so embedded. A similar method has been used

Melvin D. Bolton and Bart P.H.J. Thomma (eds.), *Plant Fungal Pathogens: Methods and Protocols*,
Methods in Molecular Biology, vol. 835, DOI 10.1007/978-1-61779-501-5_7, © Springer Science+Business Media, LLC 2012

for animal liver and kidney (2) and has been shown by transmission electron microscopy not to create subcellular damage to the prepared tissues (3).

Cutting leaves to reveal internal structure is not the best way to prepare them for SEM. Mechanical crushing and tearing are common even if tissues are cut with a sharp new razor blade (4), and this damage is quite obvious at relatively low magnifications (Fig. 1). Fracturing samples may be more successful than cutting. However, untreated critical point dried plant tissues generally break along tough cell walls and vascular bundles, which may not expose intracellular fungal elements adequately.

Frozen tissues break more cleanly, but freezing presents its own drawbacks. Biological samples are inherently poor thermal conductors, so they freeze slowly. Because they contain water, ice crystals form during slow freezing and may damage structures of interest, especially in unfixed tissue (5). Rapid freezing helps minimize crystal formation or keep crystals small enough that they may be below the resolution of the examining system at the low magnifications common in biological electron microscopy. Plunge freezing in liquid nitrogen at about –200°C would seem to freeze tissues very quickly, but it does not. Because nitrogen's melting point (–210°C) is so close to its boiling point (–196°C), introduction of any warmer materials into liquid nitrogen causes vigorous boiling.

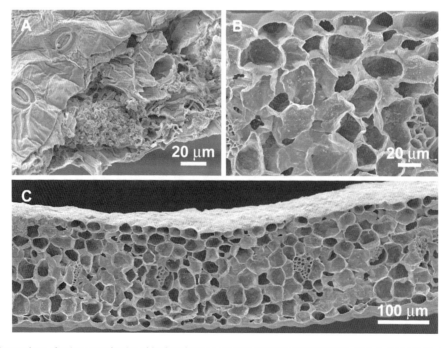

Fig. 1. Comparison of cut vs. cryofractured leaf surfaces; scanning electron micrographs. (**a**) A sugarbeet (*Beta vulgaris*) leaf cut with a new razor blade before fixation and critical point drying. Crushing and tearing artifacts are obvious. (**b**) Cells in a cryofractured edge from the same sugarbeet leaf identically fixed and dried, without crushing and tearing. (**c**) The cleanly cryofractured leaf tissue.

When samples are immersed in liquid nitrogen, the so-called Leidenfrost phenomenon occurs whereby a layer of gas surrounds the introduced tissue and insulates it. The actual cooling rate of tissue in liquid nitrogen, then, is much slower than would be expected (6), leaving ample opportunity for ice-crystal formation and subcellular damage. Another result of the Leidenfrost phenomenon is air drying (or more specifically, drying in the nitrogen-gas layer that surrounds the immersed tissue). Air drying, with passage of a gas–liquid interface through a sample, produces large surface-tension forces that damage delicate tissues and produce distortions readily visible in the SEM (1).

However, freezing tissue that has been fully infiltrated with and submerged in 100% ethanol prevents or minimizes cellular damage from ice-crystal formation. At liquid-nitrogen temperatures (colder than –196°C), ethanol freezes into a noncrystalline glass (7). Transforming a soft leaf into a uniform glass-like solid by freezing it in ethanol before fracturing also prevents mechanical crushing and tearing. Free fractures are formed randomly across the tissue surface (2), not limited to breaks along vascular bundles or cell walls in leaves of most plant species. Further, fracturing in ethanol precludes the possibility of air drying.

Freezing in ethanol is an ideal technique for use on leaves. It is less useful for stems, petioles, and roots because they generally do fracture along vascular bundles, not freely throughout the parenchyma and in cross-section.

2. Materials

1. Pink dental wax (see Note 1).
2. Double-edged razor blades (see Note 2).
3. Scintillation vials, glass, for sample processing.
4. Pasteur pipettes and rubber bulb.
5. Sodium phosphate (Millonig's) buffer, 0.1 M, pH 7.35: 11.6 g $NaH_2PO_4 \cdot H_2O$, 2.7 g NaOH, 860 mL distilled water.
6. Glutaraldehyde 2.5% in sodium phosphate (Millonig's) buffer.
7. Ethanol dilutions in distilled water: 30, 50, 70, 90, and 95% ethanol.
8. Ethanol 100% (anhydrous), stored over type 3A molecular sieves (see Note 3).
9. Fine forceps.
10. Safety glasses or face shield.
11. Cryoprotection gloves.

12. Styrofoam container, thick-walled with short sides: depth about 10 cm (see Note 4).

13. Aluminum block approximately $4 \times 5 \times 6$ cm.

14. Aluminum cylindrical SEM sample mount, 25 mm diameter.

15. Hemostat, large.

16. Hammer.

17. Round disposable aluminum weighing dish, 70 mL capacity (6.3 cm diameter $\times$ 1.9 cm deep).

18. Critical point drying containers.

19. Critical point dryer (CPD) apparatus using liquid CO_2 as the transitional fluid.

20. Small clear polystyrene storage boxes.

21. Vacuum dessicator.

3. Methods

Buffer and glutaraldehyde are stored in the refrigerator, ethanol in vented cabinets at room temperature. Use appropriate personal protective equipment and follow all institutional safety guidelines for handling and disposing of liquid nitrogen and laboratory wastes.

1. Drop a small amount of buffer on the dental wax and insert the infected leaf into the buffer drop to prevent air drying while cutting (see Note 5).

2. Cut out leaf pieces 15×15 mm or smaller with razor blade (see Note 6).

3. Immediately place excised leaf tissue in a labeled scintillation vial of 2.5% glutaraldehyde. Multiple pieces from the same sample can be placed within each vial. Volume of glutaraldehyde should be at least ten times greater than volume of leaf tissue for adequate fixation.

4. Most leaves will need to be fixed in glutaraldehyde for at least 2 h. If longer times are warranted (see Note 7), refrigerate the vials of leaf tissue in glutaraldehyde (see Note 8). If leaf pieces remain floating in glutaraldehyde after 1 h, tap the vial gently and apply light vacuum pressure (e.g., 3.33×10^4 Pa) to help remove air bubbles and aid infiltration of fixative.

5. Remove glutaraldehyde from vial with a Pasteur pipette and discard appropriately. Immediately cover leaf tissue with buffer, agitate gently, and let stand for about 15 min. Repeat two or three times depending on the thickness and volume of leaves in the vial (see Note 9).

6. After the last buffer rinse, remove and discard buffer and replace with distilled water.

7. After 15–20 min, remove water and replace with the 30% ethanol solution. Let stand for 15–30 min, remove and discard ethanol. Repeat this step with increasing concentrations of ethanol: 50, 70, 90, and 95% solutions.

8. Replace 95% ethanol with 100% (anhydrous) ethanol stored over molecular sieves (see Note 10).

9. Don the cryoprotection gloves and safety eyewear.

10. Place the aluminum block and the SEM mount into the styrofoam cooler.

11. Fill the cooler with liquid nitrogen to a depth of about 6–8 cm (Fig. 2). As the aluminum items cool, the nitrogen will boil vigorously and then subside.

12. If necessary, add liquid nitrogen to maintain the level just at the top surface of the aluminum block.

13. Put the jaws of the large hemostat into the liquid nitrogen to cool them as well (Fig. 3).

14. Place a small amount of anhydrous ethanol in an aluminum weighing dish. Using fine forceps, quickly transfer two or three leaf pieces into the dish containing the ethanol. It is important that the ethanol barely covers the leaf pieces.

15. Rest the weighing dish containing the ethanol and leaf pieces on the surface of the aluminum block (Fig. 4). The ethanol will start to freeze. Wait until it freezes completely (it should look like glass, not cloudy, although expansion cracks will appear throughout). Then use the hemostat to slide the dish off the block and submerge it in the liquid nitrogen completely for a few minutes (Fig. 5).

Fig. 2. Liquid nitrogen being added to the styrofoam shipping container used as a vessel for cooling cryofracture samples and instruments.

Fig. 3. Aluminum block, cylindrical aluminum SEM stub mount, and large hemostat being cooled in liquid nitrogen before use in leaf cryofracture.

Fig. 4. Fixed dehydrated leaf tissue samples covered by 100% ethanol in weighing dish on aluminum block. Liquid nitrogen just reaches the top of the block, and the ethanol surrounding the samples is beginning to freeze.

Fig. 5. The ethanol has frozen, and the weighing dish on top of the block will be submerged in the liquid nitrogen as has been done for the dish at right.

16. Check the liquid nitrogen level to assure that it just reaches the top of the aluminum block.

17. Using the large hemostat, retrieve the weighing dish containing the leaf tissue frozen in ethanol, empty the liquid nitrogen out of the dish, and place it on the aluminum block (see Note 11). Retrieve the aluminum stub mount and set it on the surface of the frozen ethanol over the sample (Fig. 6).

18. Sharply rap the top of the stub once with the hammer to shatter the ethanol and the embedded leaf tissue (Fig. 7; see Note 12). If you are attempting to fracture stems, petioles, or roots, consider directing the fracture with a razor blade (see Note 13).

19. Remove the stub, placing it back into the liquid nitrogen, and look at the surface of the ethanol—it should be broken, but the plant material probably will still appear intact.

20. Remove the dish from the liquid-nitrogen bath to the work surface and add fresh 100% ethanol at room temperature (Fig. 8). As the leaf thaws, it will be obvious that it has fractured (Fig. 9).

21. Repeat freezing/fracturing steps 14–20 for as many leaves as necessary, using a new dish for each batch.

22. Select a suitable holder for critical point drying (CPD). The CPD holder should have mesh fine enough to retain the smallest pieces of fractured leaf (Fig. 10a). Place the CPD holder in 100% ethanol (Fig. 10b) and transfer the leaf pieces into it (Fig. 10c), keeping them covered with ethanol at all times. Allowing samples to air dry at this point will defeat the purpose of critical point drying and obscure morphology of the fractured cell surfaces. If necessary, rinse small pieces of fractured leaf out of the dish with fresh 100% ethanol (Fig. 10d). The smaller pieces often have the best fractured edges.

Fig. 6. Aluminum SEM stub mount (previously cooled in liquid nitrogen) placed on top of the frozen ethanol that covers the leaf samples.

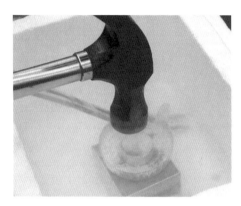

Fig. 7. Hammer striking the stub to cryofracture the frozen ethanol and contained leaf tissue.

Fig. 8. Fresh 100% ethanol is used to thaw the leaf samples after cryofracture.

Fig. 9. Thawed cryofractured leaf pieces covered by 100% ethanol in the aluminum weighing dish.

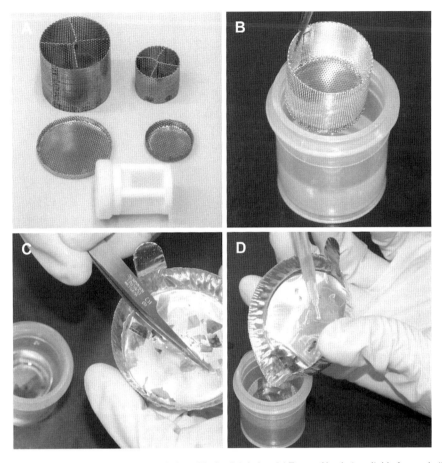

Fig. 10. The thawed cryofractured sample is ready for critical point drying. (a) Types of baskets suitable for use in the critical point dryer. (b) Critical point drying basket being submerged in a container of 100% ethanol. (c) Cryofractured leaf pieces being transferred with fine forceps from the weighing dish to the critical point drying basket in a container filled with 100% ethanol. (d) Small cryofractured pieces being rinsed from the weighing dish into the critical point drying basket with 100% ethanol.

23. When finished fracturing, close the CPD holder(s) and rinse one or two additional times with fresh 100% ethanol.

24. Quickly load the prechilled chamber of the critical point drying apparatus (Fig. 11), with minimal ethanol carryover. Close the chamber and purge with liquid CO_2 to displace the ethanol in the tissue. Seal the chamber completely and heat to increase the temperature and pressure above the critical point for liquid CO_2 (31.1°C, 7.38×10^6 Pa). Maintain heat above the critical temperature so that recondensation does not occur while slowly venting the CO_2 vapor to atmospheric pressure.

25. When the drying procedure is finished, remove leaf pieces and store in labeled clear polystyrene boxes in a vacuum dessicator.

Fig. 11. Basket of cryofractured leaf samples being transferred quickly from container of 100% ethanol to the critical point drying apparatus.

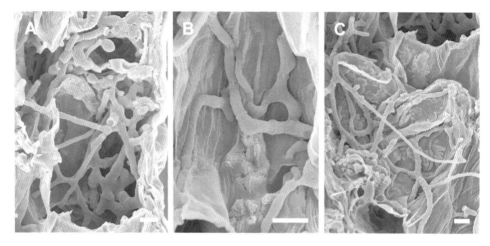

Fig. 12. (**a–c**) Hyphae of *Cercospora beticola* within cryofractured sugarbeet leaf cells. Infected leaves obtained courtesy of Melvin D. Bolton, Research Plant Pathologist, Sugarbeet and Potato Research Unit, USDA-ARS, Northern Crop Science Laboratory, Fargo, North Dakota.

26. To prepare for SEM viewing, use conductive silver paint to mount leaf pieces on appropriate sample mounts/stubs (see Note 14) so that the fractured edges (Fig. 1c; see Note 15) are visible.

27. Once the silver paint has dried, coat the sample with gold, gold-palladium, or carbon and view in the SEM (see Note 16). Depending on the amount of fungus present, you may need to examine several leaf pieces before you find fungus (Fig. 12).

4. Notes

1. Dental wax provides a yielding noncontaminating surface for cutting plant tissues; because it is hydrophobic, the aqueous buffer beads on it and forms a pool so that it is easier to keep the plant tissue submerged in the buffer.

2. Double-edged blades are thinner and sharper than single-edged blades, decreasing tearing and crushing artifact. However, they lack the reinforced back found on single-edged blades, so handle the blades with care to avoid cut fingers. Single-edged blades can be cut in half horizontally (through the central slot) with trauma shears (EMT scissors) or other sturdy scissors before removal from the protective paper to help prevent accidents.

3. Molecular sieves type 3A have a pore diameter of 30 nm, making them effective in trapping water molecules (28 nm) while excluding ethanol (44 nm). Sieves must be activated by heating in a 110°C oven for 2 h, then cooled in a dessicator before use.

4. Shipping containers, small coolers, or bait boxes work well. The 10-cm depth is important for access to facilitate the fracturing steps. If your container is too tall, you may consider cutting down the height with a serrated kitchen knife.

5. Keeping the leaf submerged in the buffer when cutting will help maintain leaf turgidity and decrease the introduction of air bubbles.

6. Draw the razor blade smoothly through the leaf tissue in a single motion for each cut without pressing down; pressing down will compress the leaf edge. Use a new blade for every one or two cuts.

7. For example, thick leaves or those with a heavy waxy cuticle require longer fixation.

8. Because the following steps of dehydration, fracturing, and critical point drying take considerable time, we commonly refrigerate samples overnight in the fixative and resume processing on the following day.

9. Each time the buffer is added it dilutes the glutaraldehyde, effectively removing it from the sample. Standing time in the buffer is as important as number of changes in order to allow time for the fixative to diffuse out of the plant tissues.

10. Be careful not to agitate the sieves when handling the 100% ethanol container and do not pipette from the area immediately above the top of the sieves in order to avoid getting particulate sieve material on the surface of the leaf tissue; the sieve debris might be visible in the SEM.

11. Change the hemostat to your nondominant hand to hold the dish. Pick up the hammer in the dominant hand.

12. Amount of force required to fracture the leaf tissue appropriately depends on the type of leaf and its turgidity. Excessive

Fig. 13. Cylindrical aluminum stub mounts that have been cut with a hacksaw to form 90° sides or parallel grooves for mounting cryofractured leaf tissue so that the cryofractured surfaces may be observed in the scanning electron microscope.

force will powder the leaf. You can't tell how well the leaf has fractured until you thaw it in step 20. Resist the urge for multiple blows to the frozen tissue!

13. Using the hemostat, hold the edge of an appropriately chilled razor blade (preferably single-edged for this purpose) on the frozen ethanol surface as the weighing dish rests on the metal block. Align the edge of the blade over the tissue along the desired fracture direction. Gently tap the back of the razor blade with the hammer to fracture the ethanol *without* driving the blade down so that it cuts the tissue.

14. If cylindrical stubs are used, they can be cut to form a 90° angle side to hold the sample, or numerous parallel grooves can be cut into the top of a stub using a hacksaw (Fig. 13); leaf pieces are mounted in the grooves formed by the saw kerf. With the latter technique, it is possible to minimize sample exchanges in the SEM and survey many fractured leaf edges in a short time.

15. Fractured surfaces will appear distinctly shiny and smoother compared to unbroken ones or those that have been cut with a razor blade.

16. Leaves usually charge if not coated. Coating may not be necessary if using a variable-pressure SEM; resolution generally is better at high vacuum.

References

1. Coleman JR (1975) Biological applications: simple preparation and quantitation. In: Goldstein JI, Yakowitz H (eds) Practical scanning welectron microscopy. Plenum Press, New York

2. Humphreys WJ, Spurlock BO, Johnson JS (1974) Critical point drying of ethanol-infiltrated, cryofractured biological specimens for scanning electron microscopy. Scanning Electron Microsc/1974 (Part I):275–282

3. Humphreys WJ, Spurlock BO, Johnson JS. (1975) Transmission electron microscopy of tissue prepared for scanning electron microscopy by ethanol-cryofracturing. Stain Technol 50:119–125

4. Dawes CJ (1988) Introduction to biological electron microscopy: theory and techniques. Ladd Research Industries, Inc., Burlington, Vermont

5. Reid N, Beesley JE (1991) Cryoultramicrotomy. In: Glauert AM (ed) Sectioning and cryosectioning for electron microscopy, vol. 13 of Practical methods in electron microscopy. Elsevier, Amsterdam

6. Echlin P (1992) Low-temperature microscopy and analysis. Plenum Press, New York

7. Kveder M et al (2006) Direct evidence for the glass-crystalline transformation in solid ethanol by means of a nitroxide spin probe. Chem Phys Letters 419:91–95

Chapter 8

Real-Time PCR for the Quantification of Fungi In Planta

Steven J. Klosterman

Abstract

Methods enabling quantification of fungi in planta can be useful for a variety of applications. In combination with information on plant disease severity, indirect quantification of fungi in planta offers an additional tool in the screening of plants that are resistant to fungal diseases. In this chapter, a method is described for the quantification of DNA from a fungus in plant leaves using real-time PCR (qPCR). Although the method described entails quantification of the fungus *Verticillium dahliae* in lettuce leaves, the methodology described would be useful for other pathosystems as well. The method utilizes primers that are specific for amplification of a β-tubulin sequence from *V. dahliae* and a lettuce actin gene sequence as a reference for normalization. This approach enabled quantification of *V. dahliae* in the amount of 2.5 fg/ng of lettuce leaf DNA at 21 days following plant inoculation.

Key words: Real-time PCR, Absolute quantification, *Verticillium dahliae*, *Lactuca sativa*

1. Introduction

Quantitative real-time PCR (qPCR) measures the amount of PCR product at each cycle during a reaction, thereby allowing real-time analysis of amplification. The fractional cycle used for qPCR is known as the quantification cycle (C_q), but also as crossing point (C_p) or threshold cycle (C_t) (1). In qPCR using absolute quantification, a standard curve is generated using known starting quantities of DNA containing the target DNA (Fig. 1). The concentrations of target DNA in unknown samples can be determined from the interpolation of C_q values with the linear equation of the standard curve.

QPCR technology has been applied for the measurement of fungal biomass in many plant–fungus interactions (2–6). The potential applications of this technology include susceptibility phenotyping for diseases, the assessment of fungal development in

Melvin D. Bolton and Bart P.H.J. Thomma (eds.), *Plant Fungal Pathogens: Methods and Protocols*,
Methods in Molecular Biology, vol. 835, DOI 10.1007/978-1-61779-501-5_8, © Springer Science+Business Media, LLC 2012

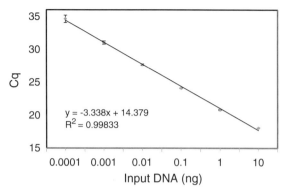

Fig. 1. Standard curve prepared by real-time PCR analysis from a tenfold dilution series of the genomic DNA of Verticillium dahliae starting at 10 ng and the primer pair VertBt-F and VertBt-R (3) specific for the amplification of *β-tubulin*. Error bars represent standard deviation from three samples of each dilution.

the host, and diagnostics. The use of qPCR has many advantages over conventional methods for quantifying fungi in plant tissue. The processing and analyses of many samples can be achieved relatively quickly and accurately using qPCR, as compared with plate assays (3). But most importantly, the qPCR method is highly sensitive, with reliable detection of 1 pg or less of fungal DNA in infected plant tissue (2, 3, 7, 8). For interactions where the amount of fungus may be low, a sensitive qPCR technique is especially important for determining the amount of fungus in the host.

A correlation between the amount of fungal biomass determined by qPCR and plant disease severity has been demonstrated in several plant–fungus interactions (9–11), suggesting applicability of qPCR to plant disease resistance screening procedures. The application of qPCR in screening plants for resistance to fungi is appealing for several reasons. Primer pairs used in PCR applications can be prepared for specific amplification of a particular fungus, and not closely related species, reducing ambiguity of detection over conventional plating assays (3). When scoring disease symptoms, those symptoms caused by fungi are not always easily distinguished from other plant disorders or senescence. Moreover, in tolerant interactions, disease symptoms are not visible, yet the plant tissues are heavily colonized by the fungus. In addition, plant cultivars can show genetic variation for partial resistance (12). Thus, the application of qPCR for fungal quantification in plant tissue may be an especially valuable tool for susceptibility phenotyping of plants.

The detailed protocol outlined in this chapter allows for the quantification of the fungal pathogen *V. dahliae* in lettuce leaves by qPCR analysis. However, with the appropriate primer pair for target DNA amplification, the methodology outlined can be useful for quantifying fungi in other pathosystems. The method employed utilizes absolute quantification, and includes materials and methods for DNA extraction and purification, and qPCR. The method takes advantage of SYBR Green 1 methodology and β-tubulin primers that allow specific amplification of target DNA from *V. dahliae*, but not other *Verticillium* spp. or fungi (3).

2. Materials

1. Fungus (see Note 1).
2. Lettuce (*Lactuca sativa* L.) seedlings.
3. Drigalski spatula.
4. Plant growth chamber with light and temperature control.
5. Hoagland's solution.
6. Microcentrifuge tubes (1.5-mL).
7. Liquid nitrogen.
8. Forceps.
9. Balance.
10. Screw-cap tubes (15-mL).
11. Micropestles.
12. Vortexer.
13. FastDNA kit (MP Biomedicals, Solon, Ohio) (see Note 2).
14. Ampure magnetic beads and SPRIplate® 96-Ring Magnet Plate (Agencourt, Beverly, MA) (see Note 3).
15. 70% Ethanol.
16. Incubator (with 37°C setting).
17. Nuclease-free water.
18. Qubit® fluorometer and Quant-iT™ dsDNA BR Assay Kit (Invitrogen, Carlsbad, CA) (see Note 4).
19. PCR tubes (0.2-mL).
20. PCR tubes (0.5-mL).
21. IQ SYBR® Green Supermix (Bio-Rad, Carlsbad, CA) (see Note 5).
22. Lightcycler® 480 II (Roche).
23. PCR plates (96-well, white, straight sides).
24. Optically clear plate seals for 96-well PCR plates.
25. Repeating pipettor.
26. Single channel pipettors (1–10, 20–200, 200–1,000 μL).
27. Multichannel pippettor (1–10 μL).
28. Low-retention filter tips for pipettors (1–10, 20–200, 200–1,000 μL).

29. DNA primers for *V. dahliae* β-tubulin (3): VertBt-F 5′AACAA
    CAGTCCGATGGATAATTC-3′ and VertBt-R 5′-GTACCGG
    GCTCGAGATCG-3′ (see Notes 6 and 7).

30. DNA primers for lettuce actin: qActPI558F, 5′-ACATAGCGG
    GAGCATTGAAC-3′ qActPI638R, 5′-ACACCCCGTTCTTC
    TCACAG 3′ (see Notes 7 and 8).

3. Methods

3.1. Fungal Growth and Plant Inoculation

1. Scrape the fungus from the Petri plates (see Note 9) using a sterile Drigalski spatula, and enumerate using a hemocytometer.

2. Inoculate plants with a 5 mL aliquot of the Hoagland's solution containing the suspension of conidia. Plant inoculations were carried out as described previously (13) with modifications (see Note 10).

3. At 21 days postinoculation, remove the entire third and fourth true leaves with a razor blade (Fig. 2 and see Notes 11 and 12), place in sterile 15-mL screw-cap tubes, and immediately freeze in liquid nitrogen (see Note 13).

3.2. DNA Extraction

1. Remove leaf samples in sterile 15-mL screw-cap tubes from the −80°C storage and pour liquid nitrogen into tubes.

Fig. 2. The lettuce line PI 251246 at 21 days after inoculation with *V. dahliae*. The *white circle* indicates the position at which the third and fourth leaves are cut from the plant.

2. Prechill a forceps and 1.5-mL microcentrifuge tube with liquid nitrogen. Remove leaf tissue from either side of the midvein with forceps and weigh 100 mg of tissue in the prechilled tube, and immediately add liquid nitrogen to the tube (see Note 14).

3. Thoroughly grind the individual leaf samples by hand in separate 1.5-mL microcentrifuge tubes using a micropestle in liquid nitrogen.

4. Suspend the ground samples in 800 μL buffer CLS-VF and 200 μL buffer PPS from the FastDNA kit (MP Biomedicals) and vortex for 30 s. Spin samples for 10 min and process for DNA extraction according to the FastDNA kit protocol (MP Biomedicals).

3.3. Purification of DNA Template

1. Add 108 μL of Ampure magnetic beads to a 60 μL aliquot of extracted DNA in 0.2-mL tubes and mix by pipetting. Incubate at room temperature 5–10 min.

2. Place on SPRIplate® and allow ring to form for 10 min.

3. Remove supernatant carefully from the bottom of the well without disturbing the ring of beads. Discard the supernatant.

4. Remove samples from the SPRIplate® plate and add 200 μL of 70% ethanol to each, mix and return to plate for 3 min. Remove ethanol and repeat once.

5. Remove from SPRIplate® and incubate at 37°C until ethanol has evaporated.

6. Resuspend beads in 25 μL of nuclease-free water by vigourous vortexing. Centrifuge very briefly.

7. Place on SPRIplate® and allow ring to form 10 min.

3.4. Quantification of DNA Template

1. Dilute Quant-iT dsDNA BR Assay Reagent 200-fold in Quant-iT™ dsDNA BR Buffer. The reagents are kept at room temperature.

2. Aliquot 198 μL of the dilution prepared in step 1 for each of the samples and 190 μL for two standards into 0.5-mL PCR tubes.

3. Remove 2 μL of sample on SPRIplate® and add to the 198 μL aliquot of diluted Quant-iT™ dsDNA BR Reagent or add 10 μL of standard; mix and incubate without agitation for 5 min (see Note 15).

4. Take the reading on the Qubit® fluorometer and multiply the result by 100 to yield ng/μL. Typical results for this plant tissue range from 5 to 20 ng/μL.

3.5. Real-Time Quantitative PCR

1. Dilute all plant DNA template samples in water to 3 ng/μL. Carefully remove the plant DNA template from the Ampure magnetic beads on the SPRIplate®, to avoid including any beads in the amplification reaction.

2. Prepare 32 μL of 3 ng/μL template for each plant sample so the same template dilution is used for all reactions. Three individual reactions (technical replicates) are run for each biological replicate sample with both VertBt and qActPI primers.

3. Prepare a standard curve of tenfold dilutions from 1 ng to 1 pg using genomic DNA from *V. dahliae* in 100 μL volumes. Aliquot 90 μL water to each dilution using a repeating pipette (see Note 16).

4. Prepare one master mix for an entire 96-well PCR plate of 20 μL reactions: the master mix contains 1X IQ SYBR® Green Supermix (Bio-Rad) (see Note 17), 200 nM of each of the primers (VertBt-F and VertBt-R), and nuclease-free water.

5. Aliquot 15 μL master mix to each well using a repeating pipette.

6. Add 5 μL of the plant sample templates (3 ng/μL) to each well, individually.

7. Add 5 μL water to three wells for the No Template Control (NTC) (see Note 18).

8. Add 5 μL of prepared standard curve (with genomic DNA of *V. dahliae*) to the plate starting with the 1 pg dilution and ending with the 1 ng dilution. Include three replicates per dilution.

9. Load the 96-well PCR plate containing samples into the Lightcycler® 480 II.

10. Adjust the Lightcycler® program on the SYBR Green I manufacturer template (see Note 19) to include a 5 min 95°C denaturation step followed by 40 cycles of 95°C for 10 s and 65°C (see Note 20) for 35 s (single data acquisition point) followed by the default melt curve analysis of heating to 95°C, cooling to 65°C for 1 min, and then heating to 97°C at a rate of 1°C/5 s (continuous data acquisition).

11. Next confirm template quality and quantity with the actin reference gene primers. Prepare a separate 96-well PCR plate for the amplification of the lettuce actin sequence (see Note 21). Prepare one master mix for the entire plate containing 1X IQ SYBR® Green Supermix (Bio-Rad) and 400 nM of each qActPI forward and reverse primer (total reaction volume of 20 μL).

12. Add 5 μL of the plant sample templates (3 ng/μL) to each well, individually.

13. Add 5 μL water to three wells for the NTC.

14. Add 5 μL of cloned actin sequence (see Note 5) to three wells as a positive control (concentration of 1,000 copies/μL).

15. Load the 96-well PCR plate containing the samples into the Lightcycler® 480 II.

16. Adjust the Lightcycler® program on the SYBR Green I manufacturer template (see Note 19) to include a 5 min denaturation

step at 95°C, followed by 40 cycles of 95°C for 10 s, and 60°C for 30 s followed by melt curve analysis (see step 11 above).

3.6. Analysis of Absolute Quantification Data

1. After the run is completed, data generated from the amplification curves are analyzed using the software bundled with the real-time PCR thermocycler (see Note 22). Typical qPCR amplification curves are shown (Fig. 3). The software varies by real-time PCR thermocycler manufacturer. The results are reported as the mean C_q value of the PCR replicates per sample (SD is also determined), for both the target and the actin reference. Data can be presented in tabular (Table 1) or graphic form (i.e., when

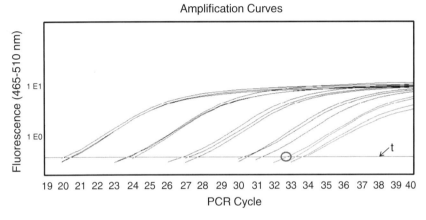

Fig. 3. Typical real-time PCR amplification curves. The *circle* depicts the fractional quantification cycle (C_q) used to calculate the amount of fungal DNA in a sample. Using the Lightcycler® 480 II (Roche) real-time PCR thermocycler, the threshold (t) determines C_q values across all samples in the log-linear phase of the amplification curve and can be adjusted.

Table 1
Results of qPCR analyses of the fungus *Verticillium dahliae*, strain VdLs.17 in lettuce line PI 251246 leaves using VertBt (for *β-tubulin*) and qActPI (for actin) primer pairs 21 days after inoculation

Samples	*β-tubulin* C_q	SD *β-tubulin*	Concentration[a] (fg/ng)	Actin C_q	SD Actin
Control[b] sample a	N/A	N/A	0.00	21.04	0.06
Control sample b	N/A	N/A	0.00	20.74	0.06
Control sample c	N/A	N/A	0.00	20.29	0.09
Inoculated sample a	34.32	0.05	7.53	21.50	0.08
Inoculated sample b	36.64	2.84	2.48	20.21	0.09
Inoculated sample c	33.30	0.53	15.93	20.63	0.35

[a]The concentration is expressed as fg *V. dahliae* DNA/ng of plant leaf DNA
[b]Control samples were prepared from leaves of uninoculated plants

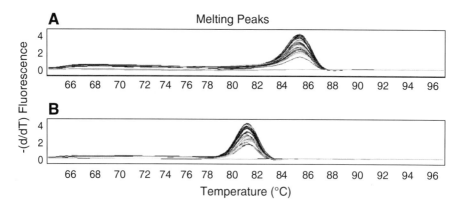

Fig. 4. Melt profiles of qPCR reactions. (a) Melt profile of the qPCR product obtained using primers VertBt-F and VertBt-R (no peaks were observed in NTCs or in control plants). (b) Melt profile of the qPCR product obtained using primers qAct-Pl558F and qActPl638R for lettuce actin (no peak was obtained in a fungus-only DNA template control).

presenting multiple time points). Additionally, the target can be reported as fungal concentration per defined unit of template (i.e., fg *V. dahliae* DNA/ng plant DNA; Table 1). The target fungal gene C_q values are used to determine the concentration of fungal DNA present from the linear equation of the amplification. The linear equation is generated from the standard curve of known starting DNA concentrations containing the target gene (Fig. 1).

2. Examine the melt peak graph to determine that a single peak is present for each sample (see Note 23). The melting temperature is 85.5°C (Fig. 4a) for Vert Bt primers and 81°C for the Actin primers (Fig. 4b). Additional melt peaks indicate non-specific amplification or primer-dimers in the reaction. This causes C_q inaccuracy since the SYBR Green I dye binds to double-strand DNA without specificity toward the target DNA sequence.

3. Ensure no amplification occurred with target primers in the control plant samples or the NTC. There should be no C_q data recorded (Table 1) and a flat line in both the amplification curves and the melt peaks. These data indicate the absence of fungal (*V. dahliae*) DNA in the noninoculated plant tissue and lack of cross-contamination or primer-dimers in the PCR reaction.

4. Check for uniformity of the template across samples by comparing the actin reference C_q values between samples (Table 1). The reference C_q data can also be used to normalize the target data using an Infection Coefficient (C_q actin reference/C_q target) (3).

4. Notes

1. The fungus used throughout these experiments was *Vertcillium dahliae*, strain VdLs.17.

2. There are numerous DNA extraction kits for plant tissues but these kits yield cleaner DNA from lettuce than other kits tested for this procedure.

3. Removal of PCR inhibitors from lettuce leaf tissue samples is important for increased qPCR sensitivity and accuracy. The Ampure magnetic beads and SPRIplate® 96-Ring Magnet Plate (Agencourt, Beverly, MA) system were successfully applied in this example but other systems are commercially available for the purpose of DNA purification.

4. While the Qubit® fluorometer (Invitrogen, Carlsbad, CA) is used for this application, other types of fluorometers and sources of picogreen dye are available for the quantification of dsDNA

5. Many companies sell different SYBR green mixes with varying proprietary formulations which can affect primer efficiency. These mixes are crosscompatible across qPCR thermal cyclers. When designing new primers it is advisable to try several different mixes to find the optimal primer efficiency and amplification of the target gene. This qPCR was optimized with IQ SYBR® Green Supermix and the use of other mixes would require re-optimization of the primers.

6. The VertBt-F and VertBt-R primers amplify a single copy gene in the race 2 strain of *V. dahliae*, VdLs.17 (http://www.broad-institute.org/annotation-/genome/verticillium_dahliae/). Additional specificity analyses, besides those previously conducted (3), by agarose gel electrophoresis revealed that the VertBt primers amplify the appropriate amplicon of 115 bp from at least four *V. dahliae* strains tested. Sequencing of the VertBt amplification product also revealed identical 115 bp sequence for VdLs.17.

7. QPCR analyses require extensive primer validation before and during qPCR runs. Initial optimization involved testing multiple annealing temperatures and primer concentrations to get the highest efficiency without primer-dimers. The dynamic range of both the ActPI and VertBt primers was determined by standard curve amplification based on copy number. Standard curves were prepared with 5×10^5 to five copies of the relevant gene and the LightCycler® 480 software Fit Points method was used to determine efficiency. Actin PI primers yielded an equation of $y = -3.5x + 34.42$ for an efficiency of 93% and an error of 0.00680. Amplification was stable through five copies

with an average C_q of 33.38 and standard deviation of 0.66 cycles. VertBt primers yielded an equation of $y = -3.6x + 38.34$ for an efficiency of 90% and an error of 0.0831. Amplification was stable through five copies with an average C_q of 35.76 and standard deviation of 0.35 cycles. In addition, genomic DNA extracted from PI 251246 was tested for the presence of PCR inhibitors by producing two standard curves based on VdLs.17 genomic DNA. Using the VertBt primers, one standard curve was obtained with the addition of 15 ng of non-inoculated PI 251246 genomic DNA to each reaction while the other curve had only VdLs.17 genomic DNA. The two standard curves produced very similar linear equations, suggesting inhibitors from PI 251246 have negligible influence on amplification of *V. dahliae* DNA from PI 251246.

8. Actin sequence specific to *Lactuca sativa* PI 251246 was initially obtained by PCR using the degenerate primer pair F-5′FATATGGARAARATHTGG 3′ and R-5′TCRTCNACR TCRCATTTCAT 3′ (John J. Weiland, USDA-ARS). The gene-specific primer set qActPI558F and qActPI638R was designed using Primer3 (v. 0.4.0) (14) and the 80 bp amplicon generated by PCR was cloned into pCR4™-TOPO TA cloning vector (Invitrogen, Carlsbad, CA). The qActPI558F and qActPI638R primers yielded a single melt peak from plant DNA (Fig. 3b). The primers did not produce any amplification from *V. dahliae* (data not shown).

9. The fungus was plated initially on Petri plates containing potato dextrose agar, and allowing growth for 4 days at 25°C. The fungus was stored long term at −80°C as a conidial suspension in 20% (v/v) glycerol/potato dextrose broth.

10. Minor modifications included the preparation of the conidia for inoculation at 2×10^7/mL suspension in Hoagland's solution. The untreated (non-inoculated) control group consisted of the application of only Hoagland's solution. The experiment was replicated three times.

11. Given the sensitivity of qPCR in detecting target DNA, gloves are used during the handling of plant tissues and all samples to prevent the introduction of contaminating DNA. Gloves and razor blades are changed between the collections of the plant tissue samples.

12. Since leaf necrosis caused by infection with *V. dahliae* may lead to decreased copy numbers of the plant actin reference gene, necrotic leaves were not used for qPCR analyses. This is the rationale for using leaves three and four, which are generally symptomatic, but not necrotic.

13. Leaf samples were stored at −80°C.

14. Work quickly to prevent thawing of the samples. The leaf tissue samples are kept in 15-mL centrifuge tubes that are filled with liquid nitrogen. Tissue is sampled with frozen forceps and placed into frozen microcentrifuge tubes on the balance for weighing. Liquid nitrogen is again poured on the tissue after weighing. It is advisable to extract no more than 3–4 samples at a time.

15. Although the Quant-iT™ dsDNA BR Assay Kit protocol indicates 2 min, reactions we routinely determine readings after 5 min.

16. A repeating pipette is used for aliquoting water to each dilution and wherever possible in the qPCR setup to minimize pipetting error. When conducting the experiments, avoid freeze/thaw cycles of the diluted standard curve, as this decreases the sensitivity and efficiency of the standard curve. Prepare standard curves with fresh dilutions from concentrated stock DNA for qPCR.

17. The Sybr Green Supermix contains Taq polymerase, dNTPS, and the appropriate buffer.

18. The NTC is a useful check for cross-contamination and primer-dimers during the qPCR run. .

19. The Lightcycler® 480 II contains a SybrGreen I Template, which is linked to the appropriate light wavelength for SYBR Green I excitation. The SybrGreen I Template can be modified and saved for future use. Samples can be identified in the Sample Editor window at any point during the qPCR experiment.

20. The VertBt primer annealing temperature was optimized using the LightCycler 480 since the primers originally were optimized on a BioRad Icycler. The optimum annealing temperature was 65°C with the LightCycler® 480 II while it was 63°C on the Icycler (Bio-Rad).

21. No standard curve is needed for the amplification of the actin reference gene during an experimental run.

22. For this experiment the Lightcycler® software version LCS480 1.5.0.39 with the Fit Points absolute quantification analysis type was used to determine C_q and linear equation of the standard curve. Note that the noise band can be determined manually or automatically by the software. Make sure all late amplification curves with the correct melt peak are assigned a C_q rather than being automatically classified as background amplification by the software. The threshold (Fig. 3) value can also be input manually. When comparing data between different qPCR runs, use the same threshold value for C_q determination. The Lightcycler® software contains an option to generate

a report containing results and charts in a PDF file format. Additionally, the results can be exported as a tab-delimited text file that can be converted to a spreadsheet for reporting purposes. Charts can be exported in multiple picture formats.

23. Select "Melting Peaks" from the chart menu displayed after clicking on the chart option toolbar in the absolute quantification analysis window to display melt peaks.

Acknowledgments

The author acknowledges the technical assistance of Amy Anchieta, USDA ARS, and John J. Weiland, USDA-ARS, for supplying degenerate primers for plant actin amplification.

Disclaimer: Mention of trade names or commercial products in this publication is solely for the purpose of providing specific information and does not imply recommendation or endorsement by the U.S. Department of Agriculture.

References

1. Bustin SA, et al (2009) QE guidelines: minimum information tation of quantitative real-time PCR ents. Clinical Chemistry 55: 611–622

2. Brouwer M, et al (2003) cation of disease progression of several pathogens on *Arabidopsis thaliana* usin me fluorescence PCR. FEMS Microbiolog s 228:241–248

3. Atallah ZK, et al (2007) ltiplex real-time quantitative PCR to and quantify *Verticillium dahliae* colon n in potato lines that differ in response Verticillium wilt. Phytopathology 97:865–8

4. Gayoso C, et al (2007) Ass ment of real-time PCR as a method for determining the presence of *Verticillium dahliae* in different Solanaceae cultivars. Eur J Plant Pathol 118:199–209

5. Li S, et al (2008) Quantification of *Fusarium solani* f. sp. *glycines* isolates in soybean roots by colony-forming unit assays and real-time quantitative PCR. Theor Appl Genet 117:343–352

6. Alaei H, et al (2009) Molecular detection of *Puccinia horiana* in *Chrysanthemum* x *morifolium* through conventional and real-time PCR. J Microbiol Methods 76:136–45

7. Cullen DW, et al (2001) Conventional PCR and real-time quantitative PCR detection of *Helminthosporium solani* in soil and on potato tubers. Eur J Plant Pathol 107:387–398

8. Pedley KF (2009) PCR-based assays for the detection of *Puccinia horiana* on chrysanthemums. Plant Dis 93:1252–1258

9. Qi M and Yang Y (2002) Quantification of *Magnaporthe grisea* during infection of rice plants using real-time polymerase chain reaction and northern blot/phosphoimaging analyses. Phytopathology 92: 870–876

10. Hietala AM, et al (2003) Multiplex real-time PCR for monitoring *Heterobasidion annosum* colonization in Norway spruce clones that differ in disease resistance. Appl Environ Microbiol 69: 4413–4420

11. Brunner K, et al (2009) A reference-gene-based quantitative PCR method as a tool to determine *Fusarium* resistance in wheat. Anal Bioanal Chem 395:1385–1394

12. Hayes RJ, et al (2007) Variation for resistance to Verticillium wilt in lettuce (*Lactuca sativa* L.). Plant Dis 91:439–445

13. Klosterman SJ, Hayes RJ (2009) A soilless Verticillium wilt assay using an early flowering lettuce line. Plant Dis 93:691–698

14. Rozen S, Skaletsky HJ (2000) Primer3 on the WWW for general users and for biologist programmers. In: Krawetz S, Misener S (eds) *Bioinformatics Methods and Protocols: Methods in Molecular Biology*. Humana Press, Totowa

Chapter 9

Using Non-homologous End-Joining-Deficient Strains for Functional Gene Analyses in Filamentous Fungi

Mark Arentshorst, Arthur F.J. Ram, and Vera Meyer

Abstract

Fungal strains deficient in the non-homologous end-joining (NHEJ) pathway are excellent recipient strains for gene targeting approaches. In addition, NHEJ-deficiency can facilitate the formation of heterokaryons which allows rapid identification of essential genes. However, the use of NHEJ-deficient strains can also pose some limitations for gene function analyses. For example, lack of the NHEJ pathway can interfere with phenotypic analyses and complicate complementation studies. Moreover, heterokaryons are difficult to propagate and re-transform. We describe here strategies and methods to circumvent these problems and to better exploit the power of NHEJ-deficient strains. We provide methods for the establishment of transiently deficient NHEJ strains, for improved complementation analyses using AMA1-based vectors and for fast identification and propagation of heterokaryons. The methods described are applicable for a wide range of filamentous fungi.

Key words: *Aspergillus*, Ku70, NHEJ, AMA1, Heterokaryon rescue, Protoplast-mediated transformation, Complementation

1. Introduction

Targeted integration of DNA into a genome of interest is a prerequisite for functional genomic studies. However, the natural frequency of homologous recombination in filamentous fungi is very low and has hampered for a long time efficient functional gene analyses. Only after the observation that disruption of the non-homologous end-joining (NHEJ) pathway in *Neurospora crassa* resulted in homologous recombination frequencies up to 100% (1) respective mutants were established in various filamentous fungi, thereby allowing genome-wide functional genomic

Melvin D. Bolton and Bart P.H.J. Thomma (eds.), *Plant Fungal Pathogens: Methods and Protocols*,
Methods in Molecular Biology, vol. 835, DOI 10.1007/978-1-61779-501-5_9, © Springer Science+Business Media, LLC 2012

studies to become more feasible (for reviews see ref. (2, 3)). In brief, the NHEJ pathway is a mechanism ensuring the repair of chromosomal DNA double-strand breaks (DSBs) and depends on the activities of the Ku heterodimer (Ku70/Ku80-protein complex) and the DNA ligase IV-Xrcc4 complex (4, 5). The NHEJ pathway competes with another repair mechanism, the homologous recombination (HR) pathway, which mediates interaction between homologous DNA sequences, whereas the NHEJ pathway ligates DSBs without the requirement of any homology (6). By deleting either *ku70*, *ku80* or *lig4* genes, the frequency of homologous recombination events has dramatically increased in numerous filamentous fungi which rendered these strains as "gold strains" for gene manipulations including gene deletion, gene tagging and promoter replacements (for reviews see ref. (2, 3) and references therein). An added value of this high efficiency of gene targeting is that essential genes can easily be identified by the so-called heterokaryon rescue technique (7). As a detailed protocol for the heterokaryon rescue technique has been published previously for *Aspergillus nidulans* (8), we direct the reader to the original reference. However, we present here the basic concept of this technique and, based on our experiences with *A. niger*, provide the reader with practical and more generalized advices to easily dissect and work with heterokaryons.

One drawback of NHEJ inactivation is that it makes fungal strains vulnerable to DNA damaging conditions (9–12), thus posing limitations in subsequent phenotypic analyses and use of these strains. To eliminate the risk that NHEJ deficiency influences or obscures phenotypic analyses, we describe here a strategy to establish strains being transiently silenced in NHEJ. This approach has been applied for *A. nidulans* (13) and *A. niger* (14) and respective strains have proven to perform as efficient as constitutively silenced NHEJ strains with respect to gene targeting. Another drawback of constitutively silenced NHEJ strains is that complementation studies for gene deletion strains are often challenging for several reasons. For example, ectopic integration events, usually aimed for in complementation approaches, are almost impossible in NHEJ-deficient backgrounds. Therefore, transformation efficiencies are generally low in NHEJ-deficient strains and the re-transformation of deletion strains with severe phenotypes can be very difficult. A straightforward approach to bypass these disadvantages is the use of autonomously replicating AMA1-based plasmids (14). The application of these vectors for rapid complementation studies will be discussed in this chapter as well. Although the methods and approaches described here are focused on *A. niger*, they can easily be adapted to other filamentous fungi.

2. Materials and Reagents

2.1. Media and Solutions

1. Glucose (50%): For 1 L: Boil 500 mL Milli-Q (MQ) in a 1,000 mL beaker on a heated magnetic stirrer. Slowly add 500 g of D(+)-glucose anhydrous. After glucose has been dissolved, let the solution cool down to RT, add MQ up to 1 L and autoclave.

2. ASPA + N (50×): For 1 L: Add 297.5 g $NaNO_3$, 26.1 g KCl and 74.8 g KH_2PO_4 to 600 mL MQ in a 1 L cylinder. When all salts are dissolved, set pH to 5.5 with KOH (use 5 M KOH). Add MQ up to 1 L and autoclave.

3. ASPA-N (50×): For 1 L: Add 26.1 and 74.8 g KH_2PO_4 to 600 mL MQ in a 1 L cylinder. When dissolved, set pH to 5.5 with KOH. Add MQ up to 1 L and autoclave.

4. $MgSO_4$ (1 M): For 1 L: Add 246.5 g $MgSO_4 \cdot 7H_2O$ to 600 mL MQ in a 1 L cylinder. When all salts are dissolved, add MQ up to 1 L and autoclave.

5. Trace element solution (1,000×): For 1 L: Add 10 g EDTA, 4.4 g $ZnSO_4 \cdot 7H_2O$, 1.01 g $MnCl_2 \cdot 4H_2O$, 0.32 g $CoCl_2 \cdot 6H_2O$, 0.315 g $CuSO_4 \cdot 5H_2O$, 0.22 g $(NH_4)_6 Mo_7O_{24} \cdot 4H_2O$, 1.11 g $CaCl_2$ and 1.0 g $FeSO_4 \cdot 7H_2O$ to 600 mL MQ. When dissolved, set pH to 4.0 with 1 M NaOH and 1 M HCl, fill MQ up to 1 L and autoclave (see Note 1).

6. Uridine (1 M): For 100 mL: Add 22.4 g of uridine to 50 mL of warm MQ (about 50–60°C) in a 100 mL cylinder. When uridine is dissolved, add MQ up to 100 mL, sterilize by filtration and store at 4°C. Final concentration in medium is 10 mM (see Note 2).

7. Acetamide (1 M): For 100 mL: Add 5.91 g of acetamide to 50 mL of warm MQ (about 50–60°C) in a 100 mL cylinder. When acetamide is dissolved, add MQ up to 100 mL, sterilize by filtration and store at 4°C. The final concentration in cultivation medium is 10 mM (see Note 3).

8. Cesium chloride (CsCl, 1.5 M): For 100 mL: Add 25.3 g of CsCl to 50 mL of warm MQ (about 50–60°C) in a 100 mL cylinder. When CsCl is dissolved, add MQ up to 100 mL, sterilize by filtration and store at 4°C. The final concentration in cultivation medium is 15 mM (see Note 4).

9. Vitamin solution (1,000×): For 100 mL: Add 100 mg thiamin-HCl, 100 mg riboflavin, 100 mg nicotinamide, 50 mg pyridoxine, 10 mg pantotenic acid, 2 mg biotin to 50 mL of warm MQ (about 50–60°C) in a 100 mL cylinder. When all vitamins are dissolved, add MQ up to 100 mL, sterilize by filtration and store at 4°C under dark conditions (see Note 5).

10. Proline (1 M): For 100 mL: Add 11.51 g of L-Proline to 50 mL of warm MQ (about 50–60°C) in a 100 mL cylinder. When proline is dissolved, add MQ up to 100 mL, sterilize by filtration and store at 4°C. The final concentration in cultivation medium is 10 mM.

11. Urea (1 M): For 100 mL: Add 6.01 g of urea to 50 mL of warm MQ (about 50–60°C) in a 100 mL cylinder. When urea is dissolved, add MQ up to 100 mL and sterilize by filtration. The final concentration in cultivation medium is 10 mM.

12. Fluoroacetamide (FAA, 20%): For 100 mL: Add 20 g of FAA to 50 mL of warm MQ (about 50–60°C) in a 100 mL cylinder. When FAA is dissolved, add MQ up to 100 mL and sterilize by filtration. The final concentration in cultivation medium is 0.2%.

13. Hygromycin (100 mg/mL): Dissolve 1 g of hygromycin in 10 mL of MQ, sterilize by filtration, make aliquots of 500 μL and store at –20°C. The final concentration in the medium is 100 μg/mL, except for transformation plates, then use 200 μg/mL (see Note 6).

14. Caffeine (50 mg/mL): For 100 mL: Add 5 g of caffeine to 50 mL of warm MQ (about 50–60°C) in a 100 mL cylinder. When caffeine is dissolved, add MQ up to 100 mL and sterilize by filtration. The final concentration in cultivation medium is 500 μg/mL.

15. Minimal medium (MM): For 500 mL: Add under sterile conditions to 480 mL of sterile MQ: 10 mL of 50% glucose, 10 mL of 50×ASPA+N, 1 mL of 1 M MgSO$_4$ and 500 μL of 1,000× trace element solution. For MM+agar, autoclave 480 mL of MQ with 7.5 g of agar (Scharlau) and add all components after autoclaving under sterile conditions (see Note 7).

16. Minimal medium+acetamide+agar (MM-AA): For 500 mL: Autoclave 470 mL of MQ with 7.5 g of agar and add afterwards under sterile conditions: 10 mL 50% glucose, 10 mL of 50× ASPA-N, 5 mL of 1 M acetamide, 5 mL of 1.5 M CsCl, 1 M MgSO$_4$ and 500 μL of 1,000× trace element solution.

17. Minimal medium+5′-fluoroorotic acid+agar (MM-FOA): For 100 mL: Autoclave 50 mL MQ with 1.5 g of agar. Meanwhile, dissolve 75 mg of 5′-fluoroorotic acid (FOA) in 45 mL of warm MQ (about 50–60°C) and filter sterilize. After autoclaving, add the warm FOA-solution to the MQ+agar under sterile conditions. Also add 2 mL of 50% glucose, 2 mL of 50× ASPA-N, 200 μL of 1 M MgSO$_4$, 100 μL of 1,000× trace element solution, 100 μL of 1,000× vitamin solution, 1 mL of 1 M uridine and 1 mL of 1 M proline. Mix well and pour on the plates immediately (see Note 8).

18. Minimal medium+fluoroacetamide+agar (MM-FAA): For 100 mL: Autoclave 94 mL of MQ with 1.5 g of agar and add

thereafter under sterile conditions: 2 mL of 50% glucose, 2 mL of 50× ASPA-N, 200 μL of 1 M $MgSO_4$, 100 μL of 1,000× trace element solution, 1 mL of 1 M urea and 1 mL of 20% FAA.

19. Minimal medium + sucrose + agar (MMS): For 500 mL: Dissolve 162.6 g of D(+)-saccharose in 480 mL of MQ, add 6 g (1.2%) of agar and autoclave (see Note 9). For *pyrG* selection, add after autoclaving under sterile conditions: 10 mL of 50× ASPA + N, 1 mL of 1 M $MgSO_4$, 500 μL of 1,000× trace element solution. For hygromycin selection, add after autoclaving under sterile conditions: 10 mL of 50× ASPA + N, 1 mL of 1 M $MgSO_4$, 500 μL of 1,000× trace element solution, 5 mL of 50 mg/mL caffeine and 1 mL of 100 mg/mL hygromycin (see Note 10). For *amdS* selection, add after autoclaving under sterile conditions: 10 mL of 50× ASPA-N, 5 mL of 1 M acetamide, 5 mL of 1.5 M CsCl, 1 mL of 1 M $MgSO_4$ and 500 μL of 1,000× trace element solution.

20. MMS top agar: For 500 mL: Dissolve 162.6 g of D(+)-saccharose in 480 mL of MQ, add 3 g of agar (final concentration is 0.6%) and autoclave (see Note 9). Store the top agar at 65°C and add under sterile conditions: 10 mL of 50× ASPA + N, 1 mL of 1 M $MgSO_4$, 500 μL of 1,000× trace element solution. Before use, transfer the top agar to a 47–50°C water bath.

21. Complete medium (CM): For 500 mL: Add 0.5 g casamino acids, 2.5 g yeast extract and if required, 7.5 g agar to 480 mL of MQ and autoclave. Afterwards, add under sterile conditions: 10 mL of 50% glucose, 10 mL of 50× ASPA + N, 1 mL of 1 M $MgSO_4$, 500 μL of 1,000× trace element solution.

22. SMC: For 1 L: Add 242.3 g D-sorbitol, 5.5 g $CaCl_2$ and 3.9 g MES hydrate to 600 mL MQ in a 1 L cylinder. When everything is dissolved, set pH to 5.8 using 1 M NaOH and 1 M HCl. Add MQ up to 1 L and autoclave.

23. TC: For 1 L: Add 5.5 g $CaCl_2$ and 1.2 g Tris to 800 mL MQ in a 1 L cylinder. When everything is dissolved, set pH to 5.8 using 1 M NaOH and 1 M HCl. Add MQ up to 1 L and autoclave.

24. STC: For 1 liter: Add 242.3 g D-sorbitol to 600 mL TC in a 1 L cylinder. When sorbitol is dissolved, add TC up to 1 L and autoclave.

25. PEG buffer: For 10 mL: Add 2.5 g of Polyethylene glycol 6000 (PEG) to a 50-mL tube, add TC up to 10 mL under sterile conditions and dissolve PEG by shaking. Use PEG solution only fresh.

26. DNA extraction buffer: For 250 mL: Add 1.25 g sodium dodecyl sulfate (SDS), 0.3 g Tris and 9.3 g EDTA to 200 mL MQ. Set pH to 8.0 with 1 M NaOH and 1 M HCl. Add MQ up to 250 mL and autoclave the solution.

27. RNase (10 µg/µL): For 10 mL: Dissolve 100 mg RNase A in 10 mL of sterile MQ, make aliquots of 500 µL and store at –20°C.

28. Phenol:chloroform:isoamyl alcohol (25:24:1 v/v).

29. Isopropanol.

30. Sodium acetate solution (3 M): For 100 mL: Add 40.8 g sodium acetate trihydrate to 50 mL of warm MQ (about 50–60°C) in a 100 mL cylinder. Set pH to 6.0 with 1 M NaOH and 1 M HCl. Add MQ up to 100 mL and autoclave.

31. Ethanol absolute.

32. Saline solution: For 1 L: Add 9 g NaCl to 900 mL MQ in a 1 L cylinder. When NaCl is dissolved, add MQ up to 1 L and autoclave.

33. Lysing enzyme (Sigma).

34. Cellophane membrane (Biorad).

35. Myracloth (Calbiochem).

36. Cotton sticks (Hecht).

3. Methods

3.1. Establishment of a Transiently Silenced NHEJ Strain

Figure 1 highlights a strategy on how one can establish a transiently disrupted NHEJ strain in a fungus of interest. In the case of *A. niger*, we used the counter-selectable *amdS* marker which was flanked by sequences derived from the *kusA* ORF (*kusA* encodes the Ku70 homologue in *A. niger*) (11).

3.1.1. Generation of a Gene Cassette to Disrupt a NHEJ Gene

1. To transiently disrupt a *ku70*, *ku80* or *lig4* gene, a counter-selectable marker has to be chosen. In the case of *A. niger*, *amdS* and *pyrG* are available which allow counter-selection on medium containing the antimetabolite FAA or FOA, respectively.

2. After deciding which selection marker will be used, flanking regions around the marker have to be generated, which should be long enough to obtain homologous transformants with a reasonable frequency in a wild-type background. For *A. niger*, we routinely use 800–1,000 bp homologous sequences and obtain ~1–20% homologous transformants depending on the genomic locus. In the case of transiently disrupting *kusA*, both flanking regions share an overlapping region of about 300 bp close to the start codon of the *kusA* ORF (Fig. 1).

3. Several approaches can be followed to obtain the final disruption cassette: cloning strategies using restriction enzymes, in vivo recombination approaches in *Escherichia coli* or *Saccharomyces*

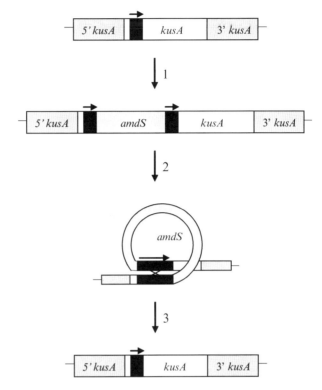

Fig. 1. Strategy for transient disruption of a *ku70* homologous gene (*kusA*) in *A. niger*. The *kusA* wild-type locus is shown on the *top cartoon*. A linear DNA fragment harbouring the counter-selectable marker *amdS*, flanked on each side with a 300 bp sequence of the *kusA* ORF (*black box* marked by an *arrow*) is transformed to an appropriate *A. niger* wild-type strain by protoplast-mediated transformation. In about 10–20% of the transformants growing on MM-AA, the construct has been integrated into the *kusA* gene locus (1). The resulting transformant (*amdS*+, *kusA*–) carries the disrupted *kusA* allele and any gene-targeting approach can be performed in this strain with high efficiency. The restoration of the disrupted *kusA* locus, can be performed by counter-selection on MM-FAA plates (2). Here, the *amdS* gene is forced to loop out via single crossover between the direct repeats (3). The figure is reproduced from ref. (14) by kind permission of Springer © 2010.

cerevisiae, Gateway technology or a fusion-PCR approach. In any case, it is essential that the DNA fragment containing the marker construct is linear when transformed to the fungus to allow *amdS* integration via the homologous *kusA* flanks. Linearization can be made by restriction (make sure that all DNAs have been digested) or by PCR (use a polymerase enzyme with a proofreading activity, e.g. Phusion from Finnzymes).

3.1.2. Transformation of a Disruption Cassette by Protoplast-Mediated Transformation

Four transformation methods have been shown to be applicable to filamentous fungi: protoplast-mediated transformation, *Agrobacterium*-mediated transformation, electroporation and biolistic transformation (for review see ref. (3) and references therein). Protoplast-mediated and *Agrobacterium*-mediated transformation, however, are the most commonly used techniques. We have recently published a detailed protocol for *Agrobacterium*-mediated transformation to *A. awamori* (15), a protocol, which for unknown reasons does not work for *A. niger* (own unpublished observations) but does for many other Ascomycetes and Basidiomycetes (16, 17). We thus provide here a protocol for protoplast-mediated transformation of *A. niger*, a method which in our hands can directly be applied to many other filamentous fungi.

1. Inoculate 1×10^8 spores in 100 mL CM, add 10 mM uridine if necessary (when using a *pyrG$^-$* strain) and incubate for 12 h at 30°C and 100 rpm.

2. Prepare protoplastation solution as follows: dissolve 200 mg Lysing enzymes in 10 mL SMC. Set pH to 5.6 with 1 M NaOH. Filter-sterilize the solution and transfer it to a sterile 50-mL tube.

3. Harvest the mycelium by filtration through sterile myracloth or a sterile coffee filter and wash one time with SMC. Add less than 0.5 g mycelium (wet weight) to the protoplastation solution and incubate for 1–2.5 h at 37°C with gentle shaking at 80 rpm, adjusting the tube in a shaker horizontally.

4. Check protoplast formation under the microscope after 1 h of incubation and then every 30 min (see Note 11).

5. Optional: Before harvesting protoplasts, mix carefully by repeatedly pipetting up and down (this additionally releases protoplasts from the mycelium).

6. Collect protoplasts through a sterile myracloth filter in a sterile 50-mL tube. Mycelial debris are not able to pass through.

7. Centrifuge at $2,000 \times g$ for 10 min at 10°C and decant the supernatant.

8. Gently re-suspend the pellet in 1 mL STC and centrifuge for 5 min at $3,000 \times g$. Discard supernatant and repeat the STC wash step.

9. For each transformation sample, add to a sterile 50-mL tube in this order: 1–20 µL DNA solution (3–10 µg), 100 µL protoplasts, 25 µL freshly made PEG buffer. Gently mix the solution by slowly pipetting up and down.

10. Add 1 mL PEG buffer and mix gently. After exactly 5 min of incubation (do not extend this incubation period because PEG is toxic to protoplasts), add 2 mL STC and mix gently by tipping.

11. Add 20 mL MMS top agar (47–50°C), gently mix by inverting the tube several times and pour onto a selective transformation agar plate (use 15 cm petri-dishes). To select for *amdS* integration in an *amdS$^-$* recipient strain, use MMS-AA agar plates.

To select for uridine prototrophy, use MMS agar plates. To select for hygromycin resistance use MMS agar plates supplemented with 200 μg/mL hygromycin and 500 μg/mL caffeine) (see Notes 6–10).

12. Incubate selective transformation plates at an appropriate temperature (25–37°C) for 3–6 days until colonies ("primary transformants") become visible.

3.1.3. Purification of A. niger Transformants and Identification of a NHEJ⁻ Strain

1. Purify primary transformants in order to obtain homokaryotic strains. In doing so, wet a sterile cotton stick with saline solution and use it to take off spores from a single colony. Streak spores on selective medium plates to get colonies from single spores.

2. Incubate plates until colonies have sporulated (3–6 days, 25–37°C).

3. Repeat steps 1 and 2.

4. Streak spores from a single colony on a CM agar plate and incubate until the plate is abundantly covered with sporulated mycelium (3–6 days, 25–37°C).

5. In order to harvest spores from CM agar plate, add 10 mL of saline solution to the plate and carefully release spores by scraping over the surface plate with a sterile cotton stick.

6. Pipet spore solution from the plate into a sterile 15-mL tube. If required, remove mycelial debris (vegetative mycelium, conidiophores) by filtration through a sterile myracloth filter.

7. Count spores using a microscope counting chamber.

8. In order to extract genomic DNA from transformants, grow selected strains overnight in 20 mL CM at 30–37°C. Use 1×10^6 spores/mL for inoculation.

9. Harvest mycelium by filtration through sterile myracloth or sterile coffee filter and transfer biomass without washing into a 2-mL Eppendorf tube (fill the tube up to the 0.5 mL line with mycelium).

10. Freeze samples at –80°C or in liquid nitrogen and dry the samples overnight in a freeze-dryer.

11. Grind the mycelium using sterile cotton sticks for 10–20 s and re-suspend the pulverized cells in 800 μL of DNA extraction buffer.

12. Add 1 μL RNase (10 μg/μL) and incubate the suspension for 30 min at 37°C, $1,000 \times g$.

13. Add 800 μL of Phenol:chloroform:isoamyl alcohol, shake vigorously for 1 min and spin down in a bench-top centrifuge for 10 min at $10,000 \times g$.

14. Transfer 700 μL of the upper DNA-containing phase to a new 2-mL Eppendorf tube, add 700 μL of Phenol:chloroform:isoamyl alcohol, shake vigorously for 1 min and spin down for 10 min at 10,000 × g.

15. Transfer 500 μL of the upper DNA-containing phase to a new 1.5-mL Eppendorf tube (see Note 12).

16. Precipitate the DNA by adding 500 μL of isopropanol (1 volume) and 50 μL of 3 M sodium acetate (1/10 volume).

17. Incubate the mixture at room temperature for a minimum of 10 min. Longer incubations up to 30 min can improve DNA recovery.

18. Centrifuge at 10,000 × g for 15 min, discard the supernatant and wash the DNA pellet with 250 μL of 70% ethanol.

19. Re-suspend the DNA pellet in 100 μL H_2O and incubate the solution at 40–60°C for 15 min. This procedure helps to dissolve the DNA completely.

20. Analyze disruption of the *NHEJ* gene either by diagnostic PCR (using 0.1 μL of the DNA solution) or by Southern hybridization, using about 5–10 μL (~5 μg) of the DNA solution (Fig. 2, strains 1 and 2).

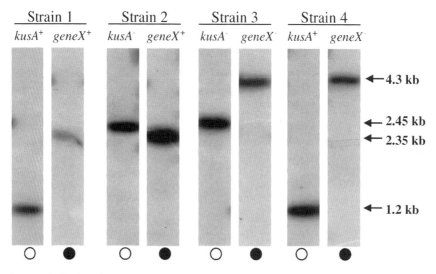

Fig. 2. Southern analysis of strains in which *kusA* has transiently been disrupted and restored after a *geneX* has been deleted. Genomic DNAs of four strains with the indicated genotype were digested with restriction enzyme I (REI, marked with *open circle*) or restriction enzyme II (REII, marked with *closed circle*). REI-restricted DNAs were hybridized with a *kusA* probe targeting the 5′ untranslated region of *kusA*. Those strains which harbour an intact *kusA* allele show a signal at 1.2 kb, whereas strains in which *kusA* has been disrupted with the *amdS* marker display a 2.45 kb signal. REII-restricted genomic DNAs were hybridized with a *geneX* probe targeting the 5′ untranslated region of *geneX*. In case of an intact *geneX* allele, a 2.35 kb band will be visible, in case *geneX* has been deleted with the *AopyrG* gene, a 4.3 kb band becomes apparent. The figure is reproduced from ref. (14) by kind permission of Springer © 2010.

3.2. Deletion of a Gene of Interest in a NHEJ⁻ Strain and Identification Whether It Is Essential or Not

In the following, we describe the targeted deletion of a hypothetical *geneX* in a NHEJ⁻ background strain as one example for a gene targeting approach. We routinely use the *A. oryzae pyrG* gene (*AopyrG*) as a selectable marker for gene deletion attempts which is flanked by 700–800 bp of 5′ and 3′ flanks of the gene of interest. The deletion cassettes can be generated by classical cloning or by fusion PCR using genomic DNA from a wild-type *A. niger* strain (N402) (14) and the *A. oryzae pyrG* containing plasmid pAO4-13 (18) as template.

1. Use a transiently silenced NHEJ⁻ strain as recipient strain (e.g. the *A. niger* strain MA169.4 displaying the genotype *kusA⁻*, *pyrG⁻*, *amdS⁺* (14)) and transform the deletion construct for *geneX* into the recipient strain by protoplast-mediated transformation (see Subheading 3.1.2).

2. Use MMS plates for selecting primary transformants (see Subheading 3.1.2). Generally, it is sufficient to select four transformants for further analysis as the rate of homologous recombination is higher than 80% in a NHEJ⁻ strain (see Note 13).

3. Purify and analyze selected transformants as described under Subheading 3.1.3 (Fig. 2, strain 3).

4. When *geneX* is an essential gene, heterokaryons containing nuclei with the genotype *geneX/pyrG⁻* and nuclei with the genotype *ΔgeneX/pyrG⁺* can be obtained at high frequency (8, 11). A first hint that *geneX* might be an essential gene can already be deduced from the growth phenotype of primary transformants on the transformation plate (Fig. 3). Heterokaryons

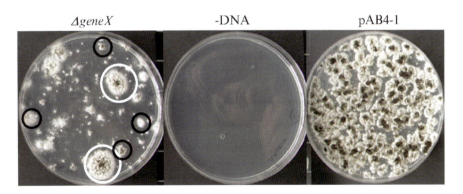

ΔgeneX   -DNA   pAB4-1

Fig. 3. Primary transformants of *A. niger* on transformation plates. A *kusA⁻* strain was transformed with a *geneX* deletion cassette, with no DNA (negative control) or with plasmid pAB4-1 containing a functional *pyrG* gene of *A. niger* (positive control; (21)). Many transformants can be detected on the positive control plate as the frequency of recombination is very high in the case of a single cross-over event. Gene *geneX* can only become successfully deleted after a double cross-over event. Hence, the transformation rate is much lower when compared to pAB4.1. Potential heterokaryons are visible (*dark circled* colonies), suggesting that *geneX* might be an essential gene. Two other colonies on the plate display a wild-type phenotype (*white circled*). Here it is likely that *geneX* has not been deleted but instead the *AopyrG* gene integrated at the *pyrG* locus, resulting in a *pyrG⁺* phenotype (*AopyrG* shares significant DNA sequence identity with the *A. niger pyrG* gene). Alternatively, the construct has been integrated randomly into the genome.

usually grow slower and more irregular compared to colonies transformed with the positive control (Fig. 3). In such a case, make sure that during purification of the transformants, not only spores but also bits of mycelium are transferred to fresh selection plates, since only the heterokaryotic mycelium is viable (see Note 14).

5. To prove that deletion of *geneX* causes a lethal phenotype, the heterokaryon rescue technique can be applied, given that the fungal species produces uninucleate spores, which breaks down the heterokaryotic state. Transfer each 100 spores of these colonies on selective and non-selective medium (e.g. for *pyrG* selection use MM agar and MM agar + 10 mM uridine). Avoid transfer of mycelium or conidiophores (e.g. filtrate spore solution through a sterile myracloth filter).

6. Incubate plates until colonies become visible (3–6 days, 25–37°C). If *geneX* is an essential gene, no growth can be observed on the selective medium plate. If *geneX* is a non-essential gene, then growth occurs on selective and non-selective plates.

3.3. Propagation of Heterokaryons

The propagation of heterokaryons to obtain sufficient biomass for genomic DNA isolation or for a next round of transformation can be challenging. When cultivated in shake flasks using CM, cells with wild-type nuclei predominate, when using MM, slow if any growth can be observed and usually mycelial pellets are formed. The cellophane membrane cultivation method (Fig. 4) circumvents these problems.

1. Put a drop of saline solution on a selective medium plate.

2. Place a cellophane membrane on top, which covers the complete plate.

3. Inoculate mycelium in the middle of the cellophane membrane using a sterile cotton stick.

4. Place a second cellophane membrane on top of the mycelium to prevent sporulation of the colony (we noticed that the

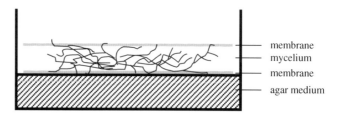

Fig. 4. Cellophane method to cultivate slow growing heterokaryons. A cellophane membrane is put on an agar plate containing selective medium. Mycelium is point-inoculated or streaked with a cotton stick onto the membrane. A second membrane is laid on *top*. During cultivation, the upper membrane prevents sporulation of the colony, whereas the lower one facilitates harvesting of the mycelium after cultivation.

presence of spores impairs DNA extraction and protoplastation efficiency).

5. Carefully remove any air bubbles between the two cellophane membranes (to avoid sporulation).

6. Incubate plates until abundant mycelial growth becomes visible (3–6 days, 25–37°C).

7. Remove the upper membrane and scrape off the mycelium from the lower membrane using a sterile cotton stick. The advantage of the lower membrane is that it prevents scrape off of agar.

3.4. Complement a Gene Deletion Strain Using AMA1-Based Vectors

pMA171 (hygromycin-based) and pMA172 (*AopyrG*-based) are autonomously replicating vectors containing either the auxotrophic selection marker *pyrG* of *A. oryzae* or the hygromycin resistance cassette as selection marker. Both markers have already been successfully used for a variety of filamentous fungi. A schematic drawing of both plasmids is given in Fig. 5a. Common to both shuttle vectors is the pBluescript backbone and the 6-kb AMA1 sequence allowing autonomous maintenance in *E. coli* and filamentous fungi, respectively. A unique rare-cutting restriction site is present in both plasmids (*Not*I) which should facilitate easy insertion of complementing genes.

1. Amplify the open reading frame of the gene of interest including flanking regions (~1,000 bp 5′ upstream and ~500 bp 3′ downstream region) from genomic DNA using PCR. Use primers which introduce *Not*I overhangs.

2. Clone the DNA fragment via *Not*I restriction into pMA171 or pMA172 and confirm DNA sequence by sequencing.

3. Transform the resulting vector into the deletion strain according to Subheading 3.1.2 using 5–10 μg circular DNA.

4. Purify and analyze transformants according to Subheading 3.1.3 (see Note 15).

5. AMA1-based complementation plasmids are rather stably maintained in *A. niger*. They become only lost after multiple rounds of cultivation under non-selective conditions. Loss of AMA1-based vectors can easily be observed by reappearance of the mutant phenotype (Fig. 5b).

3.5. Restore the NHEJ Pathway Using Counter-Selection

Both *amdS* and *pyrG* are bidirectional markers. Loss of both markers can be induced on media containing the antimetabolites fluoroacetamide (FAA) or 5′-fluoroorotic acid (FOA), respectively. Basically, mutations or loss of *amdS* (or *pyrG*) prevents the conversion of FAA (or FOA) into a toxic compound, making a fungal strain unable to grow on acetamide (auxotrophic for uridine or uracil), but resistant to FAA (or FOA). Thus *amdS*⁺ or *pyrG*⁺ strains can be cured from these markers by counter-selection.

A

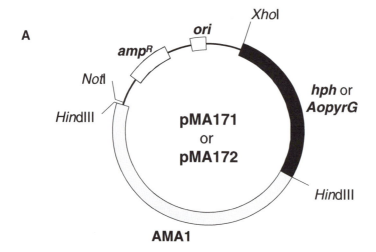

B

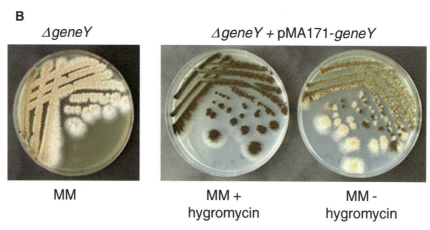

Fig. 5. (**a**) Schematic representation of the autonomously replicating plasmids containing either hygromycin (pMA171; 11,239 bp) or *AopyrG* (pAM172; 9,943 bp) as selection markers. Sites for restriction enzymes are indicated. *hph* hygromycin resistance cassette containing the *hph* gene from *E. coli* under control of the *gpdA* promoter and the *trpC* terminator of *A. nidulans*. *AopyrG A. oryzae pyrG* gene encoding an orotidine-5′-monophosphate decarboxylase flanked by its own promoter and terminator sequences. ori: origin of replication in *E. coli*; *amp^R* gene conferring ampicillin resistance in *E. coli*. Sites for restriction enzymes used for cloning are indicated. Both plasmids have been deposited at the Fungal Genetics Stock Centre. The figure is reproduced from ref. (14) by kind permission of Springer © 2010. (**b**) Complementation approach with pMA171. In this example, *geneY* important for spore color formation was deleted in *A. niger* using *pyrG* selection. The resulting strain (Δ*geneY*, *pyrG^+*, *kusA^-*) forms a fawn-colored colony (*left plate*). To prove that deletion of *geneY* is responsible for the color defect, the deletion strain was transformed with plasmid pMA171-*geneY* using hygromycin for selection. The resulting strain (Δ*geneY*, *pyrG^+*, *kusA^-*, pMA171-*geneY*, *hyg^R*) forms the typical black-colored of *A. niger* wild-type spores (*middle plate*), demonstrating that reintroduction and ectopical expression of *geneY* complements the deletion phenotype. However, pMA171-*geneY* is only stably maintained on selection plates containing hygromycin but gets lost when cultivated on non-selective medium (*right plate*, note the presence of both black and fawn colonies).

3.5.1. Curing of the amdS Marker by FAA Counter-Selection

1. To obtain *amdS^-* strains, plate 2×10^7 spores on MM-FAA agar plates.

2. Incubate plates for 1–2 weeks at 30°C.

3. Isolate FAA-resistant mutants, purify them by single streaks on MM-FAA plates.

4. Test purify strains for loss of ability to use acetamide as nitrogen source: plate strains on MM and MM-AA. *amdS* strains should not be able to grow on MM-AA.

5. Confirm loop out of the *amdS* gene and restoration of the NHEJ gene according to Subheading 3.1.3 (Fig. 2, strain 4) (see Note 16).

3.5.2. Curing of the pyrG Marker by FOA Counter-Selection

1. To obtain *pyrG⁻* strains, plate 2×10^7 spores on MM-FOA agar plates.

2. Incubate plates for 1–2 weeks at 30°C.

3. Isolate FOA-resistant mutants, purify them by single streaks on MM-FOA plates.

4. Test purify strains for uridine auxotrophy: plate strains on MM with or without 10 mM uridine. *pyrG⁻* strains should not grow on MM lacking uridine.

5. Confirm loop out of the *pyrG* gene and restoration of the NHEJ gene according to Subheading 3.1.3 (see Note 16).

4. Notes

1. The color of the 1,000× trace element solution is green when freshly made. After autoclaving, the color changes from green to purple within 2 weeks.

2. Uracil can be used instead of uridine. Uracil is less expensive, however, has a lower solubility compared to uridine. As it is not possible to make a 1 M stock solution of uracil, a 10 mM final concentration has to be made freshly by dissolving 1.12 g uracil in 1 L medium just before autoclavation.

3. Acetamide serves as nitrogen source and therefore any acetamide containing medium has to lack nitrate. In older acetamide solutions (after 3–6 months of storage), acetamide can become partially degraded. This will result in the growth of *amdS⁻* strains on MM-AA.

4. Cesium chloride is added to MM-AA to reduce background growth of *amdS⁻* strains (19).

5. Vitamin solution has to be added to MM-FOA according to (20).

6. The sensitivity towards hygromycin varies between different species. The minimal inhibitory concentration (MIC) of hygromycin can be determined when cultivating a strain of interest on MM agar plates containing serial dilutions of hygromycin. We usually apply an 8–10-fold higher MIC concentration in selective transformation medium plates and a 3–5-fold higher MIC concentration in selective purification plates.

7. For all selective media, the type and brand of agar needs to be tested. Agar might contain impurities which can inhibit growth or, alternatively, can cause background growth on selective medium (trace amounts of nitrogen are sufficient to allow background growth of an *amdS⁻* strain on MM-AA agar).

8. No stock solution can be made for FOA due to its low solubility.

9. The addition of caffeine to hygromycin selection plates strongly reduces background growth and thereby improves selectivity (C. de Bekker and H. Wösten, personal communication).

10. The yield of protoplasts depends on both the quality and quantity of the lytic enzyme batch, but also on the starting cell material. For some species (e.g., *A. nidulans*), protoplasts can be efficiently obtained when using swollen spores as starting material (8). In the case of *A. niger*, young mycelium cultivated under gently shaking usually results in highest yield of protoplasts.

11. Avoid transferring any phenol. Phenol inhibits all subsequent enzymatic reactions (PCR, restriction, etc.). When there is any doubt that phenol is still present, perform a chloroform extraction to remove traces of phenol: Add 500 μL of chloroform, mix well and spin down 10 min at $10,000 \times g$. Transfer 400 μL of the upper aqueous phase to a new 1.5-mL Eppendorf tube and continue with the DNA precipitation.

12. HR frequencies are not only dependent on the activity of NHEJ pathway, but are also strongly dependent on the gene locus. About 10% of the *A. niger* genes we have analyzed so far (>80 genes), were difficult to target—potentially because they are localized close to contig borders or localized within silenced heterochromatic DNA regions.

13. Mycelium of heterokaryons can be stored at 4°C in saline solution, or at –80°C in 50% glycerol.

14. The transformation efficiency when using a AMA1-based plasmid is usually very high (~30–100 transformants/μg DNA), as integration of a self-replicating plasmid into the genome is not required for growth.

15. Looping out of the *amdS/pyrG* gene and thereby restoration of the NHEJ gene is the most likely event which will occur. However, this needs to be confirmed by Southern blot to exclude the possibility that a mutation in the marker gene resulted in a *amdS⁻/pyrG⁻* phenotype. Such a mutation would allow growth on counter-selective medium, but does not restore the NHEJ pathway.

Acknowlegdments

We thank Charissa de Bekker and Han Wösten (Utrecht University, The Netherlands) for sharing unpublished results. Part of this work was carried out within the research program of the Kluyver Centre for Genomics of Industrial Fermentation, which is part of the Netherlands Genomics Initiative/Netherlands Organization for Scientific Research.

References

1. Ninomiya, Y., Suzuki, K., Ishii, C., and Inoue, H. (2004) Highly efficient gene replacements in *Neurospora strains* deficient for nonhomologous end-joining, Proc Natl Acad Sci USA *101*, 12248–12253.

2. Kuck, U., and Hoff, B. (2010) New tools for the genetic manipulation of filamentous fungi, Appl Microbiol Biotechnol *86*, 51–62.

3. Meyer, V. (2008) Genetic engineering of filamentous fungi - progress, obstacles and future trends, Biotechnol Adv *26*, 177–185.

4. Dudasova, Z., Dudas, A., and Chovanec, M. (2004) Non-homologous end-joining factors of *Saccharomyces cerevisiae*, FEMS Microbiol Rev *28*, 581–601.

5. Krogh, B. O., and Symington, L. S. (2004) Recombination proteins in yeast, Annu Rev Genet *38*, 233–271.

6. Shrivastav, M., De Haro, L. P., and Nickoloff, J. A. (2008) Regulation of DNA double-strand break repair pathway choice, Cell Res *18*, 134–147.

7. Nayak, T., Szewczyk, E., Oakley, C. E., Osmani, A., Ukil, L., Murray, S. L., Hynes, M. J., Osmani, S. A., and Oakley, B. R. (2006) A versatile and efficient gene-targeting system for *Aspergillus nidulans*, Genetics *172*, 1557–1566.

8. Osmani, A. H., Oakley, B. R., and Osmani, S. A. (2006) Identification and analysis of essential *Aspergillus nidulans* genes using the heterokaryon rescue technique, Nat Protoc *1*, 2517–2526.

9. Kito, H., Fujikawa, T., Moriwaki, A., Tomono, A., Izawa, M., Kamakura, T., Ohashi, M., Sato, H., Abe, K., and Nishimura, M. (2008) MgLig4, a homolog of *Neurospora crassa* Mus-53 (DNA ligase IV), is involved in, but not essential for, non-homologous end-joining events in *Magnaporthe grisea*, Fungal Genet Biol *45*, 1543–1551.

10. Malik, M., Nitiss, K. C., Enriquez-Rios, V., and Nitiss, J. L. (2006) Roles of nonhomologous end-joining pathways in surviving topoisomerase II-mediated DNA damage, Mol Cancer Ther *5*, 1405–1414.

11. Meyer, V., Arentshorst, M., El-Ghezal, A., Drews, A. C., Kooistra, R., van den Hondel, C. A., and Ram, A. F. (2007) Highly efficient gene targeting in the *Aspergillus niger kusA* mutant, J Biotechnol *128*, 770–775.

12. Snoek, I. S., van der Krogt, Z. A., Touw, H., Kerkman, R., Pronk, J. T., Bovenberg, R. A., van den Berg, M. A., and Daran, J. M. (2009) Construction of an *hdfA Penicillium chrysogenum* strain impaired in non-homologous end-joining and analysis of its potential for functional analysis studies, Fungal Genet Biol *46*, 418–426.

13. Nielsen, J. B., Nielsen, M. L., and Mortensen, U. H. (2008) Transient disruption of non-homologous end-joining facilitates targeted genome manipulations in the filamentous fungus *Aspergillus nidulans*, Fungal Genet Biol *45*, 165–170.

14. Carvalho, N. D., Arentshorst, M., Jin Kwon, M., Meyer, V., and Ram, A. F. (2010) Expanding the *ku70* toolbox for filamentous fungi: establishment of complementation vectors and recipient strains for advanced gene analyses, Appl Microbiol Biotechnol *87*, 1463–1473.

15. Michielse, C. B., Hooykaas, P. J., van den Hondel, C. A., and Ram, A. F. (2008) *Agrobacterium*-mediated transformation of the filamentous fungus *Aspergillus awamori*, Nat Protoc *3*, 1671–1678.

16. Meyer, V., Mueller, D., Strowig, T., and Stahl, U. (2003) Comparison of different transformation methods for *Aspergillus giganteus*, Curr Genet *43*, 371–377.

17. Michielse, C. B., Hooykaas, P. J., van den Hondel, C. A., and Ram, A. F. (2005) *Agrobacterium*-mediated transformation as a

tool for functional genomics in fungi, Curr Genet *48*, 1–17.

18. de Ruiter-Jacobs, Y. M., Broekhuijsen, M., Unkles, S. E., Campbell, E. I., Kinghorn, J. R., Contreras, R., Pouwels, P. H., and van den Hondel, C. A. (1989) A gene transfer system based on the homologous *pyrG* gene and efficient expression of bacterial genes in *Aspergillus oryzae*, Curr Genet *16*, 159–163.

19. Tilburn, J., Scazzocchio, C., Taylor, G. G., Zabicky-Zissman, J. H., Lockington, R. A., and Davies, R. W. (1983) Transformation by integration in *Aspergillus nidulans*, Gene *26*, 205–221.

20. Bennett, J. W., and Lasure, L. (1991) More gene manipulations in fungi, Academic Press, San Diego, pp 441–47.

21. van Hartingsveldt, W., Mattern, I. E., van Zeijl, C. M., Pouwels, P. H., and van den Hondel, C. A. (1987) Development of a homologous transformation system for *Aspergillus niger* based on the *pyrG* gene, Mol Gen Genet *206*, 71–75.

Chapter 10

Atomic Force Microscopy: A Tool for Studying Biophysical Surface Properties Underpinning Fungal Interactions with Plants and Substrates

Elizabeth Adams, Danielle Emerson, Sean Croker, Hye-Seon Kim, Shannon Modla, Seogchan Kang, and Kirk Czymmek

Abstract

One of the primary roles of the cell surface is to provide an effective barrier to various external environmental factors. Specifically, the surface properties of organisms serve as a critical obstacle to pathogen attack. Since its inception, Atomic Force Microscopy (AFM) has enabled nanoscale imaging of cell surfaces in their native state. However AFM has yet to be systematically applied toward resolving surface features and the forces underpinning plant-fungal interactions. In an effort to understand the physical forces involved at the plant-microbe interface, we describe a method for the attachment of fungal spores to AFM tips and the subsequent measurement of unbinding forces between spores with a range of substrates and plant surfaces under physiologically relevant conditions. Investigations of binding events using AFM offer an unexplored, sensitive, and quantitative method for analyzing host-pathogen/microbe-surface interactions.

Key words: AFM, *Arabidopsis thaliana*, Biofunctionalization, Fluorescent proteins, Fungi, *Fusarium oxysporum*, Plant-pathogen interactions, Silane, Unbinding forces

1. Introduction

The cell surface protects the cell and mediates interactions with various substrates, cells, other organisms as well as participates in regulating cellular and developmental activities in response to external stimuli (1–3). Elucidating the chemical and physical properties of the cell surface and how these properties change in response to various stimuli and surface property-phenotype/function relationships is essential for understanding how organisms function. Characterization of biological surfaces has been greatly aided by a diverse array of cytological and surface analysis tools

Melvin D. Bolton and Bart P.H.J. Thomma (eds.), *Plant Fungal Pathogens: Methods and Protocols*,
Methods in Molecular Biology, vol. 835, DOI 10.1007/978-1-61779-501-5_10, © Springer Science+Business Media, LLC 2012

(4–6). In this chapter, we present the application of Atomic Force Microscopy (AFM) as a powerful tool to measure the interactions of fungal spores in a targeted way with various substrates and plant surfaces.

The colonization of plant root surfaces by the soil borne fungal pathogen *Fusarium oxysporum* initiates the development of vascular wilt and root rot diseases in over 100 different plant species, including tomato, sugarcane, banana, chickpea, and potato (7–9). Several other species of *Fusarium*, including *F. graminearum* and *F. verticillioides*, are major pathogens of wheat, barley, and/or maize crops worldwide (10–12). In addition to direct crop loss, mycotoxins produced by certain *Fusarium* species pose a serious threat to human and animal health (13, 14). Because fungal spores are a main source of inoculum, understanding how spores attach to various substrates is a key step toward understanding fungal proliferation and pathogenesis. To demonstrate the utility of AFM for plant-fungal interactions, we employed this technique to examine the strength of *F. oxysporum* spore interactions with different substrates and the factors that influence their interactions. The density and location of attached fungal spores on the root surface is a critical determinant of virulence, yet the factors that contribute to spore attachment are not well understood. Spore binding events are governed by the chemical and physical properties of both the fungal cell and the binding surfaces. Plants are likely to have evolved specific surface properties that hinder pathogen attachment and/or growth as a preformed mechanism (15), and environmental factors such as hydration, ion concentrations, and charge on both the plant and fungal surfaces are likely to affect the ability of pathogen attachment to plant surfaces.

Scanning probe microscopes utilize a sharpened tip to scan across a sample and create an image of the surface. Similarly, an AFM tip with attached fungal spores (Fig. 1) or other relevant biomolecules can be scanned point-by-point and the forces between the spore on the tip and the surface can be measured and mapped. Specifically the measured forces can be used to determine surface interactions (i.e., generate force distance curves). The ability to attach various biomolecules to the tip, combined with the force sensitivity of the AFM (on the order of nanoNewtons), makes it possible to measure the strength of a single molecular interaction, a process first investigated in 1994 for measuring binding forces between streptavidin and biotin molecules (16). The AFM tip can also be functionalized by covalent attachment with a range of materials from individual proteins (17) to whole fungal cells (18). Generation of force distance curves can be achieved by bringing the tip close to the surface of interest resulting in binding events between the two surfaces. Retraction of the tip breaks these bonds, making it possible to

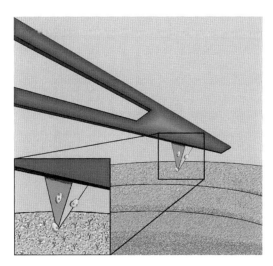

Fig. 1. Illustration of the interface between the fungal spore functionalized tip and plant surface. The AFM cantilever (dark gray) with fungal spores (yellow) attached to sharp AFM tip (light gray) in close proximity to the plant surface. Inset shows a single fungal spore at the AFM tip apex with putative surface ligands (green circles) that can bind to corresponding plant surface molecules (blue) by various molecular forces.

measure the amount of force involved and indirectly providing adhesive forces between the two surfaces. Another significant advantage of AFM is the ability to operate in liquids, which allows quantification of forces in buffers and solutions that mimic physiological conditions (19).

Force interactions in our study were measured with an AFM mounted on an inverted light microscope. This use of the combined optical/AFM microscope made it possible to precisely locate a region of interest by allowing simultaneous visualization of the tip and the target surface. In addition to measuring fungal spore interactions with host plant structures, spore interactions with several types of synthetic substrates such as glass, Teflon®, parafilm, and mica were evaluated. By studying the forces involved in spore unbinding from these materials, information was revealed about the surface properties affecting spore attachment. For example, Teflon® is extremely hydrophobic, providing a surface similar to the adaxial leaf surface. Furthermore, the AFM can be used to evaluate the strength of bacterial, fungal, or other microbe interactions with materials or surfaces of industrial, medical, and agricultural importance (e.g., sterile surfaces used in hospitals and veterinary clinics, grain and food storage containers), and this knowledge can be used for assessing the physical and environmental factors influencing contamination under various conditions.

2. Materials

2.1. Fungal Culture

1. *F. oxysporum* strain O-685 (obtained from Fusarium Research Center at Penn State).

2. Carboxymethyl Cellulose (CMC) medium (15 g CMC, 1 g Yeast Extract, 0.5 g $MgSO_4$, 1 g NH_4NO_3, and 1 g KH_2PO_4 per liter).

2.2. Plant Ecotypes

1. *Arabidopsis thaliana* ecotypes used were Cape Verde Islands (Cvi-0) and Greenville (Gre-0).

2. Arabidopsis nutrient (AN) agar: 5 mmol⁻¹.

3. KNO_3, 2.5 mmol⁻¹ KH_2PO_4, 2 mmol⁻¹ $MgSO_4$, 2 mmol⁻¹ $Ca(NO_3)_2$, 0.005% sequestrine, 70 μmol⁻¹ H_3BO_3, 14 μmol⁻¹ $MnCl_2$, 0.5 μmol⁻¹ $CuSO_4$, 1 μmol⁻¹ $ZnSO_4$, 0.2 μ⁻¹ Na_2MoO_4, 10 μmol⁻¹NaCl, 0.01 mmol⁻¹ $CoCl_2$, and 1% agar.

4. Sterilization solution: 1 mL bleach, 1 mL double distilled water.

2.3. Coating of AFM Tips

1. Piranha solution: 1 mL sulfuric acid, 2 mL hydropgen peroxide.

2. 4% 3-aminopropyldimethylethoxysilane (APDES) solution: 1.9 mL ethanol, 0.1 mL 18 MΩ ultrapure water, 0.08 mL (APDES) (Gelest Inc., Morrisvile, PA).

3. 4% 3-aminopropyltrimethylethoxysilane (3-APTES) solution: 1.9 mL ethanol, 0.1 mL 18 MΩ ultrapure water, 0.08 mL (APTES).

4. 50:50 EDC/NHS solution: 0.4 M *N*-ethyl-*N'*-(3-dimethylaminopropyl)-carbodiimide hydrochloride (EDC), 0.1 M *N*-hydroxy-succinimide (NHS), double distilled water.

5. Glutaraldehyde solution: 1.25% glutaraldehyde in distilled water.

2.4. AFM and Light Microscopes

1. Bioscope II and Nanoscope V controller (Bruker Instrument, Santa Barbara, CA, USA).

2. Zeiss Axiovert 200 M inverted light microscope (Carl Zeiss, Inc. Germany).

3. Zeiss LSM 510 VIS confocal microscope on Axiovert 200 M.

4. Silicon nitride tips (DNP-S and MLCT tips: Bruker Probes, Santa Barbara, CA, USA).

5. Vibration Isolation table.

6. SPSS version 16.0 for Windows (SPSS Inc, Chicago, Illinois, USA).

3. Methods

3.1. Growth Conditions of F. oxysporum

1. Inoculate 100 mL of CMC medium with a 1 cm diameter agar plug with mycelium from a culture of *F. oxysporum*.

2. Grow cultures on a shaker (110 rpm) at 26°C for 7 days.

3. Filter culture through several layers of cheesecloth, centrifuged at 3,000 rpm for 5 min, wash four times with double distilled water and then count using a hemocytometer.

3.2. A. thaliana Growth Conditions

1. Sterilize seeds in sterilization solution for 5 min followed by washing four times in double distilled water.

2. Place seeds in water at 4°C in the dark for 72 h.

3. Sow seeds on AN agar and incubate in a growth chamber with a constant temperature of 21°C, 13 h dark, and 11 h light cycle with a light intensity of 65 mol m^2/s.

4. Use plants for analysis between 14 and 30 days old.

3.3. Preparation of AFM Tips Prior to Spore Attachment

1. Dip the AFM tips (nominal spring constants of 0.01–0.5 N/m) several times into a fresh piranha solution in a glass petri dish.

2. Transfer the tips into a petri dish containing 5 mL 18 MΩ ultrapure water for several minutes to remove all traces of the piranha solution and dry at 50°C.

3.4. Silane Coating of AFM Tips (Amine Terminated Monolayer)

1. Silane treat AFM tips using 100 µL of 4% APDES solution or 100 µL of 4% APTES for 1 h at room temperature. The purpose of silanization was to covalently attach an amine group to the silicon nitride surface of the AFM tip.

2. Remove any weakly bound silane by transfering AFM tips into 2 mL ethanol for several minutes.

3. Dry AFM tips for 5 min at 100°C.

4. Incubate silanized AFM tips in 50 µL glutaraldehyde solution per tip containing *F. oxysporum* spores (5×10^6 spores/mL) for 1 h.

5. After incubation, wash AFM tips in 18 MΩ ultrapure water and store in ultrapure water at 4°C until use. If you are planning on using the tips within the hour, leave the tips at room temperature.

3.5. BSA and EDC/NHS Coating of AFM Tips

1. Coat silanized AFM tips with 50 µL of 1 mg/mL of BSA per tip, overnight at 4°C. The purpose of coating with BSA is to create a molecular monolayer that the fungal spores can be covalently linked to.

2. Incubate coated AFM tips in 50 µL EDC/NHS solution per tip for 1 h at room temperature, with a final spore concentration

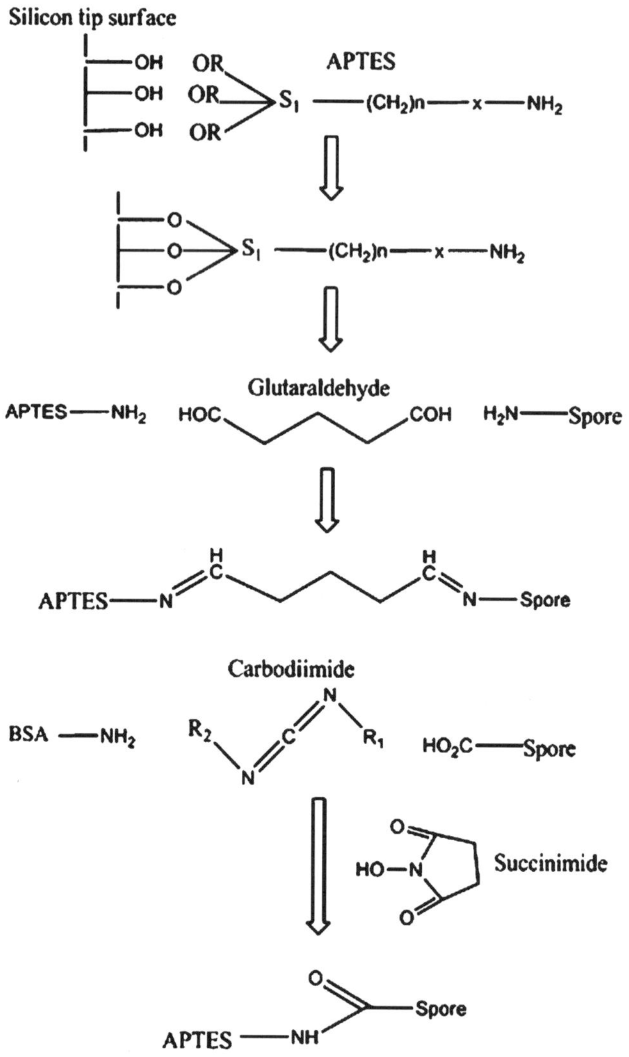

Fig. 2. Functionalization of tip surfaces. Functionalization was achieved using either (**a**) APTES or glutaraldehyde or (**b**) EDC/NHS coupling methods. The coupling chemistry used to attach the fungal spores affected the number of spores bound (see Subheading 3.6).

of 5×10^6 spores/mL. EDC and NHS act as chemical crosslinkers by covalently linking the proteins in the fungal cell wall to the BSA on the AFM tip surface.

3. After the incubation, wash the AFM tips in 18 MΩ ultrapure water and store in ultrapure water at 4°C until use.

4. Functionalize tips immediately prior to use. Chemistry underpinning both coating methods is shown in Fig. 2.

5. Fungal cells treated with EDC/NHS clump more frequently than other coating methods. Following attachment of spores, store tips in a humidity chamber (see Note 1) at 4°C.

3.6. Percentage of Cells Bound to the AFM Tips Can Be Affected by the Coupling Method

1. Test a range of spore concentrations (i.e., 1.0×10^6/mL to 1.0×10^7/mL).

2. For *F. oxysporum* the optimal concentration is 5.0×10^6/mL using APTES self-assembled monolayer (SAM) and 6×10^7/mL spores for APDES SAM (see Note 2).

3. Figure 3 illustrates the spore density on different AFM tips; Fig. 3a demonstrates the ideal density of spores while the spore density in Fig. 3b is too high.

3.7. Preparation of Artificial Surfaces

1. Wash glass, Teflon®, and parafilm with 5 mL of 10% v/w of Micro-90 (Cole-Parmer, Vernon Hills, IL). This step was performed to remove any microscopic contamination and ensure uniform cleanness of surface for reliable measurements. The surface of glass was cleaned before each experiment to maintain a consistent charge density. Glass cleaning is an important step, as it has been observed that the surface charge density of glass will vary, not only with the type of glass used, but also with the different methods used for cleaning.

2. Teflon® and parafilm are hydrophobic surfaces that are thought to resemble, to some extent, the hydrophobic nature of the waxes present on the plant leaf surface.

3. Mica and glass were used as a model for a hydrophilic surface that carries a defined charge, such as some areas present on plant surfaces. Freshly cleaved mica and glass slides can be utilized as a model for negatively charged surfaces. All carry different densities of negative charge but are structurally similar. Freshly cleave the mica before each experiment.

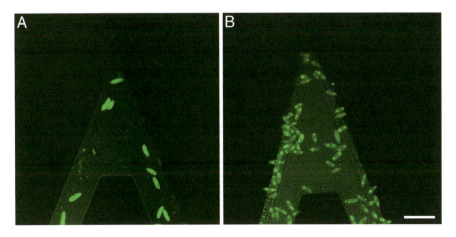

Fig. 3. Examples of an AFM tip coated with fungal spores. Spores were attached using APTES and 1.25% glutaraldehyde at two different spore densities: (**a**) is ideal while in (**b**) the spore concentration is too high. Confocal microscopy was used to acquire fluorescence from GFP labeled spores (ovoid bright green structures). Scale bar = 20 μm.

4. Wash samples with copious amounts of tap water followed by distilled water, 18 MΩ ultrapure water two propanol washes and finally 18 MΩ ultrapure water.

5. Dry samples at 60°C.

6. Grow plants for 22–25 days. The abaxial and adaxial surfaces of the plant leaf can vary in their chemical composition. For example, the adaxial and abaxial surfaces on the epicuticular layer contain 74% and 83% total wax, respectively (20).

7. Stabilize leaves by adding a very thin layer of epoxy resin to the slide prior to placement.

8. Allow epoxy resin to dry before adding 0.1 M PBS.

9. Perform measurements on both the abaxial and adaxial leaf surfaces, using the coupling chemistry described in Subheading 3.4.

10. For roots, position hydrated seedlings on a standard glass slide and target regions 2 cm from the root tip for measurements of unbinding forces.

3.8. Confocal Microscopy of AFM Tips Coated with Fungal Spores

1. Verify that spores were bound to the apex of the tip, by viewing cantilevers with a Zeiss LSM 510 VIS laser scanning microscope attached to an Axiovert 200 M using a 40× C-Apochromat (1.2 Numerical Aperture) water immersion objective lens or equivalent confocal or conventional light microscope.

2. Image AFM tips in a single-well Nalge-Nunc chambered coverslip system (Nalge-Nunc, Naper, IL).

3. Invert cantilevers and mount them on a small piece of double-sided sticky tape and cover AFM tip with deionized water to prevent spores from drying.

4. Collect images using a 488 nm laser line of a 15 mW argon laser (Coherent Enterprises, Santa Clara, CA) with a 505 nm long pass emission filter to visualize ZsGreen fluorescent protein labeled spores.

3.9. AFM Instrumentation Setup

1. Align the laser on the end of the spore coated AFM tip submerged in PBS.

2. Align the coated tip over the region of interest and engage the tip.

3. Engage in contact mode (see Note 3) and then switch to force mode (see Note 4).

3.10. AFM Measurements and Data Analysis

1. Collect the force measurements at room temperature in 0.1 M PBS using a VeecoBioScope II Atomic Force Microscope coupled to an Axiovert 200 M using PicoForce mode.

2. Collect approximately 1,000 force curves for each of the probed regions.

3. Of these, perform approximately 250 at different locations within each region, ensuring that the sampling was representative of the area investigated. This was achieved using the x and y offsets in the force mode of approximately 2 µm each offset.

4. Confirm repeatability by using three separate AFM tips for each experimental condition.

5. Review force curves to determine the presence of specific, nonspecific, no interactions, and multiple interactions (see Fig. 4 for representative curves).

6. Categorize force curves by the presence of specific features within each curve (21). Briefly, a specific interaction shows signs of polymer extension, which is presented as a curved unbinding event with a change in direction at the initial unbinding step, as the molecules are being pulled apart (Fig. 4a).

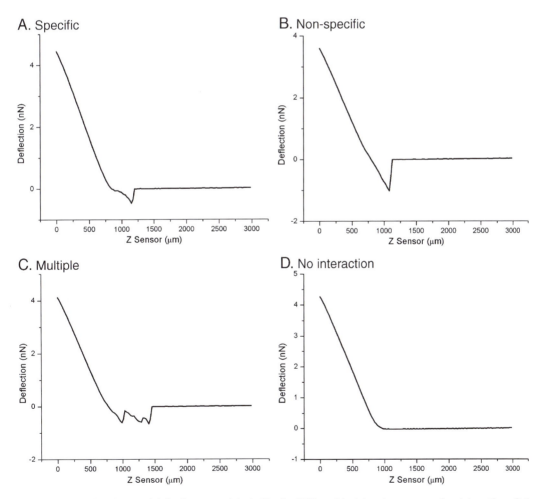

Fig. 4. Examples of each type of deflection curve detected by the AFM used to determine spore-surface interactions. Only specific single interactions were further analyzed.

7. A nonspecific interaction is characterized by the presence of an unbinding event that showed no change in the direction of unbinding (Fig. 4b). This is characteristic of an interaction driven by capillary forces. Multiple unbinding events represent several molecules being pulled apart either simultaneously or at a slight delay from each other (Fig. 4c). These were not analyzed further as their energies of individual interactions could not be separated completely. The total numbers of these different curves were analyzed to determine the distribution of unbinding events, average and standard deviation.

8. Compare several different coupling chemistries to attach the fungal spores to the AFM tips, against glass as a control substrate, to determine the most efficient system. In this case, APTES with 1.25% glutaraldehyde or BSA with 0.2 M EDC and 0.05 M NHS coupling chemistries were compared (Fig. 5).

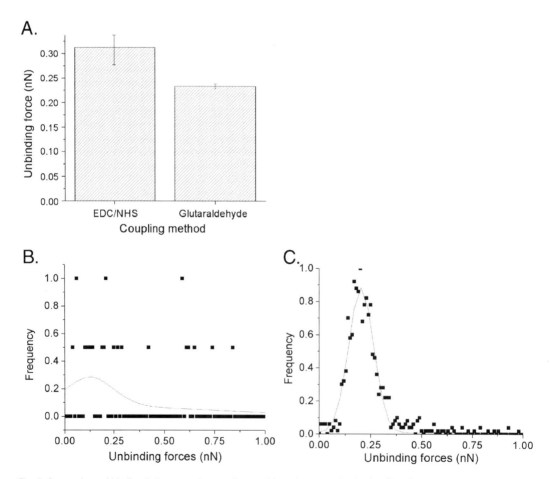

Fig. 5. Comparison of binding between a glass surface and fungal spores attached to the AFM tips using either EDC/NHS or glutaraldehyde coupling. (**a**) Average unbinding forces. (**b**) Distribution of the unbinding force curves for EDC/NHS coupling and (**c**) for glutaraldehyde coupling within each data set, demonstrating that the force was influenced by the coupling method used.

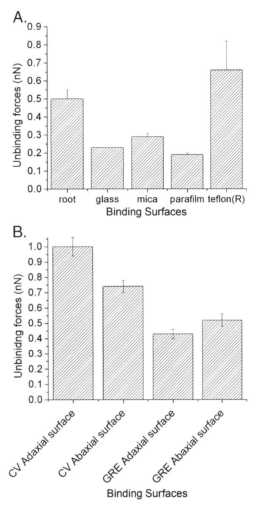

Fig. 6. Comparison of the average unbinding forces of the fungal spore functionalized AFM tips against various substrates. (**a**) Cvi-0 root comparison with control surfaces. Control surfaces were arranged in order of hydrophobicity from glass (hydrophilic) to Teflon (hydrophobic). The average unbinding forces of Teflon® were the closest to the unbinding forces of Cvi-0 adaxial leaf surface (Fig. 6b average 1.0 nN) with no significant difference between these two surfaces, while the Cvi-0 root binding forces were intermediate and averaged 0.5 nN (**b**) Unbinding forces to the abaxial and adaxial leaf surfaces of both *A. thaliana* ecotypes Gre-0 and Cvi-0 suggest chemical differences between these surfaces, potentially affecting the recognition and binding of plant fungal pathogens. Gre-0 is resistant to infection by *F. oxysporum* strain O-685, while Cvi-0 is highly susceptible.

9. Figure 6 shows the unbinding forces of the control substrates (Fig. 6a) and plant leaf surfaces (Fig. 6b). The role of surface charge and long-range electrostatic interactions can be examined by conducting force measurements in 0.1 M PBS containing 2 M NaCl, In this example, the adaxial surface showed a decrease in unbinding forces in the presence of NaCl while the abaxial surface remained unchanged (Fig. 7).

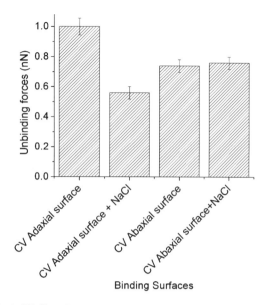

Fig. 7. The effect of NaCl on the average unbinding forces detected against different plant leaf surfaces. Differences in unbinding forces among control surfaces could be affected by charged proteins in both fungal and plant surfaces. The presence of the sodium chloride effectively masked the charged molecules on the surfaces, which yielded differences in unbinding forces when comparing the abaxial and adaxial surfaces of the leaf.

3.11. Data Analysis

1. Review all curves and select curves with a single binding event (see Fig. 4 for example).

2. Determine the forces from each selected force curve and measure using the Multiple Curve Analysis software within the AFM instrument program.

3. Export these force values into an Excel spreadsheet.

4. Determine the distribution, average and standard error measurements from these single-binding events (see Note 5). From the thousands of force curves that are collected for each data set, typically 10% will have single binding events.

5. Determine the significance between treatments by applying a student T-test of the average unbinding forces detected (21).

6. Use Levene's test for equality of variances to analyze the distribution of force values which are non-Gaussian (Alpha value of 0.05). This test will confirm if there are different types of single binding events. For example, single binding forces can be relatively weak or strong at the same location which represents different interaction types. This phenomenon will affect distribution of binding forces and typically occurs in more complicated systems where multiple components could be responsible for binding. The fungal cell wall is one such example where multiple carbohydrates can interact with different surfaces.

4. Notes

1. Humidity chamber: to create a sealed container for storing the tips, use a sealable container with a wet paper towel in the bottom. The tips can be stored in a gel pack on top of the wet paper towel under a drop of water. Usually around 50 μL per tip is enough to keep them covered until needed.

2. This step is significant primarily because the number of bound spores influences the types of forces detected (see Subheading 3.11). In addition, if the spore concentration was too high, both sides of the AFM tip tended to get coated, interfering with the amount of laser light hitting the tip and consequently the laser signal detection, resulting in the tip being unusable.

3. AFM settings are important in determining forces. Keep parameters constant between different samples, and maintain the same tip velocity. The following settings were used to engage the AFM tip on all surfaces:

   (a) Scan size 0 nm.

   (b) Scan rate 1 Hz.

   (c) Integral gain and proportional gain 0.01.

   (d) Deflection setpoint +0.5 V (this is a relative value and should be 0.5 V higher than the vertical value).

4. Once the tip is engaged, it is useful to make a note of the Z position. Using a relative trigger helps prevent the tip from applying too much pressure on the sample surface. The relative trigger determines the maximum amount of tip deflection and therefore is a measure of the amount of force applied to the sample. This value can be entered by the operator and is usually set to less than 10% of the sample approximate height. Force measurements were carried out at a constant tip velocity of 4 μm/s. The following settings can be used for PicoForce:

   (a) Ramp 1.5 μm.

   (b) Scan rate 1 Hz.

   (c) Trigger mode: Relative.

   (d) Data type: Deflection.

5. The number of curves is important to determine significant differences among samples. For example, the total number of force curves (and percentage of measured unbinding forces) was 3,096 (1.65%) for parafilm and 1,868 (35.5%) for mica. In a comparison of both synthetic hydrophobic surfaces tested, parafilm, showed significantly less unbinding forces than Teflon® with an average unbinding force of 0.19 nN.

Acknowledgments

Partial support for this work was provided by Delaware EPSCoR through the Delaware Biotechnology Institute with funds from the National Science Foundation Grant EPS-0447610 and EPS-0814251 and the State of Delaware.

References

1. Huckelhoven, R. (2007) Cell Wall-Associated Mechanisms of Disease Resistance and Susceptibility . *Ann. Rev. Phytopathol.* **45**: 101–127.

2. Kumamoto, C.A. (2008) Molecular mechanisms of mechanosensing and their roles in fungal contact sensing. *Nat. Rev. Micro.* **6**:667–673.

3. Plaine, A., Walker, L., Da Costa, G., Mora-Montes, H.M., McKinnon, A., Gow, N., Gaillardin, C., Munro, C.A., and Mathias, R.L. (2008) Functional analysis of *Candida albicans* GPI-anchored proteins: Roles in cell wall integrity and caspofungin sensitivity. *Fungal Genet. Biol.* **45**:1404–1414.

4. Heath, M.C. (2000) Advances in Imaging the Cell Biology of Plant-Microbe Interactions. *Ann. Rev. Phytopathol.* **38**:443–459.

5. Bolshakova, A.V., Kiselyova, O.I., and Yaminsky, I.V. (2004) Microbial Surfaces Investigated Using Atomic Force Microscopy. *Biotechnol. Prog.* **20**:1615–1622.

6. Czymmek, K.J., Bourett, T.M., DeZwaan, T.M., Sweigard, J.A., and Howard, R.J. (2002) Utility of cytoplasmic fluorescent proteins for live-cell imaging of *Magnaporthe grisea* in planta. *Mycologia* **94**:280–289.

7. Agrios, G.N. (2005) Plant Pathology (5th ed), Elsevier Academic Press.

8. Smith, I.M., Dunez, J., Lelliott, R.A., Phillips, D.H., and Arche S.A. eds (1988). European handbook of plant diseases. Blackwell Scientific Publications Oxford.

9. Czymmek, K.J., Fogg, M., Powel,l D.H., Sweigard, J., Park, S-Y, and Kang S. (2007) In vivo time-lapse documentation using confocal and multi-photon microscopy reveals the mechanisms of invasion into the Arabidopsis root vascular system by *Fusarium oxysporum.* *Fungal Genet. Biol.* 2007 **44**(10):1011–1023.

10. Birzele B, Meier A, Hindorf, H., Krämer, J., and Dehne, H.-W. (2002) Epidemiology of Fusarium infection and deoxynivalenol winter wheat in the Rhineland, Germany. *Eur. J. Plant Pathol.* **108**:667–673.

11. Kang, Z., and Buchenauer, H. (2000) Cytology and ultrastructure of the infection of wheat spikes by *Fusarium culmorum. Mycol. Res.* **104**:1083–1093.

12. Munkvold, G.P. (2003) Epidemiology of Fusarium diseases and their mycotoxins in maize ears. *Eur. J. Plant Pathol.* **109**:705–713.

13. Kumar, V., Basu, M.S., and T.P. Rajendran (2008). Mycotoxin research and mycoflora in some commercially important agricultural commodities. *Crop Prot.* **27**:891–905.

14. Desjardins, A.E. (2006) Fusarium Mycotoxins: Chemistry, Genetics & Biology. APS Press.

15. Berrocal-Lobo, M., and Molina, A. (2007). Arabidopsis defense response against *Fusarium oxysporum. Trends Plant Sci.* **13**:145–150.

16. Florin E-L, Rief M, Lehmann, H., Ludwig, M., Dornmair, C., Moy, V. T., and Gaub, H. E. (1994) Sensing specific molecular interactions with the atomic force microscope. *Biosens. Bioelectron.* **10**: 895–901.

17. Ebner, A., Hinterdorfer, P., and Gruber, H.J. (2007) Comparison of different aminofunctionlization strategies for attachment of single antibodies to AFM cantilevers. *Ultramicrosc.* **107**:922–927.

18. De Groot WJ, Ram AF and Klis, F.M. (2005) Features and functions of covalently linked proteins in fungal cell walls. *Fungal Genet. Biol.* **42**:657–675.

19. Willemsen, O.H., Snel, M.M.E., Cambi, A., Greve, J., De Grooth, B.G., and Figdor, C.G. (2000) Biomolecular Interactions Measured by Atomic Force Microscopy. *Biophys. J.* **79**: 3267–3281.

20. Gniwotta, F., Vogg, G., Gartmann, V. Carver, T.W.L., Riederer, M., and Jetter, R. (2005) What Do Microbes Encounter at the Plant Surface? Chemical Composition of Pea Leaf Cuticular Waxes. *Plant Physiol.* **139**:519–530.

21. Heinz, W.F. and Hoh, J.H. (1999) Spatially resolved force spectroscopy of biological surfaces using the atomic force microscope. *Tibtech* **17**:143–150.

Chapter 11

Use of the Yeast Two-Hybrid System to Identify Targets of Fungal Effectors

Shunwen Lu

Abstract

The yeast two-hybrid (Y2H) system is a binary method widely used to determine direct interactions between paired proteins. Although having certain limitations, this method has become one of the two main systemic tools (along with affinity purification/mass spectrometry) for interactome mapping in model organisms including yeast, *Arabidopsis*, and humans. It has also become the method of choice for investigating host–pathogen interactions in fungal pathosystems involving crop plants. This chapter describes general procedures to use the GAL4-based Y2H system for identification of host proteins that directly interact with proteinaceous fungal effectors, thus being their potential targets. The procedures described include cDNA library construction through in vivo recombination, library screening by yeast mating and cotransformation, as well as methods to analyze positive clones obtained from library screening. These procedures can also be adapted to confirmation of suspected interactions between characterized host and pathogen proteins or determination of interacting domains in partner proteins.

Key words: Fungal pathogens, Plant disease, Protein–protein interactions, Yeast transformation, Synthetic dropout media, Reporter genes, Autoactivation, Aureobasidin A, α-Galactosidase assay

1. Introduction

The yeast two-hybrid (Y2H) system was first pioneered by Fields and Song in 1989 (1) and has been since widely used to study protein–protein interactions in various binary systems (2–7) and, more recently, to establish interactomes in model organisms including yeast, *Arabidopsis,* and humans (8–11). This system was originally developed based on the intriguing properties of the GAL4 protein of the baker's yeast *Saccharomyces cerevisiae*. GAL4 is a master transcription activator controlling expression of several *GAL* genes encoding enzymes involved in galactose utilization (12).

Melvin D. Bolton and Bart P.H.J. Thomma (eds.), *Plant Fungal Pathogens: Methods and Protocols,*
Methods in Molecular Biology, vol. 835, DOI 10.1007/978-1-61779-501-5_11, © Springer Science+Business Media, LLC 2012

GAL4 consists of two separable and functionally essential domains: an N-terminal DNA-binding domain (BD) which binds to specific DNA sequences called upstream-activating sequences (UAS), and a C-terminal activation domain (AD) which interacts with basal transcriptional machinery (Fig. 1). In Y2H assays, the two testing proteins (the "bait" and the "prey") are fused in frame with the BD and AD domains, respectively, and coexpressed in a yeast strain which carries two or more reporter genes. A physical interaction between the bait and prey proteins would reconstitute proximity of the native GAL4 domains, thus activating the reporter genes (Fig. 1). Apart from testing the interaction between two known proteins as originally designed (1), Y2H can be also used for library screening to identify the interacting partner(s) of a characterized protein (the "bait") without prior knowledge of the nature of the "prey." This application has been adapted to identify novel targets of many functionally characterized proteins (8–11, 13) including proteinaceous fungal effectors involved in plant pathogenesis (14, 15).

Proteinaceous fungal effectors, e.g., avirulence (Avr) proteins and certain host-selective toxins (HSTs), are of diverse proteins (usually small and cysteine-rich) secreted by the fungal pathogen and recognized by the host plant, in particular host–pathogen interactions (16–18). Although specific interactions between the characterized fungal effectors and the corresponding host resistance (or susceptibility) genes may have been well characterized, the cognate host receptors or targets of most fungal effectors are still unknown. It is often difficult to assign candidate interacting partners for fungal effector proteins simply by exploring available interactomes or comparative proteomics databases because most fungal effectors have no apparent homologies to any known proteins

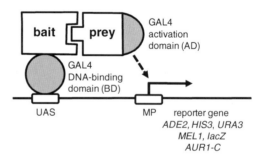

Fig. 1. Schematic diagram showing the principle of the GAL4-based yeast two-hybrid system (adapted from Fields and Song (1)). A physical interaction between the bait and prey fusion proteins brings the GAL4 domains into close proximity and activates the reporter gene. *UAS* upstream-activating sequence; *MP* minimal promoter. Enzyme activities encoded by the reporter genes: Nutritional: *ADE2* phosphoribosylaminoimidazole carboxylase (adenine biosynthesis); *HIS3* imidazoleglycerol-phosphate dehydratase (histidine biosynthesis); *URA3* orotidine-5′-phosphate decarboxylase (uracil biosynthesis). Colormetric: *MEL1* secreted α-galactosidase; *lacZ* intracellular β-galactosidase. Antibiotic resistance: *AUR-1C* mutated phosphatidylinositol: ceramide phosphoinositol transferase (conferring resistance to aureobasidin A). The specific reporter genes in related yeast strains are indicated in Table 1.

(17, 18). ToxA, the major proteinaceous HST produced by the wheat tan spot fungus *Pyrenophora tritici-repentis* (19, 20) and the leaf/glume blotch fungus *Stagonospora nodorum* (21), has been shown to interact with two chloroplast proteins, ToxABP1 (a homologue of *Arabidopsis* THYLAKOID FORMATION 1 protein) (14) and PCN (a plastocyanin) (15). However, the host specificity-determining target(s) of ToxA is still unknown because in both ToxA-sensitive and -insensitive wheat lines, *ToxABP1* and *PCN* are found to be identical (100% nucleotide identity) and expressed in the same patterns (14, 15). Furthermore, ToxA does not interact directly with *Tsn1*, a disease resistance-like gene which governs the ToxA-triggered susceptibility in wheat (22). These findings emphasize the complexity of effector–host interactions and the need for an efficient method to unravel the interacting networks.

This chapter is intended to describe general experimental procedures to use the GAL4-based Y2H system for identification of host plant proteins that directly interact with proteinaceous fungal effectors, thus being their potential targets. These procedures include generation of the bait construct and expression strain, cDNA library construction through in vivo recombination, library screening by yeast mating and cotransformation, as well as analysis of the library-derived positive clones. The methods described have been used successfully to identify two pathogenesis-related proteins that are differentially expressed in ToxA-sensitive and -insensitive wheat lines and interact directly with ToxA (Lu, S., Friesen, T.L., Faris, J.D., unpublished). In addition, the materials and methods used for library screening can also be adapted for confirmation of suspected interactions between characterized host and pathogen proteins or determination of interacting domains in identified protein partners. This chapter is not intended to give a comprehensive review on the rapidly evolving Y2H systems and the related protocols published during the last decades, but some key references are cited in related sections. A brief introduction on biochemical methods that can be used for further validation of physical interactions between paired proteins is given in the last section of the chapter.

2. Materials

Plasmids and yeast strains are described using examples from the MatchMaker Y2H system (Table 1).

2.1. Plasmids and Yeast Strains

2.1.1. Plasmids

Plasmids used in the Y2H system act as shuttle vectors that can be maintained in *E. coli* and in *S. cerevisiae*. pGBKT7 (Fig. 2a) and pGADT7 (Fig. 2b) are the two common vectors for construction of the bait and the prey fusion proteins, respectively. pGADT7-Rec (Fig. 2b) is a prey library vector which contains a multiple cloning

Table 1
GAL4-based yeast two-hybrid library screening systems[a]

Company	Y2H system[b] features	GAL4-BD plasmid[c] selection marker	GAL4-AD plasmid[d] selection marker	Yeast strain reporter genes
Clontech	Matchmaker SMART technology Allows library construction directly in yeast Antibiotic selection Epitope tags T7 promoter for in vitro transcription	pGBKT7 *TRP1* pGBKT7-p53 *TRP1* (control) pGBKT7-Lam *TRP1* (control)	pGADT7 *LEU2* pGADT7-Rec *LEU2* pGADT7-RecT *LEU2* (control)	Y2HGold *ADE2, HIS3, MEL1, AUR1-C* AH109 *ADE2, HIS3, MEL1, lacZ* Y187 *MEL1, lacZ*
Agilent Technologies (Stratagene)	HybriZAP Lambda vector to generate high-efficiency libraries and increased representation	pBD-GAL4-Cam *TRP1*	pAD-GAL4-2.1 *LEU2*	YRG-2 *HIS3, lacZ*
Life Technologies (Invitrogen)	ProQuest Low copy number vectors attR sites for Gateway cloning Allows both positive and negative selection	pDEST32 *LEU2*	pDEST22 *TRP1* pEXP-AD502 *TRP1*	MaV203 *HIS, URA3, lacZ*

[a]This table is given to show the availability of major GAL4-based Y2H systems only and is not intended to recommend a specific commercial source

[b]Detailed product information can be found at: Matchmaker, http://www.clontech.com; HybriZAP, http://www.genomics.agilent.com; ProQuest, https://www.invitrogen.com

[c]pGBKT7, the bait cloning vector (Fig. 2a); pGBKT7-p53 and pGBKT7-Lam, the controls plasmids (Clontech) that express the tumor suppressor protein 53 and lamin protein, respectively

[d]pGADT7, the prey cloning vector (Fig. 2b); pGADT7-Rec, the prey library vector (Fig. 2b); pGADT7-RecT, the control plasmid (Clontech) which expresses the SV40 virus large T-antigen known to interact with p53 but not with Lam

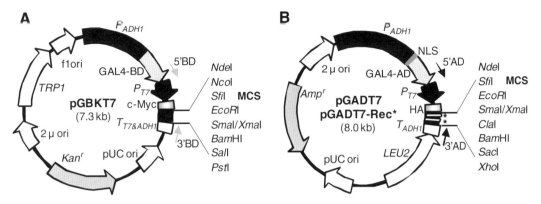

Fig. 2. Main features of pGBKT7 (**a**), pGADT7, and pGADT7-Rec (**b**) (adapted from www.Clontech.com). pGADT7 and pGADT7-Rec are the same except that the later has the SMART III and CDS III oligonucleotide sequences (*asterisks*) which are incorporated at the MCS for in vivo recombination (23). P_{ADH1} *S. cerevisiae ADH1* promoter; T_{ADH1} *Saccharomyces cerevisiae ADH1* terminator; P_{T7} T7 RNA polymerase promoter; NLS SV40 nuclear localization signal; HA HA epitope tag; *c-Myc* c-Myc epitope tag; *Amp^r* Ampicillin resistance gene; *Kan^r* kanamycin resistance gene; *5′AD and 3′AD* pGADT7-specific primers; *5′BD and 3′BD* pGBKT7-specific primers.

site (MCS) flanked by short stretches of sequence homologous to the SMART (Switching Mechanism at 5′ end of RNA Transcript) oligonuceotides that are designed for in vivo recombination (23). Control plasmids pGBKT7-p53, pGBKT7-Lam, and pGADT7-RecT are also listed in Table 1.

2.1.2. Yeast Strains

1. AH109, the recipient yeast strain for expression or coexpression of the fusion proteins: *MATa, trp1-901, leu2-3,112, ura3-52, his3-200, gal4Δ, gal80Δ, LYS2 :: GAL1$_{UAS}$-GAL1$_{TATA}$-HIS3, GAL2$_{UAS}$-GAL2$_{TATA}$-ADE2, URA3 :: MEL1$_{UAS}$-MEL1$_{TATA}$-lacZ, MEL1* (see Note 1).

2. Y2HGold, the alternative recipient strain for expression or coexpression of the fusion proteins: *MATa, trp1-901, leu2-3, 112, ura3-52, his3-200, gal4Δ, gal80Δ, LYS2 : : GAL1$_{UAS}$-Gal1$_{TATA}$-HIS3, GAL2$_{UAS}$-Gal2$_{TATA}$-ADE2, URA3 : : MEL1$_{UAS}$-Mel1$_{TATA}$-AUR1-C, MEL1* (see Note 2).

3. Y187, the library strain and a mating partner with AH109 or Y2HGold: *MATα, leu2-3,112, ura3-52, trp1-901, his3-200, ade2-101, gal4Δ, gal80Δ, met–, URA3 :: GAL1$_{UAS}$-GAL1$_{TATA}$-lacZ, MEL1*(see Note 3).

2.2. Buffers and Solutions

1. 10 N NaOH solution. Dissolve 40 g sodium hydroxide pellets in 80 mL of ddH$_2$O. Bring volume to 100 mL with ddH$_2$O. Do not autoclave.

2. 10× TE buffer: 0.1 M Tris–HCl, pH 7.5, 10 mM EDTA. Adjust pH to 7.5 and autoclave.

3. 10× LiAc: 1 M lithium acetate. Adjust pH to 7.5 with dilute acetic acid and autoclave.

4. 50% PEG: Resuspend 25 g polyethylene glycol (MW = 3,350) in ~20 mL ddH$_2$O. Bring volume up to 50 mL with ddH$_2$O and sterilize through a 0.45-μm syringe filter (see Note 4).

5. Fish sperm DNA (single-stranded DNA fragments): 10 mg/mL in 1× TE buffer, denatured by boiling for 10 min and immediately placed on ice. Divide to aliquots in 1.5-mL microcentrifuge tubes and store at –20°C.

6. Potassium phosphate solution (67 mM): Dissolve 0.91 g KH$_2$PO$_4$ in ~80 mL of ddH$_2$O. Adjust pH to 7.5, bring volume up to 100 mL with ddH$_2$O, sterilize through a 0.22-μm syringe filter, and store at 4°C for up to 1 year.

7. Lyticase stock solution: 5 units/μL lyticase in 1× TE buffer. Store at –20°C.

8. Z-buffer: Dissolve 16.1 g Na$_2$HPO$_4$·7H$_2$O, 5.5 g NaH$_2$PO$_4$·H$_2$O, 0.75 g KCl, 0.246 g MgSO$_4$·7H$_2$O in ~800 mL of ddH$_2$O. Adjust pH to 7.0, autoclave and store at room temperature for up to 1 year.

9. X-gal and X-α-gal stock solutions: Dissolve X-gal (5-bromo-4-chloro-3-indolyl-β-D-galactopyranoside) or X-α-gal (5-bromo-4-chloro-3-indolyl-α-D-galactopyranoside) in N,N dimethyl-formamide (DMF) at a final concentration of 20 mg/mL. Divide to aliquots in 1.5-mL microcentrifuge tubes and store in the dark at –20°C.

10. Aureobasidin A (AbA) stock solution: 0.1 mg/mL, dissolved in 100% ethanol, stored at 4°C.

11. Z-buffer/X-gal solution: 100 mL of Z-buffer, 0.27 mL of β-mercaptoethanol, 1.67 mL of X-gal stock solution (20 mg/mL). Mix well just before use.

12. 50% glycerol (autoclaved).

13. 100% DMSO (Dimethyl sulfoxide).

14. 3 M sodium acetate (pH 4.8).

15. 95% ethanol (chilled at –20°C).

16. 0.9% NaCl solution.

2.3. Yeast Media

1. YPD and YPDA media (for 1 L): Dissolve 10 g yeast-extract, 20 g bacto-peptone, 0.1 g adenine hemisulfate (for YPDA only) in ~800 mL of ddH$_2$O. Adjust pH to 5.6–5.8 with 1 N NaOH. Add 20 g agar (for plates only) and bring volume up to 950 mL with ddH$_2$O. Autoclave at 121°C for 15 min. Add 50 mL of 40% glucose after cooling down to 55°C in a water bath and mix well before pouring the plates (see Note 5).

2. Synthetic dropout (SD) medium (for 1 L): Dissolve 1.7 g yeast nitrogen base without amino acids and ammonium sulfate 5.0 g ammonium sulfate, 20 g D-Glucose, 0.60–0.69 g

ready-to-use dropout (DO) supplement (see Note 6), or 100 mL of appropriate 10× DO stock solution (see below) in ~800 mL ddH$_2$O. Adjust pH to 5.6–5.8 with 1 N NaOH. Add 20 g agar (for plates only) and bring volume to 1 L with ddH$_2$O. Autoclave at 121°C for 15 min.

3. 10× DO stock solution (for 1 L): Dissolve 200 mg L-Adenine hemisulfate salt, 200 mg L-Arginine HCl, 200 mg L-Histidine HCl monohydrate, 1,000 mg L-Leucine, 300 mg L-Lysine HCl, 200 mg L-Methionine, 500 mg L-Phenylalanine, 2,000 mg L-Threonine, 200 mg L-Tryptophan, 300 mg L-Tyrosine, 200 mg L-Uracil, and 1,500 mg L-Valine in 1 L ddH$_2$O (omit one or more amino acids as needed to make a specific 10× DO solution for desired SD medium, e.g., add all except for L-Leucine and L-Tryptophan to make a 10× –Leu/–Trp DO solution). Autoclave at 121°C for 15 min and store at 4°C for up to 1 year.

4. SD + X-α-gal plates: Prepare appropriate SD agar plates as described above, spread 50 μL of X-α-gal stock solution onto the surface of the solidified agar plates (100-mm) just before use. Alternatively, add 1 mL of X-α-gal stock solution to 1 L of SD agar medium after autoclave (cooled down to 55°C) before pouring the plates.

5. SD-aureobasidin A (AbA) plates: Prepare appropriate SD agar plates as described above and add 1 mL of AbA stock solution to 1 L of SD agar medium after autoclave (cooled down to 55°C) before pouring the plates (see Note 7).

Optional: Kanamycin (50–100 μg/mL) can be added to all yeast media to avoid bacterial contamination.

3. Methods

3.1. Growth and Maintenance of Yeast Strains

3.1.1. General Growth Conditions

Most strains of *S. cerevisiae* grow optimally at 30°C. On nonselective YPD or YPDA agar plates, single cells become visible colonies usually 2–3 days after plating. On selective SD media, colonies may appear 2–5 days after plating depending on the stringency of the selection and the yeast strains (see Note 8). For liquid cultures starting with a single colony (2–3 mm in diameter) with shaking at 150–200 rpm, 24–36 h of incubation is sufficient to reach stationary growth stage.

3.1.2. Storage of Yeast Strains

For a short-term storage, yeast cells can be maintained on agar plates (tightly wrapped with Parafilm) and kept at 4°C for 1–2 months without losing viability (see Note 9). For long-term storage, yeast cells can be suspended in appropriate liquid media with 25% glycerol and kept in cryogenic vials at –80°C (see Note 10). To recover a strain, streak a small amount of cells onto an YPD or YPDA plate and incubate at 30°C for 2–3 days. Before starting with transformation or mating, verify growth phenotypes of the yeast strain by streaking the cells onto appropriate SD agar plates and allow growth at 30°C for 2–3 days.

3.2. Generation of the Bait Construct and Expression Strain

3.2.1. Construction of the Bait Plasmid

1. Amplify the gene of your interest (*GYI*) encoding the fungal effector protein (not including the N-terminal signal peptide) by PCR using primers that contain 5′-end restriction sequences compatible with the MCS on the bait vector pGBKT7 (Fig. 2a). Be sure that the *GYI* sequence to be cloned does not contain the restriction sites that have been incorporated at the 5′-ends of the PCR primers and will be translated in frame with the GAL4-BD domain.

2. Purify the amplified PCR product by phenol/chloroform extraction/ethanol precipitation or column purification.

3. Digest 1.0–3.0 μg of the purified PCR product using the restriction enzymes incorporated into the primer sequences under the conditions described by the enzyme suppliers (inactivate the restriction enzymes by heating or column purification after digestion).

4. Ligate the digested PCR product into pGBKT7 that has been linearized with compatible restriction enzymes using a T4 DNA ligase under required conditions.

5. Transform an *E. coli* strain with the ligated recombinant DNA (select for *Kan^R*) and isolate the bait plasmid (pGBKT7-GYI) by column purification.

6. Confirm the sequence identity and the correct GAL4-BD-GYI fusion by DNA sequencing using pGBKT7-specific primers 5′BD (5′-TCATCGGAAGAGAGTAGT-3′) and 3′BD (5′-AGAGT-CACTTTAAAATTTGTAT-3′) (Fig. 2a).

3.2.2. Preparation of Yeast Competent Cells

1. Inoculate 5 mL of YPDA medium in a 50-mL centrifuge tube with a single colony (2–3 mm in diameter) of the yeast strain (AH109 or Y2HGold). Incubate overnight at 30°C with shaking (200–250 rpm).

2. Inoculate 50 mL of YPDA in a 200-mL flask with 0.5 mL of the overnight culture from **step 1**. Incubate at 30°C with shaking (200–250 rpm) for 3–4 h until the $OD_{600} = 0.4$–0.6.

3. Spin down the cells in a 50-mL tube using a GSA rotor (or equivalent) at $1,000 \times g$ for 5 min at room temperature. Discard the supernatants and resuspend the cell pellet in sterile ddH$_2$O.

4. Spin down the cells and resuspend the cell pellet in 1 mL of TE/LiAc solution (freshly prepared by mixing 5 mL of 10× TE and 5 mL of 10× LiAc with 40 mL of ddH$_2$O). Use competent cells immediately to obtain best transformation efficiency (see Note 11).

3.2.3. Yeast Transformation   Both AH109 and Y2HGold can be used for expression of the bait fusion construct or coexpression of the bait and the prey constructs. Include aureobasidin A (AbA) in SD/-LTHA and SD/-LTHA + X-α-gal plates if Y2HGold is used (see Note 7).

1. Combine 0.1 μg each of the bait and the prey plasmid DNA (pGADT7 or pGADT7-RecT, Table 1) in a 1.5-mL microcentrifuge tube for each transformation experiment as listed below:

   (a) pGBKT7-GYI only

   (b) pGBKT7-GYI + pGADT7

   (c) pGBKT7-p53 + pGADT7-RecT (positive control)

   (d) pGBKT7-Lam + pGADT7-RecT (negative control)

   (e) No DNA (blank control)

2. Add 0.1 mg of the denatured single-strand fish sperm DNA.

3. Add 0.1 mL of competent cells to each tube and mix well.

4. Add 0.6 mL of PEG/LiAc solution (freshly prepared by mixing 1 mL of 10× TE and 1 mL of 10× LiAc with 8 mL of 50% PEG) to each tube and vortex at high speed for at least 10 s (see Note 12).

5. Incubate at 30°C for 30 min (invert tube every 10 min).

6. Add 20 μL of DMSO. Mix well by gentle inversion. Do not vortex.

7. Incubate the tube in a 42°C water bath for 15–20 min.

8. Spin down cells at 13,000 rpm in an Eppendorf centrifuge at room temperature for 10 s and discard the supernatant.

9. Resuspend the cell pellet in 0.5 mL of sterile 1× TE buffer.

10. Spread 100 μL of the transformed cells onto SD/−Trp, SD/−Leu–Trp (-LT), and SD/−Leu–Trp–His–Ade (-LTHA) agar plates separately. Incubate the plates (face down) at 30°C for 2–3 days.

11. Compare results of the five transformations. The yeast transformants expressing the bait fusion construct alone (Expt. 1) should grow only on the SD/−Trp plate. The yeast transformants coexpressing the bait construct and the pGADT7 vector (Expt. 2) should grow on the SD/-LT plate but not on the SD/-LTHA plate (Fig. 3a), like the negative control (Fig. 3c). Any growth on SD/-LTHA plate in Expt. 2 will suggest that the bait fusion protein has "activated" the reporter genes on its own ("autoactivation," see Note 13). Also, make sure that the colonies on the SD/-LT plate (Fig. 3a, left) in Expt. 2 are similar in size to those of the controls (Fig. 3b, c, left). If the colonies are significantly smaller, the expressed bait protein is likely "toxic" to the yeast, thus not suitable for further analysis (see Note 14).

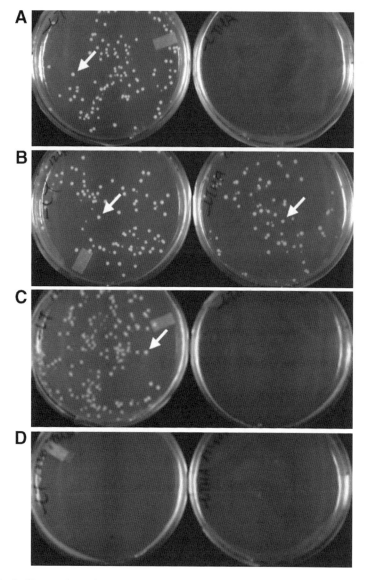

Fig. 3. SD agar plates showing the results of transformation of AH109 with a bait fusion construct and control plasmids. Plates were inoculated with 100 μL of the transformed cells and incubated at 30°C for 3 days. Transformants coexpressing a bait and prey plasmids (**a–c**, corresponding to Expts. 2–4 in Subheading 3.2.3) all grew on SD/-LT plates (*left*, *arrow* indicates a single colony), but only those expressing the positive control constructs (**b**) were able to grow on SD/-LTHA plate (*right*). All transformants in Expts. 1–4 grew equally well on SD/–Trp plates which select for *TRP1* only (not shown). The blank control did not produce any colonies as shown in (**d**).

12. If no apparent "autoactivation" or toxicity is associated with the bait construct, pick up a single colony from the SD/–Trp plate in Expt. 1 and streak onto fresh SD/–Trp plate. Incubate at 30°C for 2–3 days. Use the yeast cells from the streaked colonies to make a glycerol stock and store at –80°C (see Note 10).

13. Perform a western blot analysis to verify the expression of the fusion protein following standard procedures (see Note 15).

3.3. Construction of a cDNA Library

This section describes procedures for the construction of a "non-normalized" cDNA library. Key materials used for first-strand cDNA synthesis and double-stranded cDNA amplification are components of the "Mate & Plate" Library System (Clontech). A "normalized" cDNA library may be required if the target genes are known to be expressed at a low level. The construction of such a library involves special procedures that are not within the scope of this chapter. Existing protocols (24) may be followed if applicable.

3.3.1. Isolation of RNA from Plant Tissues

1. Grow the host plant in a greenhouse or a growth chamber under the conditions appropriate for the fungal pathosystem to be tested.

2. Inoculate the plant with an effector-producing fungal isolate or infiltrate the plant with fungal culture filtrates that contain the secreted effector protein, or with the purified effector protein (if available). Incubate the inoculated plants for a desired period of time.

3. Collect plant materials at several time points, e.g., 4, 12, 24, and 48 h after inoculation or infiltration (be sure to include the point where the induced symptoms become visible). Freeze the collected materials immediately in liquid nitrogen.

4. Extract mRNA or total RNA using standard protocols or an appropriate commercial kit (see Note 16). Keep the isolated RNA on ice and proceed to first-strand cDNA synthesis immediately.

3.3.2. First-Strand cDNA Synthesis

1. Set up the reaction in a 0.2-mL PCR tube as follows:

   RNA (Containing 0.025–1.0 μg mRNA or 0.1–2.0 μg total RNA): 1–3 μL

   CDS III oligo(dT) primer: 1 μL

   ddH$_2$O: 0–2 μL

   Total volume: 4 μL

2. Incubate at 72°C for 2 min, then immediately on ice for ≥2 min.

3. Add the following to the reaction tube and then incubate at 42°C for 10 min:

   5× First-Strand Buffer: 2.0 μL

   20 mM DTT: 1.0 μL

   10 mM dNTP Mix: 1.0 μL

   MMLV Reverse Transcriptase: 1.0 μL

4. Add 1.0 μL of SMART III oligonucleotide to the reaction mixture and incubate at 42°C for 1 h in a PCR thermalcycler (see Note 17).

5. Add 1.0 μL of RNase H and incubate at 37°C for 20 min. The synthesized first-strand cDNA can be used immediately for PCR amplification (below) or stored at –20°C for later use.

3.3.3. Amplification of Double-Stranded cDNA

1. Set up two reactions in 0.2-mL PCR tubes, each containing as follows:

   First-strand cDNA (see Subheading 3.3.2): 2 μL

   10× Advantage 2 PCR Buffer: 10 μL

   50× Advantage 2 Polymerase Mix: 2 μL

   50× dNTP Mix: 2 μL

   5′ PCR Primer: 2 μL

   3′ PCR Primer: 2 μL

   10× GC-Melt Solution: 10 μL

   ddH$_2$O: 70 μL

   Total volume: 100 μL

2. Run PCR at the following conditions: 95°C, 30 s; 30 cycles of 95°C, 10 s, 68°C, 6 min; 68°C, 5 min.

3. Run 7 μL of the PCR product from each sample on a 1.0–1.5% agarose gel. A smear of range 0.3–6.0 kb should be visible (Fig. 4, lanes 1 and 3).

3.3.4. Purification of Double-Stranded cDNA

1. Prepare two CHROMA SPIN™ TE-400 Columns (Clontech) following the manufacturer's instructions.

2. Carefully load the double-stranded (ds) cDNA sample from above to the center of the gel bed's flat surface in each column prepared in step 1 and centrifuge at $700 \times g$ for 5 min.

3. Combine the flow-through samples (containing ds cDNA) into a 1.5-mL microcentrifuge tube.

4. Add 1/10 volume of 3 M sodium acetate (pH 4.8) and 2.5 volume of 95% ethanol (prechilled at –20°C). Mix well and incubate at –20°C for 1 h to overnight.

5. Centrifuge at 13,000 rpm for 20 min at room temperature. Remove the supernatant and allow the pellet in the tube to air-dry for 10 min.

6. Resuspend the pellet in 25 μL of dd H$_2$O.

7. Run 2 μL of the purified ds cDNA on a 1.0–1.5% agarose gel. The intensity of the DNA smear (Fig. 4, lanes 2 and 4) should be similar to that of the unpurified ds DNA (lanes 1 and 3). The total amount of the purified ds cDNA should be 2.0–5.0 μg (see Note 18).

8. Proceed to in vivo recombination with pGADT7-Rec as described below or store at –20°C for later use.

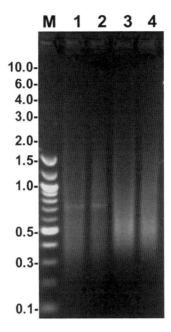

Fig. 4. Agarose gel images showing double-stranded (ds) cDNA (ranging 0.3–6 kb) generated using the procedures described in Subheading 3.3. The first-strand cDNA was synthesized from ~0.1 μg of mRNA isolated from 2-week-old wheat leaves at 12 h (lanes 1 and 2) and 48 h (lanes 3 and 4) after infiltration with a fungal effector protein. Lanes 1 and 3: unpurified ds cDNA from Subheading 3.3.3 (see step 3); lanes 2 and 4: purified ds cDNA from Subheading 3.3.4 (see step 7). M: 100 bp DNA ladder; size positions relative to 2.0–10.0 kb are indicated based on 1 kb DNA ladder in the same gel (not shown).

3.3.5. In Vivo Library Construction in Yeast

1. Prepare competent cells of the yeast strain Y187 as described in Subheading 3.2.2.

2. Combine the following in a sterile 15-mL tube:

   ds cDNA (from Subheading 3.3.4) (Include equal amounts of ds cDNA samples made from different time points): 20 μL

   pGADT7-Rec (0.5 μg/μL): 6 μL

   Single-stranded fish sperm DNA (denatured): 20 μL

3. Add 600 μL of competent cells to the tube and mix well.

4. Add 2.5 mL of PEG/LiAc Solution and mix by vortexing.

5. Incubate at 30°C for 30 min (invert tube every 10 min).

6. Add 160 μL of DMSO and mix well, and then incubate the tube in a 42°C water bath for 20 min.

7. Centrifuge at $1,000 \times g$ for 5 min and discard the supernatant.

8. Resuspend the cell pellet in 3 mL of YPD liquid medium and incubate at 30°C with slow shaking (<200 rpm) for 60–90 min.

9. Spin down the cells at $1,000 \times g$ for 5 min and discard the supernatant.

10. Resuspend in 30 mL of NaCl solution (0.9%) in a 50-mL tube.

11. Spread 300 µL of the cell suspension on SD/−Leu agar plate (150-mm, ~100 plates are needed) (see Note 19).

12. Incubate plates (upside down) at 30°C for 3–5 days.

13. Harvest cells from each plate by adding 5 mL of SD/−Leu liquid medium followed by dislodging with sterile glass spreaders. Combine all cells in a sterile flask, spin down at $1,000 \times g$ for 5 min, and adjust density to 2×10^7 with freezing medium (SD/−Leu in 25% glycerol).

14. Make prey library stocks by distributing the cells into 2-mL cryogenic vials (1 mL each) and store at −80°C for up to 1 year.

3.4. Library Screening

3.4.1. Library Screening by Yeast Mating

1. Recover the bait expression strain (see Subheading 3.2.3, step 12) from glycerol stock on SD/−Trp plate at 30°C for 2–3 days.

2. Inoculate 50 mL of SD/−Trp liquid medium with a single colony of the bait strain and incubate at 30°C with shaking at 250 rpm until the culture has a $OD_{600} \geq 0.8$ (usually, an overnight incubation is sufficient).

3. Spin down the cells at $1,000 \times g$ for 5 min at room temperature and resuspend in 5 mL of SD/−Trp liquid medium.

4. Thaw a 1-mL aliquot of the Y187 library cells (see Subheading 3.3.5, step 14) in a room temperature water bath.

5. Combine the 5-mL bait culture and the 1-mL library cells in a sterile 2-L flask, and add 45 mL of 2× YPDA containing kanamycin at 50 µg/mL (YPDA/Kan).

6. Incubate at 30°C for 20–24 h with gentle swirling (30–50 rpm) (see Note 20).

7. Transfer the mating mixture to a sterile centrifuge bottle and spin down the cells at $1,000 \times g$ for 10 min, and resuspend the cell pellet in 10 mL of 0.5× YPDA/Kan.

8. Spread 200 µL of the mating cell suspension onto 150-mm SD/-LTHA plates (or SD/-LTHA+AbA if Y2HGold is the bait strain) and incubate at 30°C for 5–7 days (see Note 21). At the same time, spread 100 µL of a 1:10 (or 1:100) dilution on SD/−Trp, SD/−Leu, and SD-LT plates separately to estimate mating efficiency (see Note 22). A successful mating should yield ~1×10^6 diploid colonies (Fig. 5a–c).

9. Pick up large colonies (2–3 mm in diameter) emerged on SD/-LTHA plates (Fig. 5d) for further analysis as described in Subheading 3.5.

3.4.2. Library Screening by Cotransformation

Use this method for a direct library screening if the "cDNA library" obtained in Subheading 3.3.4 (step 6) is not used for in vivo recombination in Y187.

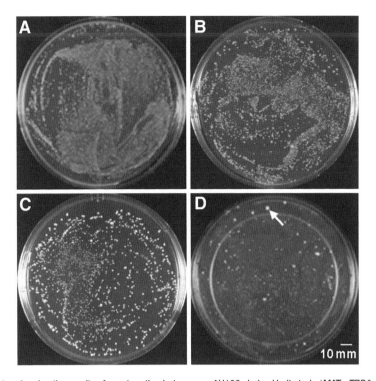

Fig. 5. SD agar plates showing the results of yeast mating between an AH109-derived bait strain (*MATa; TRP1*) and Y187-derived prey library cells (*MATα; LEU2*). Plates were inoculated with mating cells for determination of mating efficiency (**a–c**, 100-mm plates) or for selection of positive interactions (**d**, 150-mm plate) as described in Subheading 3.4.1 and incubated at 30°C for 5 days. The number of colonies on SD/−Trp plate (**a**), which allows the unmated haploid bait cells (AH109) to grow, is apparently higher than that on SD/−Leu plate (**b**) which allows the unmated haploid prey library cells (Y187) to grow. In this case, the Y187 library cells are considered the "limiting" partners in the mating. The number of colonies (~1,000) on the SD/-LT plate (**c**), which only allows the mated diploid cells to grow, appears slightly lower than that in (**b**), indicating a successful mating. The large colonies (*arrow*) on SD/-LTHA plate (**d**) are "positive" clones subjected to further analysis.

1. Prepare competent cells of the yeast strain AH109 (or Y2HGold) as described in Subheading 3.2.2.

2. Combine the following in a sterile 15-mL tube:

   ds cDNA (see Subheading 3.3.4, step 6) (Include equal amounts of ds cDNA samples made from different time points): 20 μL

   pGADT7-Rec (0.5 μg/μL): 6 μL

   pGBKT7-GYI plasmid DNA (0.5 μg/μL): 10 μL

   Single-stranded fish sperm DNA (denatured): 20 μL

3. Continue transformation following steps 3–9 described in Subheading 3.3.5.

4. Resuspend the cell pellet in 6 mL of NaCl solution (0.9%).

5. Spread 200 μL of the cell suspension on SD/-LTHA plates and incubate at 30°C for 3–5 days. At the same time, spread 150 μL of a 1:25 diluted cell suspension (30 μL of the 6-mL

transformed cells in 720 μL of 0.9% NaCl solution) separately on SD/−Leu and SD/-LT plates and incubate at 30°C for 2 days to determine transformation efficiency (see Note 23).

6. Pick up large colonies (2–3 mm in diameter) emerged on the SD/-LTHA plates for further analysis as described below.

3.5. Analysis of Positive Clones

3.5.1. Verification of Reporter Gene Phenotypes

1. Pick up single colonies from the original mating (see Subheading 3.4.1) or cotransformation (see Subheading 3.4.2) plates using 20-μL sterile pipette tips and streak on SD/-LHTA plates. Incubate the plates at 30°C. Colonies with a robust growth should be seen in 2–3 days (see Note 24)

2. Restreak the transferred colonies onto a second SD/-LTHA plate and incubate at 30°C for 1–2 days.

3. Repeat step 2 one or two more times (see Note 25).

4. Streak the colonies onto SD/-LTHA + X-α-gal plates. Incubate at 30°C for 24–36 h. Most positive clones should grow and turn blue within 12–24 h (Fig. 6a). This assay will confirm the His+, Ade+, and Mel1+ phenotypes and suggest the activation

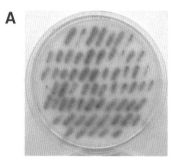

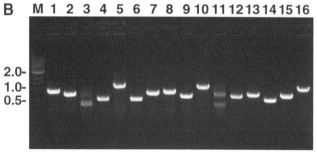

Fig. 6. Analysis of positive clones obtained from library screening. (**a**) α-galactosidase assay. Yeast cells were streaked onto SD/-LTHA + X-α-gal and incubated at 30°C for 36 h. Most clones were able to produce a blue color which indicates the activation of the *MEL1* reporter gene. (**b**) PCR screening. Agarose gel images showing the cDNA inserts (ranging from 0.5 to 1.5 kb) amplified from 16 "positive" clones identified in (**a**). The PCR products were amplified using 5′AD and 3′AD primers specific to the prey library vector pGADT7-Rec (Fig. 2b). Multiple bands observed in lanes 3 and 11 were due to the presence of more than one prey library plasmid in the yeast clone. M: 1 kb DNA ladder.

of the three corresponding reporter genes (*HIS3*, *ADE2*, and *MEL1*) due to the coexpression of (but not necessarily by the physical interaction between) the bait and prey proteins in the yeast clone (see below).

3.5.2. PCR Screening to Exclude False Positive Clones

A majority (>60% in our experiments) of the "blue" clones seen above may carry noncoding, e.g., the 3′-untranslated region (UTR) sequences only. These false positive clones must be excluded before any further analysis. In addition, some of the "blue" clones may contain coding sequences that are fused to the GAL4-activation domain in a "wrong" reading frame. These clones are also likely false positive although it is known that yeast tolerates translational frameshifts (25).

1. Scrape a few cells using a sterile pipette tip from a blue colony on SD SD/-LTHA + X-α-gal plates (see Subheading 3.5.1, step 4).

2. Resuspend the cells in a 250-μL PCR tube containing 5 μL of 67 mM potassium phosphate and 1 μL of lyticase solution. Incubate at 37°C for 30–60 min.

3. Freeze the cell suspension at –80°C for >10 min, and then thaw the tube at 95°C for 2 min.

4. Add 44 μL of PCR reaction mix as shown below to the tube.

   GoTaq Green Mix (Promega): 25 μL

   5′AD primer (10 μm) (5′AD: 5′-CTATTCGATGATGAAGATA-CCCCACCAAACCC-3′): 2 μL

   3′AD primer (10 μm) (3′AD: 5′-ATCGTAGATACTGAAAAA-CCCCGCAAGTTCACT-3′): 2 μL

   ddH$_2$O: 15 μL

5. Perform PCR at the following conditions: 95°C, 5 min; 35 cycles of 95°C, 20 s, 60°C, 30 s, 72°C, 3–5 min; 72°C, 10 min.

6. Analyze a 5-μL aliquot of the PCR product on a 1.0–1.5% agarose gel. The size of the PCR-amplified products usually ranges from 0.5 to 1.5 kb and most clones should give a single band with different sizes (Fig. 6b).

7. Purify the PCR products by column purification..

8. Confirm the identity of the prey cDNA insert and an "in-frame" fusion by direct DNA sequencing using the 5′AD (or 3′AD) primer and BLAST searches against NCBI or other databases.

9. Take positive prey clones that contain coding sequences (fused in frame with the GAL4-AD on pGADT7-Re) for further analysis (see Note 26).

Some prey fusion proteins could activate reporter genes by binding to the reporter gene-related promoter elements without physical interaction with the bait protein. These false positive clones cannot be excluded simply by DNA sequencing. Individual Y2H tests are mandatory to confirm the physical interaction between the bait and prey proteins. The Y2H testing procedures described below are similar to those used for testing of the bait construct (see Subheading 3.2.3).

1. Isolate prey plasmids from yeast library clones using a yeast plasmid isolation kit (Clontech) and transform into *E. coli* (select for ampicillin resistance).

2. Pick up a few *Amp^R* single colonies and extract plasmid DNA using standard protocols.

3. Confirm the identity of the cDNA insert-encoding putative effector-interacting protein (EIP) and the GAL4-AD-EIP fusion in each plasmid by DNA sequencing.

4. Set up five transformation experiments, each contains a bait/prey combination as follows (also see Subheading 3.2.3):

   (a) pGBKT7 + pGADT7-EIP

   (b) pGBKT7-GYI + pGADT7-EIP

   (c) pGBKT7-p53 + pGADT7-RecT (positive control)

   (d) pGBKT7-Lam + pGADT7-RecT (negative control)

   (e) No DNA (blank control)

5. Perform yeast transformation as described in steps 2–9 of Subheading 3.2.3.

6. Spread 100 μL of the transformed cells onto SD/-LT and SD/-LTHA agar plates separately. Incubate the plates (face down) at 30°C for 2–3 days.

7. Compare results of the five transformations. The yeast transformants coexpressing the prey fusion proteins and bait construct (Expt. 2) should grow on both SD/-LT and SD/-LTHA plates like those of the positive control (Expt. 3). The yeast transformants coexpressing the prey fusion protein and the bait vector (Expt. 1) should grow only on the SD/-LT and not on the SD/-LTHA plate, like those of the negative control. Any growth observed on the SD/-LTHA plate in Expt. 1 will indicate an "autoactivation" associated with the prey fusion protein.

8. Proceed to full-length protein test if the prey protein has no "autoactivation" activity (see Note 27).

Most prey plasmids rescued from yeast library clones carry cDNA sequences encoding partial proteins. It is unknown whether or not the encoded amino acid residues are likely located on the surface of the corresponding full-length protein in its native tertiary structure, thus being accessible to the effector. It is essential to test if the "positive" interaction confirmed above also occurs between the bait and the full-length prey protein.

1. Clone the full-length coding sequence of the *EIP* gene (not including any signal peptide sequences at the N-terminus) into the prey vector pGADT7 following procedures described for the construction of the bait plasmid (see Subheading 3.2.1).

2. Transform the new pGADT7-EIP construct into yeast along with the bait construct as described in Subheading 3.5.3.

3. Compare the reporter gene phenotypes of the transformants. If the full-length protein gives the same "positive" results, the identified "prey" protein can be considered as an EIP subjected to further analysis at the investigator's choice (see Subheading 3.6).

3.5.5. β-Galactosidase Filter Lift Assay

The β-galactosidase (plate, liquid, or colony-lift filter) assay can be used to detect the β-galactosidase activity in certain yeast strains, e.g., AH107 and Y187 that carry the *lacZ* reporter gene. It was widely used in Y2H experiments in early studies that depended on the selection for *HIS3* (which is a weak selection) and *lacZ* only (1, 2). More recently, this assay has become optional (or not applicable) for yeast strains carrying additional reporter genes such as *ADE2* (which is a strong nutritional selection), *AUR1-C* (which is a strong antibiotic selection), and *MEL1* (which is used for selection for the α-galactosidase activity). The colony-lift filter β-galactosidase assay (which is more stringent than plate and liquid assays) introduced in this section is intended for investigators who may use a yeast strain that relies on the selection for β-galactosidase activity. Note that some yeast strains like Y2HGold (Table 1) do not carry the *lacZ* reporter gene, and hence cannot be tested using this method.

1. Streak yeast cells on SD/-LTHA plates (100-mm) and incubate at 30°C for 2–3 days.

2. Cover the plate with a sterile Whatman filter paper (75-mm) and rub the filter gently with a spreader to allow the colonies to "stick" onto the filter.

3. Lift the filter paper carefully and submerse (colonies facing up) in liquid nitrogen for 10–15 s.

4. Remove the filter from liquid nitrogen and thaw at room temperature for 5 min.

5. Place the filter (colony side up) in a clean 100-mm petri dish containing a sterile Whatman filter paper (of the same size) that has been presoaked in 3 mL of Z buffer/X-gal solution.

6. Incubate the filters at 30°C (or room temperature) for 1–12 h. The appearance of the blue colors on the colonies will suggest a positive interaction (see Note 28).

3.6. Further Validation of Positive Interactions by Other Methods

Biochemical methods such as coimmunoprecipitation (Co-IP) (26, 27) or a glutathione-*S*-transferase (GST)-fusion pull-down assay (28, 29) are commonly used to validate the physical interactions between the bait and prey proteins that are identified in vivo by a Y2H assay. A strong interaction identified through Y2H usually turns out to be also positive in one or more biochemical assays (30). These biochemical tests provide independent evidence that supports the physical interaction observed in yeast. However, a negative result from these in vitro assays does not necessarily dismiss a positive interaction identified using Y2H assays because some factors which facilitate the interaction in vivo may not be available in the in vitro conditions. In addition, bimolecular fluorescence complementation (BiFC) assays (31) and fluorescence resonance energy transfer (FRET) (32, 33) have been developed in recent years and can be used to monitor the colocalization of two interacting proteins in the same type of cells (e.g., leaf mesophyll cells of a host plant) or the same compartment (e.g., chloroplasts) of the cell. These methods provide a useful tool to demonstrate that the interaction observed is indeed relevant to a particular biological process such as effector-trigged disease (or resistance) in a host–pathogen interaction. The detailed protocols of these methods are not within the scopes of this chapter. Readers who are interested in these aspects may follow the related references provided in the reference list of this chapter.

4. Notes

1. AH109 can be used for mating or cotransformation. You can select for His3⁺, Ade2⁺, Mel1⁺, and LacZ⁺ phenotypes in all AH109-derived stains.

2. Y2HGold is a new strain which has replaced AH109 in the current Clontech MatchMaker systems. In addition to *ADE2, HIS3,* and *MEL1*, this strain has *AUR1-C* which can be used to select for resistance to aureobasidin A (AbA). You cannot use this strain alone to test interactions based on the LacZ⁺ phenotype because it does not carry the *lacZ* reporter gene.

3. Y187 can also be used as a host strain for testing an interaction between paired proteins, but you can select for Mel1⁺ and LacZ⁺ phenotypes only because this strain does not carry the *ADE2* and *HIS3* reporter genes, although it is auxotrophic for histidine and adenine.

4. Incubate the suspension at 37°C until the powder has completely dissolved. Use a 0.45-μm syringe filter to sterilize the solution and store at room temperature.

5. The 40% glucose solution should be autoclaved separately to improve the performance (filter-sterilized glucose is usually recommended but we found that an autoclaved preparation works just fine).

6. It is more convenient to use DO supplements from a commercial source. Various ready-to-use DO supplement powders can be purchased from Clontech (www.clontech.com) under catalog numbers 630413–630431.

7. Aureobasidin A (AbA) is a cyclic depsipeptide isolated from the filamentous fungus *Aureobasidium pullulans* R106 and is toxic to yeast at low concentrations (50–100 ng/mL). AbA inhibits the yeast inositol phosphorylceramide (IPC) synthase encoded by the *AUR1* gene (34). Use this antibiotic for selection only if you use the yeast strain Y2HGold, which expresses the mutant version of the gene (*AUR1-C*) conferring resistance to AbA (35). The final concentration of AbA should not be >100 ng/mL.

8. Y187 grows slower than AH109 and Y2HGold on both non-selective and selective media. It usually takes one more day for Y187 colonies to reach the same size as those of AH109 and Y2HGold.

9. Do not use cells from an agar plate that has been kept at 4°C for more than 1 month to make competent cells for transformation. Fresh cultures streaked from glycerol stocks will work best for this purpose.

10. For frozen stocks, pick a single colony (2–3 mm in diameter) and inoculate into a 2-mL cryogenic vial containing 0.9 mL of appropriate liquid medium. Incubate at 30°C with shaking (200–250 rpm) for 24 h. Add equal volume of 50% glycerol and mix well by vortexing before freezing at –80°C.

11. Yeast competent cells prepared in this method can be left at room temperature for a few hours without losing viability. Always use freshly prepared competent cells for transformation.

12. Be sure to vortex the tube vigorously at this stage. Mixing thoroughly is critical for transformation efficiency.

13. "Autoactivation" activity associated with the bait construct must be eliminated before starting any library screening projects or testing a specific interaction. One can try to modify the bait construct by changing the size of the fused effector protein on the same bait vector if such change does not affect the activity of the effector. If the problem remains, the LexA-based DUALhybrid (DualsystemsBiotech, www.dualsystems.com) or other equivalent systems may be chosen to fuse to the effector

protein to the BD domain in two opposite directions. The autoactivation caused by one direction may be eliminated by switching to the other direction. Follow the manufacturer's instructions if these systems are used.

14. Try to modify the bait construct as suggested in Note 13. If the toxicity remains, switch to a different bait vector that has a lower expression level. The Clontech pGBT9 vector can be used for this purpose, but the expression level of the bait fusion protein may be too low to be detectable by Western blot (see Note 14 below).

15. A Western blot analysis is preferred to verify the expression of the bait fusion protein, but a negative result from such a test does not always indicate a failure of the expression. "Toxic" proteins or a degradation of the expressed fusion protein upon extraction may result in an expression level that is too low to be detectable. A bait fusion protein that is not detected by a Western blot analysis may still give positive clones when used for library screening. This is actually true if the bait protein is expressed using a low-copy plasmid, e.g., pGBT9. It is the investigator's choice to determine how to proceed at this point. Standard protocols (30, 36) can be followed to perform this analysis. For convenience, the Y-PER Yeast Protein Extraction Reagent (Pierce) can be used for yeast protein extraction. The bait fusion protein can be detected using the anti-cMyc or anti-GAL4 DNA-BD antibodies (Clontech).

16. In general, a minimum of 100 ng of total RNA or 25 ng of mRNA is required for cDNA synthesis. We have been using the Dynabeads mRNA DIRECT Kit (Invitrogen) for mRNA isolation from wheat leaf tissues (usually, 100–200 mg) and have obtained satisfying results.

17. Incubate at 25–30°C for 10 min at room temperature before incubation at 42°C if you use the CDSIII/6 random primer for first-strand cDNA synthesis.

18. The quality and quantity of double-stranded cDNA are critical for library construction. Repeat first-stand cDNA synthesis and/or ds cDNA amplification until the desired size range and the amount are obtained before going to the next step.

19. To estimate the total number of colonies in the library, spread 100 μL of a 1:100 dilution on SD/−Leu plates (100-mm). Incubate at 30°C for 3 days. Count the number of colonies on the plate and multiply by 30,000.

20. The speed of swirling must be kept below 50 rpm. Shaking at a speed >50 rpm will significantly reduce mating efficiency.

21. SD/-LTHA plates are used for selection for both *HIS3* and *ADE2* (high stringency). If weak interactions are also looked for, first spread the cells on SD/−Leu–Trp–His (-LTH) plates

(medium stringency). Remember to add 5–60 mM 3-amino-1,2,4-triazole (3-AT) to the SD/-LTH plates to suppress the background growth. 3-AT is used to titrate the GAL4-independent minimum level of *HIS3* expression in the yeast cells (it is not necessary to add 3-AT to SD/-LTHA plates). Copy the diploid colonies emerged on SD/-LTH plates to SD/-LTHA plates for further analysis.

22. The mating efficiency can be calculated using the equation: % diploids = the number of colonies (#cfu)/mL of diploids/#cfu/mL of limiting partner × 100. A low mating efficiency will greatly reduce the chance to obtain positive clones. Repeat the mating experiment if the mating efficiency is <10%.

23. Transformation efficiency (transformants per μg of pGADT7-Rec DNA) = #cfu on the SD-Leu plate × 1,000/3. The total number of colonies screened = #cfu on the SD-LT plate × 1,000.

24. Always pick up large (>2 mm in diameter) colonies. Small colonies (<1 mm in diameter) that appear after 2 days are usually "false" positives and will not grow (or grow very poorly) when transferred onto fresh SD/-LTHA plates.

25. The yeast transformants obtained from library screening may carry more than one prey library plasmids, thus complicating the downstream analyses. A repeated (2–3 times) streaking of the same transformant on SD/-LTHA media will segregate multiple prey plasmids into different cell populations.

26. In yeast and other organisms, certain genes may encode proteins whose translation involves programmed ribosomal frameshifting which depends on specific "signals" embedded in the mRNA sequences (25). It is difficult to predict whether or not a protein is subjected to ribosomal frameshifting simply based on its coding sequences. If you are interested in a cDNA sequence identified from a positive clone that is not fused in frame with the GAL4-AD domain, you can reclone it "in frame" with the GAL4-AD on pGADT7, then coexpress it with the bait protein in yeast. If the cotransformant retains the positive interaction, the prey clone may be a true "positive."

27. A full-length protein test is not always possible using a typical GAL-4-based Y2H system in which the activation of the reporter genes by the bait/prey protein complex occurs in the nucleus. The presence of any transmembrane domain in the prey would result in a cytoplasmic localization of the protein complex, thus complicating the test results. The Split-Ubiquitin Y2H system (37) can be applied if the prey protein is likely membrane-associated.

28. The time it takes a "positive" colony to turn blue may vary from 30 min to 12 h depending on the strength of the interactions. Prolonged incubation (>12 h) is not recommended because it may result in false positives.

Acknowledgments

The author thanks Dr. Timothy Friesen and Dr. Melvin Bolton for reviewing the manuscript. All yeast two-hybrid work performed in the author's laboratory is funded by the Agricultural Research Service, the United States Department of Agriculture under the CRIS project 5442-21000-033-00D.

References

1. Fields S., and Song O. (1989). A novel genetic system to detect protein-protein interactions. *Nature* 340, 245–246.

2. Chien C.T., Bartel P.L., Sternglanz R., and Fields S. (1991). The two-hybrid system: A method to identify and clone genes for proteins that interact with a protein of interest. *Proc Natl Acad Sci USA* 88, 9578–9582.

3. Fields S., and Sternglanz R. (1994). The two-hybrid system: An assay for protein-protein interactions. *Trends Genet* 10, 286–292.

4. Van Criekinge W., and Beyaert R. (1999). Yeast Two-Hybrid: State of the art. *Biol Proced Online* 2, 1–38.

5. Walhout A.J., Boulton S.J., and Vidal M. (2000). Yeast two-hybrid systems and protein interaction mapping projects for yeast and worm. *Yeast* 17, 88–94.

6. Fields S., and Bartel P.L. (2001). The two-hybrid system. A personal view. *Methods Mol Biol* 177, 3–8.

7. Causier B., and Davies B. (2002). Analysing protein-protein interactions with the yeast two-hybrid system. *Plant Mol Biol* 50, 855–870.

8. Parrish J.R., Gulyas K.D., and Finley R.L., Jr. (2006). Yeast two-hybrid contributions to interactome mapping. *Curr Opin Biotechnol* 17, 387–393.

9. Yu, Haiyuan, et al (2008). High-quality binary protein interaction map of the yeast interactome network. *Science* 322, 104–110.

10. Bruckner A., Polge, C., Lentze N., Auerbach D., and Schlattner U. (2009). Yeast two-hybrid, a powerful tool for systems biology. *Int J Mol Sci* 10, 2763–2788.

11. De Las Rivas J., and Fontanillo C. (2010). Protein-protein interactions essentials: Key concepts to building and analyzing interactome networks. *PLoS Comput Biol* 6, e1000807.

12. Johnston S.A., Salmeron J.M., Jr., and Dincher S.S. (1987). Interaction of positive and negative regulatory proteins in the galactose regulon of yeast. *Cell* 50, 143–146.

13. Deane C.M., Salwinski L., Xenarios I., and Eisenberg D. (2002). Protein interactions: Two methods for assessment of the reliability of high throughput observations. *Mol Cell Proteomics* 1, 349–356.

14. Manning V.A., Hardison L.K., and Ciuffetti L.M. (2007). Ptr ToxA interacts with a chloroplast-localized protein. *Mol Plant Microbe Interact* 20, 168–177.

15. Tai Y., Bragg J., and Meinhardt S. (2007). Functional characterization of ToxA and molecular identification of its intracellular targeting protein in wheat *American J. Plant Physiol.* 2, 76–89.

16. Wolpert T.J., Dunkle L.D., and Ciuffetti L.M. (2002). Host-selective toxins and avirulence determinants: What's in a name? *Annu Rev Phytopathol* 40, 251–285.

17. Friesen T.L., Meinhardt S.W., and Faris J.D. (2007) The *Stagonospora nodorum*-wheat pathosystem involves multiple proteinaceous host-selective toxins and corresponding host sensitivity genes that interact in an inverse gene-for-gene manner. *Plant J* 51, 681–92.

18. Stergiopoulos I., and de Wit P.J. (2009). Fungal effector proteins. *Annu Rev Phytopathol* 47, 233–263.

19. Ballance G.M., Lamari L., Kowatsch R., and Bernier C.C. (1996). Cloning, expression and occurrence of a gene encoding the Ptr necrosis toxin from *Pyrenophora tritici-repentis Mol. Plant Path.* On-Line (http://www.bspp.org.uk/mppol/) 1996/1209ballance).

20. Ciuffetti L.M., Tuori R.P., and Gaventa J.M. (1997). A single gene encodes a selective toxin causal to the development of tan spot of wheat. *Plant Cell* 9, 135–144.

21. Friesen T.L., Stukenbrock E.H., Liu Z., Meinhardt S., Ling H., Faris J.D., Rasmussen J.B., Solomon P.S., McDonald B.A., and Oliver R.P. (2006). Emergence of a new disease as a result of interspecific virulence gene transfer. *Nat Genet* 38, 953–956.

22. Faris, J.D. Zhang, Z. Lu H., Lu S., Reddy L., Cloutier S., Fellers J.P., Meinhardt S.W., Rasmussen J.B., Xu S.S., Oliver R.P., Simons K.J., and Friesen T.L. (2010). A unique wheat disease resistance-like gene governs effector-triggered susceptibility to necrotrophic pathogens. *Proc Natl Acad Sci USA* 107, 13544–13549.

23. Zhu Y.Y., Machleder E.M., Chenchik A., Li, R., and Siebert P.D. (2001). Reverse transcriptase template switching: a SMART approach for full-length cDNA library construction. *Biotechniques* 30, 892–897.

24. Shcheglov A., Zhulidov P, Bogdanova E. and Shagin D. (2007) Generation of normalized cDNA libraries. *In* Nucleic acids hybridization modern applications. Buzdin A. and Lukyanov S. Eds. Springer, Dordrecht, The Netherlands. pp. 316.

25. Jacobs J.L., Belew A.T., Rakauskaite R. and Dinman J.D. (2007) Identification of functional, endogenous programmed-1 ribosomal frameshift signals in the genome of *Saccharomyces cerevisiae. Nucleic Acids Res* 35, 165–174.

26. Naumovski L. (2001). Two-hybrid interactions confirmed by coimmunoprecipitation of epitope-tagged clones. *Methods Mol Biol* 177, 151–159.

27. Lee C. (2007). Coimmunoprecipitation assay. *Methods Mol Biol* 362, 401–406.

28. Kraichely D.M., and MacDonald P.N. (2001). Confirming yeast two-hybrid protein interactions using in vitro glutathione-S-transferase pulldowns. *Methods Mol Biol* 177, 135–150.

29. Vikis H.G., and Guan K.L. (2004). Glutathione-S-transferase-fusion based assays for studying protein-protein interactions. *Methods Mol Biol* 261, 175–186.

30. Garcia-Cuellar M.P., Mederer D., and Slany R.K. (2009). Identification of protein interaction partners by the yeast two-hybrid system. *Methods Mol Biol* 538, 347–367.

31. Barnard E., McFerran N.V., Trudgett A., Nelson J., and Timson D.J. (2008). Development and implementation of split-GFP-based bimolecular fluorescence complementation (BiFC) assays in yeast. *Biochem Soc Trans* 36, 479–482.

32. Arndt-Jovin D.J., and Jovin T.M. (1989). Fluorescence labeling and microscopy of DNA. *Methods Cell Biol* 30, 417–448.

33. Sekar R.B., and Periasamy A. (2003). Fluorescence resonance energy transfer (FRET) microscopy imaging of live cell protein localizations. *J Cell Biol* 160, 629–633.

34. Takesako K., Kuroda H., Inoue T., Haruna F., Yoshikawa Y., Kato I., Uchida K., Hiratani T., and Yamaguchi H. (1993). Biological properties of aureobasidin A, a cyclic depsipeptide antifungal antibiotic. *J Antibiot* (Tokyo) 46, 1414–1420.

35. Hashida-Okado T., Ogawa A., Endo M., Yasumoto R., Takesako K., and Kato I. (1996). *AUR1*, a novel gene conferring aureobasidin resistance on *Saccharomyces cerevisiae*: a study of defective morphologies in Aur1p-depleted cells. *Mol Gen Genet* 251, 236–244.

36. Guo D., Rajamaki M.L., and Valkonen J. (2008). Protein-protein interactions: The yeast two-hybrid system. *Methods Mol Biol* 451, 421–439.

37. Iyer K., Burkle L., Auerbach D., Thaminy S., Dinkel M., Engels K., and Stagljar I. (2005). Utilizing the split-ubiquitin membrane yeast two-hybrid system to identify protein-protein interactions of integral membrane proteins. *Sci STKE* 2005, p. l3.

Chapter 12

Characterization of Plant-Fungal Interactions Involving Necrotrophic Effector-Producing Plant Pathogens

Timothy L. Friesen and Justin D. Faris

Abstract

Recently, great strides have been made in the area of host-pathogen interactions involving necrotrophic fungi. In this article we describe a method to identify, produce, and characterize effectors that are important in host–necrotrophic fungal pathogen interactions, and to genetically characterize the interactions. The main strength of this method is the combined use of pathogen inoculation, a pathogen culture filtrate bioassay, and genetic analysis of susceptibility and sensitivity in segregating host-mapping populations. These methods have been successfully used to identify several *Stagonospora nodorum* necrotrophic effectors and to characterize the genetic and phenotypic effects of individual host–effector interactions in the wheat-*S. nodorum* system. *S. nodorum* isolates that induce a differential response on two lines are used to produce culture filtrates that contain necrotrophic effectors while the wheat lines differing in reaction to the pathogen are used to develop a mapping population. The wheat population is used to develop DNA marker-based genetic linkage maps and culture filtrates are infiltrated across the mapping population. Linkage and quantitative trait loci (QTL) analysis is used to identify regions of the wheat genome harboring genes that govern sensitivity to necrotrophic effectors. The same populations are inoculated with the effector-producing isolate to determine the significance and proportion of disease explained by individual host gene–effector interactions. Additionally, from this information, differential lines that are sensitive to single effectors are developed for further purification and characterization of the effectors, eventually resulting in the identification, molecular cloning, and characterization of the effector genes.

Key words: DNA marker, Dothideomycete, Effector, Fungus, Host-selective toxin, Host-specific toxin, Molecular mapping, Necrotroph, Necrotrophic effector, Necrotrophic pathogen, Quantitative trait loci, *Stagonospora nodorum*

1. Introduction

Necrotrophic fungal pathogens have long been thought to be much more nonspecific and quantitative in their attack on plants than their biotrophic counterparts. Numerous studies to determine the inheritance of resistance and to quantify the number of

Melvin D. Bolton and Bart P.H.J. Thomma (eds.), *Plant Fungal Pathogens: Methods and Protocols*,
Methods in Molecular Biology, vol. 835, DOI 10.1007/978-1-61779-501-5_12, © Springer Science+Business Media, LLC 2012

resistance genes involved in Stagonospora nodorum blotch of wheat reported multiple quantitative trait loci (QTLs), or genomic regions, associated with resistance, which provided much support for the notion that wheat-*Stagonospora nodorum* interactions were complex, quantitatively controlled, and did not follow a gene-for-gene scenario (see ref. (1) for review). However, recent work in the wheat-*S. nodorum* system, and several other host–pathogen systems, has shown that this is not the case and that many of the necrotrophic pathogens have evolved mechanisms to attack plants in very specific ways, in some cases using resistance pathways (2–4). The *Stagonospora nodorum*-wheat system is a good example of a host-pathogen system where necrotrophic effectors are employed by the fungal pathogen to induce the host to kill itself and then proliferate and sporulate on the necrotic tissue (see refs. (5, 6) for review).

These discoveries were primarily the result of dissecting the genetic nature of wheat-*S. nodorum* interactions by essentially "Mendelizing" the system, i.e., characterizing the host and pathogen components of a single interaction using microbiology and molecular genetics. Conventional genetic approaches involve inoculating a population of wheat lines with isolate(s) of the pathogen followed by the identification of host QTLs statistically associated with resistance and further, rather laborious, efforts to decipher the identity of the pathogen and host factors responsible for the effects of a given QTL. In our approach, we first employ methods to produce and purify individual effectors in culture. Infiltration of an individual effector into wheat lines of a segregating mapping population, followed by analysis of the phenotypic response to the effector, i.e., scoring each member of the population as sensitive or insensitive to the effector, provides a basis for understanding the genetics of the host–effector interaction and allows the chromosomal location of the host gene conferring sensitivity to the effector to be determined using molecular linkage mapping in conjunction with DNA markers (7). Then, the same wheat mapping population can be inoculated with spores of the effector-producing isolate and data on disease development for each line is collected followed by QTL analysis to determine the significance and effects of the host–effector interaction (7–12). This method differs from a conventional method in that the genetics of individual host–effector interactions and the chromosomal positions of effector sensitivity loci are known prior to conducting disease surveys. Therefore, the genetic dissection of this system in this way provides knowledge regarding the genetic factors responsible for the effects of putative QTL. This method also allows analysis of additive and epistatic effects among interactions, provides information useful for the identification of single host–effector differential lines, and leads to the identification of DNA markers suitable for genotyping and marker-assisted selection of lines with enhanced resistance.

Here, we describe the method whereby we produce and purify these effectors in culture and then use these individual effectors to characterize the genetics of the effector–host interaction.

2. Materials

2.1. Plant Growth

1. SC10 plant growth containers and racks (Stuewe & Sons, Inc.).
2. SB100 bedding plant soil (Sungro Horticulture).

2.2. Fungal Growth, Inoculation, Infiltration, and Yeast Expression

1. Fries liquid media—5 g ammonium tartrate $((NH_4)_2C_4H_4O_6)$, 1 g ammonium nitrate (NH_4NO_3), 0.5 g magnesium sulfate $(MgSO_4 * 7H_2O)$, 1.3 g potassium phosphate (KH_2PO_4), 2.6 g potassium phosphate (K_2HPO_4), 30 g sucrose, 1 g yeast extract, 2 mL trace element stock solution (167 mg LiCl, 107 mg $CuCl-H_2O$, 34 mg H_2MoO_4, 72 mg $MnCl_2-4H_2O$, 80 mg $CoCl_2-4H_2O$, 1 L water), and 1,000 mL water.
2. V8-PDA—150 mL V8 juice, 10 g Difco PDA, 3 g $CaCO_3$, 10 g agar, and 850 mL distilled H_2O.
3. YPD liquid media—1% yeast extract, 2% peptone, and 2% dextrose.
4. 250-mL Erlenmeyer flasks.
5. 100×15-mm Petri plates.
6. Inoculating loops, pipettes, pipette tips, 1-mL syringes.
7. High volume low pressure (HVLP) paint sprayer apparatus for inoculations.
8. Humidity chamber humidified with Ultra sonic humidifier (Venta-sonic VS205).
9. Light and temperature controlled growth chamber for plant growth.
10. Temperature controlled rotary shaker.
11. Fluorescent light banks for fungal growth.

2.3. Proteinaceous Effector Separation and Characterization

1. Miracloth (EMD Chemicals, 475855-1R).
2. Whatman 6 filters (1006 090).
3. Büchner funnel.
4. 1-L side-armed Erlenmeyer flask.
5. Millipore Durapore membrane 0.45 μm (HVLP04700).
6. Kontes ultraware 47-mm microfiltration assembly (953750-5347).

7. Amicon ultra ultracell regenerated cellulose spin filters, 3 kDa molecular weight cut-off (MWCO; UFC800308), 10 kDa MWCO (UFC801008), 30 kDa MWCO (UFC803008).

8. Fisherbrand regenerated cellulose dialysis tubing, 3,500 kDa MWCO (21-152-9).

9. ÄKTA prime plus liquid chromatography system (GE Healthcare).

10. Varian Pro Star high pressure liquid chromatography (HPLC) system (Varian Inc.).

11. Prep grade liquid chromatography columns.

    Cation exchange—HiTrap SPXL, HiTrap CMFF (GE Healthcare).

    Anion exchange—HiTrap QFF, HiTrap DEAE FF, HiTrap ANX FF (GE Healthcare).

    Affinity—Bioscale Mini CHT type II (BioRad)

    Size exclusion—HiLoad 16/60 Superdex 30 prep grade (GE Healthcare).

12. HPLC columns.

    Ion exchange—IEC CM-825 75×8 mm (Shodex), IEC SP-825 75×8 mm (Shodex).

    Size exclusion chromatography—BioSep SEC-S2000, 600×7.8 mm, 5 μm (Phenomonex).

13. Pronase (EMD Chemicals, 537088).

14. Bio-Rad Protean II and SDS-PAGE electrophoresis system (BioRad).

2.4. Genetic Linkage Mapping

1. Segregating population of >100 homozygous lines derived from polymorphic parents.

2. DNA isolation protocol.

3. Source of DNA markers (see Note 1).

4. DNA marker visualization technology (see Note 2).

5. Linkage mapping software, e.g., Mapmaker (13).

2.5. Identification of Effector Sensitivity Loci

1. Raw DNA marker data collected from genotyping of the segregating population.

2. Phenotypic data collected from culture filtrate infiltrations of the segregating population.

3. Linkage data from molecular mapping analysis with mapping software.

4. QTL analysis software, e.g., QGene (14), Map Manager QTX (15), QTL Cartographer (16), Mapmaker QTL (17).

3. Methods

3.1. Growth of the Fungus

1. *Stagonospora nodorum* is transferred as dried plugs (see Note 3) to 100×15-mm Petri plates containing V8-PDA medium.

2. Cultures are grown under 24 h fluorescent light until pycnidia are produced and pycnidiospores are released (Approximately 7 days) (Fig. 1e).

3. Pycnidiospores are used for all host inoculations as well as for inoculating Fries medium for culture filtrate production.

Fig. 1. (**a**) Three-day-old liquid shake cultures of *Stagonospora nodorum*. (**b**) Three-week-old stationary cultures of *S. nodorum*. (**c**) Infiltration of culture filtrate into wheat leaves. Water soaked area is where liquid has gone into the leaf intercellularly. (**d**) 0–3 reaction scale used to evaluate reaction to culture filtrates on wheat lines. From top to bottom 0 = no reaction, 1 = mottled chlorosis, 2 = chlorosis/necrosis without tissue collapse, 3 = necrosis with complete tissue collapse. (**e**) *S. nodorum* pycnidia producing pycnidiospores (*pink*) on V8-PDA media.

3.2. Disease Assay by Fungal Inoculation

1. Wheat plants are grown to the 2–3 leaf stage in SC10 containers using SB100 potting soil mix and watered and fertilized as appropriate.

2. Plants inoculations are done with *S. nodorum* at spore concentrations of 1×10^6 spores/mL.

3. Plants are inoculated until runoff using an HVLP paint sprayer apparatus.

4. Once inoculated, plants are moved to humidity chambers supplied with 100% relative humidity using cool mist humidifiers (Venta-sonic VS205). While in the humidity chamber, plants are kept under constant light. Humidifier settings should be set high enough so that free moisture remains on the leaves but low enough to avoid excessive runoff.

5. After 24 h at 100% relative humidity, plants are moved to a controlled growth chamber at 21°C (day and night temperature) with a 12-h photoperiod for 6 days.

6. The plants are rated for disease 7 days post-inoculation using a 0–5 disease rating scale based on lesion type (10).

3.3. Production of Culture Filtrate and Identification of Necrotrophic Effector Proteins

1. Proteinaceous necrotrophic effectors of *S. nodorum* are produced in culture by inoculating 60 mL of Fries medium with pycnidiospores (Fig. 1e) in 250-mL Erlenmeyer flasks.

2. Flasks are placed on a shaker with constant shaking (100 rpm) at 27°C for 3 days.

3. After 3 days of growth under shaking conditions, the cultures, which now contain a large amount of mycelium (Fig. 1a), are transferred to stationary growth in a dark growth chamber at 21°C.

4. After the transfer to stationary growth (Figs. 1b and 2a), cultures are checked weekly for activity by removing 1 mL of cell free culture filtrate with a sterile 1-mL syringe and filter sterilizing by pushing the 1 mL of filtrate through a 0.45 syringe filter (Millipore) followed by infiltration of approximately 50 μL of filter sterilized culture filtrate into 2–3 leaves of each of the susceptible wheat lines that have been characterized for necrotrophic effector sensitivities. Once a visible level of activity (necrosis) is identified based on the plant bioassay (Fig. 1c, d), cultures are processed and used. See Subheading 3.4 for details on the infiltration bioassay.

5. Cultures identified to contain necrosis producing effectors based on the plant bioassay (Fig. 1d) are harvested and filtered.

6. Cultures are filtered through Miracloth followed by vacuum filtration through a Whatman 6 filter using a Büchner funnel and a side arm Erlenmeyer flask.

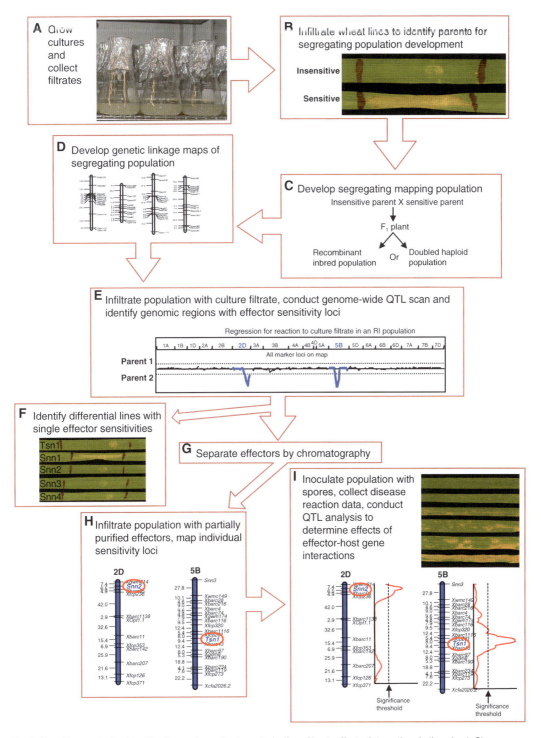

Fig. 2. Flow diagram for the identification and genetic characterization of host–effector interactions in the wheat-*Stagonospora nodorum* pathosystem.

7. Filtrates are then subjected to vacuum filtration using a 47-mm Millipore, Durapor, 0.45-μm membrane, and a Kontes ultra-ware 47-mm microfiltration assembly to remove any mycelial fragments (see Note 4).

3.4. Infiltration Assay Using Processed Fungal Culture Filtrates

1. Filter sterilized culture filtrates are infiltrated into susceptible wheat lines as well as any differential wheat lines that are known to harbor single effector sensitivities.

2. Plants to be infiltrated are placed under fluorescent lights for 20 min prior to infiltrations to enhance stomata opening and easing the infiltration process. Infiltrations are done using a needleless 1-mL syringe (Fig. 1c).

3. Culture filtrate is infiltrated into the leaf by pressing the open end of the syringe onto a secondary leaf and pushing the liquid inter-cellularly through the stomatal openings leaving a water soaked appearance where the liquid is present within the leaf (Fig. 1c).

4. Infiltration reactions are evaluated at 3 days post-infiltration on a scale of 0–3 where 0 shows no reaction, 1 shows a slight but visible chlorosis, 2 shows a visible chlorosis without tissue collapse, and 3 shows complete tissue collapse (Fig. 1d).

5. Protease sensitivity is done at this stage to evaluate whether the effectors are likely proteins. Two proteases, Proteinase K and Pronase, are used (EMD Biosciences, Inc). Pronase is a mixture of endo- and exo-proteases and due to the lower optimal temperature range, Pronase has proven to be the best option for protease sensitivity testing (see Note 5).

6. Culture filtrates are subjected to 1 mg/mL Pronase in a final concentration of 50 mM MOPS buffer for 4 h at room temperature. Controls include (1) culture filtrates in MOPS buffer alone and (2) Pronase in 50 mM MOPS buffer alone (see Note 6).

7. Pronase treated culture filtrates and controls are infiltrated into sensitive wheat lines. Significant reduction or elimination of necrosis activity as compared to the untreated controls indicates an effector is proteinaceous in nature.

3.5. Population Development and Genome Mapping

1. Identify two wheat lines that differ for sensitivity to the culture filtrates and susceptibility to the fungal inoculations (Fig. 2b).

2. Cross-hybridize the two parental wheat lines.

3. Develop a recombinant inbred (RI) or doubled haploid (DH) population (Fig. 2c; see Note 7).

4. Isolate DNA from leaf tissue of the parents and each progeny line of the segregating mapping population.

5. Screen the parents of the mapping population with DNA markers to identify markers that reveal polymorphisms between the parents (see Note 8).

6. Genotype the segregating population with DNA markers that revealed polymorphisms between the parents.

7. Use the marker genotyping data to assemble markers into linkage groups using linkage analysis software such as Mapmaker (13).

8. Determine the most plausible order of markers within each linkage group and the linkage distances between markers using the linkage analysis software (Fig. 2d; see Note 9).

3.6. Identification of Genomic Regions Harboring Necrotrophic Effector Sensitivities

1. Infiltrate culture filtrates of the necrotrophic effector-producing isolate into leaves of each line of the wheat mapping population (see Note 10).

2. After infiltration, move the plants to a growth chamber with a 12-h photoperiod at 21°C.

3. Evaluate reactions of the plants to the culture filtrates 3 days after infiltration and score the reactions using the 0–3 scale as described above (Fig. 1d).

4. Regress the phenotypic data on the molecular mapping data using interval regression analysis in a QTL analysis software package to identify significant marker-trait associations, which result in the identification of genomic regions harboring loci that confer sensitivity to effectors present in the culture filtrates (Fig. 2e).

3.7. Determine the Number of Effector–Host Gene Interactions Involved

1. Count the number of genomic regions significantly associated with the phenotypic reactions to the culture filtrate (Fig. 2e; see Note 11).

3.8. Development of Differential Lines Sensitive to Single Necrotrophic Effectors

1. Visually inspect the DNA marker data to identify progeny lines that harbor marker alleles from the sensitive parent at one genomic region significantly associated with reaction to the culture filtrate, and concomitantly harbors marker alleles from the insensitive parent at all other genomic regions associated with reaction to the culture filtrate (Fig. 2f; see Note 12).

2. Propagate additional seed of the selected lines for future use in characterizing the host–effector interactions.

3.9. Chromatographic Separation of Necrotrophic Effectors

Once differential lines are identified, necrotrophic effectors can be purified and characterized. Isolation of these proteins is performed by both low pressure and HPLC. An ÄKTA prime plus system (GE Healthcare) is used in the initial stages of purification to do preparatory separation. Preparatory separation is followed by analytical separation using HPLC.

1. Crude culture filtrates that have been prepared according to 3.3 are first dialyzed overnight against either water or running buffer using 3,500 MWCO dialysis tubing.

2. Initially, strong cation and anion exchange columns are used along with the appropriate buffers (see Note 13) to identify a column that binds and separates the effector proteins (see Note 14).

3. Appropriate running buffers are chosen based on which column is used i.e., a buffer with buffering capacity at a low and high pH are used for the cation and anion exchange columns, respectively.

4. Columns are loaded with dialyzed active culture filtrate and can be loaded with 1 mL to 1 L of culture filtrate depending on column size.

5. Typically a linear gradient of 0–300 mM NaCl over 20–100 mL (depending on the column size) is used for elution of the proteins.

6. Fractions of 1–5 mL are collected and fractions under peaks are assayed by infiltration (see steps 2–4, Subheading 3.4 above) on the appropriate differential lines, parental lines, and resistant checks. Active fractions are identified based on an infiltration bioassay (Subheading 3.4) where fractions inducing necrosis are considered active. On a typical prep grade column, three to five fractions will contain activity and the center of this set of fractions normally contains the highest quantity of the effector.

7. Using the active fractions from previous steps, additional methods of prep grade separation are then attempted to gain a higher level of purity. These methods include additional ion exchange, affinity, and size exclusion chromatography.

8. Using active fractions from the prep grade columns, the analytical grade HPLC columns are employed in the final stages of purification. These include ion exchange and size exclusion columns. Typically ion exchange is a good final step due to the benefit of a final concentrated sample.

9. Purified and possibly isolated protein effectors are then used in mapping of the host receptors, characterization of the interactions, and eventual cloning of the effector genes.

3.10. Analysis of Fractions Containing Necrotrophic Effectors

1. Proteins within the active fractions are visualized using 1-dimensional polyacrylamide gel electrophoresis (1D-PAGE) (see Note 15).

2. Protein bands are cut out of the 1D-PAGE gels, digested with trypsin and subjected to mass spectrometry (MS) analysis following normal procedures.

3. MS-MS spectra are analyzed against a protein database generated from the *S. nodorum* genome sequence to identify peptides associated with candidate effector proteins in order to identify the candidate effector genes.

4. Candidate genes are initially subjected to 5′ and 3′ RACE to define gene boundaries. Several kits are available for generating 5′ and 3′ RACE ready cDNA. We had success with the SMART RACE cDNA amplification kit from Clontech. RACE ready cDNA is amplified as described in the Clonetech user manual.

5. Once gene boundaries are identified, primers are designed for PCR amplification of the entire coding region from the start codon to the termination site from cDNA of the corresponding virulent *S. nodorum* isolate. Amplicons are cloned into a PCR cloning vector where the fragment is sequenced and verified.

6. Candidate genes that come out of this procedure are then initially verified using the *Pichia pastoris* heterologous expression system (Invitrogen). Candidate genes are cloned into a *P. pastoris* expression vector (Invitrogen) and transformed into one of the *P. pastoris* strains (Invitrogen) as described in the user manual supplied by Invitrogen (see Note 16). Various yeast strains and yeast vectors are available. We have successfully used the *P. pastoris* strain X33 and the expression vector pGAPZ for gene expression (Invitrogen).

7. Positively transformed *P. pastoris* colonies are grown in YPD liquid medium.

8. Cell free culture filtrates of the *P. pastoris* strain expressing the candidate gene are infiltrated into the wheat differential lines to verify that this is the necrotrophic effector of interest.

3.11. Mapping Effector Sensitivities as Mendelian Loci

Effectors that have been separated from all other effectors with activity in a given host population (Fig. 2g) can then be used for more specific mapping of the sensitivity gene in the host.

1. Infiltrate fully or partially purified effectors into leaves of each line of the wheat mapping population (see Note 17).

2. After infiltration, move the plants to a growth chamber with a 12-h photoperiod at 21°C.

3. Evaluate reactions of the plants to the partially purified effectors 3 days after infiltration and score the reactions as either sensitive or insensitive.

4. Convert scores for reaction to the partially purified effector infiltrations to genotypic scores (see Note 18).

5. Determine the position of the effector sensitivity gene on the genetic linkage map relative to previously mapped DNA markers (Fig. 2h) using linkage analysis software (e.g., Mapmaker).

3.12. Determine the Role of a Compatible Effector–Host Gene Interaction in the Development of Disease

1. Using a QTL analysis software package, evaluate the disease data obtained from spore inoculation experiments (Subheading 3.2) along with the raw marker data of the entire marker data set (including effector sensitivity loci).

2. Determine the significance of the effector sensitivity locus and the amount of variation explained by the host–effector interaction in disease caused by the isolate used for spore inoculations (Fig. 2i; see Note 19).

4. Notes

1. In wheat, the most widely used and user-friendly types of markers are PCR-based, which may include simple sequence repeat (SSR) markers (18–24), amplified fragment length polymorphism (AFLP) markers (25–27), target region amplified polymorphism (TRAP) markers (28, 29), expressed sequence tag (EST)-based PCR markers (30–32), and single nucleotide polymorphism (SNP) markers (33). See the corresponding marker references for information regarding primer sequences, PCR protocols, and application.

2. Standard protocols for marker visualization include the separation of amplified fragments by agarose or polyacrylamide gel electrophoresis followed by staining of the DNA fragments and documentation by photography or digital imaging (28, 30, 31, 34). High-throughput technologies involve the use of dye-labeled PCR primers and the separation of amplified fragments by capillary gel electrophoresis such as described in Liu et al. (35).

3. For storage of *S. nodorum*, single-spored isolates are grown on V8-PDA until the mycelium reaches the edge of a 10-cm Petri plate. A 5-mm cork borer is used to generate plugs that are removed from the plate and dried overnight in a laminar flow hood. Dried plugs can be stored at −80°C for more than 10 years.

4. Some filters tend to bind certain proteins. We have found that the durapore material has worked best for *S. nodorum* effectors but different material may work better for other effectors.

5. Pronase activity can be different for different cultures. Diluting the culture filtrates to the threshold of activity provides better success rates when testing individual effectors.

6. Additional time and temperatures can be used as needed to make this experiment more effective, as Pronase works differently on culture filtrates with different protein components and concentrations.

7. RI populations are developed by crossing two parents, growing the F_1 plants to maturity and forcing the F_1 plants to self-pollinate to produce F_2 seeds. A number of F_2 seeds (usually 100–200) are grown individually and allowed to self-pollinate to produce F_3 seed. From this point, a single seed is selected from each F_2-derived progeny at each generation and self-pollinated. This procedure, known as single seed descent, is repeated until the plants reach the F_6 or F_7 generation at which time the seed for each progeny line is bulked. Wheat DH populations are usually developed using the methods described by Matzk and Mahn (36). The development of a DH population is faster than an RI population, but requires much more labor and dedicated facilities. Both types of populations segregate for parental traits but each progeny line is genetically fixed (homozygous). Therefore, these populations can be used in replicated experiments. These populations should consist of 100–200 progeny lines.

8. DNA markers are based on differences (polymorphisms) in the DNA sequences of the two parents of a mapping population. Polymorphisms can arise through mutation, errors in DNA replication, or insertions or deletions of larger tracts of DNA. The detection of polymorphisms, or the visualization of DNA markers, can be done in a variety of ways, most of which are based on polymerase chain reaction (PCR) technology. See refs. Paterson (37) and Faris et al. (34) for reviews.

9. Once the data for a large number of DNA markers has been collected for all members of the population, the marker data is used to assemble linkage maps. Linkage mapping software such as Mapmaker (13) is used to identify marker linkage groups, assemble the markers into the most plausible order, and to identify genetic linkage distances between markers, which is based on recombination frequency between the markers and expressed in centiMorgans (cM) (see refs. (34, 37) for reviews). The number of DNA markers needed to assemble robust genetic linkage maps depends on the size of the genome and total number of chromosomes. Ideally, it is best to strive for one marker every 10 cM. For example in wheat, which has 21 chromosome pairs and a genome size based on recombination frequency of about 3,600 cM, it requires at least 300–400 evenly spaced markers to obtain acceptable coverage.

10. The culture filtrates are infiltrated into the secondary leaf of 2-week-old seedlings that have been grown in a greenhouse at approximately room temperature. Before infiltration, plants are removed from the greenhouse and placed under fluorescent lights for 15 min.

11. The number of genomic regions significantly associated with reaction to the culture filtrate is indicative of the number of

effectors present in the culture filtrate that are recognized by corresponding effector sensitivity genes segregating in the population. In other words, if the QTL analysis shows that two genomic regions are strongly associated with reaction to the culture filtrate, this would suggest that two host gene–effector interactions are operating in this particular system (Fig. 2e).

12. If more than one effector–host gene interaction is involved (i.e., multiple genomic regions associated with culture filtrate reactions are identified), the molecular marker data is used to identify progeny lines that harbor single effector sensitivity loci. This is done by evaluation of the raw marker data to identify progeny lines that harbor marker alleles derived from the insensitive parent at all regions of the genome associated with reaction to the culture filtrate except for one of interest, for which the marker alleles from the sensitive parent would be selected. These lines can then be used in purification and characterization of the corresponding pathogen produced necrotrophic effector.

13. Depending on the chromatography column to be used, different running buffers may be appropriate. Columns are typically supplied with user manuals that will provide suggested buffers for a range of applications.

14. Most often the column of choice for crude separation and concentration is a strong cation or anion exchange column such as the SPXL or QXL from GE Healthcare or the UNOsphere Rapid S or UNOsphere Q columns from Biorad, Hercules CA, respectively. These prepacked columns come in various sizes and different sizes can be used depending on the starting volume of culture filtrates. Many additional prepacked columns are available and may be useful including both additional strong and weak anion and cation exchange columns and affinity columns.

15. When separating proteins of less than 30 kDa, precast 16–16.5% Tris-tricine SDS-PAGE gels (38) have given the best results and are available from several vendors.

16. Invitrogen provides several kits that simplify the process of heterologous expression. For our expression we found the best results came from the X33 *P. pastoris* strain using the pGAPZ vector and the native signal sequence rather than the pGAPZα vector which contains a *P. pastoris* signal sequence.

17. The effectors are infiltrated into the secondary leaf of 2-week-old seedlings that have been grown in a greenhouse at approximately room temperature. Before infiltration, plants are removed from the greenhouse and placed under fluorescent lights for 15 min.

18. The conversion of phenotypic scores to genotypic scores is needed to determine the chromosomal position of the gene

controlling effector sensitivity relative to the DNA markers. For example, if DNA marker alleles derived from the sensitive parent are designated with an "A" and marker alleles from the insensitive parent are designated with a "B," then sensitive and insensitive lines of the mapping population should be given "A" and "B" designations, respectively, for the gene controlling sensitivity, which is a morphological marker that can then be evaluated together with the DNA markers in linkage analysis.

19. QTL analysis software programs have the ability to assess single marker-trait associations (single factor regression) and can also perform more sophisticated analyses such as interval regression (39) to determine the most plausible position of a QTL between markers and account for potential masking effects of other QTLs. The logarithm of the odds (LOD) and coefficient of determination (R^2) values for the effector sensitivity gene locus indicate the significance of the locus in the development of disease, and the proportion of total variation in disease explained by the locus, respectfully. Therefore, these two values indicate the significance and the degree of effects conferred by the compatible effector–host gene interaction in the mapping population. In addition to this information, whole genome scans can potentially reveal additional genomic regions associated with disease conferred by the spore inoculations. See Liu et al. (11), Friesen et al. (7, 9), and Abeysekara et al. (12) for examples.

References

1. Xu SS, Friesen TL, Cai XW (2004) Sources and genetic control of resistance to Stagonospora nodorum blotch in wheat. Recent Res Devel Genet Breeding 1:449–469

2. Faris JD, Zhang Z, Lu H, Lu S, Reddy L, Cloutier S, Fellers J P, Meinhardt SW, Rasmussen JB, Xu SS, Oliver RP, Simons KJ, and Friesen TL (2010) A unique wheat disease resistance-like gene governs effector-triggered susceptibility to necrotrophic pathogens. Proc Natl Acad Sci USA 107:13544–13549

3. Nagy ED, and Bennetzen JL (2008) Pathogen corruption and site-directed recombination at a plant disease resistance gene cluster. Genome Res 18:1918–1923

4. Lorang JM, Sweat TA, Wolpert TJ (2007) Plant disease susceptibility conferred by a "resistance" gene. Proc Natl Acad Sci USA 104:14861–14866

5. Friesen TL, Faris JD (2010) Characterization of the wheat-*Stagonospora nodorum* disease system: What is the molecular basis of this quantitative necrotrophic disease interaction? Can J Plant Pathol 32:20–28

6. Friesen TL, Faris JD, Solomon PS, Oliver RP (2008b) Host specific toxins: effectors of necrotrophic pathogenicity. Cell Micro 10:1421–1428

7. Friesen, TL, Zhang, Z, Solomon, PS, Oliver, RP and Faris, JD (2008a) Characterization of the interaction of a novel *Stagonospora nodorum* host-selective toxin with a wheat susceptibility gene. Plant Physiol 146:682–693

8. Friesen TL, Stukenbrock EH, Liu ZH, Meinhardt SW, Ling H, Faris JD, Rasmussen JB, Solomon PS, McDonald BA, Oliver RP (2006) Emergence of a new disease as a result of interspecific virulence gene transfer. Nat Genet 38: 953–956

9. Friesen TL, Meinhardt SW, Faris JD (2007) The *Stagonospora nodorum*–wheat pathosystem involves multiple proteinaceous host selective toxins and corresponding host sensitivity genes that interact in an inverse gene-for-gene manner. Plant J 51: 681–692

10. Liu ZH, Friesen TL, Meinhardt SW, Ali S, Rasmussen JB, Faris JD (2004) QTL analysis and mapping of resistance to Stagonospora

nodorum leaf blotch in wheat. Phytopathology 94:1061–1067

11. Liu ZH, Friesen TL, Ling H, Meinhardt SW, Rasmussen JB, Faris JD (2006) The *Tsn1*-ToxA interaction in the wheat-*Stagonospora nodorum* pathosystem parallels that of the wheat-tan spot system. Genome 49:1265–1273

12. Abeysekara NS, Friesen TL, Keller B, Faris JD (2009) Identification and characterization of a novel host-toxin interaction in the wheat - *Stagonospora nodorum* pathosystem. Theor Appl Genet 120:117–126

13. Lander ES, Green P, Abrahamson J, Barlow A, Daly MJ, Lincoln SE, Newburg L (1987) Mapmaker: an interactive computer package for constructing primary genetic linkage maps of experimental and natural populations. Genomics 1:174–181

14. Joehanes R, Nelson JC (2008) QGene 40, an extensible Java QTL-analysis platform. Bioinformatics 24: 2788–2789

15. Manly KF, Cudmore RH Jr, Meer JM (2001) MAP MANAGER QTX, cross-platform software for genetic mapping. Mamm Genome 12:930–932

16. Wang S, Basten CJ, Zeng Z-B (2010) Windows QTL Cartographer 25 Department of Statistics, North Carolina State University, Raleigh, NC (http://statgenncsuedu/qtlcart/WQTLCarthtm)

17. Lincoln S, Daly M, Lander E 1992 Mapping genes controlling quantitative traits with MAPMAKER/QTL11. Whitehead Institute Technical Report, Cambridge, Massachusetts

18. Röder MS, Korzun V, Wendehake K, Plaschke J, Tixier M-H, Leroy P, Ganal MW (1998) A microsatellite map of wheat. Genetics 149:2007–2023

19. Pestsova E, Ganal MW, Röder MS (2000) Isolation and mapping of microsatellite markers specific for the D genome of bread wheat. Genome 43:689–697

20. Sourdille P, Singh S, Cadalen T, Brown-Guedira GL, Gay G, Qi L, Gill BS, Dufour P, Murigneux A, Bernard M (2004) Microsatellite-based deletion bin system for the establishment of genetic-physical map relationships in wheat (*Triticum aestivum* L). Funct Intergr Genomics 4:12–25

21. Somers DJ, Isaac P, Edwards K (2004) A high-density microsatellite consensus map for bread wheat (*Triticum aestivum* L). Theor Appl Genet 109:1105–1114

22. Song QJ, Shi JR, Singh S, Fickus EW, Costa JM, Lewis J, Gill BS, Ward R, Cregan PB (2005) Development and mapping of microsatellite (SSR) markers in wheat. Theor Appl Genet 110:550–560

23. Torada A, Koike M, Mochida K, Ogihara Y (2006) SSR-based linkage map with new markers using an intraspecific population of common wheat. Theor Appl Genet 112: 1042–1051

24. Xue S, Zhang Z, Lin F, Kong Z, Cao Y, Li C, Yi H, Mei M, Zhu H, Wu J, Xu H, Zhao D, Tian D, Zhang C, Ma Z (2008) A high-density intervarietal map of the wheat genome enriched with markers derived from expressed sequence tags. Theor Appl Genet 117:181–189

25. Vos P, Hogers R, Bleeker M, Reijans M, Van de Lee T, Hornes M, Frijters A, Pot J, Peleman J, Kuiper M, Zabeau M (1995) AFLP: A new technique for DNA fingerprinting. Nucleic Acids Res 23:4407–4414

26. Faris JD, Gill BS (2002) Genomic targeting and high-resolution mapping of the domestication gene *Q* in wheat. Genome 45: 706–718

27. Haen KM, Lu HJ, Friesen TL, and Faris, JD (2004) Genomic targeting and high-resolution mapping of the *Tsn1* gene in wheat. Crop Sci 44:951–962

28. Liu ZH, Anderson JA, Hu JG, Friesen TL, Rasmussen JB, Faris JD (2005) A wheat intervarietal genetic linkage map based on microsatellite and target region amplified polymorphism markers and its utility for detecting quantitative trait loci. Theor Appl Genet 111:782–794

29. Chu C-G, Xu SS, Friesen TL, Faris JD (2008) Whole genome mapping in a wheat doubled haploid population using SSRs and TRAPs and the identification of QTL for agronomic traits. Mol Breeding 22:251–266

30. Lu HJ, Fellers JP, Friesen TL, Meinhardt SW, Faris, JD (2006) Genomic analysis and marker development for the *Tsn1* locus in wheat using bin-mapped ESTs and flanking BAC contigs. Theor Appl Genet 112: 1132– 1142

31. Reddy L, Friesen TL, Meinhardt SW, Chao S, Faris JD (2008) Genomic analysis of the *Snn1* locus on wheat chromosome arm 1BS and the identification of candidate genes. Plant Genome 1:55–66

32. Zhang Z, Friesen TL, Simons KJ, Xu SS, Faris JD (2009) Development, identification, and validation of markers for marker assisted selection against *Stagonospora nodorum* toxin sensitivity genes *Tsn1* and *Snn2* in wheat. Mol Breeding 23:35–49

33. Chao S, Zhang W, Akhunov E, Sherman J, Ma Y, Luo M-C, Dubcovsky J (2008) Analysis of gene-derived SNP marker polymorphism in US wheat (*Triticum aestivum* L) cultivars. Mol Breeding 23:23–33

34. Faris JD, Friebe B, Gill BS (2004) Genome Mapping In: Encyclopedia of Grain Science

Edited by C Wrigley Elsevier, San Diego, CA pp 7–16

35. Liu S, Chao S, Anderson JA (2008) New DNA markers for high molecular weight glutenin subunits in wheat. Theor Appl Genet 118:177–183

36. Matzk F, Mahn A (1994) Improved techniques for haploid production in wheat using chromosome elimination. Plant Breed 113:125–129

37. Paterson AH (1996) Making Genetic Maps In: Paterson, AH (ed) in *Genome Mapping in Plants,* pp 23-39 R G Landes Company, Austin, TX

38. Schägger H (2006) Tricine-SDS-PAGE. *Nature Protocols* 1: 16–22

39. Haley CS, Knott SA (1992) A simple regression method for mapping quantitative trait loci in line crosses using flanking markers. Heredity 69:315–324

Chapter 13

Heterologous Production of Fungal Effectors in *Pichia pastoris*

Anja Kombrink

Abstract

In this chapter a method for the heterologous production of fungal proteins in the yeast *Pichia pastoris* is described. Starting with cloning of the sequence encoding the gene of interest into the expression vector, this protocol describes *P. pastoris* transformation, production of the protein in a fermentor, and purification of the protein. This method has successfully been used for the production of a number of fungal effector proteins.

Key words: *P. Pastoris*, Heterologous protein production, Fungal effector, Yeast

1. Introduction

The methylotrophic yeast *Pichia pastoris* has become an important system for the production of heterologous proteins. The *P. pastoris* expression system has several features that make it useful for the production of heterologous proteins. In contrast to the prokaryote *E. coli*, *P. pastoris* has the machinery for posttranslational modification of proteins, which in many cases is needed for the production of correctly folded and biologically active eukaryotic proteins. The expression of the gene of interest can easily be induced via the alcohol oxidase 1 (*AOX1*) gene promoter that is activated upon changing the carbon source of *P. pastoris* from glycerol to methanol. Furthermore, the *P. pastoris* transformation is efficient and cultures with high cell densities can be obtained. Since the *Pichia* expression system has been used for the production of a wide range of proteins, it has been described in several reviews (1–3). Multiple *P. pastoris* strains and expression vectors for transformation with the gene of interest are available for protein production (1, 2).

Melvin D. Bolton and Bart P.H.J. Thomma (eds.), *Plant Fungal Pathogens: Methods and Protocols*,
Methods in Molecular Biology, vol. 835, DOI 10.1007/978-1-61779-501-5_13, © Springer Science+Business Media, LLC 2012

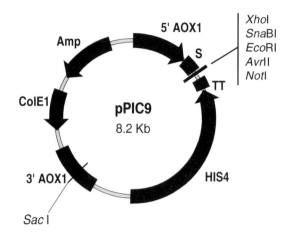

Fig. 1. Schematic representation of the pPic9 vector used for *Pichia pastoris* transformation. The gene of interest is cloned in the multiple cloning site, preferably using the *Eco*RI and *Not*I restriction sites. *S* signal peptide sequence; *TT* transcription termination; *ColE1 E. coli* replication origin; *Amp* ampicillin resistance gene.

Here we describe the particular method that was successfully used for the production of fungal effector proteins. More information and DNA sequences of the multiple cloning sites of the expression vectors can be found in the Invitrogen Instruction manual (4, 5).

We use *P. pastoris* strain GS115 which is defective in the histidine dehydrogenase gene. The expression vector, pPic9 (Fig. 1), contains the *HIS4* gene which allows selection of transformants that are able to grow on medium without histidine. Transformation of *P. pastoris* is based on recombination in the native *AOX1* locus in the *P. pastoris* genome and the *AOX1* sequence in linearized pPic9. The vector pPic9 contains the *HIS4* gene, the *AOX1* promoter, a transcription termination signal and further downstream *AOX1* sequences.

For the expression of fungal effectors by *P. pastoris*, the sequence encoding the mature protein of interest, lacking the endogenous signal peptide, should be used, as pPic9 contains the sequence encoding a *P. pastoris* signal peptide for optimal secretion. The cloning site of pPic9 contains restriction sites allowing in frame directional cloning. For purification of the produced protein a histidine-tag is fused to the N-terminus of the protein. Furthermore, an N-terminal FLAG-tag is attached to enable specific detection of the tagged protein on western blot.

For the production of heterologous proteins we use a fermentor to achieve a high density cell culture. Additional information to this protocol about protein production by *Pichia pastoris* in fermentors, can be found in the protocol written by Stratton et al. (6), and conditions for optimal protein production in fermentor cultures are reviewed by Cereghino et al. (7). Small scale production of proteins in flasks is less efficient since the conditions in the cultures cannot be adjusted continuously for optimal growth. Furthermore, yeast cells require much oxygen and the continuous addition of substrate (methanol induction) is more difficult in small scale flask production. However, depending on the protein of interest and the amount that is needed for experiments, it might be possible to obtain enough protein in a flask culture.

2. Materials

2.1. Construct Preparation for P. pastoris Transformation

1. pPic9 vector (Invitrogen, Carlsbad, USA).
2. Restriction endonucleases and buffers.
3. T4 ligase and ligation buffer.
4. DH5α competent cells.
5. LB/ampicillin (100 μg/mL) plates.
6. LB/ampicillin (100 μg/mL) liquid medium.
7. *AOX1* primers: 5'GACTGGTTCCAATTGACAAGC, 3'GCA AATGGCATTCTGACATCC.
8. DNA purification kit.

2.2. Pichia Transformation

1. *P. pastoris* strain GS115.
2. Sterile 300 mL flasks.
3. YPD medium: 1% yeast extract, 2% peptone, 2% dextrose (glucose).
4. 28–30°C incubator with shaker.
5. Ice cold sterile water.
6. Ice cold 1 M sorbitol.
7. Electroporation device.
8. MD (minimal dextrose) plates: 1.34% YNB, 4×10^{-5}% biotin, 2% dextrose.

   Prepare 10×YNB: 134 g yeast nitrogen base with ammonium sulfate and without amino acids, add water to 1 L and filter sterilize. 500× biotin: 20 mg biotin add water to 100 mL and filter sterilize. 10× dextrose: 200 g of D-glucose, water to 1 L and autoclave. After autoclaving 800 mL water with 15 g agar add 100 mL of 10×YNB, 2 mL of 500× biotin, and 100 mL of 10× dextrose.
9. 25 mM NaOH.

2.3. Fermentor Run

1. YPD medium (see above).
2. Three sterile 300 mL flasks.

3. FM22 medium: 42.9 g KH_2PO_4, 5 g $(NH_4)_2SO_4$, 1 g $CaSO_4 \times 2H_2O$, 14.3 g K_2SO_4, 11.7 g $MgSO_4 \times 7H_2O$, 40 g glycerol, add water to 1 L. Adjust pH to 4.5. After autoclaving for 20 min, add 500× biotin.

4. FM22 medium (for fermentor): 80.1 mL H_3PO_4-(ortho)85, 3 g $CaSO_4 \times 2H_2O$, 42.9 g K_2SO_4, 35.1 g $MgSO_4 \times 7H_2O$, 120 g glycerol, add water to 3 L.

5. Fermentor (3000 bench-top fermentor, New Brunswick).

6. 25% NH_3 solution.

7. 50% glycerol: 500 mL 50% glycerol, after autoclaving add 2 mL PMT4 + 2 mL 500× biotin.

8. Methanol + 12 mL PMT4 + 2 mL 500× biotin.

9. PMT4 trace element solution, for 1 L: 2.0 g $CuSO_4$, 0.08 g NaI, 3.0 g $MnSO_4 \times H_2O$, 0.2 g $Na_2MoO_4 \times 2H_2O$, 0.02 g H_3BO_3, 0.5 g $CaSO_4 \times 2H_2O$, 0.5 g $CoCl_2$, 7 g $ZnCl_2$, 22 g $FeSO_4 \times 7H_2O$, 1 mL H_2SO_4. Filter sterilize.

10. Antifoam 204, antifoam B (Sigma).

2.4. Concentration and Purification

1. Amicon device for protein concentration (400 mL, stirred cell, Millipore) + membrane.

2. Lysis buffer: 6.9 g $NaH_2PO_4 \times H_2O$, 17.54 g NaCl, 0.68 g immidazole, add water to 1 L.

3. Dialysis membrane that has pore sizes at least half the size of the protein of interest or smaller.

4. Ni-NTA Superflow resin (Qiagen).

5. Column (xk 16, pharmacia biotech) with peristaltic pump.

6. Wash buffer: 6.9 g $NaH_2PO_4 \times H_2O$, 17.54 g NaCl, 1.36 g imidazole, add water to 1 L.

7. Elution buffer: 6.9 g $NaH_2PO_4 \times H_2O$, 17.54 g NaCl, 17.00 g imidazole.

3. Methods

3.1. Construct Preparation for P. pastoris Transformation

1. Determine the sequence encoding the signal peptide of the gene of interest to allow cloning of the sequence that encodes the mature peptide.

2. Design primers. The forward primer contains the sequence for the restriction site which is followed by the histidine- and FLAG-tag sequences (His: CATCATCATCATCATCAT, Flag: CCCGACTACAAGGACGACGATGACAAG). The reverse primer contains the second restriction site. If *Eco*RI and *Not*I are not present in the gene, use these restriction sites in the

forward and reverse primer respectively. Alternative restriction sites in the multiple cloning site of pPic9 can be found in Fig. 1 and in the Invitrogen Instruction manual (4).

3. Use these primers to amplify the gene of interest by PCR and purify the PCR product.

4. Ligate the amplified fragment into pPic9 and sequence the final constructs (see Note 1).

5. Transform chemical competent DH5α cells with 10 μL of the ligation mix. Plate the cells on LB-ampicillin plates. Select positive colonies with colony PCR the next day; use the forward or reverse primer AOX1on pPIC9 in combination with the gene specific forward or reverse primer.

6. Inoculate 3 mL LB/ampicillin with positive colonies and grow them over night at 37°C.

7. After miniprep, pPic9 is linearized by digestion with *Sac*I. Check whether endogenous *Sac*I restriction sites are present in the gene of interest. If *Sac*I is present, the Invitrogen Instruction manual gives alternative restriction sites (see Note 2). For *Pichia* transformation 5–20 μg DNA is needed (see Note 3).

8. Heat inactivate the restriction enzyme by putting the reaction at 65°C for 15 min. Purify the DNA and elute the DNA in 15 μL demineralized water or TE buffer.

3.2. Pichia Transformation

1. Grow yeast strain GS115 in 5 mL YPD in a 50-mL conical tube over night at 28–30°C and shaking at 225–250 rpm (see Note 4).

2. Inoculate 75 mL fresh YPD with the preculture in a 300-mL flask. Grow over night until the cultures has an OD of 1.3–1.5 (see Note 5).

3. Centrifuge the culture for 5 min at 4°C at $1,500 \times g$. Discard the supernatant and resuspend the pellet in 50 mL ice cold sterile water.

4. Centrifuge as in step 3 and resuspend the pellet in 25 mL of ice cold water.

5. Centrifuge as in step 3 and resuspend the pellet in 10 mL of ice cold 1 M sorbitol.

6. Centrifuge as in step 3 and resuspend the pellet in 1 mL of ice cold 1 M sorbitol.

7. Mix 80 μL of the cells with the linearized DNA and transfer the mix to a precooled electroporation cuvette with 0.2 cm gap size. Incubate on ice for 5 min.

8. Pulse the cells with the settings for yeast electroporation (Charging voltage 1,500 V, resistance 200 Ω).

9. Add 1 mL of ice cold 1 M sorbitol to the cells and transfer the content of the cuvette to an Eppendorf tube.

10. Plate 200 µL of the cells on a MD plate. Centrifuge the remainder of the cells and plate them after resuspending the pellet in a small volume. After 2–4 days colonies appear on the plate.

11. Check the presence of your gene by colony PCR. Using sterile toothpicks or sterile pipette tips, transfer yeast cells to 30 µL of 25 mM NaOH. Boil the cells (in a PCR machine) for 10 min and use 1 µL of the suspension in the PCR reaction.

3.3. Fermentor Bioflo 3000 Run

Instructions for operation of the fermentor can be found in the manual (8) and online (9).

1. Grow the *P. pastoris* transformant in 3×5 mL YPD in a 50-mL conical tube overnight in 28°C shaking at 225–250 rpm.

2. Centrifuge the cells for 10 min at $1,300 \times g$. Resuspend the pellet in a few mL of FM22 medium and inoculate 3×100 mL FM22 medium in a 300-mL flask. Grow 36–60 h at 28°C (see Note 6).

3. Autoclave the fermentor filled with 3 L FM22 medium at least one day prior to the start of the fermentor run. Close the open tubes with clamps to keep the inside sterile.

4. Preferably, start the fermentor in the morning. After calibration of the pH meter, connect the fermentor to air, water, thermometer, DO_2, and pH electrode. The temperature should be set at 30°C. Connect and start the NH_3 feed to increase the pH to 4.5.

5. Add 12 mL PMT4, 3 mL 500× biotin, 1 mL antifoam 204, and 0.1 mL antifoam B to the medium.

6. Calibrate the DO electrode; set "span" to 100 and unplug the cable to set "zero."

7. Start the rotor speed (agit) at 200 rpm and set "control" to DO_2. Agit "set" should be 1,000 rpm. The rotor starts at 200 rpm, but the speed is connected to the measured DO_2 value, which is set to 30 P.I.D. When the DO_2 reaches 30 (30%), the rotor speed will go up till maximum 1,000 rpm. Air is set to 8.0 P.I.D.

8. Collect the cells by centrifugation for 10 min at $1,300 \times g$. Remove the supernatant and resuspend the pellet in a few mL of FM22. Add the culture to the fermentor. After 24–36 h of growth the glycerol will be finished (see Note 7). At this point, start the (50%) glycerol feed at speed 10, according to the fermentor used for this protocol. For 4 h the culture will grow on minimal glycerol. Figure 2 shows an example of a fermentor run, showing the continuous changes in rotor speed and DO_2.

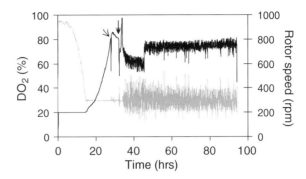

Fig. 2. Fermentor run of 95 h. The open arrow indicates the time at which the glycerol is finished and the cells stop growing. This results in a sudden drop of the rotor speed and an increase of the DO_2. After 4 h of feed on minimal glycerol, the carbon source is switched to methanol (*closed arrow*).

9. After 4 h, stop the glycerol feed (see Note 8). Start the methanol feed at speed 2 over night.

10. Next day, check the fluctuation of DO_2 and the rotor speed. To decrease the fluctuations you might adjust the glycerol feed. The DO_2 usually fluctuates between 20 and 40% and the rotor speed between 600 and 800 rpm.

11. After 4 days, (e.g., start on Monday and harvest on Friday) harvest the medium.

12. Centrifuge the medium for 15 min at $3,000 \times g$ to pellet the yeast cells. Store the supernatant at 4°C or for long-term storage at –20°C until purification.

3.4. Concentration and Purification of the Protein

Perform the concentration and purification of the protein at 4°C.

1. Concentrate the medium to less than 50 mL (see Note 9) using the appropriate membrane that has pore sizes at least half the size of the protein of interest or smaller.

2. Dialyze the concentrated protein solution 2 times against 5 L lysis buffer at 4°C for 8–12 h. Use dialysis membrane with appropriate pore size, which is at least half the size of your protein.

3. Pour a Ni-NTA column. A 10-cm column is usually sufficient. Overnight packing of the column is recommended.

4. Wash the column with 10× its volume of lysis buffer (1 mL/min).

5. Apply your sample at 0.5 mL/min. Collect the flow through to check on gel whether the protein of interest has indeed bound to the column.

6. Wash the column with 10× its volume of wash buffer. Collect the wash buffer flow through to check the absence of your protein that should still be bound to the column.

7. Elute the protein with elution buffer and collect fractions of approximately 1 mL (see Note 10).

8. Run a protein gel with the collected fractions. Include wash buffer flow through and sample flow through. Stain the gels with coomassie staining. Because of the high imidazole concentration in the samples, a smear will be visible.

9. Pool the samples which give a band on the protein gel and dialyze 2 times against 5 L demineralized water at 4°C for 8–12 h.

4. Notes

1. If ligation into pPic9 is not efficient, you could consider cloning into pPic9 via a pGEMT cloning. Sequence the construct in pGEMT or in pPic9.

2. Alternative restriction sites include *Pme* I, *Nsi* I, and *Nsi* I, more information can be found in the Invitrogen Instruction manual.

3. Transformation of *P. pastoris* via electroporation is very efficient. Measure the DNA concentration after miniprep and digest 10–20 µg in total, divided in 2–4 digestion reactions. Pool the digestion reactions and purify the DNA. Use the obtained amount for transformation.

4. The amount of cells you will obtain using these amounts is enough for up to 12 transformations. It is not recommended to store the competent cells.

5. Possibly the culture has grown too far and needs to be diluted in the morning; in this case it should be grown for an additional 3–5 h until an OD of 1.3–1.5 is reached.

6. The growth speed of the transformants might differ. When using fresh yeast from plate for inoculation, 24 h is enough to obtain a thick culture. The 100 mL preculture can be grown for 60 h (over the weekend), but the suspension will be very thick, and most likely contains a high percentage of dead cells.

7. When the culture is growing as expected for Mut⁺ transformants, the glycerol from the medium will be finished after 24–36 h. During the night, the DO_2 will drop till 30%. The rotor speed will go up to provide oxygen in order to maintain DO_2 30. This continues until the glycerol is finished and the cells stop growing. The growth stop will result in a sudden increase of the DO_2 because the cells do not use oxygen anymore. At the same time the rotor speed will drop, since there is no need for oxygen (see Fig. 2). Keep an eye on the fermentor to notice this moment and immediately start minimal glycerol feed.

8. Upon turning off the glycerol feed, the DO_2 should immediately go up, which means that the feed was minimal and no accumulation in the medium occurred because of a glycerol surplus.

9. Usually we concentrate 1 L medium, which is half the yield of one fermentor run, to approximately 50 mL.

10. Collect the fractions, elute at least until the brown/green color has disappeared from the column. Usually 36 fractions of ~1 mL are enough, but you might continue the elution to be sure all protein has been eluted.

References

1. Gregg JM, Cereghino JL, Shi J, Higgins D R (2000) Recombinant protein expression in *Pichia pastoris*. Molecular Biotechnology. 16:23–52.

2. Daly R, Hearn MTW (2005) Expression of heterologues proteins in *Pichia pastoris*: a useful experimental tool in protein engineering and production. Journal of Molecular Recognition. 18:119–138.

3. Macualey-Patrick S et al (2005) Heterologues protein production using the *Pichia pastoris* expression system. Yeast. 22:249–270.

4. Invitrogen, *Pichia* Expression Kit; A manual of methods for the expression of recombinant proteins in *Pichia pastoris*. Catalog no. K1710–01.

5. http://tools.invitrogen.com/content/sfs/manuals/pich_man.pdf.

6. Stratton J, Chiruvolu V, Meagher M (1998) High Cell-density fermentation. In: Higgins andCregg (ed) *Pichia* protocols, 103. Totowa, New Jersey.

7. Cereghino et al (2002) Production of recombinant proteins in fermenter cultures of the yeast *Pichia pastoris*. Current Opinion in Biotechnology. 13:329–332.

8. Guide to Operations, Bioflo 3000 bench-top fermentor. Manual no: M1227-0051/F. Edison, New Jersey.

9. http://www.igz.ch/f/puebersicht_f.asp?action=download&fileid=6123.

Chapter 14

The Application of Laser Microdissection to Profiling Fungal Pathogen Gene Expression *in planta*

Wei-Hua Tang, Yan Zhang, and Jon Duvick

Abstract

Laser microdissection (LM) of plant tissues infected with a fluorescent protein-tagged fungus is a useful method for obtaining samples highly enriched in fungal RNA for downstream analysis such as hybridization to a microarray. This paper outlines the requirements for successful LM of infected tissues and details a set of protocols for (1) preparing and sectioning infected tissue samples under conditions that preserve both RNA integrity and cytological features; (2) capturing fungal structures *via* LM; and (3) extraction and amplification of transcripts for further analysis.

Key words: Laser microdissection, Laser capture, Fungal infection, Transcript profiling, Fluorescent protein

1. Introduction

The ability to monitor fungal pathogen gene expression *in planta* has long been a goal for plant pathologists seeking to understand pathogenesis. This goal has been significantly advanced by the availability of complete fungal genome transcriptomes and robust transcript profiling methods (1, 2). However, except in a few cases (3–5), it is difficult to separate the pathogen infection structures from surrounding plant tissue once penetration has occurred. Although bulk infected plant tissues can be used directly for RNA extraction and profiling, these experiments inevitably suffer from a low signal/background ratio and lack control over the developmental stage of infection structures sampled. Several technologies developed in the past decade have made it possible, at least in part, to overcome these difficulties: transgenic fungal strains that

Melvin D. Bolton and Bart P.H.J. Thomma (eds.), *Plant Fungal Pathogens: Methods and Protocols*,
Methods in Molecular Biology, vol. 835, DOI 10.1007/978-1-61779-501-5_14, © Springer Science+Business Media, LLC 2012

constitutively express fluorescent proteins in the fungal cytosol to visualize fungal infection structures with the aid of a fluorescent microscope (6) and laser microdissection (LM) of infected tissues to facilitate the harvest of specific fungal cells during infection of plant tissues (7). These techniques can be used to provide relatively uniform, pathogen-rich samples for downstream genome-wide gene profiling via microarray hybridization or high throughput sequencing, when combined with rigorously monitored cDNA synthesis and amplification. Although not addressed specifically in this paper, these methods also apply to experiments aimed at obtaining plant host transcripts in cells undergoing infection (e.g., (8)) or interacting with symbiotic organisms (9).

2. Materials

2.1. Construction of Fluorescent-Tagged Fungal Strain

Choice of fluorescent protein will depend on the properties of the tissue being evaluated after fixation as well as other factors (see Notes 1–5). The AmCyan gene used in our experiments (7) is available from BD Biosciences, San Jose, CA, U.S.A. Optimized filter set for AmCyan is Excitation D440/40x, Dichroic 470dcxr, Emission D500/40m. A standard filter set for ECFP can also be used for AmCyan. However, to observe fungi *in planta*, a wide-band GFP filter (Exciter HQ470/40, Dichroic Q495LP, Emitter HQ500LP) (Chroma Technology Corp, Rockingham, VT U.S.A.) is preferable for distinguishing the fungal signal from plant tissue autofluorescence.

2.2. Sample Fixation and Embedding

Scintillation vials, Fisher Scientific.

Glass vacuum chamber.

Acetone.

Xylene.

Paraplast plus, Sigma.

Tissue Tek metal molds (EMS, Hatfield, PA USA).

Embedding machine BMJ-1, AIHUA, Tianjing Tianli Inc., Tianjing, China.

BP-111RS Research Laboratory Microwave Tissue Processor, Microwave Research & Applications, Inc., Laurel, MD, U.S.A.

2.3. Sectioning and Tape Transfer

Leica RM2235 Microtome.

Paraffin Tape-Transfer system, Instrumedics, Richmond, IL, U.S.A.

2.4. Laser Microdissection

Arcturus Veritas Microdissection System, including CapSure® LCM Caps (part no. LCM0211), Applied Biosystems, Silicon Valley, CA, U.S.A, or PALM Microbeam system (P.A.L.M.

Microlaser Technologies, Carl Zeiss, Germany) and 0.5-mL RNase-free microfuge tube (Ambion, Austin, TX U.S.A.).

2.5. RNA Isolation and Quality Assessment

PicoPure RNA isolation kit, Arcturus Bioscience, Mountain View, CA, U.S.A.

RNase-Free DNase Set, Qiagen, Valencia, CA, U.S.A.

Agilent RNA 6000 Pico kit, Agilent Technologies, Waldbronn, Germany.

Agilent Bioanalyzer 2100, Agilent Technologies, Waldbronn, Germany.

RNA6000 ladder, Ambion.

2.6. Linear Amplification of RNA

TargetAmp 2-round aminoallyl-aRNA amplification kit, Epicentre Biotechnologies, Madison, WI, U.S.A.

SuperScript III Reverse Transcriptase (200 U/µL), Invitrogen, Carlsbad, CA, U.S.A.

SuperScript II Reverse Transcriptase, Invitrogen, Carlsbad, CA, U.S.A.

RNeasy MinElute Cleanup Kit, Qiagen, Valencia, CA, U.S.A.

Ethanol 100%.

PCR machine, PTC-100 Peltier Thermal Cycler, U.S.A.

0.5 mL tube for cap (thin-walled reaction tube with domed cap), N8010611 GeneAmp.

Speedvac concentrator.

3. Methods

The following section outlines methods for fixation, embedding, and sectioning of tissues; visualizing and recovering cells by LM; and RNA extraction, quality control, and amplification. Researchers are encouraged to monitor each step carefully for quality of outcome, and to modify the methods as appropriate for the system under study. Refer to Fig. 1 for overview of the LM process (see Note 6).

3.1. Fixation and Embedding of Infected Tissues

For this step, preservation of both RNA quality and cytological features is critical. Use of a lab microwave with precise temperature control facilitates the rapid dehydration and embedding that are critical for preserving RNA integrity (10). Optimal conditions that also preserve cytological detail and protein fluorescence may require trial-and-error experimentation. The methods outlined here are adapted from Ohtsu and Schnable (11) with additions from Inada and Wildermuth (12), and were used successfully for

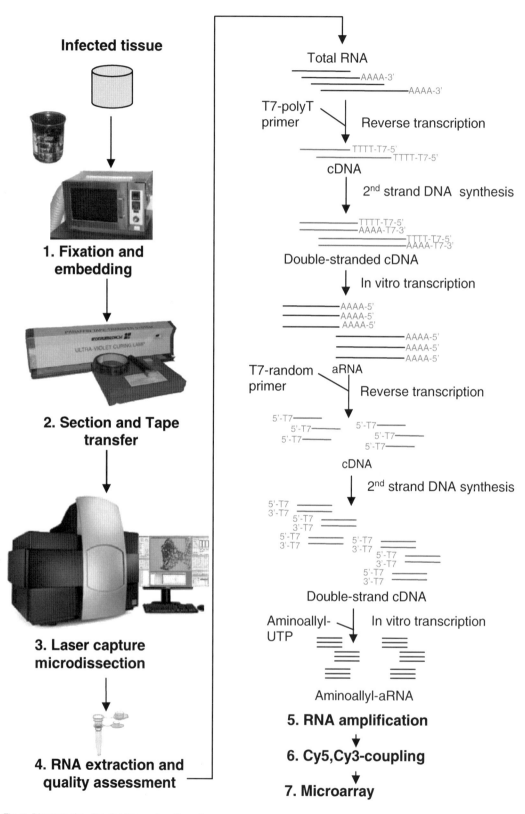

Fig. 1. Diagram showing the laser microdissection process.

preparing maize stalk tissue for fluorescence microscopy and LCM of fungal mRNA (7). At each step involving a change in reagents, be sure to pour off previous reagent as completely as possible.

1. At the scheduled time post inoculation, move the infected plants and mock-inoculated plants to a location close to a dissecting fluorescent microscope (see Note 1 for biosafety considerations).

2. Harvest one of the inoculated samples quickly using scissors and/or blades to expose the infection sites, and check under fluorescent microscope with a suitable filter set (a Longpass GFP filter is good) to confirm that the infection process is within the expected range.

3. Trim the inoculated sample to small blocks (no larger than $1 \times 1 \times 0.5$ cm each) using blades, and immediately put into a vial containing about 15 mL of 100% acetone (4°C). Up to six blocks from the same sample may be combined into one vial, and this should be done within 5 min (as quickly as possible). Total sample volume should not exceed 5% of the acetone volume.

4. Put the vial in a vacuum chamber, subject it to 15–30 min of vacuum infiltration (~400 mmHg) at room temperature, and slowly bring back to normal pressure. (Note: you should see air bubbles move slowly up from the bottom sample blocks, but the surface of acetone in the vial should not appear as if boiling.) If there are more samples, trim another six sample blocks during this time to catch up.

5. Change to fresh 100% acetone (4°C), continue to apply vacuum for 15–30 min, and slowly bring back to normal pressure. Meanwhile, prepare about 1 L water and heat to 37°C.

6. Change to fresh 100% acetone (room temperature), close the vial, and place it in a ~300-mL glass beaker with 37°C water. The surface of the water should be about half the height of the vial. Put the whole setup into a BP-111RS Research Laboratory Microwave Tissue Processor, insert the thermocouple/sensor into the water in the beaker, set for 37°C and microwave for 15 min. Pause the microwave every 5 min and shake the vial a little bit to mix the sample blocks inside the vial. Make sure that the microwave setting does not boil the acetone at all, and that the vial temperature is similar to the surrounding water temperature.

7. Swipe off the water from the outside surface of the vial, change the contents to fresh 100% acetone (room temperature), and continue vacuum infiltration (400 mmHg) for 15–30 min.

8. Change to fresh 100% acetone (37°C), put into the beaker setup again, and microwave at 37°C for 15 min. Again, pause

the microwave every 5 min to shake the vial and check the water temperature.

9. Repeat steps 7 and 8 until the total microwave time at 37°C reaches 45 min.

10. Change to acetone:xylene (1:1 volume) prewarmed to 60°C. Heat the water in the beaker to 60°C without the vial. Put the vial into the beaker and microwave at 60°C for 1 min 15 s.

11. Change to 100% xylene prewarmed to 60°C, microwave at 60°C for 1 min 15 s.

12. Change to xylene:paraplast (1:1 volume) prewarmed to 60°C, and microwave at 65–68°C for 10 min. Pause at 5 min to shake the vial.

13. Change to fresh 100% paraplast prewarmed to 70°C, microwave at 70°C for 30 min. Pause once every 10 min to shake the vial. Turn on the embedding machine.

14. Repeat step 13 four times or until the total microwave time with 100% paraplast is over 2.5 h.

15. Change to fresh 100% paraplast (70°C). Samples can be kept at this stage for 5 min up to 4 h while other samples are being processed.

16. Tissue blocks and paraplast are next poured into Tissue Tek base molds (metal). It is important to adjust the position and orientation of tissue blocks appropriately, taking into consideration the eventual direction of sectioning, which should, if possible, allow the lengthwise profile of fungal hyphae to be visible. Use a prewarmed needle for this purpose, and slowly add the liquid paraplast from the dispenser to fill the mold. The tissue blocks should sink in the bottom, not float. Add the Tissue Tek embedding ring and dispense more liquid paraplast. Carefully move the metal mold with tissue blocks to the cold plate (left side of the embedding machine). Wait until the blocks harden, whereupon the tissue in solid paraplast in the mold can be stored at 4°C for up to a month.

3.2. Sectioning and Tape Transfer

1. Paraplast blocks are dug out from the mold using a blade, and trimmed to a trapezoidal shape with the tissue in the center.

2. With a Leica microtome, set the section thickness to greater than 10 μm, and rotate the wheel to trim off the surface paraplast until the tissue is exposed. Then section at 10 μm thickness to get a "rainbow" serial (see Note 7).

3. Get the transfer tape ready for paraffin. Peel a piece of cover tape and lay it on a flat surface, and transfer the "rainbow serial" to the surface of the cover tape. Use the small hand-roller to adhere and flatten it (see Note 8).

4. Place the cover tape (tissue side down) on a slide sitting under the UV chamber, and use the hand-roller to affix and flatten it.

5. Cure the adhesive with 1 min of UV light, and then peel the cove taper off. The tissue rainbow should remain on the slide.

6. The slide is deparaffinized twice for 10 min each in 100% xylene. After air-drying in a fume hood at room temperature for 2–3 h, the sample is ready for LM.

3.3. Laser Microdissection

Outlined here are methods optimized for two distinct LM platforms: laser capture microdissection (using a Veritas Microdissection System from Arcturus/Molecular Devices), and the LM and pressure catapulting system (using a PALM Microbeam from P.A.L.M. Microlaser Technologies). The primary difference between these platforms is the way samples are ejected into the cap as illustrated in Fig. 2. The recovery percentage for targeted cells is higher with laser capture microdissection system (Veritas), but the prepared sample needs to be of optimum quality (completeness of wax embedding, tissue firmness, and flatness/evenness of the section) for successful capture. Uneven, loose sections raise the risk of contamination from unwanted tissues touching the capture membrane. For the LM and pressure catapulting system (PALM) the minimum quality threshold for sample sections is lower and consequently tissue preparation is easier (see Note 9).

3.3.1. Veritas System

1. After starting the Veritas machine, load the slides with infected tissues, up to three at a time. The machine automatically closes the door to the sample chamber, and scans the slides to create a global "roadmap" on the computer screen.

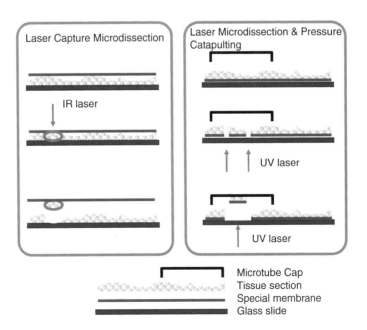

Fig. 2. Comparison of the two major types of laser microdissection.

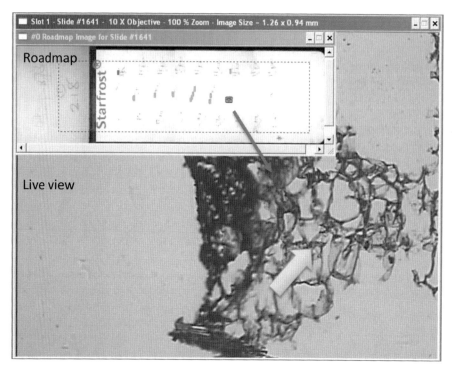

Fig. 3. Screenshot of laser capture microdissection system showing the "roadmap" (inset) and "live video" window. The large arrow points to the laser target position (*small circle*).

2. Move the red box of "live video" to show detailed image for a region of interest.

3. Select objective (2×, 10×, 20×, or 60×) from microscope window, and adjust the light intensity for observation under light field. Use the shutter to switch to fluorescence observation (such as Red, Blue, UV, or Green). Fungal structures can be distinguished from their surrounding cells under a UV filter with AmCyan fluorescence. The areas that have been selected for capture (dark grey squares in Fig. 3) can be stored in the computer, and then captured all at once (Fig. 3).

4. Move cap to the slide by selecting "Place Cap at Region Center." Enable the capture laser function and, while under light field, reduce the light intensity to a minimum, then click "Locate Capture Laser Position" to view the position of capture laser.

5. Move the capture laser to a blank region without any sample on the slide, double-click to test fire the capture laser.

6. Regulate "power" (ranging from 80 to 95 mW based on our experience) and "pulse" (around 2,500 μs) to adjust focus of capture laser (Fig. 4). Then label the target cells on the monitor screen. In a previous study (Zhang., Y. and Tang, W.H. unpublished), a fine UV-laser beam was first used to cut around

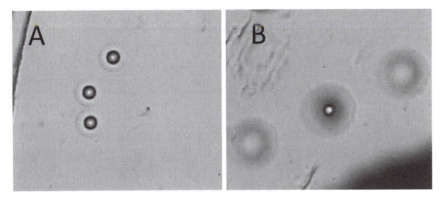

Fig. 4. Test firing the capture laser to check focus. (**a**) Capture laser in focus; (**b**) capture laser not focused.

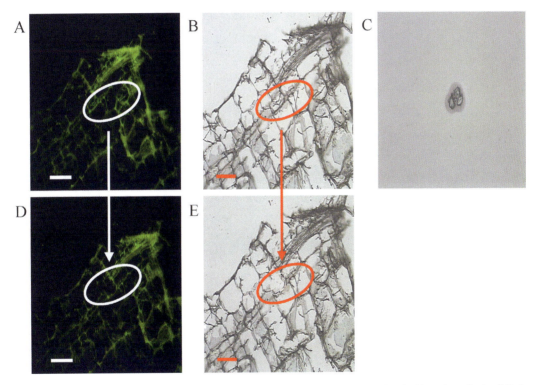

Fig. 5. Laser capture microdissection of fluorescent-tagged *Fusarium graminearum* hyphae inside maize cells. (**a, d**) Under UV filter; (**b, e**) light field; (**a, b**) before laser capture; (**c**) captured in the cap; (**d, e**) after laser capture.

the target (maize stalk parenchyma cells containing fluorescent-tagged *Fusarium graminearum* hyphae) to disconnect the target cells from surrounding cells, and an infrared (IR) laser beam was targeted to the specific area of the thermoplastic polymer film of the transfer cap just above the target cells, to activate the film to fuse with the target cells (Fig. 5). In this way, the target cells were captured by the cap and could be transferred to RNA extraction buffer.

7. The areas of collected sample are tracked by Veritas software as a reference for calculating captured cell number. In our study, two biological replicates, each composed of around 1,000 cells, were captured for RNA isolation.

8. After collecting sufficient amount of cells, adjust the microscope objective to inspect the cap for captured cells, and use "ablation" to destroy individual unwanted tissues attached to the cap if there are any (see Note 9).

3.3.2. PALM Microbeam System

The following steps are based on our previous studies using maize parenchyma cells (7):

1. Deparaffinized slides are placed on the microscope stage face up, where they can be visualized under brightfield illumination on the computer monitor through a video camera.

2. Fungal hyphae expressing AmCyan fluorescent protein are identified under a GFP longpass emission filter (HQ485/25x, HQ515LP, chroma).

3. Host cells containing fungal hyphae are first selected using the graphic tools of PALM RoboSoftware and then cut along the margins using a low-heat UV (337 nm nitrogen) laser.

4. The cells are then collected by laser pressure, which catapults them into the lid of a 0.5-mL RNase-free reaction tube filled with 40 μL of extraction buffer (from the PicoPure RNA isolation kit) placed in a holder immediately above the slide.

5. The collection time per 40 μL of extraction buffer should be no longer than 30 min. The cell numbers are estimated based on the collected area calculated by the PALM RoboSoftware.

In our experiments, we collected two samples, approximately 1 mm² in area (approximately 800 cells) per sample (see Note 9).

3.4. RNA Isolation and Quality Assessment

1. If using Veritas system, pipette 50 μL extraction buffer (XB) into a 0.5-mL microfuge tube. Insert the Capsure Macro LCM Cap (containing collected cells) onto the microfuge tube and incubate for 30 min at 42°C. Centrifuge at $800 \times g$ for 2 min to collect cell extract into the microfuge tube. Remove the CapSure Macro LCM Cap.

2. If using the PALM system, the extraction buffer is spun down to the bottom of tubes after collection and then briefly vortcxcd and incubated at 42° C for half an hour. If the 800 cells cannot be captured in one microfuge tube cap within half an hour, up to four caps of 50 μL extraction buffer collection can be pooled into one tube.

3. The sample (obtained in step 1a or 1b above) is then quickly frozen and stored at −80° C before RNA extraction, or proceed directly to the next step.

4. Isolate RNA from the LC sample from step 1a or 1b above, using PicoPure RNA isolation kit. Carefully follow the manufacturer's instructions (see Note 10). The purified RNA should be eluted into a 0.5-mL microfuge tube in 11 μL of buffer.

5. Remove 1 μL for RNA quality assessment and store the rest at −80°C (preferably aliquoted into to several tubes).

6. Using the Agilent Bioanalyzer 2100 and Agilent RNA 6000 Pico kit, carefully follow the manufacturer's instructions to obtain RNA quality and quantity estimates (see Note 11). You will want to accumulate 10–11 samples for a single run on the Bioanalyzer.

7. Check the electropherogram display on a computer monitor; if the 28S:18S ribosomal RNA peak height ratio is higher than 1 (as in (7); Fig. 5), the RNA quality can be considered sufficient for later analysis.

3.5. Linear Amplification of RNA

For a typical microarray experiment, anywhere from 1 to 20 μg of cRNA will be required per sample. If the total yield of LM RNA is less than 10 ng, we recommend the TargetAmp 2-round Aminoallyl-aRNA amplification kit to prepare the sample for downstream Agilent microarray hybridization. If the yield of LM-total RNA is greater than 50 ng, the TargetAmp 1-round Aminoallyl-aRNA amplification kit may suffice. If the yield of LM-total RNA is between 10 and 50 ng, either use the TargetAmp 2-round Aminoallyl-aRNA amplification kit from a fraction of LM-derived RNA or use the TargetAmp 1-round Aminoallyl-aRNA amplification kit based on your own optimization. The authors slightly favor the former. With practice, up to four reactions can be performed together, while for first time user, no more than two reactions should be processed at a time.

1. Round-One, first-strand cDNA synthesis: mix 1 μL TargetAmp T7-Oligo(dT) Primer B and about 0.5 ng of LCM-derived total RNA (in no more than 2 μL volume), add RNase-free water to bring up the total volume to 3 μL, and incubate at 65°C for 5 min in a PCR machine. Chill on ice for 1 min, then centrifuge at $10,000 \times g$ for 15 s. Prepare the Round-One first-strand cDNA synthesis master mix in a separate microfuge tube with 2 μL per reaction on ice as follows: 1.5 μL TargetAmp Reverse Transcription PreMix-SS, 0.25 μL DTT (from the TargetAmp kit), and 0.25 μL SuperScript III Reverse Transcriptase (200 U/μL, from Invitrogen). After gently mixing, add 2 μL of the Round-One first-strand cDNA synthesis master mix into each reaction, gently mix again, and incubate at 50°C for 30 min in a PCR machine with lid also heated to 50°C.

2. Round-One, second-strand cDNA synthesis: Prepare the Round-One, second-strand cDNA Synthesis master mix (5 μL per reaction mix) by combining 4.5 μL DNA Polymerase

PreMix-SS 1 and 0.5 µL DNA Polymerase-SS 1 gently on ice. Gently mix and then incubate at 65°C for 10 min. Centrifuge at $10,000 \times g$ for 15 s, then incubate at 80°C for 3 min. Centrifuge at $10,000 \times g$ for 15 s, and put the reaction on ice. Add 1 µL of cDNA Finishing Solution-SS to each reaction, gently mix, incubate at 37°C for 10 min, and then immediately incubate at 80°C for 3 min. Centrifuge at $10,000 \times g$ for 15 s, then chill on ice.

Warm the T7 RNA polymerase and T7 transcription buffer to 37°C. Prepare the Round-One, In Vitro Transcription Master Mix (39 µL per reaction) at room temperature by combining 4 µL T7 transcription buffer, 27 µL In Vitro Transcription PreMix A, 4 µL DTT, and 4 µL T7 RNA polymerase, gently mix, then add 39 µL to each reaction, mix gently, then incubate at 42°C for 4 h in a PCR machine. Then add 2 µL of RNase-Free DNase I to each reaction, mix gently, and incubate at 37°C for 15 min.

3. Round-One, In Vitro Transcription of aRNA: Warm the T7 RNA polymerase and T7 transcription buffer to 37°C. Prepare the Round-One, In Vitro Transcription Master Mix (39 µL per reaction) at room temperature by combining 4 µL T7 transcription buffer, 27 µL In Vitro Transcription PreMix A, 4 µL DTT, and 4 µL T7 RNA polymerase, gently mix, then add 39 µL to each reaction, mix gently, then incubate at 42°C for 4 h in a PCR machine. Then add 2 µL of RNase-Free DNase I to each reaction, mix gently, and incubate at 37°C for 15 min.

4. Round-One, RNA purification: Use Qiagen's RNeasy MinElute Cleanup Kit and carefully follow the manufacturer's instructions. The resulting ~14 µL aRNA elution can be stored overnight at −80°C or directly proceed to next step.

5. Round-Two, first-strand cDNA synthesis: Transfer the ~14 µL aRNA elution to a PCR tube, add 2 µL Random Primer-SS, and adjust the volume of each aRNA sample to 3 µL by speed vacuum centrifugation without heat (see Note 12). Incubate at 65°C for 5 min in a PCR machine. Chill on ice for 1 min and then centrifuge at $10,000 \times g$ for 15 s. Prepare the Round-Two, first-strand cDNA synthesis Master Mix on ice (2 µL per reaction) by combining 1.5 µL Reverse Transcription PreMix-SS, 0.25 µL DTT, and 0.25 µL SuperScript II Reverse Transcriptase (200 U/µL). Add 2 µL Master Mix into each sample and gently mix. Incubate at room temperature (20–25°C) for 10 min, and then incubate at 37°C for 1 h in a PCR machine. To each sample, add 0.5 µL of RNase H-SS, gently mix, incubate at 95°C for 2 min, chill on ice for 1 min, and then centrifuge at $10,000 \times g$ for 15 s.

6. Round-Two, second-strand cDNA Synthesis: Add 1 μL T7-Oligo(dT) Primer C into each reaction, gently mix and incubate at 70°C for 5 min, then incubate at 42°C for 10 min. Prepare the Round-Two, second-strand cDNA Synthesis Master Mix (13.5 μL per reaction) on ice by combining 13 μL DNA Polymerase PreMix-SS 2 and 0.5 μL DNA Polymerase-SS 2. Add 13.5 μL into each reaction, gently mix, and incubate at 37°C for 10 min. Centrifuge at $10,000 \times g$ for 15 s, and then incubate at 80°C for 3 min. Centrifuge at $10,000 \times g$ for 15 s, and then chill on ice.

7. Round-Two In Vitro Transcription of Aminoallyl-aRNA: Warm the T7 RNA polymerase and T7 Transcription Buffer to 37°C. Prepare the Round-Two, In Vitro Transcription Master Mix (40 μL per reaction) at room temperature by combining 9.6 μL RNase-Free Water, 4 μL T7 Transcription Buffer, 16 μL In Vitro Transcription PreMix B, 2.4 μL Aminoallyl-UTP, 4 μL DTT, and 4 μL T7 RNA polymerase. Add 40 μL Master Mix into each reaction. Gently mix, and then incubate at 42°C for 9 h. Incubation for more than 9 h is not recommended. If the reaction is left overnight, set the PCR machine to 4°C after 9 h incubation. Add 2 μL of RNase-Free DNase I to each reaction, mix gently, and then incubate at 37°C for 15 min.

8. Aminoallyl-aRNA purification: Use Qiagen's RNeasy MinElute Cleanup kit and follow the manufacturer's instructions (see Note 13).

9. Determining concentration of the aminoallyl-aRNA: 1 μL of the amplified aminoallyl-aRNA is diluted to around 1 ng/μL for quality assessment using an Agilent Bioanalyzer 2100 and RNA 6000 Pico Kit as in 4.5 steps 10–14. Check the electropherogram from a computer monitor; if it is a "bell" shape with a size peak of more than 250 nucleotides, the RNA quality can be considered sufficient for later analysis.

3.6. Aminoallyl cRNA Labeling (for Microarray Hybridization) (see Note 14)

For each comparison, equal amounts of aminoallyl cRNA are coupled with Cy3 or Cy5 monoreactive NHS esters for 30 min in the dark according to Ambion's dye-coupling protocol (Amino-Allyl MessageAmp II Kit, Ambion). After the reactions are quenched with 4 M hydroxylamine HCl, RNeasy columns (Qiagen) are eluted with water and quantitated on a spectrophotometer to measure sample concentration and dye incorporation.

3.7. Recommendations for Downstream Microarray Analysis

1. Fluorescence Labeling
The amplified aminoallyl-aRNA can be coupled with either biotin for Affymetrix gene chip hybridization or Cy3/Cy5 for Agilent microarray hybridization. We recommend Cy3/Cy5 coupling if it is an option, because the efficiency of Cy3/Cy5 coupling can be directly determined by fluorescence measurements

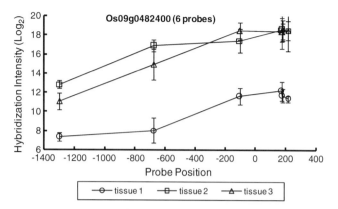

Fig. 6. Effect of probe position on hybridization intensities for a single gene in an Agilent microarray analysis of LM-derived samples, using rice as an example. The y-axis shows hybridization intensities for multiple probes of a rice gene, Os09g0482400 in three types of rice tissue. The x-axis (probe position) represents the distance (number of nucleotides) from the 5′ end of the 60-mer probe target sequence to the translation termination codon.

after reaction and before hybridization, while efficiency of biotin coupling cannot be examined directly.

2. Probe Preparation

The amplified aminoallyl-aRNAs range in size from 100 to 500 bp, shorter than the original mRNA, so we advise skipping the fragmentation step in the probe preparation procedure. This small aRNA fragment size also has implications for Agilent microarray performance, since any 60mer probes having a target more than 500 bp away from polyA site will have disproportionately low signal relative to the actual expression level (Fig. 6). Since most commercial Agilent microarray 60mer probes fall within the 3′UTR due to the higher specificity relative to coding regions, this may not be a major issue, but this caveat should be kept in mind, especially if a new array design is contemplated.

3.8. Data Analysis

Microarray analysis of LM-derived RNA also needs specific consideration. The transcriptome from a cell population that is obtained from LM can be very different from that of a mixed tissue sample or another LM-picked cell population. Therefore these samples may not meet the assumption for quantile normalization that most of the genes are constantly expressed over the range of intensities or that the numbers of up- and downregulated genes over the intensity range are equal. Therefore, we do not recommend using quantile normalization and its derived methods, such as LOWESS method, for this type of microarray. Using an intensity-based normalization method (13) with a set of reference genes is more suitable.

4. Notes

1. Biosafety considerations: Depending on the containment conditions mandated by regulatory agencies for the specific pathogen being studied, researchers will need to put together a containment plan for each stage of the experiment to insure that the recombinant fungal pathogen is not released to the environment.

2. Ability to visualize the pathogen inside the plant tissue of choice is critical. Many plant cells can produce phenolic compounds upon fungal infection, contributing to autofluorescence (14, 15). Autofluorescence at wavelengths that overlap with the fluorescent protein may obscure the fungal signal, so careful preexperimentation is essential. Preliminary experiments should be carried out using a transgenic fungal strain expressing a fluorescent protein driven by a strong constitutive promoter (such as the RP27 promoter from *Magnaporthe oryzae*) (16). The suspension of conidia at the appropriate concentration should be able to provide a reproducible infection pattern in host tissues based on prior experiments. In our studies (7), a conidial suspension of the transgenic fungal strain (10^6 per milliliter) in sterile water was freshly prepared, and was used for inoculation as soon as possible.

3. To ensure that sufficient sample numbers (i.e., sites of infection at a similar stage) can be obtained by LM, the method of inoculation should be carefully selected based on the following criteria: (A) The fungal infection sites need to be sufficiently numerous to allow final LM of about 1,000 cells. (B) The plant tissue selected for inoculation should be reasonably uniform (i.e., not too complex). If the inoculated tissue is too complex, the fungal pathogen in different parts of this tissue may not grow synchronously, making it difficult to achieve a uniform infection stage. (C) The tissue that the pathogen will infect under natural conditions should be considered first, as this will best reflect the molecular basis underlying the natural situation. However, this point might be compromised by point C. (D) Multiple individual plants should be inoculated to get independent biological replicates.

4. Maximizing the specificity and homogeneity of LM samples: Pilot experiments should be done to observe the infection pattern of the fluorescent-tagged fungal pathogen inside the plant tissue. Harvesting infected tissues at a fixed time postinoculation is the first step in obtaining uniformly-staged fungal samples. But this may not be sufficient because each fungal propagule infects host cells independently and the process of infection cannot be assumed to be uniform, especially if the

fungus encounters host cells with different cell wall thickness and different degree of viability. Therefore, the features of host cells the pathogen associates with can also be used as a sampling criterion. For example, in infection of maize stem, fungal invasion of vascular bundles, parenchyma cells, and epidermal cells can be distinguished microscopically, so these tissues can be isolated separately by LM. The morphology of the fungal pathogen (either spores, hyphae, or sexual reproduction structures) can also be considered as a feature to help distinguish different infection stages. In addition, considering the lag between gene expression and phenotypic expression, it may be advantageous to harvest the infected tissue at a time slightly earlier than the stage identified by fluorescent microscopy. Finally, the time of inoculation should be fixed in the day/ night cycle, to avoid the expression differences caused by circadian gene expression (17).

5. Identifying appropriate control samples: The type(s) of control samples used for evaluating pathogenesis-related gene expression will depend on the goals of the experiment and the specific profiling technology to be used, but in many cases it will be important to contrast pathogenesis-related gene expression to a different growth condition, e.g., growth *in vitro*. In addition, the researcher may want to evaluate mock-inoculated host tissue transcripts to establish background fungal signal levels in a typical LM-retrieved sample. While these details are beyond the scope of this paper, numerous reviews address the range of approaches to pathogen profiling and analysis (18).

6. The procedure from sample preparation to RNA amplification (Subheading 3.1–3.5; Fig. 1) requires a minimum of 5 days for one sample. Optimizing sample preparation and preexperiments will take additional time.

7. In Subheading 3.2, step 2, the choice of section thickness depends on the target sample. Because the fungal hyphae are usually less than 10 μm thick, a thinner section is better in order to avoid capturing too many plant cell components along with target fungal hyphae. However, thinner sections may contain shorter fragments of hyphae if the hyphae are not growing exactly parallel to the section angle.

8. In Subheading 3.2, step 3: A "rainbow" set of serial sections is ideal because it can contain more tissue samples on a single slide to ease further LM. However, if it is difficult to get a "rainbow" serial, it is also fine to use the tape-transfer system, one section per slide.

9. In Subheading 3.3, we recommend capturing the plant cell containing the fungal hyphae (area >8 μm in diameter) instead of capturing the hyphae (around 2–3 μm in width) only. In the

Veritas system, it is technically difficult to capture any target area smaller than 5 µm in diameter, because the minimal size of the capture laser at 40× objective is around 5 µm in diameter. In PALM microbeam system, because the cutting UV laser will destroy the RNA within its cutting lane (1–2 µm in width), cutting exactly along the hyphae increases the possibility of degrading the RNA in the hyphae. Probably because catapulting uses a lower power laser, catapulting directly targeting the hyphae didn't cause RNA degradation in our experience.

10. In Subheading 3.4 step 4, the volume of PicoPure Conditioning Buffer to be added should be 5 times of that total volume of sample collected in step 1a or 1b. For example, if two caps of 50 µL extraction buffer collection were pooled into one tube in step 1b, then the total sample volume is 100 µL, so add 500 µL of PicoPure Conditioning Buffer.

11. In Subheading 3.4 step 6, prepare the RNA ladder in advance, dilute the ladder to 1 ng/µL, and aliquot to 2 µL per tube. Instead of preparing 550 µL of RNA 600 Pico gel matrix, one can choose to prepare smaller amount (such as 150 µL) of RNA 600 Pico gel matrix. The prepared gel matrix needs to be used within 4 weeks (preferably within 1 week).

12. In Subheading 3.5, step 5 is a critical step, as in our experience overdrying can cause nonspecific amplification. A preexperiment is recommended. Use a tube with the exact amount of water as the RNA solution to test the time needed for vacuum drying.

13. In Subheading 3.5 step 8, use of Qiagen's RNeasy MinElute Cleanup Kit instead of RNeasy Mini Kit is recommended because the expected yield of aminoallyl-aRNA from laser microdissected sample usually is smaller than 20 µg.

14. In Subheading 3.6, the aRNA amplification is sufficient for Agilent microarrays; for Affymetrix arrays, the amplification method needs to be optimized. We recommend the use of Affymetrix's GeneChip Expression 3, Amplification Two-Cycle Target Labeling and Control Reagents (900494), combined with the MEGAscript High Yield Transcription Kit (Ambion, AM1333).

Acknowledgments

We thank K. Ohtsu and P. Schnabel for discussions of LM technique, and Barbara Valent for stimulating discussions. We also thank Xiang Tang, Xiao-Wei Zhang, Zhi-Yong Zhang, Dong Zhang, and Ting-Lu Yuan for help in organizing the lab protocols. Wei-Hua

Tang's research is supported by the Ministry of Science and Technology of China (grant nos. 2007CB108700 and 2006CB101901).

References

1. Shaw KJ and Morrow BJ (2003) Transcriptional profiling and drug discovery. Curr Opin Pharmacol 3: 508–512

2. McDonagh A, Fedorova ND, Crabtree J et al. (2008) Sub-telomere directed gene expression during initiation of invasive aspergillosis. PLoS Pathog 4: e1000154

3. Hahn M and Mendgen K (1997) Characterization of *in planta*-induced rust genes isolated from a haustorium-specific cDNA library. Mol Plant Microbe Interact 10: 427–437

4. Godfrey D, Zhang Z, Saalbach G et al. (2009) A proteomics study of barley powdery mildew haustoria. Proteomics 9: 3222–3232

5. Mosquera G, Giraldo MC, Khang CH et al. (2009) Interaction Transcriptome Analysis Identifies *Magnaporthe oryzae* BAS1-4 as Biotrophy-Associated Secreted Proteins in Rice Blast Disease. The Plant Cell 21: 1273–1290

6. Czymmek KJ, Bourett TM and Howard RJ (2004) Fluorescent Protein Probes in Fungi. Methods in Microbiology 34: 27–62

7. Tang WH, Coughlan S, Crane E et al. (2006) The application of laser microdissection to *in planta* gene expression profiling of the maize anthracnose stalk rot fungus *Colletotrichum graminicola*. Mol Plant Microbe Interact 19: 1240–1250

8. Chandra D, Inada N, Hather G et al. (2010) Laser microdissection of Arabidopsis cells at the powdery mildew infection site reveals site-specific processes and regulators. Proc Natl Acad Sci USA 107: 460–465

9. Balestrini R, Gomez-Ariza J, Lanfranco L et al. (2007) Laser microdissection reveals that transcripts for five plant and one fungal phosphate transporter genes are contemporaneously present in arbusculated cells. Mol Plant–Microbe Interact 20: 1055–1062

10. Takahashi H, Kamakura H, Sato Y et al (2010). A method for obtaining high quality RNA from paraffin sections of plant tissues by laser microdissection. J Plant Research DOI: 10.1007/s10265-010-0319-4

11. Ohtsu K and Schnabel PS (2007) Maize Tissue Preparation and Extraction of RNA from Target Cells for Genotyping. Cold Spring Harb Protoc; 2007; doi:10.1101/pdb.prot4784

12. Inada N and Wildermuth MC (2005) Novel tissue preparation method and cell-specific marker for laser microdissection of Arabidopsis mature leaf. Planta 221: 9–16

13. 't Hoen PA, Turk R, Boer JM et al. (2004) Intensity-based analysis of two-colour microarrays enables efficient and flexible hybridization designs. Nucleic Acids Res 32: e41

14. Siranidou E, Kang Z, and Buchenauer H (2002) Studies on symptom development, phenolic compounds and morphological defence responses in wheat cultivars differing in resistance to Fusarium head blight. J Phytopathol 150: 200–208

15. Philippe S, Tranquet O, Utille JP et al. (2007) Investigation of ferulate deposition in endosperm cell walls of mature and developing wheat grains by using a polyclonal antibody. Planta 225: 1287–1299

16. Bourett, TM, Sweigard JA, Czymmek KJ, et al. (2002) Reef coral fluorescent proteins for visualizing fungal pathogens. Fungal Genet Biol 37: 211–220

17. Dong W, Tang X, Yu Y et al. (2008) Systems biology of the clock in *Neurospora crassa*. PLoS One 3: e3105

18. Wise RP, Moscou, MJ, Bogdanove AJ, and Whitham SA (2007) Transcript Profiling in Host–Pathogen Interactions. Annu Rev Phytopathol 2007. 45: 329–69

Chapter 15

Metabolomics Protocols for Filamentous Fungi

Joel P.A. Gummer, Christian Krill, Lauren Du Fall, Ormonde D.C. Waters, Robert D. Trengove, Richard P. Oliver, and Peter S. Solomon

Abstract

Proteomics and transcriptomics are established functional genomics tools commonly used to study filamentous fungi. Metabolomics has recently emerged as another option to complement existing techniques and provide detailed information on metabolic regulation and secondary metabolism. Here, we describe broad generic protocols that can be used to undertake metabolomics studies in filamentous fungi.

Key words: Systems biology, Metabolomics, Metabolome, Metabolite, Fungi, Derivatisation, GC-MS, LC-MS, NMR, SPE

1. Introduction

Metabolomics is the most recent of the "omics techniques to be embraced by functional genomics" (1–4). The metabolome is defined as the quantitative complement of low-molecular-weight metabolites present in a cell under a given set of physiological conditions (5, 6). Whilst often viewed as a complementary technique to transcriptomics and proteomics, metabolomics does offer distinct advantages (5–7). First, the metabolome is directly related to the phenotype; metabolic fluxes are regulated not only by gene expression, but also by post-transcriptional and post-translational events (Fig. 1). Second, changes in the metabolome are amplified relative to changes in the transcriptome or proteome. It is also true that the technology involved in metabolomics is generic, as a given metabolite, unlike a transcript or protein, is the same in every organism that contains it. Consequently, the metabolomics data will directly highlight metabolic pathways and processes differentially regulated,

Melvin D. Bolton and Bart P.H.J. Thomma (eds.), *Plant Fungal Pathogens: Methods and Protocols*,
Methods in Molecular Biology, vol. 835, DOI 10.1007/978-1-61779-501-5_15, © Springer Science+Business Media, LLC 2012

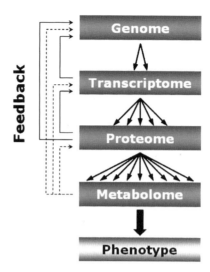

Fig. 1. General schematic of the omics pathway.

and by consequence, the genes and proteins responsible for their synthesis. In broad terms, the field can be broken into two distinct forms, targeted and non-targeted. In targeted metabolomics, we know the metabolite we are looking for and thus can optimise the extraction methods and instrumentation accordingly. Good examples of targeted metabolomics include the analysis of mycotoxins in food samples. The primary focus of non-targeted metabolomics is to determine the relative concentration of as many metabolites as possible in the sample at the time of sampling. Unlike DNA or proteins that are comprised of a defined set of building blocks, the variation of metabolites in a living cell in terms of chemical properties is enormous ranging from small and polar (e.g. hexoses) and to large and non-polar (e.g. lipids). To maximise the coverage of metabolites, different technologies are often employed and these are described briefly below.

Gas chromatography-mass spectrometry (GC-MS) is a versatile technique that is well suited to non-targeted metabolite profiling of volatile and thermally stable polar and non-polar metabolites (8). Classes of metabolites routinely measured by GC-MS include sugars, amino acids, phosphorylated metabolites, organic acids, lipids and amines. Typically, most of these compounds are non-volatile and thus require comparatively complex derivatisation stages to be included during sample preparation to confer volatility and thermal stability. There are several configurations of mass spectrometers that are typically used in combination with GC, including linear quadrupole, ion trap and time-of-flight. Throck Watson and David Sparkman (2007) provide comprehensive details of each of these platforms, and Dunn (9) provides an elegant overview of each with a focus on metabolomics. The protocol listed below describes the use of a single linear quadrupole mass spectrometer (10–13).

Liquid Chromatography-Mass Spectrometry (LC-MS) is another technique commonly used in metabolomics. Sample preparation for LC-MS is simple and it caters well for compounds that are not well suited for GC-MS (large, thermally labile, non-volatile etc.). However, LC is technically demanding, requiring the use of large amounts of MS-grade solvents. These are not only expensive to purchase, but also to dispose of. Technical reproducibility is not trivial, particularly compared with GC, and retention time drift is also problematic. There is also the issue that no single column is suitable for a wide range of analyte classes. Although this is true for GC, it is significantly more so for LC. To date in fungi, LC-MS has been typically used to examine and screen for secondary metabolites (14) rather than non-targeted scanning of primary metabolites. Secondary metabolites are typically unsuited to GC-MS as they are often too large, particularly after derivatisation. The method below outlines a robust technique for screening secondary metabolites using liquid chromatography with an ion trap mass spectrometer.

The third metabolomics technique to be considered in this chapter is Nuclear Magnetic Resonance (NMR). NMR is a spectroscopic technique that identifies, particularly, atom–atom bonds, and so differs fundamentally from the methods reviewed above which are MS-based. Like all techniques, there are strengths and weaknesses to consider when using NMR for metabolomics (15, 16). NMR is a non-biased and non-destructive technique that requires very little sample preparation. This lessens the likelihood of sample loss or the introduction of artefacts into the preparation. NMR also has the potential for high-throughput fingerprinting, allowing for the analysis of hundreds of samples per day. It is only paramagnetic nuclei that can be detected, but as these include ^1H, ^{13}C, ^2H, ^{14}N, ^{15}N, ^{17}O, ^{19}F, ^{23}Na, ^{29}Si, ^{31}P and ^{35}Cl, all biomolecules are well covered. The significant drawback with NMR compared with MS-based methods, however, is sensitivity. It has been reported that NMR is only half as sensitive as MS techniques, thus precluding the detection of low-abundance metabolites (17). These strengths and weaknesses aside, NMR has been extensively used for metabolic fingerprinting, particularly in the medical fields. Studies are now starting to emerge exploiting its strengths in the fungal field.

2. Materials

Use only analytical grade solvents or higher grade if available. Always use high purity milli-Q filtered water. The volume of solutions and number of each consumable required will depend on the number of replicates/extracts used. All of the required safety protocols and MSDSs must be followed

2.1. Materials for GC Analysis

1. 5- or 10-mL Pipette with tip, one per replicate.

2. 50-mL Falcon tubes, one per replicate.

2.1.1. Harvesting of Fungal Tissue and Quenching of Metabolism

3. Forceps.

4. Quenching solution; −40°C 75% methanol in water.

Flask Culture

5. Rinsing solution; −40°C 50% methanol in water.

6. Dry ice, enough to submerge tubes of quenching & rinsing solutions.

7. 50-mL Syringes, one per replicate.

8. Cotton gauze 100 cm^2, one per replicate.

9. Spatula.

10. Liquid nitrogen, enough to submerge falcon tubes.

11. Freeze dryer.

Plate Culture

1. Scalpel, handle size 3.

2. Round edge scalpel blade, size 10 (to fit handle).

3. 2-mL Snap-lock microcentrifuge tubes (see Note 1).

4. Liquid nitrogen, enough to submerge 2-mL microcentrifuge tubes.

5. Freeze dryer.

2.1.2. Extraction of Metabolites from Fungal Tissue

1. Spatula.

2. 2-mL Safe-lock microcentrifuge tubes (see Note 1).

Cell Disruption and Metabolite Isolation

3. Weight scale—capable of measuring mg quantities.

4. Methanol.

5. Milli-Q water.

6. Ribitol (>99% purity).

7. Tissue lyser—equipped with capsule to hold 2-mL tubes (QIAGEN; see Note 2).

8. Liquid nitrogen.

9. Ice.

10. Vortex.

11. Microcentrifuge tubes (1.5–2 mL).

12. Bench-top microcentrifuge.

13. Pipettes.

Drying Metabolites

1. Speedivac vacuum concentrator.

2. Freeze dryer (see Note 3).

2.1.3. Derivatisation and GC-MS Analysis of Fungal Metabolites

1. Methoxyamine HCl, 20 mg per sample.

2. Pyridine, 10 mL would be sufficient.

3. Thermomixer.

4. *N*-Trimethylsilyl-*N*-methyl   trifluoroacetamide   (MSTFA), 30 μL per sample.

5. Hexane.

6. Alkanes (see Note 4).

7. Glass vials; 1 mL.

8. Glass insert; 100 μL.

9. Crimp caps.

10. Crimper.

11. GC-MS equipped with an autosampler and syringe capable of 1 μL injections.

2.1.4. Deconvolution and Identification of Metabolites from GC/ LC-MS Data

1. Deconvolution and analysis software.

2. Multivariate analysis software, capable of PCA.

3. Statistical analysis software.

2.2. Materials for LC Analysis

2.2.1. Extraction of Fungal Secondary Metabolites

1. 100% Methanol, AR grade (or MS grade).

2. Milli-Q water.

3. 5% Methanol in Milli-Q water.

4. Hexane, LC-MS grade or better.

5. Sodium chloride, AR grade.

6. Griseofulvin (internal standard).

7. 50-mL Falcon tubes.

8. Bench-top centrifuge and 50-mL tube adaptors.

9. Vortex.

10. Sonic waterbath (optional).

11. 40-mL amber glass vials.

12. Bulk C18 Solid Phase Extraction (SPE) material.

13. 8-mL empty SPE reservoirs with frits.

14. 500 mg C18 SPE columns or cartridges.

15. 50-mL syringe.

16. Glass wool.

17. Nitrogen drying manifold.

18. Pipettes and filter tips.

2.2.2. LC-MS Analysis of Fungal Metabolites

1. Microcentrifuge tubes (1.5 mL).

2. Bench-top microcentrifuge.

3. Amber glass autosampler vials with caps.

4. 100-μL limited volume glass inserts.

5. Water, LC-MS grade.

6. Acetonitrile, LC-MS grade.

7. Formic acid, (>99% pure).

8. HPLC-MS system with autosampler.

9. RP18 column with guard column.

2.3. Materials for NMR Analysis

2.3.1. Culturing and Culture Harvesting

1. Carbon substrate (^{13}C-Labelled optional (see Note 5)) to add to fungal growth medium. For natural abundance spectra add e.g. 40 mM glucose.

2. Make a 250-mM solution of the required standard in 1 mL deuterium oxide. Lyophilise to dryness. Prior to NMR spectrum acquisition, resuspend thoroughly in 1 mL deuterium oxide, centrifuge at maximum speed in a bench-top microcentrifuge to pellet any insoluble matter, and add 700 µL to a NMR tube.

3. Sterile mortar and pestle (1 per sample).

4. Liquid nitrogen.

5. Methanol:water Solution: make a 70:30 methanol:water solution and store at –20°C. Each sample requires 30 mL.

6. Sterile glass funnel lined with Whatman #1 paper filter, placed in a 50-mL Falcon tube standing in ice (1 per sample).

7. Rotary evaporator.

8. One 10 mL tube per sample.

9. Freeze dryer.

10. Deuterium oxide.

11. Internal reference compound (optional but recommended) (see Note 6).

12. Two 1.5 mL eppendorf tubes per sample.

13. Bench-top microcentrifuge.

14. Clean NMR tubes (1 per sample).

15. There are a number of NMR analysis software platforms for viewing and annotating spectra.

3. Methods

The following procedures were performed on *in vitro* cultures of the filamentous ascomycete *Stagonospora nodorum*. Briefly, the fungus was grown by inoculation of both liquid minimal medium and solid agar plate medium with 25 mM glucose as the carbon source. Flask cultures were grown in the dark for 1 week at 22°C with shaking at 150 rpm. Plate cultures were grown for 1–2 weeks with a 12-h white-light regimen at 22°C (18). For statistical significance of the metabolomic data, the preparation of eight biological replicates is recommended. The appropriate controls should be included (see Note 7). It is not the purpose of this protocol to include operation of LC/GC-MS or NMR equipment. Platform-specific training will be required for any of those instruments.

3.1. GC Methods

3.1.1. Harvesting Fungal Tissue and Quenching of Metabolism (see Note 8)

Flask Culture

1. Pipette 10 mL of culture (cut tapered tip off pipette tip if required) into a 50-mL falcon tube containing 20 mL Quenching Solution.

2. (In advance) Remove the plunger from a 50-mL syringe, fold a 100-cm^2 square of gauze 3 times and line the eluting end of the syringe with the folded gauze. Pour the entire contents of the falcon tube into the syringe, place the plunger back in the end and elute the solution into a fresh falcon tube (if the analysis of extracellular metabolites/media components is required).

3. Carefully remove the plunger, pour 10 mL rinsing solution over the mycelia resting on the gauze, return the plunger and elute the washing solution for discarding.

4. Using forceps carefully remove the gauze from the syringe and scrape the mycelia into a fresh falcon tube with a spatula.

5. Freeze the mycelia in liquid nitrogen.

6. Lyophilise the fungal tissue in a freeze dryer.

Plate Culture

1. Using a scalpel equipped with a round-edged blade, scrape the mycelia from the plate whilst avoiding disruption of the growth medium.

2. Scrape the mycelia from the scalpel blade into a microcentrifuge tube.

3. Freeze the mycelia in liquid nitrogen.

4. Lyophilise the fungal tissue in a freeze dryer.

3.1.2. Extraction of Metabolites from Fungal Tissue

The following protocol describes the extraction of metabolites from lyophilised fungal tissue regardless of the culture conditions or harvest method. The data analysed in this protocol uses metabolites extracted from 2 mg of lyophilised fungal tissue (see Note 8). The extraction steps should be performed at low temperatures wherever possible.

1. Transfer at least 2 mg of dried fungal tissue to a 2-mL safe-lock microcentrifuge tube and record the weight of the transferred tissue.

2. To the tube, add 685 μL of methanol and shake vigorously for 2 min in a tissue lyser at 30 Hz (see Note 2).

3. Remove from the lyser and to the tube add 75 μL of water containing 1.25 μg ribitol (internal standard) per 1 mg of fungal tissue and return the tube to the tissue lyser for a further 2 min.

4. Freeze the tube in liquid nitrogen by gripping the tube with forceps in an upright position until the suspension is frozen.

5. Remove from the liquid nitrogen and gently open the lid to the tube as soon as possible, to avoid the lid bursting off as the tube warms. Allow the suspension to thaw on ice.

6. Vortex the tube briefly and centrifuge at $20,000 \times g$ for 2 min to collect the cell debris.

7. Transfer the entire supernatant to a fresh microcentrifuge tube being careful not to disturb the pellet.

8. To the remaining pellet, add 250 μL 90% methanol. Place the tube back in the shaker for a further 2 min.

9. Remove the tube from the tissue lyser and again centrifuge at $20,000 \times g$ for 2 min.

10. Add the supernatant to that previously collected and discard the pellet.

11. Mix the combined supernatants by vortex and transfer a 2 mg equivalent (see Note 9) of supernatant to a fresh tube.

12. Dry the extracted samples in 200–300-μL glass inserts placed inside 1.5-mL plastic tubes using a centrifugal evaporator (see Note 3).

13. If the analysis of extracellular components is required, the eluate collected during the harvest should be vortexed and 100 μL (the amount used may need to be adjusted depending on the concentration of media components remaining in the extracellular eluate) transferred to a fresh microcentrifuge tube containing 2.5 μg ribitol and drying the contents in a centrifugal evaporator (see Note 3). Follow the steps as per the metabolite extracts from this point.

3.1.3. Derivatisation and GC-MS Analysis of Fungal Metabolites

Various options are available for derivatisation. The method below uses MSTFA which is suitable for most classes of compounds. Other derivatisation techniques are described in detail in Halket and Zaiken (19).

1. Dry the extracted samples (approximately 2 mg equivalent of tissue) in 200–300-μL glass inserts placed inside 1.5-mL plastic tubes using a centrifugal evaporator.

2. Add 20 μL of methoxyamine hydrochloride and shake for 2 h (1,250 rpm, 37°C) (see Note 10 for online derivatisation procedure provided a Gerstel or CTC autosampler is available).

3. Add 30 μL of MSTFA and continue shaking (1,250 rpm, 30°C) for 30 min.

4. Finally add 5 μL of alkanes (6.2 μg/mL) to enable retention indices to be calculated in the data analysis stage.

5. Transfer glass inserts into autosampler vials and seal with crimp caps. Derivatised samples should be analysed within 2 days.

6. Place derivatised samples into the autosampler tray on the GC-MS and set up a sequence table specifying vial positions and sample names.

7. 1 μL injections are most commonly carried out in splitless and split injection modes depending upon sample concentration (see Note 11).

8. It is advised that methanol blanks be run every three samples in splitless injection mode and every six in split mode to minimise possible carryover of compounds remaining on the column to the subsequent analyses.

9. For comprehensive metabolomic profiling we suggest using the 5.6°C/min oven temperature gradient (see Table 1 for parameters and mass spectrometer settings).

10. Tuning of the instrument and air/water checks should be undertaken before each experiment.

3.1.4. Deconvolution and Identification of Metabolites from the GC-MS Data

The acquired GC-MS data can be analysed using the freely available AMDIS package (http://chemdata.nist.gov/mass-spc/amdis/) or a commercial software package such as AnalyzerPro™ (SpectralWorks Ltd., Runcorn, United Kingdom). AnalyzerPro identifies individual analytes from the GC-MS data which includes deconvoluting co-eluting and partially co-eluting peaks using the mass spectral

Table 1
GC-MS parameters for non-targeted metabolomics using a single-quad mass spectrometer

GC column	Varian factor 4 column (VF-5ms; 30 m × 0.25 mm × 0.25 μm + 10 m Ezi-guard) (see Note 12)
Autosampler	1 pre-wash with hexane 5 sample pumps prior to injection 5 post washes with methanol, 5 with hexane
GC oven gradient (5.6°C/min)	70°C (hold 1 min), ramp to 76°C at 1°C/min, ramp to 325°C at 5.63°C/min (hold for 3 min)
GC oven gradient (15°C/min)	70°C (hold 1 min), ramp to 325°C at 15°C/min (hold 3 min) (see Note 13)
Other GC parameters	Carrier gas flow (Helium) 1 mL/min at constant flow (see Note 14) Inlet temperature 250°C
Mass spectrometer parameters	Transfer line (auxiliary) 280°C Quadrupole mass analyser 150°C Source 250°C Solvent delay before MS turned on approximately 7.5 min (will require some adjustment with different instruments/columns) Scan m/z range 50–600, sample rate of 2^2 for the 5.6°C/min method (2^1 for the 15°C/min method)

information collected. With the construction of a "metabolite library" (see Note 15), analytes can be cross-referenced against it and identified as metabolites. An output of metabolite peak areas representing the abundance of each metabolite can be used for the subsequent analysis. The following describes this process of analysis.

1. Check the chromatography for overloading peaks. If required, individual analytes may need to be quantified on a less abundant ion.

2. Analytes that are not matched against a reference from the library should be identified as "unknowns" and labelled accordingly, incorporating identifiers such as the retention time, retention index and base peak in the label.

3. Identify analytes common to the test samples and the controls. Some of these are likely to have been introduced during preparation and should be eliminated from the data set.

4. Normalise the metabolite abundances to the ribitol (internal standard) by dividing the peak area of each metabolite by the peak area of the ribitol.

5. Scaling the data by $x = \log(x+1)$ transformation of the ribitol-normalised data to reduce the impact of a large dynamic range of metabolite abundances within the data set.

6. Perform a principal component analysis (PCA) of the scaled data to identify major sources of variance in the data set, and thus the most influential metabolites.

7. Influential metabolites should be validated for statistical significance using the ribitol-normalised data.

3.2. LC Methods

3.2.1. Extraction Protocol for Secondary Metabolites

Culture Filtrate

1. For each sample, weigh 0.5 g of bulk SPE material (preferably water wettable, see Note 16) to a 50-mL tube. Activate with 3 mL methanol, remove excess, then wash and equilibrate twice with 5 mL milli-Q water (can be prepared beforehand, see Note 17). Carefully remove excess water and introduce the culture filtrate in ~40 mL batches. Shake horizontally for 20–30 min each.

2. Allow to stand for 1 min for SPE material to settle between renewing culture filtrate. Repeat until the desired volume of culture filtrate is processed.

3. While the culture filtrate is extracting, prepare one empty SPE reservoir with frit per sample.

4. Remove all culture filtrate from tubes. Add 5 mL milli-Q water, resuspend the SPE material and transfer to an empty SPE reservoir with frit.

5. Wash the SPE material with 5 mL 5% methanol. Elute with 3 mL 100% methanol (see Notes 16 and 17).

3.2.2. Mycelium

1. Weigh freeze-dried mycelium in a 50-mL tube, add 100 µg Griseofulvin, 30 mL 100% methanol and shake for 2 h.

2. Filter the extract into a weighed 40-mL amber glass vial using a 50-mL syringe with ca 1 cm high glass wool stuffed into the eluting end (can be prepared beforehand). Squeeze the mycelium thoroughly. Dry the extracts under a stream of nitrogen and gentle heating (see Note 18).

3. Reconstitute dried extracts in 10 mL milli-Q water. Vortex thoroughly (or use a sonic water bath) to completely dissolve extracts.

4. Add sodium chloride to the extracts to facilitate phase separation (optional, see Note 19). Overlay with 10 mL hexane and vortex or shake vigorously. Centrifuge at maximum speed for 5–15 min.

5. Remove the hexane completely (see Note 20).

6. Load the cleared aqueous extracts onto 500 mg C18 SPE columns at a flow rate of 1–5 mL/min. Wash with 5 mL 5% methanol in milli-Q water. Elute with 3 mL 100% methanol into weighed 4-mL amber glass vials (see Note 21).

3.2.3. LC-MS Analysis of Metabolites

1. Dry the methanol extracts obtained from the SPE on the nitrogen drying manifold (see Note 18).

2. Resuspend in 300 µL 5% acetonitrile, transfer to a reaction tube and centrifuge at maximum speed for 5–10 min (see Notes 16 and 21).

3. Transfer a 50-µL aliquot into a glass vial containing a 100-µL limited volume glass insert.

4. The parameters for an optimal HPLC separation and MS detection are highly platform dependent and need to be optimised for every standard, analyte or combination of those. We recommend using standard method development settings as recommended by the manufacturer of your instrument(s) and refer to further publications (14, 20–25).

5. Our laboratory typically uses an Agilent 1100 LC-MSD IonTrap. The HPLC mobile phase is water:acetonitrile 95:5 (both with 0.1% formic acid), flow rate 0.2 mL/min. We commonly use a 150 × 2.1-mm, 3-µm GraceSmart C18 column with guard column and inject 1–5 µL of sample, running non-injection blanks at the beginning of each sequence and every 3–4 samples to check for column contamination (Table 2).

3.3. NMR Methods

3.3.1. Methanol: Water Extraction of Polar Metabolites

1. Grow fungal culture in flask in preferred media and growth conditions.

2. Pre-cool a sterile mortar and pestle with liquid nitrogen.

3. Add the lyophilised fungal culture and grind under liquid nitrogen until it is reduced to a fine powder (see Note 22).

Table 2
Typical chromatography parameters for LC separation of fungal secondary metabolites

Step	Time (min)	% A (water)	%B (acetonitrile)
Start	0.0	95	5
Initial hold	10.0	95	5
Gradient	60	0	100
Top hold	75	0	100

Post time: 15 min
Electrospray ionisation conditions "smart conditions" were activated with needle voltage 3,500 kV, drying gas temperature 325°C, drying gas flow 12 L/min
Trap parameters typically used: trap drive 100%, Ion count 30,000, maximum accumulation time 50 ms

4. Add 20 mL –20°C 70:30 methanol:water solution and continue grinding until the sample is resuspended.

5. Transfer the resuspended sample to a #1 Whatman paper-lined glass funnel and collect filtrate in a 50-mL Falcon tube standing in ice.

6. Wash the mortar and pestle twice with 5 mL methanol:water solution, and transfer to the funnel as in the previous step.

7. Set the water bath on the rotary evaporator to 45°C. Apply the sample and reduce the sample volume to 2–3 mL.

8. Transfer the sample to a pre-weighed 10 mL tube and snap freeze in liquid nitrogen.

9. With the lid loosened, removed or punctured, lyophilise the sample overnight in a freeze dryer.

10. Reweigh the 10 mL tube to determine the dry weight of the lyophilised sample and store at –80°C until ready for NMR spectroscopy.

3.3.2. Preparation for NMR Spectroscopy

1. Resuspend the sample thoroughly in 1 mL deuterium oxide by pipetting and vortexing as necessary and transfer to a pre-weighed 1.5-mL Eppendorf tube.

2. Centrifuge the sample at maximum speed in a bench-top microcentrifuge to pellet any particulate debris. Remove the supernatant to a second 1.5-mL tube.

3. Dry the first Eppendorf in a heating block and re-weigh to determine the weight of the pellet. The pellet weight is deducted from the dry weight of the sample determined from the 10-mL tube in step 9 above to give the true dry weight of the polar metabolites.

4. Centrifuge the sample in the second Eppendorf at maximum speed as before (see Note 23).

5. Transfer 700 μL supernatant to a clean NMR tube (see Note 24).

6. Load the NMR tube into a NMR spectrophotometer and acquire a ^{13}C spectrum (see Note 25).

7. Once a spectrum has been acquired, the sample may be retrieved and stored for further analysis (see Notes 26 and 27).

4. Notes

1. Other microcentrifuge tubes can be used; however, some brands are prone to leaking during the metabolite extraction that follows.

2. If a tissue lyser is unavailable, steps should be taken to ensure thorough homogenisation of the fungal mycelia. Effectively lyophilised flask-cultured fungus may be more suitable than plate cultures, as the mycelia forms a finer suspension with the methanol more readily during metabolite extraction.

3. Highly concentrated metabolite extracts may require the combined use of a vacuum concentrator and a lyophiliser to dry. The lyophiliser is recommended as the extract will be more readily dissolved during later derivatisation steps. The extract however will not remain frozen in the lyophiliser if the methanol content is too high. Vacuum concentrating to remove the methanol is therefore required. The fungal extracts are likely to contain a high proportion of sugars; this method of drying the extract can also be used to prevent "caramelising" the metabolite extract.

4. The alkanes comprise a mixture of n-Decane, n-Dodecane, n-Pentadecane, n-Nonadecane, n-Docosane, n-Octacosane, n-Dotriacontane and n-Hexatriacontane in hexane, as internal retention index controls.

5. There is a wide range of ^{13}C labelled compounds available from e.g. Omicron Biochemicals, Sigma-Aldrich, Cambridge Isotope Laboratories etc. For labelled glucose we added 40 mM D-[1-^{13}C]-glucose to the MM-C medium. In the case of labelled mannitol, which was a poor growth substrate for some mutant strains which we investigated, cultures were first grown for 2 days on MM-C with 40 mM glucose. The samples were harvested and washed to remove all medium and then resuspended in 50 mL MM-C with 20 mM D-[1-^{13}C]-mannitol and cultured for a further 24 h

6. There have been a number of internal reference compounds used in fungal ^{13}C-NMR studies in order to act as a point of

reference for the chemical shifts of resonance peaks. These include acetone (26), acetate (27), EDTA (28–30), dioxane (31, 32), hexamethyldisiloxane (33), TMS (29, 34–38) and trimethylsilyl propionate (TMSP) (39). Deficiencies have been noted with each of these and the recommendation has been made that 2,2-dimethyl-2-silapentane-5-sulfonic acid (DSS) be used as the universal standard, due to its being insensitive to variations in temperature and pH, and the fact that it is chemically inert and has a single, sharp, unambiguous highfield resonance peak (40).

7. At various steps during the harvesting and extraction of metabolites, some thought must be given to the preparation of controls, which can be analysed by GC-MS along with the experimental samples. The subsequent data analysis will be made easier if analytes introduced during the preparation of samples for analysis can be quickly identified and excluded from the data set. Such controls include quenching solution, rinsing solution, media, and extraction solvents. These can be dried down and derivatised alongside the metabolite extracts for GC-MS analysis.

8. Harvesting and quenching by these methods allow the separation of mycelia from the growth media and snap freezing in less than 10 s.

9. The 2 mg equivalent of supernatant is calculated from the recorded weight of the fungal mycelia and a total extraction volume of 1,000 μL. For example, if the initial amount of fungal tissue was 2.5 mg, 800 μL of the extract would contain an equivalent of a 2 mg metabolite extract and 2.5 μg of ribitol. The amount of tissue used in the metabolite extraction will depend on the concentration of metabolites required for analysis, which will depend on the limits of detection of the individual analytical instrument used in the study.

10. Provided that a capable autosampler is available, online derivatisation can be used. This has the advantage of consistent derivatisation times before injection for all samples and is particularly important for large experiments. It should be noted however that it is important to keep reagents fresh (i.e. no older than 3 days).

11. The optimal split ratio has been described as the smallest ratio at which the compound present at highest concentration does not saturate the mass spectrometer detector (41). Overloading is evident when the bell shape curve of a peak becomes asymmetrical. In many biological samples, sugars are often present at concentrations many fold higher than the majority of other compounds. In such cases, overloading of peaks sometimes cannot be avoided if quantification of the less abundant metabolites is to be achieved; samples should therefore be run using a higher split ratio to accurately quantify the more abundant compounds.

12. A range of columns are available with various bonded phases designed for specific purposes. The Varian VF-5ms column is suitable for a wide range of applications however there are a number of columns with similar properties available from alternative suppliers.

13. The 15°C/min oven gradient can be used for higher-throughput analyses however this will be to the detriment of compound separation.

14. In order to obtain reproducible retention times for each compound across experiments on a single column, a standard (i.e. mannitol) can be used to lock the instrument method so that column pressure is adjusted to ensure elution of the standard at a fixed time. This enables the method to be re-locked following column maintenance. A lock time should be chosen to maintain the column flow close to 1 mL/min for optimal chromatographic resolution.

15. An efficient method for rapidly identifying observed analytes is using a metabolite library. The subsequent analysis of the generated metabolite profile will be aided by the use of a metabolite library generated by analysing purchased "standards" on the analytical instrument and under the same conditions as is being used for the metabolomic study. Some software such as AnalyzerPro will allow the creation of an "in-house" library which can be used to easily identify metabolites in the data.

16. The volume of solvent depends on the amount as well as the solubility of compounds and therefore needs to be determined empirically. Make sure SPE columns and bulk material never run dry as this will negatively affect recovery. Add water to SPE material slowly and carefully to excessive frothing.

17. Using water wettable C18 material (e.g. Alltech Prevail®) will reduce frothing and allows for omitting the preconditioning step with methanol. SPE can either be performed per individual sample using cartridges or columns with disposable syringes or in parallel on a vacuum manifold. Most manufacturers recommend an optimal flow rate of 1–5 mL/min for sample binding, but since binding of analytes occurred during shaking, the flow rate here can be faster.

18. This may take several hours, but allows for a great number of samples to be processed in parallel. Always adjust the Nitrogen nozzles to approximately 1–2 cm above extract surface and flow rate low enough that the extracts do not splash. Gentle heat (~40°C) may be applied to speed up evaporation. Alternatively, the methanol extracts can be filtered into 100-mL glass bulbs and dried on a rotary evaporator.

19. We found that methanol extracts of *S. nodorum* contain a large amount of lipids and fatty acids. These need to be removed

or they will interfere with the SPE and/or LC. If needed, they can be recovered from the hexane easily and analysed subsequently, e.g. by GC-MS (42). Adding sodium chloride to the extract facilitates phase separation in case foam comprised of extract/hexane/air or any other sort of interphase forms between hexane and the extract. The whole step may be omitted if the reconstituted extract is clear (i.e. coloured, yet translucent).

20. Make sure not to carry over any hexane to the SPE, as this will cause the analytes to elute prematurely from the SPE material.

21. Alternatively to removing particulates by centrifuging, Whatman Mini-UniPrep™ Syringeless Filters, which also act as autosampler vials, therefore speeding up the processing and reducing sample loss, can be used.

22. Fungal mycelium is amenable to grinding and should not require any grinding accessory. If the sample comprises a large proportion of spores or is otherwise difficult to reduce to a fine powder, then glass beads or acid washed sand may be added to assist in grinding.

23. Samples should always be centrifuged prior to loading into NMR tubes since any particulate matter will interfere with the spectrum acquisition. Samples may be coloured, but they should be translucent.

24. To clean NMR tubes, rinse several times with de-ionised water, followed by several rinses with 100% acetone. Dry the tubes in an oven at 70°C for 30 min. Do not leave the plastic lids on as these may melt.

25. It is not the purpose of this protocol to include operation of an NMR spectrophotometer. Platform-specific training will be required for the spectrophotometer to be used (as well, of course for any LC/GC-MS equipment).

26. To remove and retain a sample from an NMR tube, place a 1.5-mL eppendorf tube over the open end of the tube. Hold the eppendorf in your hand with the NMR tube oriented upside down along your wrist and forearm. Swing your arm once over your shoulder and the centrifugal force should be sufficient to transfer the sample into the eppendorf. It is advisable to make sure you have sufficient space to perform this manoeuvre, ensure other personnel are at a safe distance, and practice with water first to avoid losing a precious sample.

27. NMR is a non-destructive method and samples may be stored indefinitely for further analysis by NMR or other techniques. Samples may be stored at −80°C or may be lyophilised and stored at room temperature. If it is likely that the sample may be contaminated, 0.01% sodium azide may be added to prevent growth of contaminants as it is NMR invisible.

References

1. Fiehn, O. (2002) Metabolomics - The link between genotypes and phenotypes. *Plant Mol. Biol.* 48, 155–171.

2. Goodacre, R., Vaidyanathan, S., Dunn, W.B., Harrigan, G.G. & Kell, D.B. (2004) Metabolomics by numbers: Acquiring and understanding global metabolite data. *Trends Biotechnol.* 22, 245–252.

3. Hall, R.D. (2006) Plant metabolomics: From holistic hope, to hype, to hot topic. *New Phytol.* 169, 453–468.

4. Weckwerth, W. & Morgenthal, K. (2005) Metabolomics: From pattern recognition to biological interpretation. *Drug Discov. Today* 10, 1551–1558.

5. Kell, D.B., Brown, M., Davey, H.M., Dunn, W.B., Spasic, I. & Oliver, S.G. (2005) Metabolic footprinting and systems biology: The medium is the message. *Nat. Rev. Microbiol.* 3, 557–565.

6. Oliver, S.G., Winson, M.K., Kell, D.B. & Baganz, F. (1998) Systematic functional analysis of the yeast genome. *Trends Biotechnol.* 16, 373–378.

7. Hollywood, K., Brison, D.R. & Goodacre, R. (2006) Metabolomics: Current technologies and future trends. *Proteomics* 6, 4716–4723.

8. Throck Watson, J. & David Sparkman, O. (2007) Introduction to Mass Spectrometry: Instrumentation, applications and strategies for data interpretation. Wiley, Chichester.

9. Dunn, W.B. (2008) Current trends and future requirements for the mass spectrometric investigation of microbial, mammalian and plant metabolomes. *Phys. Biol.* 5, 1–24.

10. Lowe, R.G., Lord, M., Rybak, K., Trengove, R.D., Oliver, R.P. & Solomon, P.S. (2008) A metabolomic approach to dissecting osmotic stress in the wheat pathogen *Stagonospora nodorum*. *Fungal Genet. Biol.* 45, 1479–1486.

11. Solomon, P.S., Wilson, T.J.G., Rybak, K., Parker, K., Lowe, R.G.T. & Oliver, R.P. (2006) Structural characterisation of the interaction between *Triticum aestivum* and the dothideomycete pathogen Stagonospora nodorum. *Eur. J. Plant Pathol.* 114, 275–282.

12. Solomon, P.S., Rybak, K., Trengove, R.D. & Oliver, R.P. (2006) Investigating the role of calcium/calmodulin-dependent protein kinases in *Stagonospora nodorum*. *Mol. Microbiol.* 62, 367–381.

13. Tan, K.C., Trengove, R.D., Maker, G.L., Oliver, R.P. & Solomon, P.S. (2009) Metabolite profiling identifies the mycotoxin alternariol in the pathogen *Stagonospora nodorum*. *Metabolomics* 5, 330–335.

14. Nielsen, K.F. & Smedsgaard, J. (2003) Fungal metabolite screening: Database of 474 mycotoxins and fungal metabolites for dereplication by standardised liquid chromatography-UV-mass spectrometry methodology. *J. Chromatogr. A* 1002, 111–136.

15. Dettmer, K., Aronov, P.A. & Hammock, B.D. (2007) Mass spectrometry-based metabolomics. *Mass Spectrom. Rev.* 26, 51–78.

16. Nicholson, J.K. & Wilson, I.D. (2003) Understanding 'global' systems biology: Metabonomics and the continuum of metabolism. *Nat. Rev. Drug Discov.* 2, 668–676.

17. Sumner, L.W., Mendes, P. & Dixon, R.A. (2003) Plant metabolomics: Large-scale phytochemistry in the functional genomics era. *Phytochemistry* 62, 817–836.

18. IpCho, S.V.S., Tan, K.-C., Koh, G., Gummer, J., Oliver, R.P., Trengove, R.D. & Solomon, P.S. (2010) The transcription factor StuA regulates central carbon metabolism, mycotoxin production, and effector gene expression in the wheat pathogen *Stagonospora nodorum*. *Eukaryot. Cell* 9, 1100–1108.

19. Halket, J.M. & Zaikin, V.G. (2003) Review: Derivatization in mass spectrometry-1. Silylation. *Eur. J. Mass Spectrom.* 9, 1–21.

20. di Mavungu, J.D., Monbaliu, S., Scippo, M.L., Maghuin-Rogister, G., Schneider, Y.J., Larondelle, Y., Callebaut, A., Robbens, J., C., v.P. & S., d.S. (2009) LC-MS/MS multi-analyte method for mycotoxin determination in food supplements. *Food Addit. Contam. A* 26, 885–895.

21. Ren, Y., Zhang, Y., Shao, S., Cai, Z., Feng, L., Pan, H. & Wang, Z. Simultaneous determination of multi-component mycotoxin contaminants in foods and feeds by ultra-performance liquid chromatography tandem mass spectrometry. *J. Chromatogr. A* 1143, 48–64.

22. Senyuva, H.Z., Gilbert, J. & Ozturkoglu, S. (2008) Rapid analysis of fungal cultures and dried figs for secondary metabolites by LC/TOF-MS. *Anal. Chim. Acta* 617, 97–106.

23. Spanjer, M.C., Rensen, P.M. & Scholten, J.M. (2008) LC-MS/MS multi-method for mycotoxins after single extraction, with validation data for peanut, pistachio, wheat, maize, cornflakes, raisins and figs. *Food Addit. Contam. A* 25, 472–489.

24. Sulyok, M., Krska, R. & Schuhmacher, R. (2007) A liquid chromatography/tandem mass spectrometric multi-mycotoxin method for the quantification of 87 analytes and its application to semi-quantitative screening of moldy food samples. *Anal. Bioanal. Chem.* 389, 1505–1523.

25. Vishwanath, V., Sulyok, M., Labuda, R., Bicker, W. & Krska, R. (2009) Simultaneous determination of 186 fungal and bacterial metabolites in indoor matrices by liquid chromatography/ tandem mass spectrometry. *Anal. Bioanal. Chem.* 5, 1355–1372.

26. de Koker, T.H., Mozuch, M.D., Cullen, D., Gaskell, J. & Kersten, P.J. (2004) Isolation and purification of pyranose 2-oxidase from *Phanerochaete chrysosporium* and characterization of gene structure and regulation. *Appl. Environ. Microbiol.* 70, 5794–5800.

27. Donker, H.C.W. & Braaksma, A. (1997) Changes in metabolite concentrations detected by ^{13}C-NMR in the senescing mushroom (*Agaricus bisporus*). *Postharvest Biol. Tec.* 10, 127–134.

28. Ceccaroli, P., Saltarelli, R., Cesari, P., Pierleoni, R., Sacconi, C., Vallorani, L., Rubini, P., Stocchi, V. & Martin, F. (2003) Carbohydrate and amino acid metabolism in *Tuber borchii* mycelium during glucose utilization: a ^{13}C NMR study. *Fungal Genet. and Biol.* 39, 168–175.

29. Martin, F., Ramstedt, M., Söderhäll, K. & Canet, D. (1988) Carbohydrate and amino acid metabolism in the ectomycorrhizal ascomycete *Sphaerosporella brunnea* during glucose utilization. *Plant Physiol.* 86, 935–940.

30. Ramstedt, M., Martin, F. & Söderhäll, K. (1989) Mannitol metabolism in the ectomycorrhizal basidiomycete *Piloderma croceum* during glucose utilization. A ^{13}C-NMR study. *Agric. Ecosys. Environ.* 28, 409–414.

31. Thomas, G.H. & Baxter, R.L. (1987) Analysis of mutational lesions of acetate metabolism in *Neurospora crassa* by ^{13}C nuclear magnetic resonance. *J. of Bacteriol.* 169, 359–366.

32. Yoshida, M., Murai, T. & Moriya, S. (1984) ^{13}C NMR spectra of plant pathogenic fungi. *Agric. Biol. Chem.* 48, 909–914.

33. Jobic, C., Boisson, A.M., Gout, E., Rascle, C., Fèvre, M., Cotton, P. & Bligny, R. (2007) Metabolic processes and carbon nutrient exchanges between host and pathogen sustain the disease development during sunflower infection by *Sclerotinia sclerotiorum*. *Planta* 226, 251–265.

34. Martin, F., Boiffin, V.V. & Pfeffer, P.E. (1998) Carbohydrate and amino acid metabolism in the *Eucalyptus globulus-Pisolithus tinctorius* ectomycorrhiza during glucose utilization. *Plant Physiol.* 118, 627–635.

35. Martin, F., Canet, D. & Marchal, J.P. (1985) ^{13}C nuclear magnetic resonance study of mannitol cycle and trehalose synthesis during glucose utilization by the ectomycorrhizal ascomycete *Cenococcum graniforme*. *Plant Physiol.* 77, 499–502.

36. Martin, F., Canet, D., Marchal, J.-P. & Brondeau, J. (1984) *In vivo* natural-abundance ^{13}C nuclear magnetic resonance studies of living ectomycorrhizal fungi : observation of fatty acids in *Cenococcum graniforme* and *Hebeloma crustuliniforme*. *Plant Physiol.* 75, 151–153.

37. Shachar-Hill, Y., Pfeffer, P.E., Douds, D., Osman, S.F., Doner, L.W. & Ratcliffe, R.G. (1995) Partitioning of Intermediary carbon metabolism in vesicular-arbuscular mycorrhizal leek. *Plant Physiol.* 108, 7–15.

38. Bago, B., Pfeffer, P.E., Douds, D.D., Jr., Brouillette, J., Bécard, G. & Shachar-Hill, Y. (1999) Carbon metabolism in spores of the arbuscular mycorrhizal fungus *Glomus intraradices* as revealed by nuclear magnetic resonance spectroscopy. *Plant Physiol.* 121, 263–272.

39. Forgue, P., Halouska, S., Werth, M., Xu, K., Harris, S. & Powers, R. (2006) NMR metabolic profiling of *Aspergillus nidulans* to monitor drug and protein activity. *J. Proteome Res.* 5, 1916–1923.

40. Wishart, D.S. & Sykes, B.D. (1994) Chemical shifts as a tool for structure determination. *Methods Enzymol.* 239, 363–392.

41. Kanani, H., Chrysanthopoulos, P.K. & Klapa, M.I. (2008) Standardizing GC-MS metabolomics. *J. Chromatogr. B* 871, 191–201.

42. Indarti, E., Majid, M.I.A., Hashim, R. & Chong, A. (2005) Direct FAME synthesis for rapid total lipid analysis from fish oil and cod liver oil. *J. Food Compos. Anal.* 18, 161–170.

Chapter 16

Targeted Gene Replacement in Fungi Using a Split-Marker Approach

Rubella S. Goswami

Abstract

Targeted gene replacement is one of the primary strategies for functional characterization of fungal genes and several methods have been developed for this purpose over the years. The increased availability of genome sequence information in the present times has enabled wider adoption of protocols based on the knowledge of the gene sequence and its surrounding region. Among such targeted gene replacement approaches, the spilt-marker method has gained popularity in filamentous fungi. This method involves only two rounds of PCR and does not require any subcloning. It is based on the availability of a marker gene (e.g., the hygromycin gene) and sequences of the gene of interest, as well as around 1 kb long regions flanking the gene on either side. The technique includes PCR amplification of the flanking regions of the gene of interest and the marker gene followed by a fusion PCR which leads to the creation of two molecular cassettes, each containing a part of the marker gene fused to one flanking region. These molecular cassettes are then simultaneously used for transformation of protoplasts. Three homologous recombination events, one within each flanking region and one in the marker gene, lead to the replacement of the gene of interest with a functional marker gene. The transformants are then grown on selective media and emerging colonies can be screened for presence of the marker and absence of the gene being replaced using various methods.

Key words: Split-marker, Gene replacement, Fungi, Gene deletion, Fungal transformation, *Fusarium graminearum*, *Gibberella zeae*

1. Introduction

Gene replacement has been one of the fundamental approaches used for functional characterization of fungal genes or open reading frames. Over the years, several techniques have been developed and adopted for this purpose. With the increased availability of whole genome sequences, the split-marker deletion approach is becoming more widely adopted for functional genomic studies.

Melvin D. Bolton and Bart P.H.J. Thomma (eds.), *Plant Fungal Pathogens: Methods and Protocols*,
Methods in Molecular Biology, vol. 835, DOI 10.1007/978-1-61779-501-5_16, © Springer Science+Business Media, LLC 2012

This approach relies on the availability of the sequence of the gene of interest and portions flanking it on both sides prior to replacement. This strategy was initially developed for *Saccharomyces cerevisiae* by Fairhead et al. (1, 2) and is reported to have been used for *Cochliobolus heterostrophus* and *Gibberella zeae* (3, 4). Split-marker is a PCR-based method that does not require subcloning and only involves two rounds of PCR. The hygromycin B phosphotransferase (*hph*) gene conferring resistance to the antibiotic hygromycin or any other marker is used, and it depends upon homologous recombination for gene replacement in the fungus. The technique involves PCR amplification of regions flanking the gene of interest on either side and the marker gene followed by a fusion PCR which leads to the creation of two molecular cassettes each containing a part of the marker (*hph*) gene fused to one flanking region. These two cassettes are simultaneously used for transformation. Three homologous recombination events occur, one within each flanking region and one in the marker (*hph*) gene, to successfully replace the gene of interest with a functional marker (*hph*) gene. The methods described here were developed for *Fusarium graminearum* and the *hph* gene as a selectable marker; however, the marker can be easily changed. Modifications that may be required based on the fungal species being transformed, source construct for the marker gene, and other factors are mentioned in the text.

2. Materials

Solutions, specific media, supplies, and specific equipment used at various stages of the protocol:

2.1. DNA Extraction

1. *Nuclei Lysis Buffer*: 100 mL 1 M Tris pH 7.5, 50 mL of 0.5 M EDTA pH 8, 10 g of Cetyl Trimethyl Ammonium Bromide (CTAB) are added to 500 mL with distilled water and autoclaved.

2. *DNA Isolation Buffer*: 31.89 g Sorbitol, 6.05 g Tris Base/THAM, and 0.84 g EDTA are added to 500 mL distilled water, the pH is adjusted to 7.5 and autoclaved.

3. *5% Sarkosyl*: 12.5 g N-laurylsarcosine is added to 250 mL distilled water and autoclaved.

4. *DNA extraction buffer*: Mix 1:1:0.4 volume of the Nuclei lysis buffer, DNA isolation buffer, and 5% Sarkosyl to prepare this buffer.

5. *TE (pH 8)*: 4.0 mL of 1 M Tris pH 8 and 0.8 mL of 0.5 M EDTA pH 8 are added to 400 mL distilled water and autoclaved.

6. *Complete medium*: Vitamin and trace element stock solutions need to be prepared first. The vitamin stock solution is constituted

by adding 4 g inositol, 200 mg Calcium pantothenate, 200 mg choline. chloride, 100 mg thiamine, 75 mg pyridoxine, 75 mg nicotinamide, 50 mg ascorbic acid, 30 mg riboflavin, 5 mg *p*-aminobenzoic acid, 5 mg folic acid, 5 mg biotin to a 50:50 ethanol:H_2O mixture to bring up the volume to 1 L. The trace element stock solution is prepared by adding 5 g citric acid, 5 g $ZnSO_4.6H_2O$, 1 g $Fe(NH_4)_2(SO_4)_2.6H_2O$, 250 mg $CuSO_4.5H_2O$, 50 mg $MnSO_4$, 50 mg H_3BO_3, 50 mg $Na_2MoO_4.2H_2O$ to 95 mL distilled water. Both the trace element solution and vitamin stock solution can be filter-sterilized and stored. A final medium is prepared by adding 1 g KH_2PO_4, 0.5 g $MgSO_4.7H_2O$, 0.5 g KCl, 30 g sucrose, 2 g $NaNO_3$, 2.5 g N-Z Amine, 1 g yeast extract, 0.2 mL trace element solution, 10 mL vitamin stock solution, and 20 g agar (Bacto) to distilled water, the volume adjusted to 1 L, and autoclaved (5).

2.2. Preparation of Protoplasts

1. *CMC medium*:15 g Carboxmethylcellulose (low viscosity from Sigma), 1.0 g NH_4NO_3, 1.0 g KH_2PO_4, 0.5 g $MgSO_4.7H_2O$, and 1.0 g yeast extract are added to 750 mL prewarmed distilled water with continuous stirring. Once the solutes are dissolved, the volume is adjusted to 1 L with distilled water and autoclaved (see Note 1).

2. *YEPD Broth*: 3.0 g Yeast Extract, 10.0 g Bacto peptone, and 20.0 g Dextrose are added to 500 mL distilled water while stirring and the volume is subsequently adjusted to 1 L with distilled water and autoclaved.

3. *Protoplasting solution*: 500 mg Driselase (Sigma, St. Louis, MO) and 100 mg Lysing enzyme (Sigma, St. Louis, MO) are dissolved in 20 mL of 1.2 M KCl by stirring for 30 min at room temperature. The solution is subsequently filter-sterilized using a 0.45-μm filter.

2.3. Transformation

1. *TB3 medium*: 3 g yeast extract, 3 g casamino acids, and 20% sucrose are added to 500 mL distilled water, the volume is adjusted to 1 L and autoclaved. For solid medium, 0.7% low melting point agarose is added prior to autoclaving.

2. *V8 juice medium*: 200 mL V8 juice, 2 g $CaCO_3$, and 15 g Bacto agar are added to 800 mL distilled water, the volume is adjusted to 1 L and autoclaved.

3. *Mung bean agar medium*: 40 g of mung bean is placed in 750 mL of boiling distilled water for 23 min and filtered through two layers of cheese cloth. The volume is then adjusted to 1 L, 15 g of agar is added and autoclaved (6).

4. *1× STC*: This solution contains 1.2 M Sorbitol, 10 mM Tris-HCl (pH 8.0), and 50 mM $CaCl_2$. It can be prepared by

adding 120 mL 2 M sorbitol, 2 mL 1 M Tris-HCl (pH 8.0), and 2 mL 5 M CaCl$_2$, bringing the volume up to 200 mL with the distilled water followed by autoclaving.

5. *1× PTC*: 40% (40 g in 100 mL) PEG 4000 or 8000 is added to 1× STC. The volume is adjusted to 50 mL with STC and filter-sterilized using a 0.45-μm filter (see Note 2).

2.4. Other Supplies

In addition to standard laboratory supplies, a few specific items that are required for this protocol are listed below:

1. 24-well blocks or 15-mL snap cap tubes.
2. Miracloth (Calbiochem Corporation, La Jolla, CA).
3. Glass beads (0.5 mm) and Bead beater (BioSpec Products Inc., Bartlesville, OK).
4. DyNA Quant 200 fluorometer (Hoefer, San Francisco, CA), Nanodrop ND-1000 spectrophotometer (Thermo Scientific, Wilmington, DE), or any spectrophomometer for DNA quantification.
5. Kits for probe labeling (radioactive or nonradioactive).
6. DNAse-free RNAse A 10 mg/mL.
7. Qiagen's QIAquick PCR Purification kit (Valencia, CA).
8. Funnels.
9. 500-mL centrifuge bottles.
10. 50-mL Sorvall tubes.
11. 15-mL screw-cap falcon tubes.
12. 1.5-mL microfuge tubes.
13. Hygromycin or other antibiotic for selection of transformants.
14. Chloroform.
15. Isoamyl alcohol.
16. Isopropanol.
17. DMSO.
18. Ammonium acetate.
19. Potassium chloride.

3. Methods

3.1. DNA Extraction

DNA can be extracted from fungal cultures using several methods, including commercial kits. The CTAB method commonly used and adapted for *F. graminearum* (7) is described below:

1. Add 5 mL liquid complete medium (5) in 24-well blocks or 15 mL snap cap tubes, inoculate with plugs from the growing

edge of 7-day-old *F. graminearum* cultures, and incubate for 3 days at room temperature with continuous shaking at 150 rpm.

2. Harvest fungal mycelium using Miracloth and transfer into 2.2-mL microfuge tubes.

3. Freeze–dry the mycelia overnight. Add glass beads (0.5 mm) of approximately half the quantity of the mycelium to each tube and beat in a bead beater for 10 s to disrupt the mycelia.

4. Centrifuge the contents at $16,000 \times g$ for 30 s, and add 0.8 mL of DNA Extraction Buffer (DEB), containing Nuclei Lysis Buffer, DNA Isolation Buffer, and 5% Sarkosyl (*N*-lauryl-sarkosine) in the ratio of 1:1:0.4, to each tube.

5. Vortex the tubes thoroughly, and incubate at 65°C for 60 min with occasional gentle mixing at 15-min intervals.

6. Add chloroform:isoamyl alcohol (750 μL) mixed at a standard ratio of 24:1, to each tube, mix thoroughly, and centrifuge for 15 min at $14,000 \times g$.

7. Transfer the aqueous phase into a fresh microfuge tube, add 3 μL DNAse-free RNAse A to each tube, and incubate at room temperature for 30 min.

8. To precipitate DNA, add 600 μL isopropanol and 150 μL of 3 M ammonium acetate to the contents of the tubes, mix thoroughly, incubate at room temperature for 10 min, and centrifuge for 10 min at 15,000 rpm.

9. Remove the supernatant and air-dry the pellet at room temperature.

10. Wash the DNA pellet with 100 μL of 70% ethanol, air-dry, and dissolve in 50 μL TE. Quantify the DNA (This can be done using a DyNA Quant 200 fluorometer or Nanodrop ND-1000 spectrophotometer) at 260 nm or any other method.

3.2. Preparation of Gene Replacement Constructs

This method involves preparation of two linear constructs, which are developed through fusion PCR. Each of these constructs contains a flank of the target gene and two thirds of the marker gene or the entire marker gene. These linear constructs are used simultaneously for transformation of the protoplasts and replacement takes place through homologous recombination at two locations, one within the marker gene and another between the flanks. An illustration of the procedure is provided in Fig. 1.

3.2.1. Primer Design

Four universal/selectable marker primers and four gene-specific primers are required.

1. *Universal/selectable marker primers*: Two universal primers to amplify the entire marker gene from the source construct and two primers to amplify two thirds of the marker gene when

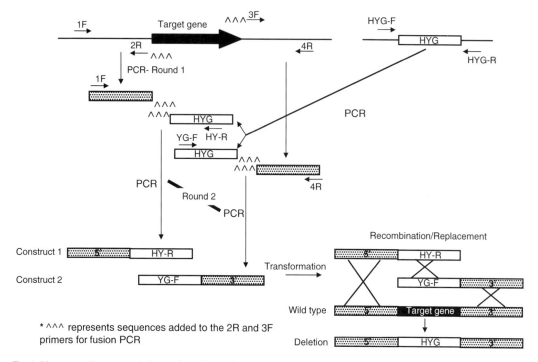

Fig. 1. Diagrammatic representation of the split-marker-targeted replacement strategy including location of primers, PCRs, constructs, and homologous recombination.

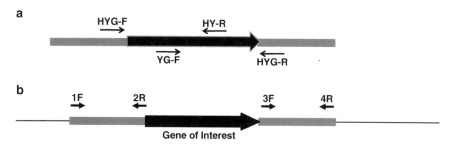

Fig. 2. (a) Diagrammatic representation of the location of the primers for amplification of the marker gene (HYG) and the internal primers for amplification of the 2/3rd portion of the gene. (b) Diagrammatic representation of the location of the forward and reverse primers for the flanking regions of the gene of interest.

used along with the forward and the reverse primer, respectively, need to be designed (Fig. 2a). To design primers for the complete marker gene (HYG-F and HYG-R in our example), use sequences of the regions immediately flanking the *hph* in your source construct. If your marker gene is flanked by a universal primer such as M13 forward and reverse, then these primers can be used for amplification of the whole gene fragments (i.e., as HYG-F and HYG-R) (see Note 3). For the internal primers (HY-R and YG-F in our example), use the sequence of the marker gene. If you are using the hygromycin gene (*hph*) as your marker, you can use the internal forward and reverse primer sequences (HY-R and YG-F) given below along with the HYG-F and HYG-R for amplification of two thirds of the gene sequence.

(a) HY-R GTA TTG ACC GAT TCC TTG CGG TCC GAA

(b) YG-F GAT GTA GGA GGG CGT GGA TAT GTC CT

2. *Gene-specific primers*: Select one upstream and one downstream region 0.6–1.1 kb long, flanking your gene of interest. Use a standard primer design program such as WebPrimer (http://www.yeastgenome.org/cgi-bin/web-primer) or Primer 3 (http://frodo.wi.mit.edu/primer3/) to select one forward (1F and 3F) and one reverse (2R and 4R) primer for each flank (Fig. 2b). Add extensions consisting of sequences that are complementary to the marker primers HYG-F and HYG-R to the 5′ end of the 2R primer and 5′ end of the 3F primer, respectively. This extension is required for fusion of the flanks to the marker sequence (*hph*) and can vary in design depending on the source construct for the marker gene.

If your marker gene is in a vector construct with universal primer sequences flanking the gene—add sequences complementary to the universal primers to the 2R and 3F primers. For example, if you have a construct with M13F and M13R flanking the HYG gene, you will need to add sequences complementary to M13F and M13R to the 5′ end of your 3F and 2R primers, to add the common region for fusion PCR to your flanking regions.

M13F 5′ CGCCAGGGTTTTCCCAGTCACGAC 3′.

M13R 5′ AGCGGATAACAATTTCACACAGGA 3′.

2R (M13)—region complementary to the 3′ end of the first flank (i.e., reverse primer in your primer set for flank 1, i.e., 2R primer designed previously) + reverse complement of the M13F.

3F (M13)—region complementary to the 5′ end of the second flank (i.e., forward primer in your primer set for flank 2, i.e., 3F primer designed previously) + reverse complement of the M13R (see Note 4).

If your marker gene is not in a vector construct with universal primer sequences flanking the gene—you will have to add a sequence approximately 21 bp in length complementary to the vector region at the 5′ end of your marker gene to the 5′ end of the 2R primer. Likewise, you will have to add a sequence complementary to the vector region at the 3′ end of the marker gene to the 5′ end of the 3F primer.

3.2.2. PCR Amplification for Preparation of Constructs

1. *PCR Amplification (Round 1)*: Set up PCR reactions for amplification of the flanking regions and the marker gene. PCR conditions, particularly annealing temperatures, may have to be

Table 1
Primer combinations and templates to be used for the first round of PCR as well as the expected amplicons

Reaction number	Primer combinations	Template	Amplicon
1	1F and 2R	Fungal genomic DNA	5′ flank of gene
2	3F and 4R	Fungal genomic DNA	3′ flank of gene
3	HYG-F and HY-R	Marker gene cassette (hygromycin)	2/3rd of marker gene from the 5′ end
4	YG-F and HYG-R	Marker gene cassette (hygromycin)	2/3rd of marker gene from the 3′ end
5	HYG-F and HYG-R	Marker gene cassette (hygromycin)	Entire marker (*hph*) gene

optimized depending on the primer sequence. Example of a generic PCR program to be used may include: 1 cycle at 94°C followed by 25 cycles of 30 s at 94°C, 30 s at 55°C, and 2 min at 72°C. Finally, add an extension cycle of 5 min at 72°C. The primer–template combinations and respective amplicons are listed in Table 1.

Purify the PCR products prior to the next round of PCR to remove excess primers. This can be done using any standard commercial purification kit.

2. *PCR Amplification (Round 2)*: During this PCR reaction, each flank from round 1 is fused to the marker through PCR splicing by overlap extension (7, 8). Either 2/3rd of the marker gene amplified from each end or the entire marker gene can be used as template along with the gene-specific primers. If the marker primer HYG-F and HYG-R combination is used in round 1 and the entire gene is amplified, then this fragment can be used for fusion PCR with both 3′ and 5′ flanks in the second round. The primer–template combinations which are referred to in Table 1 and respective amplicons for the second round of PCR are listed below (Table 2):

Set up two 50 μL PCR reactions for amplification of adequate amounts of each construct. Add 20 ng of each flank and the marker gene (*hph*) as template. Use primer and template combinations mentioned above (see Notes 6 and 7).

PCR reaction conditions will vary based on the primers used. A protocol that has been used in *F. graminearum* while using the entire *hph* gene as a marker and the HY-R and YG-F primers includes 1 cycle at 94°C followed by 30 cycles of 30 s at 94°C, 30 s at 55°C, and 2 min 30 s at 68°C with a final extension cycle of 10 min at 68°C.

Table 2
Primer combinations and templates to be used for the second round of PCR (fusion PCR) as well as the expected amplicons (see Note 5)

Reaction number	Primer combinations	Template	Amplicon
1	1F and HY-R	5′ flanking fragment from reaction 1 and 2/3rd of marker gene from the 5′ end from reaction 3	Construct 1 for transformation with the 5′ end flank and 2/3rd of the marker gene
2	4R and YG-F	3′ flanking fragment from reaction 2 and 2/3rd of marker gene from the 3′ end from reaction 4	Construct 2 for transformation with the 3′ end flank and 2/3rd of the marker gene

Purify and concentrate the products from the fusion PCR (constructs 1 and 2) using a PCR Purification kit. Each construct can be purified separately or both the constructs can be purified and concentrated together using the PCR Purification kit.

3.3. Protoplast Isolation

3.3.1. Preparation of Spore Suspension

1. Inoculate 100 mL CMC medium with the *F. graminearum* isolate being studied from culture plates or glycerol stock and incubate for 5 days at room temperature with shaking at 150 rpm.

2. Filter the cultures through one layer of Miracloth placed in a funnel and collect the filtrate in 500 mL centrifuge bottles. Wash the Miracloth with 50 mL sterile distilled water and collect the washes with the filtrate.

3. Distribute the filtrates and washes into 50 mL Sorvall tubes and centrifuge at $2,500 \times g$ for 10 min at 4°C. Subsequently, discard the supernatant and wash the pellet twice with sterile distilled water.

4. After the final wash, the pellet (consisting of pure spores) should be resuspended in 15% glycerol in order to bring the spores to a final concentration of about 10^8 spores/mL and stored at −80°C for future use.

3.3.2. Protoplast Preparation

1. Inoculate 100 mL YEPD medium with 1 mL of the spore suspension (from glycerol stocks of 10^8 spores/mL) and incubate for 12–16 h at room temperature (until germ tubes are about 1.5 times the length of the spore) with shaking at 150 rpm.

2. Harvest fungal mycelia by filtering through two layers of Miracloth and wash once with 50 mL sterile distilled water and once with 30 mL 1.2 M potassium chloride (KCl). Then transfer

the mycelia into 20-mL filter-sterilized protoplasting solution, and incubate for 30 min at 30°C with shaking at 80 rpm to mix the contents well.

3. After 30 min incubation, monitor protoplasting at regular intervals, until more than 50% of the mycelia are digested and protoplasts released (see Note 8).

4. Filter the protoplasts through two layers of Miracloth and collect the filtrate in 50 mL conical tubes. To maximize protoplast harvest, the Miracloth can be washed 5 times with 20 mL 1.2 M KCl. For this step, transfer funnel with Miracloth to a sterile 250 mL Erlenmeyer flask. Wash the Miracloth with 5×20 mL 1.2 M KCl. The filtrate and washes should be combined, divided in equal aliquots into 50 mL Sorvall tubes, and centrifuged at $2,500 \times g$ for 10 min at 4°C (see Note 9).

5. Aliquot washes into 50-mL conical tubes and centrifuge at $2,500 \times g$ for 10 min at 4°C (see Note 10).

6. Wash the pellet with protoplasts twice with 30 mL 1.2 M KCl with intermittent centrifugation (centrifuge and discard) at $2,500 \times g$ for 10 min at 4°C.

7. Resuspend the pellet (protoplasts) in 1× STC to bring it to a final concentration of 10^8 protoplasts/mL. Lastly, add dimethyl sulfoxide (DMSO) to a final concentration of 7%. The protoplasts can then be stored in 200 μL aliquots at –80°C for subsequent use in transformation.

3.4. Transformation

1. Prepare plates containing 20 mL TB3 medium supplemented with 150 mg/L hygromycin (see Note 11).

2. Thaw two aliquots (tubes containing 200 μL of the protoplasts previously stored at –80°C) on ice.

3. Add 14–18 μL containing 1–2 μg of each construct, obtained from the second round of PCR amplifications which contain the flanking regions and portions of the marker gene, to one tube. Use a second tube as a control to check viability of the protoplasts by adding 14–18 μL of STC (used to resuspend the protoplasts).

4. Incubate both tubes on ice for 20 min. Subsequently, add 1 mL PTC solution to each tube, mix the contents well, and incubate at room temperature for 20 min.

5. Transfer the protoplasts into 15-mL screw-cap falcon tubes and add 5 mL TB3 liquid medium. Incubate the tubes in a slanted position for 12–16 h at room temperature with shaking at 150 rpm.

6. Centrifuge the tubes at $2,500 \times g$ for 10 min at 4°C. Discard the supernatant and resuspend the mycelia in each tube in 1 mL STC.

7. Divide the resuspended mycelia into two or three aliquots of approximately 350 µL each. Add one aliquot of the resuspended mycelia to 10 mL TB3 media containing 0.7% agarose. Mix well and use this to overlay a previously prepared 150-mm petri dish containing 20 mL of the same medium supplemented with 150 mg/L hygromycin. You can repeat this for the other aliquot(s).

8. Incubate these plates overnight in darkness at room temperature. The next day, overlay the plates again with 15 mL of TB3 medium containing 250 µg/mL hygromycin and 0.7% low melting temperature agarose (see Note 11).

9. After 4–5 days of incubation, transfer the hygromycin-resistant colonies to V8 juice medium containing 250 mg/L hygromycin and 1.5% agarose.

10. Prepare pure cultures by single spore isolation from these putative transformant colonies growing on the V8 juice medium and plate them on mung bean agar medium containing 250 mg/L hygromycin. Macroconidia can then be harvested from these plates and used for pathogenicity tests.

3.5. Confirmation of Gene Replacement Using PCR

3.5.1. Initial Confirmation of Gene Replacement

Successful replacement of the gene of interest can be verified using PCR and further confirmed by Southern hybridization. Several combinations of primers internal to the marker and the gene of interest can be designed and used for this purpose. An easy and quick PCR-based screen is briefly described below:

1. Design gene-specific and marker-specific primers based on the complete sequence of the gene of interest using a standard primer design software and standardize amplification conditions.

2. Grow mycelia of the mutant and wild-type isolates and extract DNA as described earlier.

3. Obtain vector DNA from the source cassette for the marker gene.

4. Set up PCR reactions with the primer and template combinations listed in Table 3.

5. Standard PCR conditions can generally be used, but may have to be modified depending on the primer pairs. A possible protocol could be as follows: One cycle of 94°C for 2 min followed by 30 cycles of 94°C for 1 min, 62°C for 1 min, 68°C for 3 min, and one cycle at 68°C for 10 min. Adjustment of the annealing temperature is the most common modification required (see Note 12).

6. Amplification of both the gene and the marker from the mutants is indicative of an ectopic integration event.

Table 3
Primer combinations and templates to be used for the PCR-based screening of transformants as well as the expected amplicons

Primers	Template	Amplicon
Gene-specific forward and reverse primers	Genomic DNA from the wild-type strain Genomic DNA from the mutant strain	For successful replacement, the gene should amplify from the wild type but not from the mutant strain. Compare size of amplification product if there is an amplification in the mutant
Marker-specific forward and reverse primers	Genomic DNA from the mutant strain DNA from vector with the marker cassette (positive control) Genomic DNA from the wild-type strain (negative control)	For successful replacement, the gene should amplify from templates 1 and 2
1F and HY/R	Genomic DNA from wild-type strain Genomic DNA from the mutant strain	This is a positive screen for replacement of the upstream end of the gene. For successful replacement, the gene should amplify from the mutant strain but not from the wild type
4R and YG/F	Genomic DNA from wild-type strain Genomic DNA from the mutant strain	This is a positive screen for replacement of the downstream end of the gene. For successful replacement, the gene should amplify from the mutant strain but not from the wild type

3.5.2. Southern Hybridization

Any standard Southern hybridization protocol can be used. A published protocol from Rosewich et al. has been used successfully for *F. graminearum* (9). Certain strategies found to work well in preparing the Southern blots and probes are briefly mentioned below:

1. *DNA amount:* Generally, 5 µg of genomic DNA from each of the mutants and the wild type is required. However, amounts as low as 1.5 µg have been used when there is a scarcity of DNA.

2. *Restriction enzymes:* The selection of restriction enzymes for digestion of the DNA is important for screening knock-outs. Restriction enzymes that cut once in the regions flanking the genes but have no restriction sites in either in the native gene or the marker (*hph*) gene should be used.

3. *Probes*: Gene-specific fragments and the marker gene amplified during the PCR conducted to assess replacement of the genes can be used as probes. Both radioactive and nonradioactive methods for hybridization of probes to Southern blots have been used successfully for *F. graminearum* (see Note 13).

4. Notes

1. Heating the water prior to addition of CMC and stirring on a warm hotplate aids in getting CMC into solution.

2. A tip for easy preparation is to take a graduated cylinder, add a small amount of STC, then add PEG, stir until dissolved, adding more STC as necessary, but not going over 50 mL. When PEG has dissolved, bring to 50 mL with STC. Microwave the solution to warm before filtering so that the solution is less viscous.

3. The use of primers such as M13F and M13R that fuse to a standard vector sequence allows one to reuse them in making constructs for deletion using different resistance genes (3). Additionally, M13 primers have been found to work well in fusion PCR.

4. In some cases, researchers include a portion of the marker gene in one or both of their primers. This may vary from lab to lab.

5. Instead of 2/3rd of the marker gene, the entire marker gene can be used as the template. In this case, HY-F and HY-R will be replaced by HYG-F and HYG-R, respectively. In *F. graminearum*, using the entire *hph* gene appears to work better.

6. Reducing the primer concentration to 50 nM has been found to give cleaner results.

7. Though earlier reports state that DNA concentration is not critical and 1–2 μL of the purified PCR product can be used for the next PCR, we prefer to quantify the DNA amount using a spectrophotometer and electrophoresis, and add 20 ng of the marker gene (*hph*) and each flanking region.

8. After 30 min of incubation, start monitoring the protoplast formation every 10 min until half of the mycelia is digested and you can see sufficient protoplasts. Total incubation time can range from 60 min to 2.25 h, depending on the batch of enzymes. Harvest the protoplasts when you see that most of the hyphae have turned into protoplasts (most hyphae will be gone but some new protoplasts will still be forming).

9. Protoplasts tend to stick to the Miracloth. To harvest as much as possible, it helps to keep stirring with an autoclaved spatula during the washing process.

10. It helps to count the number of protoplasts before centrifugation and after the second wash to estimate the amount of STC and DMSO to add in the next step.

11. The sensitivity to hygromycin varies; therefore, hygromycin concentration to be used for selection of mutants should be standardized for each fungal species using wild-type isolates prior to starting transformation.

12. While conducting PCR using gene-specific primers on genomic DNA, a positive control can be added using primers for a house-keeping gene. For example, in *F. graminearum*, primers for the translation elongation factor 1α are often used.

13. We have used the Decaprime DNA™ II Random Priming DNA labeling kit from Life Technologies (Carlsbad, CA) for radioactive labeling and the Amersham AlkPhos Direct Labeling and Detection System with CDP-*Star* from GE Healthcare (Piscataway, NJ) for nonradioactive labeling of *F. graminearum* probes. A few changes made to the manufacturer's protocol that have worked well include the hybridization and the subsequent wash being carried out in tubes, elevation of hybridization temperature to 65°C for increased stringency, reduction in the volume of the first wash to 200 mL, and an increase in the wash time to 30 min.

Acknowledgments

I would like to thank Dr. H.C. Kistler in whose laboratory the protocol mentioned here was initially standardized for *F. graminearum* and members of his lab, particularly Karen Hilburn for her input in protocol development and manuscript preparation. Thanks are also due to Dr. Jin-Rong Xu for assisting with standardization of the method and providing additional information for the chapter.

References

1. Fairhead C et al (1996) New vectors for combinatorial deletions in yeast chromosomes and for gap-repair cloning using 'split-marker' recombination. Yeast 12: 1439–1457.

2. Fairhead C et al (1998) 'Mass-murder' of ORFs from three regions of chromosome XI from *Saccharomyces cerevisiae*. Gene 223: 33–46.

3. Catlett N et al (2003) Split-marker recombination for efficient targeted deletion of fungal genes. *Fungal Genet. Newsl* 50: 9–11.

4. Goswami RS et al (2006). Genomic analysis of host-pathogen inaction during early stages of disease development. Microbiology 152: 1877–1890.

5. Correll JC et al (1987) Nitrate nonutilizing mutants of *Fusarium oxysporum* and their use in vegetative compatibility tests. Phytopathology 77: 1640–1646.

6. Evans CK et al (2000) Biosynthesis of deoxynivalenol in spikelets of barley inoculated with macroconidia of *Fusarium graminearum*. Plant Dis 84: 654–660.

7. Gale LR et al (201) Nivalenol-type populations of *Fusarium graminearum* and F. asiaticum are

prevalent on wheat in southern Louisiana. Phytopathology 101: 124–134.

8. Ho SN et al (1989) Site-directed mutagenesis by overlap extension using the polymerase chain reaction. Gene 77: 51–59.

9. Rosewich UL et al (1999) Population genetic analysis corroborates dispersal of *Fusarium oxysporum* f. sp. *radicislycopersici* from Florida to Europe. Phytopathology 89: 623–630.

Chapter 17

Massively Parallel Sequencing Technology in Pathogenic Microbes

Sucheta Tripathy and Rays H.Y. Jiang

Abstract

Next-Generation Sequencing (NGS) methods have revolutionized various aspects of genomics including transcriptome analysis. Digital expression analysis is all set to replace analog expression analysis that uses microarray chips through their cost-effectiveness, reproducibility, accuracy, and speed. The last 2 years have seen a surge in the development of statistical methods and software tools for analysis and visualization of NGS data. Large amounts of NGS data are available for pathogenic fungi and oomycetes. As the analysis results start pouring in, it brings about a paradigm shift in the understanding of host pathogen interactions with discovery of new transcripts, splice variants, mutations, regulatory elements, and epigenetic controls. Here we describe the core technology of the new sequencing platforms, the methodology of data analysis, and different aspects of applications.

Key words: Next-generation sequencing (NGS), Short reads analysis, Fungi, Oomycetes

1. Introduction

The era of sequencing-driven research was announced in the 1970 by two landmark papers in which Sanger and Gilbert described the methodology of Sanger sequencing (1, 2). Sanger sequencing technology refers to the use of chain-terminating dideoxynucle-otide analogs that caused base-specific termination of DNA synthesis. To date, almost all sequenced eukaryotic genomes have been obtained by conventional Sanger sequencing technology, including the finishing of the human genome project in 2000. However, there are inherent limitations in the Sanger Sequencing Technology that has been in place for the last 3 decades. This conventional technology requires higher cost and longer time that cannot meet the increasing demand of large-scale sequencing in individual laboratories.

Melvin D. Bolton and Bart P.H.J. Thomma (eds.), *Plant Fungal Pathogens: Methods and Protocols*,
Methods in Molecular Biology, vol. 835, DOI 10.1007/978-1-61779-501-5_17, © Springer Science+Business Media, LLC 2012

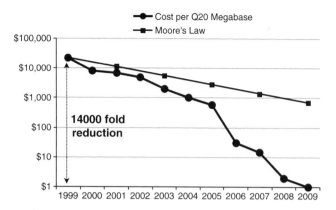

Fig. 1. Reduced cost of sequencing over the decade (adapted from 2010 data analysis from the Broad Institute). The sequence cost is estimated with Magabase sequence Phred Scores large than 20 (Q20). The actual cost reduction has outpaced the reduction predicted by the Moore's Law.

Table 1
Comparison of current next-generation sequencing technologies

	454/Roche system	Illumina/Solexa	ABI SOLiD	HeliScope
Sequencing chemistry[a]	Pyrosequencing	Synthesis with reversible terminators	Ligation-based sequencing	Synthesis by asynchronous extensions
Amplification approach	Emulsion PCR	Bridge amplification	Emulsion PCR	None
Output/run	80–120 Mb	1 Gb	1–3 Gb	1 Gb
Read length	250 bp	32–72 bp	35 bp	30 bp
Cost per Mb	$60	$2	$2	$1

[a]More detailed description of the chemistry can be found at the website of the respective companies, i.e., http://www.454.com; http://www.illumina.com; http://www.appliedbiosystems.com; http://www.helicosbio.com

Pathogenic filamentous microbes represent a large diversity of lifestyles covering divergent phylogenic lineages within the eukaryotic kingdoms. An estimated 90% of plant diseases are caused by pathogenic fungi species (3). Sequencing data have pushed forward almost every aspect of plant pathology research, such as the mechanism of virulence, population structure of virulence genes, and overview of host resistance responses. The Next-Generation Sequencing (NGS) technology holds the promise to usher a revolution in the phytopathology field by greatly increased throughput and significantly reduced costs (Fig. 1).

The strength of these new sequencing technologies lies with the fact that they are massively parallel processes rather than 96

Sanger sequencing reactions. NGS generates hundreds of megabases to gigabases of output in a single instrument run. These processes share a common feature: DNA molecules are spatially separated in a reaction chamber called a flow cell, enabling a large quantity of sequencing reactions running in parallel (4). Currently, there are three commonly used platforms, i.e., Roche (454) sequencer, Illumina genome analyzer, and Applied Biosystems SOLiD sequencer (Table 1). There are also a few single-molecule-based technologies on the horizon that are the so-called "next next generation" sequencing methods.

NGS methods have almost replaced existing gene expression studies using microarrays. This is because NGS data are more accurate, more reproducible, and free from background noises. It has been found that NGS produces better accuracy than microarray on studying absolute estimation of transcript expression levels (5). In many other fields like metagenomics, epigentics, and noncoding RNA study, NGS has also offered unprecedented opportunities.

2. Methods and Protocols

A set of core procedures underlies all the current new sequencing technologies and enables the massively parallel processes (Fig. 2). All the technologies start with library construction, amplification with emulsion PCR, or on solid surface (except the single-molecule sequencing) and are subjected to sequencing by synthesis in the flow cells.

2.1. Library Preparation

The sample prep method used by Illumina is slightly different from that of Applied Biosystems SOLiD system, but the basic goal is the same (i.e., generating large number of POLONIS). POLONIS are polymerase-generated colonies equivalent to bacterial colonies that were used in sequencing earlier.

DNA samples are fragmented by nebulization or sonication. The randomly sheared DNA sequences are used as template and further selected by sizes. The fragments of several 100 base pairs in length are isolated and enzymatically end-repaired. Adapter oligonucleotides are subsequently ligated to the templates.

2.2. Emulsion PCR

In the Roche 454 and Applied Biosystems SOLiD sequencing platforms, the amplification of template DNA is performed by Emulsion PCR (6, 7). The DNA library is diluted to single-molecule concentration. After the DNA is denatured, the single-strand template is hybridized to adapter oligonucleotides that are attached to individual beads. By emulsion, the beads are compartmentalized into water-in-oil microvesicles. These vesicles serve as individual chambers for the clonal amplification of single DNA molecules

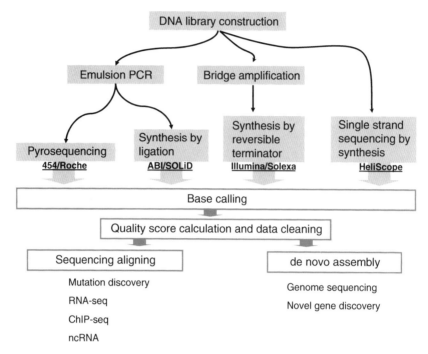

Fig. 2. Methodology of massively parallel sequencing technologies and bioinformatic analysis pipeline. In-depth descriptions are in the sections of experimental protocol and data analysis. The laboratory protocol parts are shaded with *gray boxes.*

bound to the beads. After PCR amplification in these discrete volumes, the emulsion is disrupted to release the beads. Each bead surface contains up to one million copies of the original fragment to produce detectable signal from the sequencing reaction.

2.3. "Bridge" Amplification

The PCR amplification in the Illumina genome analyzer is carried out by "bridge" amplification on solid surfaces (8). The PCR primers are covalently linked to a solid-support surface. DNA strands are denatured and captured by hybridization. The amplification is performed by DNA template "arching" over and hybridizing to an adjacent anchor oligonucleotide. A clonally amplified arching "cluster" is generated by multiple rounds of amplification. Each cluster contains approximately 1,000 clonal molecules for the subsequent sequence reaction.

2.4. Flow Cell Sequencing

High density and miniaturization of the sequencing reaction process are central to all ultra-high-throughput sequencing platforms. Capillary sequencing is no longer used for the large number of parallel sequencing processes. All the new sequencing platforms (i.e., the Roche 454 GS-FLX, the Illumina analyzer, the Applied Biosystems SOLiD, and Helicos HelioScope single-strand sequencer) utilize flow cells that contain templates tethered to a solid support.

Within this space, nucleotides and reactive reagents are iteratively applied and washed away.

2.5. Sequencing by Synthesis (SBS)

The core chemistry of all current commercial sequencing platforms uses DNA polymerase or ligase enzymes to extend many DNA strands in parallel. As DNA synthesis extension proceeds, the base type of the incorporated nucleotide or oligonucleotide can be determined in each cycle. Conventional Sanger sequencing utilizes fluorescently labeled ddNTPs with unlabeled dNTPs for polymerization. For Roche 454 platform, the substrate dNTPs are added and luminescence is used for detection of the released pyrophosphate. For Illumina platform, reversibly terminating dNTPs are fluorescently labeled. Unlike the Sanger ddNTPs, the termination of the synthesis is temporary and the emitted fluorescence enables detection as the DNA strand elongates. For Applied Biosystems SOLiD, ligation-based sequencing is performed by repeated ligation and hybridization of a limited set of semi-degenerate 8-mer oligonucleotides (7). When a matching 8-mer hybridizes to the DNA fragment sequence, DNA ligase ligates the phosphate backbone and a fluorescent signal identifies the base.

2.6. Target Sequence Enrichment

High-throughput sequencing requires very large amount of reads to get into "transcriptome depth"—in other words, obtaining low abundance RNA takes an overwhelming number of sequencing of highly abundant RNA with this technology. In order to get around this problem, several groups have focused on developing enrichment approaches where the highly abundant ribosomal RNAs are eliminated from the sequencing mixture by binding them to "locked nucleic acids," which are then removed from the mixture. Another enrichment approach is developed by Evrogen of Moscow, Russia, using a duplex-specific nuclease for normalization of RNA transcript levels. Following complementary DNA generation, the templates are denatured and a duplex-specific nuclease is added to the reaction. Abundant transcripts find matches and become double-stranded, thereby acting as targets for the nuclease. However, less abundant transcripts take longer to find their partners and so are degraded less frequently (9).

2.7. Artifacts of NGS Methods

Although NGS methods have revolutionized transcriptome analysis and genomics to a large extent, NGS is not free from artifacts. Pyrosequencing technologies such as Roche's 454 method depend on light intensities and have some problems deciphering repetitive sequences such as distinguishing between TAAAA and AAAAA (10). Applied Biosystems SOLiD sequencing technology is similar to that of 454 methods, but here the beads used for emulsion PCR are much smaller resulting in much denser packing and hence much higher throughput. However, this method produces colors for each overlapping dibase, which need to be deciphered computationally.

In this case, if there is a miscall of a base or a low-quality color call, then all the bases called thereafter are flawed. Illumina sequencing methods use cluster-based sequencing where a cluster represents a single DNA fragment. Because of the cluster-based phasing approach, DNA can be up to a certain length. Beyond that there is high substitution error.

From the perspective of practical applications, both Illumina and SOLiD platforms have their respective benefits. In high-throughput resequencing of large genomes, SOLiD is more accurate than Illumina. However, for RNA sequence analysis (RNA-seq), Illumina is more suitable.

In case of Solexa reads, about 0.3% error rate is found in the beginning of reads and 3.8% at the end of reads. It is also known that wrong base calls precede most often a G. The A to C substitution error is 10 times more than C to G substitution error (11). With Illumina ChIP-seq experiment, A to T miscall is most common where the quality scores were inconclusive (12). A strong correlation between accuracy of calling differentially expressed genes and transcript length was reported (13).

Short reads generated from NGS technologies are much smaller than the transcripts from which they are derived. Reads map at multiple locations because of paralogous gene families, low-complexity sequences, and high sequence similarity between alternatively spliced isoforms of the same gene. In addition, polymorphisms, reference sequence errors, and sequencing errors require that mismatches and indels are allowed in read alignments and further contribute to lower the confidence in mappings. Due to these factors, a significant number of reads are *multireads*: reads that have high-scoring alignments to multiple positions in a reference genome or transcript set (14). However, there is a problem with reads mapping to multiple isoforms of a gene, causing isoform expression to not be predicted accurately. Previous computational methods have either rejected reads that mapped into multiple locations or assigned reads heuristically to genes and both these methods are not free from errors.

3. Data Analysis

3.1. Base Calling

The original data generated from all NGS platforms are large sets of tiled fluorescence or luminescence images. These images are recorded during sequencing process on the flow-cell surface after each iterative sequencing step. Due to their large sizes, these image data require substantial investments alone for data storage, management, and processing. A single run of a NGS instruments generates 15GB to 15TB of data. Base calling is the computationally intensive process of identifying a nucleobase sequence from image files.

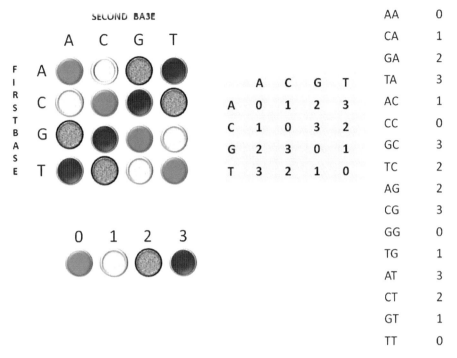

Fig. 3. Visual representations of colorspace data. In the SOLiD platform, a dinucleotide represents a color. There are only four colors, which are represented as numbers from 0 to 3. To represent 16 dinucleotides in four colors, every overlapping dinucleotide is given a color.

The computational program first finds individual beads or clusters to precisely locate each sequencing reaction. The intensity, background, and noise of images are then used to produce read sequences.

Colorspace data generated from Applied Biosystems SOLiD sequencing system are slightly different from the base calling methods where a single base is called at a time. In case of SOLiD platform, a dinucleotide represents a color and there are just four colors. The colors are represented as numbers from 0 to 3, and any number other than these represents an error. In order to portray 16 dinucleotides in four colors effectively, every overlapping dinucleotide is given a color (Fig. 3).

3.2. Quality Scoring

For each nucleobase produced from an image file, a quality value is calculated. Quality values play vital roles in all downstream analysis, from removing low-quality reads, trimming low-quality bases to aligning reads, and determining consensus-sequence. Computationally, several factors are used to estimate the quality of a base, including signal-to-noise levels, cross-talk from nearby beads or clusters, and dephasing (15). Each platform has its own tendency of generating errors (11). The 454 platform tends to produce inaccuracies in homopolymer run lengths and the Illumina platform has base-substitution error biases. Sequences coming out of sequencers are usually represented in Fastq format. Fastq is an extension of fasta format where sequence name begins with "@" followed by the sequence. The third line begins with a "+" that may have additional information. The fourth line represents quality value that is usually the ascii code of the numeric phred quality value. Since ascii code for 32 represents a [SPACE], usually 33 is added to phred value before converting it to ascii code. Fastq notation has lot of advantages and the most important one is in reducing the storage space where each phred quality score is reduced to a single character compared to three characters in case of qual files.

3.3. Data Cleaning

Raw sequence reads will contain adaptor sequences that have been ligated during library construction. Some sequencing platforms also add barcode sequences to facilitate identification of individual sample runs. Adaptor and barcode sequences can be removed using various tools such as Fastx_clipper or GATK online tools. Due to the sequencing chemistry and detection limit, the sequence quality tends to deteriorate towards the end of the read. Based on quality scores, each read is controlled and trimmed from the end and low-quality sequence reads are eliminated from the reads pool.

3.4. Aligning Sequences to the Reference

It is crucial to align sequence reads against a reference sequence, typically a previously sequenced genome, for a wide range of analysis such as polymorphism identification, transcript profiling, protein–DNA interaction site identification, and noncoding RNA discovery. Short query reads and very large amounts of query sequences pose challenges to the conventional pair-wise alignment programs. The most widely used algorithm, BLAST, uses a seed-and-extend strategy. BLAST is not suitable and very slow for handling millions of short reads. However, the modified algorithms BLAT and SSAHA speed up the process by holding an index of the query in random access memory, allowing their use for mapping 454 sequencing reads to the reference.

To handle large number of short reads with degenerate sites, many open-source solutions recently became available including MAQ, GMAP, SHRiMP, and SOAP. These programs can save computing memory and speed up searches (16–22). Apart from these, recently a way of indexing reference sequences into Burrows-Wheeler-Transformed (BWT) index has been developed (23). BWT keeps the memory footprint quite small and hence aids in analyzing huge amounts of data in a small work station. Bowtie is now available for colorspace reads and is the backbone of several other expression profiling softwares like crossbow (available for cloud computing) (24), cufflink (25), and myRNA (26). A list of all available softwares is given in Table 2.

Table 2
List of available softwares for NGS data analysis

Softwares	URL	Description
Short read alignment		
Bfast	http://sourceforge.net/apps/mediawiki/bfast/index.php?title=Main_Page	Blat-like fast accurate search tool
Bowtie	http://bowtie-bio.sourceforge.net/index.shtml	ultrafast, memory-efficient short read aligner
BWA	http://bio-bwa.sourceforge.net/bwa.shtml	Burrows-wheeler alignment tool. Gapped global alignment wrt queries
ELAND	http://bioit.dbi.udel.edu/howto/eland	Efficient local alignment of nucleotide data, runs on a single processor and allows up to two mismatches. Comes with illumine analyzer prepackaged
Exonerate	http://www.ebi.ac.uk/~guy/exoncrate/	Generic sequence alignment tool
GenomeMapper	http://1001genomes.org/downloads/genomemapper.html	Does both gapped and ungapped alignment and maps against multiple reference
GMAP and GSNAP	http://research-pub.gene.com/gmap/	Short read aligner
GNUMap	http://dna.cs.byu.edu/gnumap/	Genomic next-generation universal MAPper (gnumap)
MAQ	http://maq.sourceforge.net/	Mapping and assembly with quality
MOSAIK	http://bioinformatics.bc.edu/marthlab/Mosaik	Reference-guided assembler
MrFast and MrsFast	http://mrfast.sourceforge.net/ http://mrsfast.sourceforge.net/	Microread fast alignment search tool and microread substitution-only fast alignment search tool
MUMmer	http://mummer.sourceforge.net/	Ultrafast alignment of large-scale DNA and protein sequences
Novoalign and NovoalignCS	http://www.novocraft.com/main/downloadpage.php	Highly accurate programs for mapping next-generation sequencing reads to a reference database
PASS	http://pass.cribi.unipd.it/cgi-bin/pass.pl	Gapped and ungapped alignment
RMAP	http://rulai.cshl.edu/rmap/	Map reads with or without error probability information (quality scores) and supports paired-end reads or bisulfite-treated reads mapping
SeqMap	http://biogibbs.stanford.edu/~jiangh/SeqMap/	Supports up to 5 or more bp mismatches/INDELs. Highly tunable
SHRiMP	http://compbio.cs.toronto.edu/shrimp/	It was primarily developed with the multitudinous short reads of next-generation sequencing machines in mind
SOAP	http://soap.genomics.org.cn/	Short oligonucleotide analysis package
Slider	http://www.bcgsc.ca/platform/bioinfo/software/slider	Analyzer that uses the probability files instead of the sequence files as an input for alignment
SSAHA2	http://www.sanger.ac.uk/resources/software/ssaha2/	Sequence search and alignment by hashing algorithm
SOCS	http://solidsoftwaretools.com/gf/project/socs/	Ungapped alignment tool designed for mapping both standard SOLiD data and bisulfite transformed SOLiD data

(continued)

Table 2
(continued)

Softwares	URL	Description
Swift	http://bibiserv.techfak. uni-bielefeld.de/swift/ welcome.html	fast local alignment search, guaranteeing to find *epsilon*-matches between two sequences
SXOligoSearch	http://www.synamatix.com/ secondGenSoftware.html	Is a commercial software
VMatch	http://www.vmatch.de/	More suitable for repeat finding and miRNA search
Zoom	http://www.bioinformaticssolutions. com/products/zoom/index.php	Zillions of oligos mapped
De novo assembly		
ABySS	http://www.bcgsc.ca/platform/ bioinfo/software/abyss	Assembly by short sequences. Can be used in MPI mode
ALLPATHS	ftp://ftp.broadinstitute.org/pub/ crd/ALLPATHS/Release-LG/	De novo assembly of whole-genome shotgun microreads
Edena	http://www.genomic.ch/edena.php	Exact de novo assembler
EULER-SR	http://euler-assembler.ucsd.edu/ portal/	Contrary to the overlap-layout approach, EULER-SR uses a de Bruijn graph to construct an assembly. The assembly of a genome corresponds to an Eulerian path in the de Bruijn graph. Long (possibly erroneous) reads and mate-pairs are used to determine parts of the correct Eulerian traversal in the assembly
MIRA3	http://chevreux.org/ projects_mira.html	Mimicking intelligent read assembly
SHARCGS	http://sharcgs.molgen.mpg.de/	Short read assembler based on Robust Contig extension for genome sequencing
SSAKE	http://www.bcgsc.ca/platform/ bioinfo/software/ssake	The short sequence assembly by K-mer search and 3′ read extension
SoapDeNovo	http://soap.genomics.org.cn/soapde-novo.html	Novel short read assembly method that can build a de novo draft assembly for the human-sized genomes
VCake	http://sourceforge.net/projects/ vcake/	De novo assembly of short reads with robust error correction. An improvement on early versions of SSAKE
Velvet	http://www.ebi.ac.uk/~zerbino/ velvet/	Velvet is a de novo genomic assembler specially designed for short read sequencing technologies
SNP/Indel discovery		
ssahaSNP	http://www.sanger.ac.uk/resources/ software/ssahasnp/	ssahaSNP is a polymorphism detection tool. It detects homozygous SNPs and indels by aligning shotgun reads to the finished genome sequence
PolyBayesShort	http://bioinformatics.bc.edu/marth-lab/PbShort	A reincarnation of the PolyBayes SNP discovery tool

(continued)

Table 2
(continued)

Softwares	URL	Description
Genome browsers		
EagleView	http://bioinformatics.bc.edu/marth-lab/EagleView	Reads ACE files and is standalone
LookSeq	http://www.sanger.ac.uk/resources/software/lookseq/	LookSeq is a web-based application for alignment visualization, browsing, and analysis of genome sequence data
MapView	http://evolution.sysu.edu.cn/mapview/	Is an ACE viewer works on standalone machines
SAM	http://www.bcgsc.ca/platform/bioinfo/software/sam	Sequence assembly manager and viewer. Stores data in mysql tables
STADEN	http://staden.sourceforge.net/	Nextgen sequence analysis data recently added
XMATCHVIEW	http://www.bcgsc.ca/platform/bioinfo/software/xmatchview	Is a cross-match alignment viewer
Transcriptomics		
Cufflink	http://cufflinks.cbcb.umd.edu/	Transcript mapping, assembly, and quantification
ERANGE	http://woldlab.caltech.edu/rnaseq/	Mapping and quantifying mammalian transcriptomes by RNA-Seq. supports bowtie, BLAT and ELAND
G-Mo.R-Se	http://www.genoscope.cns.fr/externe/gmorse/	Is a gene modeler using RNAseq data
MAPNEXT	http://evolution.sysu.edu.cn/english/software/mapnext.htm	A software tool for spliced and unspliced alignments and SNP detection of short sequence reads
QPALMA	http://www.fml.tuebingen.mpg.de/raetsch/suppl/qpalma	Optimal spliced alignments of short sequence reads
RSAT	http://biogibbs.stanford.edu/~jiangh/rsat/	RNA-Seq analysis tools
TOPHAT	http://tophat.cbcb.umd.edu/	Is a fast splice junction mapper for RNA-Seq reads
Counting, e.g., CHiP-Seq, Bis-Seq, CNV-Seq		
BS-Seq	http://epigenomics.mcdb.ucla.edu/BS-Seq/download.html	Shotgun bisulphite sequencing of the Arabidopsis genome reveals DNA methylation patterning
CHIP-Seq	http://ccg.vital-it.ch/chipseq/	ChipSeq analysis server
CNV-Seq	http://tiger.dbs.nus.edu.sg/cnv-seq/	A new method to detect copy number variation using high-throughput sequencing
Findpeak	http://www.bcgsc.ca/platform/bioinfo/software/findpeaks	Perform analysis of ChIP-Seq experiments
MACS	http://liulab.dfci.harvard.edu/MACS/	Model-based analysis of ChipSeq
PeakSeq	http://archive.gersteinlab.org/proj/PeakSeq/	Systematic scoring of ChIP-Seq experiments relative to controls. A two-pass approach for scoring ChIP-Seq data relative to controls

(continued)

**Table 2
(continued)**

Softwares	URL	Description
QuEST	http://mendel.stanford.edu/sidowlab/downloads/quest/	Quantitative enrichment of sequence tags
SISSRs	http://dir.nhlbi.nih.gov/papers/lmi/epigenomes/sissrs/	Site identification from short sequence reads takes BED files as input
Alternate base calling		
Rolexa	http://svitsrv25.epfl.ch/R-doc/library/Rolexa/html/00Index.html	Is an R package for base calling
Alta-cyclic	http://hannonlab.cshl.edu/Alta-Cyclic/main.html	Is a novel illumina genome-analyzer (Solexa) base caller
PyroBayes	http://bioinformatics.bc.edu/marthlab/PyroBayes	PyroBayes is a novel base caller for pyrosequences
Integrated solutions		
CLCbio genomics workbench	http://www.clcbio.com/index.php?id=1240	Commercial software for integrated data analysis
Galaxy	http://main.g2.bx.psu.edu/	An integrated data management and analysis tool
Genomatix	http://www.genomatix.de/en/produkte/genomatix-software-suite.html#2	Integrated solutions for next-generation sequencing data analysis
JMP genomics	http://www.jmp.com/software/genomics/	Next gen visualization and statistics tool from SAS
NextGENe	http://softgenetics.com/NextGENe.html	De novo and reference assembly of illumina, SOLiD, and Roche FLX data
SeqMan genome analyzer	http://www.dnastar.com/t-products-seqman-ngen.aspx	Part of DNASTAR package. Commercial software
SHORE	http://1001genomes.org/downloads/shore.html	Weighted and gapped alignment is allowed, so the reference genome does not need to be from the same species
SlimSearch	http://www.realtimegenomics.com/	Commercial tool

As sequencing technologies are getting more mature, the read length is getting longer. With longer reads (>35 bases), there is a higher probability that the read spans across an intron/exon boundary. Most of the alignment programs have a mismatch threshold of two bases, and hence many correct alignments are lost due to this threshold. Recently, a software program called Tophat (27) was developed to tackle this problem. Tophat splits the read (A) into n fragments where n is length of "A" divided by a user-defined parameter k which defaults at 25. The fragments are mapped to the reference as seed using bowtie and a database of all the intron-exon boundaries are created from the reference sequence that falls around the read alignment. The seeded aligned fragments are extended on both sides until it meets the junction. Novel intron-exon boundaries are found using this approach.

3.5. De Novo Sequence Assembly

In the absence of any other reference sequence, the construction of contigs from a set of partially overlapping sequence reads is called de novo sequence assembly. In absence of the reference genome sequence belonging to the same species, a closely related species may be used as reference with certain acceptable error rate. Assembly of short reads with de novo sequencing strategy is a challenging option when the genome itself is large and is marred with large amount of repeats.

For the assembly of the conventional Sanger sequence reads, there are several established algorithms. However, these algorithms depend on long overlaps between sequences and they are unable to handle millions of reads. To tackle the problem of assembly of large quantity of short reads, a group of new assembly algorithms has been developed (28–33). All of these new assembly algorithms use the mathematical concept of a graph. A graph is defined as a set of nodes that can be connected by edges. In an assembly algorithm, every node in the graph is a sequence read, whereas the edges represent overlaps between read sequences. A weight is assigned to each edge, which represents the abundance of the sequence read. A single linear path through the graph will represent the "true" assembled sequence. And loops and branches indicate ambiguities that result from repeat sequences and sequencing errors. The de Bruijn graph is a refinement of this approach. In a de Bruijn graph, each edge is a k-mer instead of a read overlap. A de Bruijn graph is computationally more efficient when dealing with large numbers of reads. It is the core algorithm used in the widely used short reads assemblers such as Velvet, Euler, Edena, and ALLPATHS.

3.6. Handling Gene Multireads and Isoform Multireads

When reads map to multiple genes, it is termed as gene multireads. When reads match to multiple isoforms, it is called as isoform multireads. Two strategies have been used to handle gene multireads. One is to simply discard them (34) and the second strategy is to "rescue" multireads by allocating fractions of them to genes in proportion to coverage by uniquely mapping reads (14–35). The second strategy has proved to be in better agreement with microarray data.

The uniquely mapped reads are computed for expression analysis using the following formula:

$$v_i^{\text{uni}} = \frac{c_i^{\text{uni}}}{c^{\text{uni}}},$$

Where c_i^{uni} stands for number of reads aligning to gene i and c^{uni} is the total number of uniquely mapped reads. The transcript fraction then computed by:

$$\tau_i^{uni} = \frac{v_i^{uni}}{\tilde{l}_i} \left(\sum_j \frac{v_j^{uni}}{\tilde{l}_j} \right)^{-1},$$

Where $\tilde{l}_i$ is the effective length of gene i. For genes with a single isoform, $\tilde{l}_i$ is simply the length of that isoform. For alternatively spliced genes, we take $\tilde{l}_i$ to be the length of the union of all genomic intervals corresponding to exons of isoforms of gene i, as is done in the ERANGE software package (36).

Coming to the second method that is called as the rescue method, the abundance of gene i is computed as:

$$c_i^{rescue} = \sum_{n:i \in \pi_n} \frac{\tau_i^{uni}}{\sum_{j \in \pi_n} \tau_j^{uni}} :$$

Where π_n is the set of indices of genes to which read n maps. A multiread that maps only to genes for which $\tau^{uni} = 0$ is divided evenly among those genes. The values of v_i^{rescue} and τ_i^{rescue} are then calculated similarly to previous equations. RPKM (reads per kilobase of exon per million mapped reads) method based on this model calculates transcript abundance by $R = 10^9 \times c_i / (N_m l_i)$, where c_i is the number of reads mapping to isoform i, N_m is the total number of mappable reads, and l_i is the length of isoform i. Conversely, copy number of a gene is calculated as $X = T * C/NL = R/10^9 * T$; where T is the length of the entire transcriptome and R is RPKM value.

There are two natural forms of expression: fraction of transcripts or transcripts per million (Ti or TPM) and fraction of nucleotides or nucleotides per million (Ni or NPM) where TPM and NPM are obtained by multiplying Ti and Ni by 10^6, respectively. The fundamental assumption is that the fraction of reads obtained from an isoform is a function of fraction of nucleotides. So, as total number of reads approached infinity, Ni equals reads mapped to isoform/total number of mappable reads (Ni = Ci/N as N→∞). Under the assumption of uniformly distributed reads, it is noted that RPKM measures are estimates of $10^9 \times N_i / l_i$, which is an unnormalized value of τ_i (37).

A more recent method handles isoform multireads by explicitly estimating isoform expression levels, but does not handle gene multireads (38). Based on this estimation, FPKM (expected number of fragments per kilobase of transcript sequence per millions base pairs sequenced) is defined. The abundance of a transcript t g (all genes) in FPKM units is proportional to transcript abundance (αt).

$$106 * 103 * \alpha t \diagup 1(t) = 106 * 103 * \beta g * \Upsilon t \diagup 1(t)$$

βg and Υt are the parameters estimated from the likelihood function.

A newer strategy handles gene multireads as well as isoform multireads especially when gene multireads are significant (36).

3.7. Downstream Data Manipulation and Visualization

The read alignment programs most often produce output in generic SAM (Sequence Alignment and Mapping) or in MAQ format. However, SAM format is gaining popularity because it contains all basic information about alignments with its 11 mandatory fields and a number of optional fields. SAM files are compressed to a binary format called BAM format that uses a library similar to zlib for random fast access of data. Compressed bam files occupy only 25% of space of that of sam files. Samtools (19) performs stream-based processing on specific genomic regions without loading the entire file into memory. The SAM/BAM format, together with SAMtools, separates the alignment step from downstream analyses, enabling a generic and modular approach to the analysis of genomic sequencing data. Samtools can convert data from other alignment formats, sort and merge alignments, remove PCR duplicates, generate per-position information in the pileup format, call SNPs and short indel variants, and show alignments in a text-based viewer. Apart from samtools, other relatively new software applications such as bowtie and cufflink facilitate downstream data analysis such as finding novel intron-exon junctions, transcript discovery, and expression analysis.

Visualization of alignments is an integral part of data analysis. Browser-based visualization is very common, where one uploads sam/bam files into browser where already a preloaded annotated reference genome exists. The read alignments are displayed as tiled alignments or their RPKM values are plotted as histograms. Among the popular softwares, UCSC genome browser, Iintegrated Genomics Viewer (IGV), and MagicViewer are the popular ones.

4. Applications

4.1. Genome Sequencing

For sequencing new genomes and highly rearranged genome segments such as repeat rich regions and loci harboring virulence genes, new sequencing technology still pose a challenge. One major disadvantage of these techniques is shorter read length as compared to that of the conventional Sanger technology. The longest next-generation reads obtained are in the range of ~250 nt for 454 pyrosequencing, and Illumina sequencing has read length of 36–72 nt. As the new assembly algorithms develop, many smaller eukaryotic genomes become feasible to be sequenced entirely with short reads sequencing.

Most filamentous fungi contain few repetitive regions, and their genome sizes are typically 30–90 Mb. These features make pathogenic fungi an attractive candidate for de novo genome sequencing by NGS. Recently, a draft genome sequence of the fungus *Sordaria macrospora* has been obtained by a combination of Illumina/Solexa and Roche/454 sequencing (39). Another filamentous fungus, *Grosmannia clavigera*, is a forest pathogen with a genome of 32.5 Mb. Its genome sequences were also assembled from a combination of Sanger, 454, and Solexa sequence data (40). The small size of typical fungal genomes is suited for de novo assembly approaches to utilize the benefits of drastically reduced costs (Fig. 1). To systematically characterize all the strains stored in major fungal centers, it is conceivable to sequence all the genomes in collection by the massively parallel sequencing technologies (Table 3).

4.2. Variant Discovery

Variants can be of genetic, epigenetic, transcript, metagenomic variation in a genome. For plant pathology research, it is a central theme to discover the variations underlying the pathogenesis-related phenotypes in natural pathogen populations as well as in a laboratory-derived mutant. The conventional Sanger sequencing uses directed PCR to amplify selected genomic regions from individual strains and compares the sequence of the PCR product with the reference genome. The incentive to circumvent this expensive approach for resequencing genomes has enabled the first wave of NGS technologies.

To analyze a large and variable population, new sequencing technology for the first time offers affordable and sensitive solutions. So far, the new sequencing technologies have been used in viral and bacterial pathogens for the discovery of mutation strains with low frequency in the population. For example, with ultradeep pyrosequencing, the viral population mutation spectra have been identified in HIV clinical isolate. And the rare members can be found in relations to drug resistance (41). Due to the larger size of fungal and oomycetes genomes, similar approach will be restricted to samples with reduced complexity; however, it offers the possibility of monitoring large population with great sensitivity.

The other strength of the new technologies is the genome-wide overview of comparison between reference and query genomes. Because many pathogens evolve rapidly by mutation and by exchanging sequences among each other, sequencing virulent isolates will yield large amount of data about their mechanisms in circumventing host resistance or control chemicals. By using the 454 platform, the genomes of drug-resistant *Staphylococcus aureus* isolates and *Mycobacterium tuberculosis* isolates were compared with susceptible strains in different studies to identify drug targets and the mechanisms of resistance (29). In pathogenic fungi and oomycetes, next-generation sequencing technologies are likely to deepen our understanding of the spectrum of genome variation

Table 3
Applications of new sequencing technologies in pathogenic fungi and oomycetes

	Applications in plant pathogens	Strength as compared to conventional technologies	Drawbacks and technical developmental directions
Genome sequencing	For species with typical fungal genome sizes (30–40 Mb)	Low cost and no conventional cloning needed	Difficultly with large or repeated genomes such as oomycetes
Variant discovery	Monitoring large and complex field population with great sensitivity Genome-wide comparison of closely related sister species	Genome-wide study and high resolution	Reference genome sequence needed for optimal results
Transcriptional profiling	Expression profiling of developmental stages Gene discovery of lower expressed effectors or enzyme genes	Low cost Large dynamic range	De novo transcript assembly is under active development
Metagenome	Study community in complex samples like suppressive soil or multiple infection tissues	Handling complex sample and material that cannot be cultured	Shorter reads bring difficulties in identifying large genes
Pan-genome	Cataloging all strain variants, such as a complete effector reservoir in a given species	Sequencing large number of strains	
Regulome	Genome-wide mapping of regulation for virulence factors or components of secondary metabolite pathway	Affordable genome-wide analysis with high resolution	
Epigenome	Investigate the role of DNA methylation and histone modification in development and infection process	Affordable genome-wide analysis with high resolution	
Noncoding RNAs discovery	Study fungal development and pathogenesis	Genome-wide analysis	

within field isolates, as well as within closely related pathogenic and saprotrophic species.

4.3. Transcriptional Profiling

The transcriptome is defined as the complete set of transcripts in a cell for a specific developmental stage or physiological condition. To understand a transcriptome of a cell, it is necessary to catalog all transcript species, to capture transcript structures, and to quantify the expression levels. For a plant-pathogenic organism, transcriptome analysis has enabled us to understand how pathogen progresses through its life cycle and how host pathogen interacts at a molecular level (42).

During the last 10 years, transcriptional profiling strategies primarily rely on Expressed Sequence Tag (EST) and DNA chip and hybridization experiments. Array-based methods are high throughput and allow comparing of the levels of transcription for each transcript. However, array-based methods have several inherent limitations. First, to design an array, existing knowledge about genome sequence must be utilized. Second, the method relies on hybridization which always generates relatively high background levels due to cross-hybridization. Finally, there is a limited dynamic range of detection because of saturation of hybridization signals.

The technique of RNA-seq utilizes NGS for transcriptional profiling (43). The technique starts with a population of RNA isolated from controls and treatments. The pool containing polyA+or fractionated RNA is then converted to a library of cDNA fragments, followed by the ligation of adaptors to the end. Unlike the array-based hybridization approach, RNA-seq technology does not suffer from cross-hybridization of probes and has no inherent limitation on the number of transcripts that can be identified. The dynamic range of RNA-seq is 2–3 orders of magnitude higher than array-based experiment.

Many key questions in plant pathology research, such as mechanisms underlying gene-for-gene resistance, host vs. nonhost resistance, and biotrophy vs. necrotrophy, can be addressed by RNA-seq. For example, characterization of virulence factors has been accelerated by genomic technologies available for several pathogens. Many of these factors show a characteristic expression pattern and some of them as in the oomycetes are found to be expressed at very low level. RNA-seq technology will facilitate the identification of early expression events and rare but critical transcripts. In *Phytophthora sojae*, an oomycete pathogen, we have analyzed around 60 million reads per experimental replicate in infected sample. We compared the genes perturbed during infection process with normal condition. Interestingly, about 25% of the entire transcriptome has shown differential expression during infection (our unpublished result).

RNA-seq has offered an attractive way to discover novel transcripts. In *Hyaloperonospora arabidopsidis*, we have used the NGS in discovering novel transcripts and predicting repetitive regions in the genome. Typically after shotgun sequences are assembled into contigs, genes are predicted using various methods based on *ab initio* and sequence similarity approaches. These approaches often miss out a large number of genes that differ either in their overall GC content or in donor-acceptor sites or compositions. From our recent work with RNA-seq data on *Phytophthora sojae*, we found that around 60% of the genes are missing, 20% exons are wrong,

90% introns are missing, 20% introns are wrong, 45% loci are missing, and 20% loci are incorrect (our unpublished data). Moreover, this result was congruent in all four experimental replicates. With our present analysis, we discovered around 2% new isoforms, 20% UTRs, 40% junctions that could possibly have new genes.

4.4. Metagenome

Many, if not most, microbial organisms live in communities. As a consequence of the lifestyle of community living, the genetic material in a given environment is derived from individual microbes from different species, which is referred to as a metagenome. For example, humans live in symbiosis with many different microbial genera that inhabit the skin surface, digestive tract, and body cavities. All these bacteria are collectively called "human microbiome." There are many other complex metagenomic populations in soil, marine habitat, deep mine, and on the site of multiple infections of a given host plants (44, 45).

There are two major aspects of a megagenome that can be studied by the new technology: the identity and abundance of the microbes. To investigate the types of microbes present in a soil sample or an infection material, it is required to compare the metagenomic sequences with all other known sequenced species and isolates. To survey the relative abundance of the individual species, the sequence reads redundancy can be used to estimate the population frequency of each microbe. Metagenomic approaches will provide a molecular analysis of complex samples like "suppressive" soil or an infected field, without the requirement of culture of the organism. Comparison between different complex samples will help us to understand the role of a particular virulent strain and the consequences of change in strain frequency in fungal and oomycetes populations.

4.5. Pan-Genome

For pathogenic microbes, there is often large genetic diversity within a species. Individual strains may have overlapping yet different sets of virulence factors, mobile elements, and metabolic enzymes. The total gene repertoire in a given species is referred to as pan-genome (32, 46). Notably, it is a very different concept than metagenome, because pan-genome refers to diversity within a single species across different environment. The pan-genome concept includes the "core genome," which is shared by all individuals, and the "dispensable genome," which is shared by some individuals or unique to an individual. The size of core and dispensable genomes is relative, depending on the population diversity of a species. The pathogen *Bacillus anthracis* seems to have limited number of polymorphic nucleotides among the clinical isolates (47). By contrast, extreme levels of plasticity have been observed in Streptococcus genomes with high levels of gene gain and loss (46). Highly diverse microbes like Streptococcus may need hundreds or more strains to survey their pan-genomes.

The pan-genome concept so far is used only in bacterial species, but it will be a very useful concept for highly variable pathogenic fungi and oomycetes. For example, recent genomic analysis has shown that oomycetes possess hundreds of effectors and a subset only presents a few strains. Further Illumina sequencing has shown that most of the effectors show copy number variation among strains (our unpublished results). Therefore, the entire repertoire of virulence factors is collectively represented by isolates in a species. Another example is the discovery of mobile chromosomes in *Fusarium sp.* that accounts for pathogenicity (48). In both cases, a single reference genome will only give a snapshot of the genetic makeup, and only partly explains the pathogenesis potential as a species. The application of the next-generation sequencing technology and de novo assembly of reads will greatly enhance our understanding of microbial pan-genomes.

4.6. Regulome

The entire set of regulation components in a cell is called a regulome. Crucial regulation of gene expression activity is performed by transcription factors through binding to specific sites of the DNA sequences. Transcription factors typically bind to a promoter or other cis-elements in the genome, thereby promoting or blocking the recruitment of RNA polymerase to specific genes.

The interaction between DNA and proteins lies at the heart of transcriptional regulations. Chromatin immunoprecipitation (ChIP) is a technique for elucidating this type of interaction. First, a cross-linking agent like formaldehyde links any protein in close association with DNA. Second, a specific antibody is used to precipitate the protein along the attached DNA fragment. Finally, these DNA fragments are subsequently released by reversing the cross-linking. To identify these DNA fragments, either of the two following high-throughput methods is used: hybridization of fluorescently labeled DNA fragments to an appropriate microarray (ChIP-chip) or direct sequencing with NGS technology (ChIP-seq). The technology of ChIP-seq in general offers higher resolution than array-based method. And the large amount of data generated in a single instrument run can provide enough sequences for all sites typical of large mammalian genomes (49, 50).

Genome-wide transcription factor-binding sites have been recently discovered by "ChIP-seq" approach. Our basic understanding of transcriptional regulation has been advanced with the access of large datasets. For example, the text-book example of TATA-driven promoters has been shown to be only one of a diverse networks of promoters. Pathogenic fungi and oomycetes have complex gene regulation networks; some developmental processes are closely tied to the advance of pathogenesis. In many infections, the deployment of effectors is also closely timed by the pathogens and the large battery of oomycetes effectors are mostly induced in planta, but the exact regulations mechanisms remain to be elucidated (51).

A typical fungal species has a few 100 proteins with transcription regulation activity based on GO annotation. The genome-wide approach with next-gene sequencing technology will help us to understand the network of regulation in pathogens.

4.7. Epigenome

Epigenome refers to the form of the genome that has been biochemically modified without alteration in DNA sequence. Epigenetic changes like DNA methylation, histone modifications, and high order organization of chromatin in cell nuclei are the components of the epigenome. Depending on the different developmental stage and various environmental conditions, one genome can have several different epigenomes. Epigenetic changes play vital roles in developmental biology, cancer development, and psychiatric diseases (52). In the malaria parasite *Plasmodium falciparum*, epigenetic control serves as the predominant form of gene regulation in different life stages and infection processes (53). We will discuss the three levels of epigenetic modifications.

The first level is DNA methylation. It refers to the covalent modification by cytosine-5' methylation. It is an ancient mechanism that occurs in the common ancestor of major eukaryotic kingdoms (54). The technique of bisulfite sequencing is used to detect the DNA methylation sites. To catalog genome-wide DNA methylation patterns, NGS technology is used for the sequencing step and referred to as ultradeep bisulfite sequencing. In all major fungal groups, conserved methyltransferases and substantial amount of CG methylation have been found. Fungal DNA methylation is concentrated in transcriptionally silent, repetitive loci and appears to be used to silence mobile elements. Many virulence factor genes in pathogenic fungi and oomycetes are located in repeat rich region (48, 51) and perhaps subjected to epigenetic regulation. It is of great interest to study the pathogenic fungal methylation pattern systematically.

The second level refers to the packing of DNA into histones, which largely determines the availability of genes for transcription. Alternative histone use and covalent histone modifications such as acetylation, ubiquitination, and methylation influence the accessibility of DNA sequences by transcriptional complexes. The technology of ChIP-seq can be used to characterize histone modification in a genome-wide manner. Among a broad range of fungal genera, experimental evidences showed that histone deacetylases function in the regulation of various secondary metabolite clusters (55). To investigate the regulation of the polyketide, nonribosomal peptide, and other pathogenesis-related molecules, a genome-wide investigation in fungal species is needed.

The final level of epigenetic regulation is the higher-order of chromatin structuring. The spatial organization of chromosomes in a cell's natural state regulates gene expression, DNA replication and repair, and recombination. To investigate the special interaction of chromosomes, the technique of chromosome conformation capture (3C) has been developed. The method uses formaldehyde cross-linking followed by restriction enzyme digestion and intramolecular ligation to detect physically interacting genomic loci that are presumably important for regulating gene expression. These loci are detected by NGS sequencing. In the malaria parasite, where most of the regulation is on the epigenetic level, the physical location of the virulence gene is often at the peripheral of nuclear membrane. Similar mechanism could well impact other eukaryotic pathogens regulation.

4.8. Noncoding RNAs

Noncoding RNA (ncRNA) describes a diverse group of regulatory RNA molecules that are not transcribed from protein-encoding genes. ncRNA plays a wide range of roles in tissue development, diseases progresses, and even human cognition (52, 53, 56). Because ncRNA is relatively small in size, NGS technology is ideal for genome-wide ncRNA discovery. To characterize ncRNA bioinformatically, the sequence is further examined for secondary structure formation, genomic location, and putative function. Four hundred and fifty four technologies have been utilized to discover new and different ncRNA classes in a variety of species from worm, fruit-fly, plants, and human. In the malaria parasite Plasmodium, ncRNAs may be components of chromatin structure that regulates interaction with hosts (53). In the fungal kingdom, the role of ncRNA is gradually being discovered. One class of ncRNA is snoRNA, which has been found to be well conserved among yeasts and multicellular fungi, implying their functional importance for the fungus cells (56). With the advance of new sequencing technology, studying distinct groups of ncRNA will open a new field in plant pathology.

References

1. Sanger F, Nicklen S, Coulson AR: DNA sequencing with chain-terminating inhibitors. *Proc Natl Acad Sci USA* 1977, 74(12):5463–5467.

2. Maxam AM, Gilbert W: A new method for sequencing DNA. *Proc Natl Acad Sci USA* 1977, 74(2):560–564.

3. Hawksworth DL: The magnitude of fungal diversity: the 1.5 million species estimate revisited. *Mycological Research* 2001, 105(12):1422–1432.

4. Holt RA, Jones SJ: The new paradigm of flow cell sequencing. *Genome Res* 2008, 18(6):839–846.

5. Fu X, Fu N, Guo S, Yan Z, Xu Y ea: Estimating accuracy of RNASequencing and microarray with proteomics. *BMC Genomics* 2009, 10.

6. Margulies M, Egholm M, Altman WE, Attiya S, Bader JS, Bemben LA, Berka J, Braverman MS, Chen YJ, Chen Z et al: Genome sequencing in microfabricated high-density picolitre reactors. *Nature* 2005, 437(7057):376–380.

7. Shendure J, Porreca GJ, Reppas NB, Lin X, McCutcheon JP, Rosenbaum AM, Wang MD, Zhang K, Mitra RD, Church GM: Accurate multiplex polony sequencing of an evolved bacterial genome. *Science* 2005, 309(5741):1728–1732.

8. Bentley DR, Balasubramanian S, Swerdlow HP, Smith GP, Milton J, Brown CG, Hall KP, Evers DJ, Barnes CL, Bignell HR et al: Accurate whole human genome sequencing using reversible terminator chemistry. *Nature* 2008, 456(7218):53–59.

9. Blow N: Transcriptomics: The digital generation. *Nature* 2009, 458:4.

10. Susmita Datta, Somnath Datta, Seongho Kim, Sutirtha Chakraborty, Gill RS: Statistical Analyses of Next Generation Sequence Data: A Partial Overview *Journal of Proteomics and Bioinformatics* 2010, 3:8.

11. Dohm JC, Lottaz C, Borodina T, Himmelbauer H: Substantial biases in ultra-short read data

sets from high throughput DNA sequencing. *Nucleic Acids Res* 2008, 36(16):e105.

12. Bravo HC, Irizarry RA: Model-based quality assessment and base-calling for second-generation sequencing *Biometrics* 2010, 66(3):10.

13. Oshlack A, Wakefield M: Transcript length bias in RNA-sequencing data confounds systems biology. *Biol Direct* 2009, 4.

14. Mortazavi A, Williams BA, McCue K, Schaeffer L, Wold B: Mapping and quantifying mammalian transcriptomes by RNA-Seq. *Nat Methods* 2008, 5(7):621–628.

15. Brockman W, Alvarez P, Young S, Garber M, Giannoukos G, Lee WL, Russ C, Lander ES, Nusbaum C, Jaffe DB: Quality scores and SNP detection in sequencing-by-synthesis systems. *Genome Res* 2008, 18(5):763–770.

16. Li H, Ruan J, Durbin R: Mapping short DNA sequencing reads and calling variants using mapping quality scores. *Genome Res* 2008, 18(11):1851–1858.

17. Lin H, Zhang Z, Zhang MQ, Ma B, Li M: ZOOM! Zillions of oligos mapped. *Bioinformatics* 2008, 24(21):2431–2437.

18. Li R, Yu C, Li Y, Lam TW, Yiu SM, Kristiansen K, Wang J: SOAP2: an improved ultrafast tool for short read alignment. *Bioinformatics* 2009, 25(15):1966–1967.

19. Li H: The Sequence Alignment/Map format and SAMtools. *Bioinformatics* 2009, 25:2078–2079.

20. Li R: SOAP2: an improved ultrafast tool for short read alignment. *Bioinformatics* 2009, 25:1966–1967.

21. Homer N, Merriman B, Nelson SF: BFAST: an alignment tool for large scale genome resequencing. *PLoS ONE* 2009, 4:e7767.

22. Li R, Li Y, Kristiansen K, Wang J: SOAP: short oligonucleotide alignment program. *Bioinformatics* 2008, 24:713–714.

23. Langmead B, Trapnell C, Pop M, Salzberg SL: Ultrafast and memory-efficient alignment of short DNA sequences to the human genome. *Genome Biol* 2009, 10:R25.

24. Ben Langmead, Michael C Schatz, Jimmy Lin, Mihai Pop, Salzberg SL: Searching for SNPs with cloud computing. *Genome Biology* 2009, 10(11):134.

25. Trapnell C, Williams B, Pertea G, Mortazavi A, Kwan G, van Baren M, Salzberg S, Wold B, Pachter L: Transcript assembly and quantification by RNA-Seq reveals unannotated transcripts and isoform switching during cell differentiation. *Nat Biotech* 2010, 28(5):511–515.

26. Ben Langmead, Kasper D Hansen, Leek JT: Cloud-scale RNA-sequencing differential expression analysis with Myrna. *Genome Biol* 2010, 11(8).

27. TopHat: discovering splice junctions with RNA-Seq. *101093/bioinformatics/bth101* 2009, 25(9):1105–1111.

28. Warren RL, Sutton GG, Jones SJ, Holt RA: Assembling millions of short DNA sequences using SSAKE. *Bioinformatics* 2007, 23(4): 500–501.

29. Hernandez D, Francois P, Farinelli L, Osteras M, Schrenzel J: De novo bacterial genome sequencing: millions of very short reads assembled on a desktop computer. *Genome Res* 2008, 18(5):802–809.

30. Dohm JC, Lottaz C, Borodina T, Himmelbauer H: SHARCGS, a fast and highly accurate short-read assembly algorithm for de novo genomic sequencing. *Genome Res* 2007, 17(11): 1697–1706.

31. Zerbino DR, Birney E: Velvet: algorithms for de novo short read assembly using de Bruijn graphs. *Genome Res* 2008, 18(5):821–829.

32. Reinhardt JA, Baltrus DA, Nishimura MT, Jeck WR, Jones CD, Dangl JL: De novo assembly using low-coverage short read sequence data from the rice pathogen Pseudomonas syringae pv. oryzae. *Genome Res* 2009, 19(2):294–305.

33. Pop M: Genome assembly reborn: recent computational challenges. *Brief Bioinform* 2009, 10(4):354–366.

34. Nagalakshmi U: The transcriptional landscape of the yeast genome defined by RNA sequencing. *Science* 2008, 320:1344–1349.

35. Cloonan N, Forrest AR, Kolle G, Gardiner BB, Faulkner GJ, Brown MK, Taylor DF, Steptoe AL, Wani S, Bethel G *et al*: Stem cell transcriptome profiling via massive-scale mRNA sequencing. *Nat Methods* 2008, 5(7):613–619.

36. Mortazavi A, Williams B, McCue K, Schaeffer L, Wold B: Mapping and quantifying mammalian transcriptomes by RNA-Seq. *Nature Methods* 2008, advanced online publication(7): 621–628.

37. B Li, V Ruotti, RM Stewart, JA Thomson, Dewey C: RNA-Seq gene expression estimation with read mapping uncertainty. *Bioinformatics* 2009, 26:8.

38. Jiang H, W W: Statistical inferences for isoform expression in RNASeq. *Bioinformatics* 2009, 25:7.

39. Nowrousian M, Stajich JE, Chu M, Engh I, Espagne E, Halliday K, Kamerewerd J, Kempken F, Knab B, Kuo HC *et al*: De novo assembly of a 40 Mb eukaryotic genome from short sequence reads: Sordaria macrospora, a model organism for fungal morphogenesis. *PLoS Genet*, 6(4):e1000891.

40. DiGuistini S, Liao N, Platt D, Robertson G, Seidel M, Chan S, Docking TR, Birol I, Holt R,

Hirst M *et al*: De novo genome sequence assembly of a filamentous fungus using Sanger, 454 and Illumina sequence data. *Genome Biology* 2009, 10(9):R94.

41. Wang C, Mitsuya Y, Gharizadeh B, Ronaghi M, Shafer RW: Characterization of mutation spectra with ultra-deep pyrosequencing: application to HIV-1 drug resistance. *Genome Res* 2007, 17(8):1195–1201.

42. Bhadauria V, Popescu L, Zhao WS, Peng YL: Fungal transcriptomics. *Microbiol Res* 2007, 162(4):285–298.

43. Marioni JC, Mason CE, Mane SM, Stephens M, Gilad Y: RNA-seq: an assessment of technical reproducibility and comparison with gene expression arrays. *Genome Res* 2008, 18(9): 1509–1517.

44. Turnbaugh PJ, Hamady M, Yatsunenko T, Cantarel BL, Duncan A, Ley RE, Sogin ML, Jones WJ, Roe BA, Affourtit JP *et al*: A core gut microbiome in obese and lean twins. *Nature* 2009, 457(7228):480–484.

45. Dinsdale EA, Edwards RA, Hall D, Angly F, Breitbart M, Brulc JM, Furlan M, Desnues C, Haynes M, Li L *et al*: Functional metagenomic profiling of nine biomes. *Nature* 2008, 452(7187):629–632.

46. Tettelin H, Masignani V, Cieslewicz MJ, Donati C, Medini D, Ward NL, Angiuoli SV, Crabtree J, Jones AL, Durkin AS *et al*: Genome analysis of multiple pathogenic isolates of Streptococcus agalactiae: implications for the microbial "pan-genome". *Proc Natl Acad Sci USA* 2005, 102(39):13950–13955.

47. Ko KS, Kim J-W, Kim J-M, Kim W, Chung S-i, Kim IJ, Kook Y-H: Population Structure of the Bacillus cereus Group as Determined by Sequence Analysis of Six Housekeeping Genes and the plcR Gene. *Infect Immun* 2004, 72(9):5253–5261.

48. Venter JC, Remington K, Heidelberg JF, Halpern AL, Rusch D, Eisen JA, Wu D, Paulsen I, Nelson KE, Nelson W *et al*: Environmental genome shotgun sequencing of the Sargasso Sea. *Science* 2004, 304(5667):66–74.

49. Robertson G, Hirst M, Bainbridge M, Bilenky M, Zhao Y, Zeng T, Euskirchen G, Bernier B, Varhol R, Delaney A *et al*: Genome-wide profiles of STAT1 DNA association using chromatin immunoprecipitation and massively parallel sequencing. *Nat Methods* 2007, 4(8): 651–657.

50. Valouev A, Johnson DS, Sundquist A, Medina C, Anton E, Batzoglou S, Myers RM, Sidow A: Genome-wide analysis of transcription factor binding sites based on ChIP-Seq data. *Nat Methods* 2008, 5(9):829–834.

51. Haas BJ, Kamoun S, Zody MC, Jiang RH, Handsaker RE, Cano LM, Grabherr M, Kodira CD, Raffaele S, Torto-Alalibo T *et al*: Genome sequence and analysis of the Irish potato famine pathogen Phytophthora infestans. *Nature* 2009, 461(7262):393–398.

52. Schones DE, Zhao K: Genome-wide approaches to studying chromatin modifications. *Nat Rev Genet* 2008, 9(3):179-191.

53. Scherf A, Lopez-Rubio JJ, Riviere L: Antigenic variation in Plasmodium falciparum. *Annu Rev Microbiol* 2008, 62:445–470.

54. Pomraning KR, Smith KM, Freitag M: Genome-wide high throughput analysis of DNA methylation in eukaryotes. *Methods* 2009, 47:142–150.

55. Reyes-Dominguez Y, Bok JW, Berger H, Shwab EK, Basheer A, Gallmetzer A, Scazzocchio C, Keller N, Strauss J: Heterochromatic marks are associated with the repression of secondary metabolism clusters in Aspergillus nidulans. *Mol Microbiol*, 76(6):1376–1386.

56. Liu N, Xiao Z-D, Yu C-H, Shao P, Liang Y-T, Guan D-G, Yang J-H, Chen C-L, Qu L-H, Zhou H: SnoRNAs from the filamentous fungus Neurospora crassa: structural, functional and evolutionary insights. *BMC Genomics* 2009, 10(1):515.

Chapter 18

Confocal Microscopy in Plant–Pathogen Interactions

Adrienne R. Hardham

Abstract

The development of confocal microscopy and its application to studies of plant–pathogen interactions have revolutionised research into the role of selected molecules and cell components in pathogen infection strategies and plant defence responses. Confocal microscopy allows high-resolution visualisation of a variety of fluorescent and fluorescently tagged molecules in both fixed and living cells, not only in single cells but also in intact tissues. Confocal microscopes greatly improve image quality by reducing interference by out-of-focus light and can capture high-resolution serial optical sections through samples in the z-axis. In combination with a range of computational image analysis techniques, confocal microscopy provides a powerful tool by which molecules, molecular interactions, and cell components can be localised and studied.

Key words: Butyl methyl methacrylate, Confocal microscopy, Fluorophores, Fungi, Oomycetes, Photobleaching, Photoconversion, Plant–pathogen interactions, Protein–protein interactions, Reactive oxygen species, Scanning parameters

1. Introduction

The advent of fluorescence microscopy, in particular immunofluorescence microscopy, has had an enormous impact on studies of plant and animal molecular, cellular and developmental biology. The subsequent development of fluorescent reporter proteins, of molecular techniques for the fusion of their genes to those of proteins of interest and of methods for transfecting and expressing the fluorophore-tagged proteins in transgenic cells has further revolutionised the field, allowing studies of molecular cell biology that were previously not possible. Much high quality and informative fluorescence microscopy has been conducted with what are now often referred to as conventional widefield epifluorescence microscopes. However, these microscopes have their limitations, in particular the difficulty in obtaining information in the z-axis due

Melvin D. Bolton and Bart P.H.J. Thomma (eds.), *Plant Fungal Pathogens: Methods and Protocols*,
Methods in Molecular Biology, vol. 835, DOI 10.1007/978-1-61779-501-5_18, © Springer Science+Business Media, LLC 2012

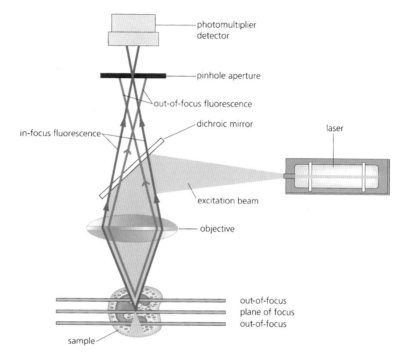

photomultiplier detector

pinhole aperture

out-of-focus fluorescence

dichroic mirror

laser

in-focus fluorescence

excitation beam

objective

out-of-focus
plane of focus
out-of-focus

sample

Fig. 1. Diagram depicting the light path in a confocal microscope (reproduced with permission from (39) in amended form).

to interference from out-of-focus light from above and below the desired focal plane. This problem is often encountered in plants because plant cells are usually thicker than animal cells and it is desirable to study them in the context of their surrounding cells in intact plant tissues.

One of the major advantages of confocal laser scanning microscopes over widefield microscopes is their ability to obtain high-quality optical sections—that is, they can collect sharp, well-resolved images from selected planes along the z-axis within thick samples. The key feature that enables confocal microscopes to reduce interference from out-of-focus light is incorporation of a pinhole aperture in front of the light detection devise (Fig. 1). In confocal microscopes, the ability to collect serial optical sections is coupled with computer software that allows the manipulation of the image data to generate three-dimensional views of the sample. The resolution of confocal microscopes in x- and y-axes is about 1.4 times better than in widefield epifluorescence microscopes (1). Resolution in the z-axis, while being about 3 times poorer than that in the x- and y-directions, is still much better than in conventional microscopes. While it is not possible to specifically prescribe materials and methods required for confocal microscopy as is done in other chapters in this volume, the following sections aim to highlight important instrument features and operational parameters that need to be taken into account and to exemplify the great potential of the technique in studies of plant–pathogen interactions.

2. Materials

2.1. Confocal Microscopes

Confocal microscopes are now commercially available from many manufacturers. In most cases, a number of models incorporating a range of different hardware and software options are offered. Major items to be considered include the microscope (upright, inverted, fixed stage), scanning module (galvanometer or resonant scanner, scan speeds), detectors (standard or high-sensitivity photomultiplier tubes (PMTs), single channel or spectral detection), and software packages (standard image analysis, region of interest analysis, fluorescence correlation spectroscopy [FCS], fluorescence resonant energy transfer [FRET], fluorescence recovery after photobleaching [FRAP] or fluorescence lifetime imaging microscopy [FLIM]).

2.2. Objective Lenses

There are three main parameters to consider when choosing objectives—their numerical aperture, magnification, and working distance. The numerical aperture is a function of the diameter of the front lens of the objective and is a measure of its light-gathering capability and its resolution. Resolution of a lens is inversely proportional to its numerical aperture, so, in general, the higher the numerical aperture the better. However, objectives with high numerical apertures have high magnifications and hence a reduced field of view. The working distance of the objective is a measure of the distance between the front lens element and the focal plane. Long working distance lenses provide space between the objective and the sample for microdissection or microinjection. In conjunction with a multiphoton laser, they also allow imaging at depth within the sample.

2.3. Lasers and Fluorophores

A range of lasers is available for use in confocal microscopy and a selection of those that are commonly used is included in Table 1. Argon ion lasers are undoubtedly the first choice. They are gas lasers and produce high power levels in visible wavelengths, with up to 10 laser lines in the blue-green region of the spectrum. Of these, the most prominent and useful are those at 488 nm (blue) and 514 nm (green). At shorter wavelengths, ultraviolet lasers emit strongly at about 350 nm, exciting not only a selection of fluorophores (Table 1) but also many endogenous autofluorescent molecules in plant cells (2). The violet laser, producing a high power line at 405 nm, is a diode-pumped solid state laser, a type of laser that is compact and efficient. Helium:neon gas lasers are often chosen for provision of green excitation wavelengths and krypton:-argon gas lasers for yellow and orange wavelengths. As indicated in Table 1, the choice of lasers should be guided by the requirements of the fluorophores that will be used.

Table 1
Excitation and emission data for commonly used fluorophores

Laser/laserline (nm)	Fluorophore	Excitation (nm)	Emission (nm)	Target
Ultraviolet/355	Hoechst 33258	352	460	DNA
	DAPI	355	461	DNA
Violet/405	Cascade blue	400	420	General label
	DyLight 405	400	421	General label
	Alexa Fluor 405	402	421	General label
	Pacific orange	405	551	General label
	Pacific blue	410	455	General label
Helium:cadmium/ 442 (blue)	Lucifer yellow CH	428	528	General label
	CFP & eCFP	435	475	General label
	Fura red	436 (high Ca^{2+})	637	Ca^{2+}
	BCECF	440	530	pH
	Cerulean	433	475	General label
	SYTOX blue	444	480	General label
Argon/488 (blue)	Fura red	472 (low Ca^{2+})	657	Ca^{2+}
	FM1-43	479	598	Membrane
	GFP & eGFP	480	507	General label
	$DiOC_6$	484	501	Membranes
	Alexa 488	488	520	General label
	H_2DCFDA	488	515	ROS
	BCECF	490	530	pH
	MitoTracker green	490	516	Mitochondria
	FITC	492	520	General label
	Cy2	492	506	General label
	ZsGreen	493	505	General label
	DyLight 488	493	518	General label
	Oregon green 488	496	524	General label
	Bodipy FL	503	512	General label
	SYTOX green	504	523	General label
Argon/514 (green)	Calcium green	506	533	Ca^{2+}
	Magnesium green	506	531	Mg^{2+}
	Rhodamine 123	510	534	Mitochondria
	Oregon green 514	511	530	General label
	FM4-64	515	740	Membranes
	YFP and eYFP	515	526	General label
	mCitrine	516	529	General label
	Alexa 514	517	542	General label
	Ethidium bromide	518	600	Nucleic acids
Helium:neon/543 (green)	Propidium iodide	536	617	Nucleic acids, plant cell walls
	Rhodamine-phalloidin	540	580	
	TRITC	550	572	
	Cy3	550	570	Actin
	Nile red	550	628	General label
	DyLight 549	555	568	General label
				Lipid
				General label

(continued)

**Table 1
(continued)**

Laser/laserline (nm)	Fluorophore	Excitation (nm)	Emission (nm)	Target
Krypton: Argon/568 (yellow)	SYTOX orange	547	570	General label
	dTomato	555	580	General label
	dsRed	560	585	General label
	Lissamine rhodamine	570	590	General label
	Alexa 568	580	605	General label
	mCherry	585	610	General label
	Texas red	596	615	General label
Krypton:Argon/647 (orange)	TOTO-3	642	660	DNA
	Cy5	650	670	General label
	Alexa 647	650	670	General label
	DyLight 649	652	670	General label

CFP cyan fluorescent protein; *eCFP* enhanced CFP; *GFP* green fluorescent protein; *eGFP* enhanced GFP; *ROS* reactive oxygen species

Fluorescent molecules that are used in epifluorescence and confocal microscopy can be divided into four main categories: (1) autofluorescent molecules (e.g. chlorophyll, phenolics, flavonoids, cellulose), (2) dyes that react with specific molecules or cell components (e.g. MitoTracker Green, DAPI, FM4-64), (3) fluorescently labelled antibodies for use in immunofluorescence microscopy (e.g. FITC-conjugated sheep anti-mouse immunoglobulin), and (4) fluorescent proteins, such as the green fluorescent protein (GFP) and its variants, which allow genetic tagging of selected proteins with fluorophores with desired spectral characteristics (see Notes 1–5).

3. Methods

Every confocal microscope has its own specific set-up and operational requirements. The following sections address a number of the important factors that need to be considered when setting the scanning parameters for image acquisition.

3.1. Scan Format

This parameter refers to the number of points (or pixels) of data in the x- and y-axes that will be collected during image acquisition. Together with the numerical aperture of the objective lens, the wavelengths of illumination and emission light, and the electronic zoom, the scan format will determine the spatial resolution of the image. A standard starting point is to select the 512×512 setting. This means that 512 pixels of data will be collected in both x- and y-directions.

3.2. Pinhole

The detection pinhole is a key element of confocal microscopes. Depending on its aperture, placement of this pinhole in front of the detector helps stop out-of-focus light reaching the detector, a major factor in improving image quality. When opened widely, the pinhole will allow both focused and out-of-focus light to pass through to the detector and the image will be similar to that obtained by a conventional widefield microscope. When closed down, the pinhole aperture will block out-of-focus light, but will also reduce the amount of light reaching the detector so that the signal becomes weak and noisy. Between these extremes, there will be a range of apertures at which the best compromises are reached, with an optimal setting typically close to one Airy disc unit, the diffraction-limited illumination volume.

3.3. Scan Mode

The scan mode parameters usually offer a choice of three spatial directions (x, y, and z), time, and wavelength. The basic set-up is to capture an image in the xy horizontal plane, but in many cases there is much to be gained by taking serial optical sections in the z-axis to acquire three-dimensional data which can later be manipulated in a variety of ways. It is also possible to record xy and xyz scans at set time intervals or at different wavelengths.

3.4. Scan Speed

Confocal microscopes allow images to be captured at different scan speeds, typically at 200, 400, 800, and 1,000 lines per second. Higher scan speeds may be desirable when viewing moving objects or to reduce dwell time and concomitant photobleaching (see Note 6). On the other hand, the higher scan speeds with conventional galvanometer scanners often also require electronic zoom which can increase the rate of photobleaching. Observation of rapidly moving objects can be achieved with resonant scanners which can scan at speeds in the range of 30 frames per second.

3.5. Electronic Zoom

In order to obtain images with maximum resolution, the digital parameters used for image capture must match the resolution capabilities of the optical system. In general, the electronic zoom needs to be set so that the smallest resolvable volume is represented by two or three pixels. This means that each pixel in the image will be about 0.07 μm, a set-up that satisfies the Nyquist sampling criterion (1).

3.6. Collecting Images

The procedure for configuring the beam path in order to capture images varies in details from one microscope to another. The main processes to be completed include adjusting the brightness levels of the laser lines that are illuminating the sample, choosing dichroic mirrors and excitation and emission filters to match the fluorescent probes being used, selecting the region of the emission spectrum that will be collected by each PMT, assigning PMT signals to different channels with selected display colours, and adjusting gain and offset

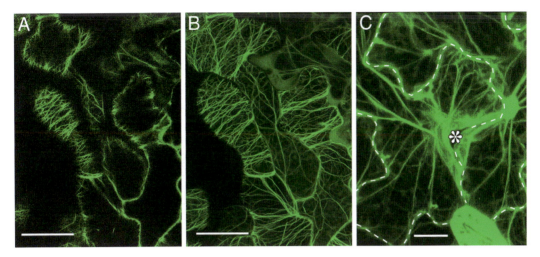

Fig. 2. *Arabidopsis* expressing the actin-binding protein, talin, fused to GFP. (**a**) One of 31 optical sections taken 0.572 μm apart through a leaf epidermal cell. (**b**) This image shows the maximum projection of all 31 sections encompassing a total z-axis depth of 17.5 μm. Images in (**a**, **b**) taken with a Zeiss LSM710 confocal microscope. (**c**) Image of actin in *Arabidopsis* plants expressing hTalin-GFP. Actin microfilaments and cables in the epidermal cell (outlined by the *white dashed line*) have been reorganised to focus on the site of attempted penetration (*asterisk*) by the avirulent pathogen, *Hyaloperonospora parasitica* CaLa2. Image taken with a Leica TCS SP2 confocal microscope (image in (**c**) courtesy of Dr Daigo Takemoto; reproduced with permission from (40)). Bar in (**a**, **b**) is 50 μm; bar in (**c**) is 20 μm.

levels for different PMTs. Once the configuration is optimal, either a single image (Fig. 2a) or a z-series of images (Fig. 2b) can be captured. The latter will require the setting of the positions for the upper and lower levels and the number of sections to be collected in the z-series scan. The number of sections will depend on a variety of factors, including the total scan depth and the end-purpose of the scan. Collection of multiple optical sections in the z-plane is one of the powerful features of confocal microscopes. It will allow combination of the multiple images into a composite image that shows the total fluorescence throughout the depth of the sampled tissue. It will also allow the generation of three-dimensional reconstructions of the fluorescent components viewed either as red-green stereo images or through image rotation or animation (3).

3.7. Sample Preparation and Analysis

In some cases, the interactions between plants and their pathogens occur on the plant surface or within the epidermal cell layer. In these situations, little prior preparation of the sample may be needed. In other cases, interactions occur within the plant tissues and it may be difficult to obtain clear images of the fluorescently labelled structures. Multiphoton lasers may provide greater penetrating power of illuminating light. Alternatively, the sample may be sectioned either in living or fixed state (see Note 7). In combination with appropriate fluorophores, confocal microscopy enables dynamic structural information to be obtained in time and space, both at the organelle and molecular levels (see Notes 8–10).

4. Notes

1. *Membrane stains:* A number of dyes will stain plant and plant pathogen membranes. The FM dyes, which are water-soluble and non-toxic and essentially non-fluorescent in aqueous solution, are especially useful for studies of endocytosis. When living cells are incubated in a solution of FM4-64 or FM1-43, the dye inserts into the outer leaflet of the plasma membrane causing it to emit bright red fluorescence when illuminated at 488 nm. With time, the process of endocytosis internalises the plasma membrane leading to the progressive fluorescence of different endomembrane compartments. FM4-64 has been used to study endocytosis in fungal hyphae (4) and endocytic events at the plant plasma membrane surrounding haustoria during infection of rice by *Magnaporthe oryzae* (5).

2. *Mitochondrial localisation:* Mitochondria can be specifically stained in plant tissues by Rhodamine 123 and MitoTracker Green. An advantage of MitoTracker Green over Rhodamine 123 is that it survives chemical fixation while Rhodamine 123 does not. Mitochondria can also be localised in transgenic plants expressing GFP-tagged nitric oxide synthase1, a mitochondrial enzyme responsible for the synthesis of nitric oxide during the plant defence response (6).

3. *Endoplasmic reticulum staining:* The fluorescent dye, 3,3'-dihexyloxacarbocyanine iodide ($DiOC_6$) (7), has received widespread use to stain the endoplasmic reticulum (ER), including studies of plant–pathogen interactions (8). However, $DiOC_6$ also stains vesicle membranes and mitochondria and causes phototoxicity. More recently, red and green BODIPY conjugates to phosphocholine (9) and glibenclamide derivatives have been reported to be more specific and less phototoxity than $DiOC_6$. In transgenic plants or pathogens, GFP can be targeted specifically to the ER by addition of HDEL or KDEL tetra-amino acid ER-retention signals. Host plants expressing GFP-labelled ER have facilitated a number of studies of ER re-organisation in response to pathogen attack (8, 10, 11).

4. *Viability staining:* Plant–pathogen interactions often involve cell death, either of the plant—as cells undergo hypersensitive cell death during effector-triggered immunity—or of the pathogen—as a result of contact with plant antimicrobial compounds. Green, blue and orange conjugates of the nucleic acid-binding dye, SYTOX, selectively stain non-viable cells with damaged membranes (12). SYTOX Green has been used to demonstrate the *in vitro* toxicity of a wheat anti-fungal peptide against eight fungal pathogens (13). The dual-excitation ratiometric pH-sensitive dye, BCECF, has also been used to

show changes in vacuolar dynamics during elicitor-induced hypersensitive cell death in tobacco BY-2 cells (14).

5. *Localising reactive oxygen species:* Derivatives of dichlorodihydrofluorescein diacetate (DCFDA) are often used to localise reactive oxygen species in plants. H_2DCFDA does not fluoresce until the acetate groups are removed by intracellular esterases. DCFDA dyes have been used to study the production of reactive oxygen species, including H_2O_2, and O_2^-, in tobacco in response to treatment with the *Phytophthora cryptogea* elicitin, cryptogein (15), and in *Penicillium expansum* in response to treatment with farnesol (16).

6. *Photobleaching:* The intensity of incident light and duration of illumination should be kept as low as possible in order to minimise photobleaching and phototoxicity. Photobleaching can cause a progressive decrease in fluorescence signal strength during image collection and can interfere with quantitation. Photobleaching is especially problematical during time-course experiments. Parameters that affect photobleaching include laser power, frame averaging, and scan speeds. In many cases, a balance needs to be achieved between limiting photobleaching and reducing noise (i.e. increase the signal-to-noise ratio).

7. *Imaging at depth in large cells and plant tissues:* It is generally not difficult to obtain good images of fluorescently labelled cellular components in isolated flat cells, including fungal hyphae. However, capturing high-quality images of components at depth in large plant cells or in cells inside plant organs is considerably more challenging. In the case of leaf samples, for example, it is usually possible to visualise fluorescent structures within epidermal cells on the side of the leaf close to the objective (10, 17–20). However, it is much more difficult to obtain clear images of structures in the underlying mesophyll cells. Chloroplasts and other structures such as cell walls both block light transmission and contribute autofluorescence that overlaps with emission spectra of the fluorophore marker proteins (21). Multiphoton lasers with their increased penetration power and small excitation volume may help address this problem (22). Another potential limitation is the working distance of the objective—it may not be possible to focus the lens at the desired depth within the tissue. Special purpose, long working distance lenses can address this problem. Nevertheless, despite these potential limitations, confocal microscopy has been used successfully to localise fluorescently tagged plant or pathogen components within inner layers of plant leaves and roots in studies of plant–fungal and plant–bacterial interactions (21, 22).

Another potential solution to obtaining high-quality images of structures inside tissues is to section the samples.

For live material, relatively thick hand or vibratome sections may provide an improvement. If it is not necessary to work with living material, the samples can be fixed and cryosectioned, typically in the range of 10–20 μm in thickness. Alternatively, after fixation, the sample may be embedded in butyl methyl methacylate resin (BMM) and sectioned on a microtome. Although there are a range of resins used for embedding and sectioning plant material for light and electron microscopy, the BMM resin is especially useful for immunolabelling and confocal microscopy because after sectioning, the resin can be removed by immersion in acetone (23, 24). This means that cellular components throughout the depth of the section are accessible to the antibodies rather than only those on the surface of the resin-embedded section. Confocal microscopy imaging of BMM sections has, for example, been used to visualise re-organisation of microtubule and actin cytoskeletons (25) and fungal effector protein transport in flax leaf mesophyll cells (Fig. 3) during fungal infection and haustorium formation.

8. *Investigating protein–protein interactions:* The exciting advances in the application of functional genomics approaches to studies of plant–pathogen interactions include bimolecular fluorescent complementation (BiFC), a microscopic technique that allows visualisation of protein–protein interactions in living plant or pathogen cells, thereby not only demonstrating the interaction between two proteins of interest but also revealing the cellular

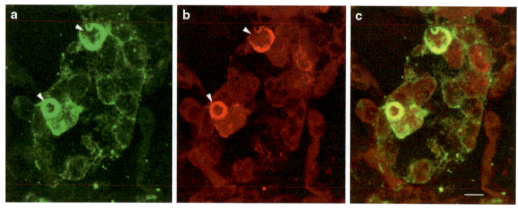

Fig. 3. Butyl methyl methacrylate (BMM) section of a flax leaf sampled 5 days after inoculation with the flax rust fungus, *Melampsora lini.* The fungus has formed two haustoria (*arrowheads*) within the leaf mesophyll cell. After removing the methacrylate resin, the section has been labelled with rabbit polyclonal anti-AvrM (41) and mouse monoclonal anti-haustorial antibody ML1 (42) followed by co-incubation in goat anti-rabbit conjugated to Alexa 488 ((a); *green*) and goat anti-mouse conjugated to TRITC ((b); *red*). (c) shows the combined signals in *green* and *red* channels. The images are maximum projections of 25 optical sections taken on a Leica TCS SP2 confocal microscope every 0.35 μm through a total depth of 8.76 μm through the leaf mesophyll. The immunolabelling shows that the AvrM effector has been transported into the plant cell cytoplasm (seen around the chloroplasts and cell periphery). Bar = 25 μm (micrographs courtesy of Dr Pamela H.P. Gan).

and subcellular location in which that interaction occurs. In the BiFC technique, DNA sequences encoding N-terminal or C-terminal portions of a GFP variant protein, typically yellow fluorescent protein (YFP), are each fused to a gene encoding one of the two putatively interacting proteins. When expressed individually in transgenic plants, neither fusion protein fluoresces; however, when co-expressed in a transgenic plant, interaction of the two tagged proteins will bring the two portions of the YFP protein together, allowing it to fluorescence. There are potential problems with the technique and a variety of safeguards and controls should be undertaken (26, 27). The approach has been applied to a number of studies of plant–pathogen interactions, in particular to the interactions of plant and viral proteins (28). The technique is illustrated in Fig. 4 where the 14-3-3 protein is separately attached to the N-terminal half and the C-terminal half of YFP. When the 14-3-3 protein dimerizes in the plant cell cytoplasm, the two halves of the YFP fluorophore are brought together, allowing the YFP molecule to fluoresce.

Protein–protein interactions can also be studied quantitatively using confocal microscopes equipped with FRET software. In this technique, both association and dissociation of proteins can be monitored in space and real-time. Proteins of interest are tagged with fluorophores that have overlapping emission and excitation spectra and interactions are detected through

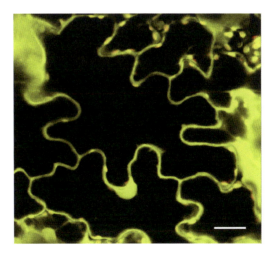

Fig. 4. Confocal microscope image of tobacco leaf epidermal cells expressing the N-terminal half of YFP and the C-terminal half of YFP each fused to the 14-3-3 protein. The 14-3-3 protein forms dimers and thus brings the two halves of the YFP fluorophore together to restore a functional fluorescent protein. The image was taken with a Leica TCS SP2 confocal microscope. Bar = 25 μm (micrograph courtesy Dr Maryam Rafiqi; reproduced with permission from (39)).

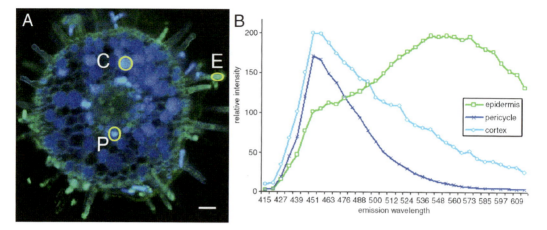

Fig. 5. (**a**) Confocal image of autofluorescent compounds in a cross-section of a *Medicago truncatula* root illuminated with a UV laser. Cells in which spectral profiles were taken are circled. *C* cortex; *P* pericycle; *E* epidermis. (**b**) Spectral emission profiles from each of the three cells circled in (**a**). Images and spectral data were obtained using a Leica TCS SP2 confocal microscope. Bar = 100 μm (images courtesy of Dr Ulrike Mathesius).

the transfer of energy from a donor fluorophore on one protein to the acceptor fluorophore on the second protein (29, 30). As with the BiFC approach, considerable care is needed when planning and conducting FRET experiments in order to check for and avoid artefacts (27). FRET can also be used to monitor protein activity through the generation of FRET activity sensors (31).

9. *Photoconversion and photoactivation:* In addition to the fluorescent proteins listed in Table 1 and in other publications (32, 33), there are two other categories of fluorescent proteins that have great potential for use in studies of plant–pathogen interactions. These are fluorescent proteins that undergo photoconversion or photoactivation when illuminated with short-wave length light. EosFP, for example, is a fluorescent protein from the coral, *Lobophyllia hemprichii*, which emits strong fluorescence at 515 nm (green) when illuminated at 506 nm, but changes its fluorescence emission to 581 nm (red) after near-UV irradiation at about 390 nm (34). Similarly, some fluorescent proteins that initially have little or no fluorescence within a defined visible spectrum can be induced to fluoresce by irradiation with violet light (35, 36). Among other purposes, photoconversion or photoactivation can be used to study dynamic events in cells such as endocytosis and exocytosis (37).

10. *Spectral imaging:* A number of confocal microscopes currently available are able to collect emitted light at multiple points across the spectrum (typically across about 300 nm) with resolutions of less than 5 nm. The emission signals at each wavelength window are collected into about 30 detection channels

in a photomultiplier. In conjunction with mathematical unmixing software, it is possible to decipher spectral emission profiles for the various fluorescent molecules in the sample. This has invaluable applications in imaging multiple fluorophores, in helping to distinguish between background autofluorescence and specific fluorophore signals and in obtaining spectral profile data from autofluorescent compounds within a sample (Fig. 5) (38).

Acknowledgement

This work was supported by Australian Research Council grants DP0771374, DP0880206, DP1093850 and LE100100078.

References

1. Hepler PK, Gunning BES (1998) Confocal fluorescence microscopy of plant cells. Protoplasma **201**: 121–157

2. Mathesius U, Bayliss C, Weinman JJ, Schlaman HRM, Spaink HP, Rolfe BG, McCully ME, Djordjevic MA (1998) Flavonoids synthesized in cortical cells during nodule initiation are early developmental markers in white clover. Mol Plant-Microbe Interact **11**: 1223–1232

3. Gunning BES (2008) Plant Cell Biology on DVD.

4. Fischer-Parton S, Parton RM, Hickey PC, Dijksterhuis J, Atkinson HA, Read ND (2000) Confocal microscopy of FM4-64 as a tool for analysing endocytosis and vesicle trafficking in living fungal hyphae. J Microsc **198**: 246–259

5. Kankanala P, Czymmek K, Valent B (2007) Roles for rice membrane dynamics and plasmodesmata during biotrophic invasion by the blast fungus. Plant Cell **19**: 706–724

6. Guo FQ, Crawford NM (2005) *Arabidopsis* nitric oxide synthase1 is targeted to mitochondria and protects against oxidative damage and dark-induced senescence. Plant Cell **17**: 3436–3450

7. Koning AJ, Lum PY, Williams JM, Wright R (1993) DiOC$_6$ staining reveals organelle structure and dynamics in living yeast cells. Cell Motil Cytoskeleton **25**: 111–128

8. Leckie CP, Callow JA, Green JR (1995) Reorganization of the endoplasmic reticulum in pea leaf epidermal cells infected by the powdery mildew fungus *Erysiphe pisi*. New Phytol **131**: 211–221

9. Foissner I (2009) Fluorescent phosphocholine-a specific marker for the endoplasmic reticulum and for lipid droplets in *Chara* internodal cells. Protoplasma **238**: 47–58

10. Takemoto D, Jones DA, Hardham AR (2003) GFP-tagging of cell components reveals the dynamics of subcellular re-organization in response to infection of *Arabidopsis* by oomycete pathogens. Plant J **33**: 775–792

11. Takemoto D, Jones DA, Hardham AR (2006) Re-organization of the cytoskeleton and endoplasmic reticulum in the *Arabidopsis pen1-1* mutant inoculated with the non-adapted powdery mildew pathogen, *Blumeria graminis* f. sp. *hordei*. Mol Plant Pathol 7: 553–563

12. Truernit E, Haseloff J (2008) A simple way to identify non-viable cells within living plant tissue using confocal microscopy. Plant Methods **4**: 15–20

13. Sun JY, Gaudet DA, Lu ZX, Frick M, Puchalski B, Laroche A (2008) Characterization and antifungal properties of wheat nonspecific lipid transfer proteins. Mol Plant-Microbe Interact **21**: 346–360

14. Higaki T, Goh T, Hayashi T, Kutsuna N, Kadota Y, Hasezawa S, Sano T, Kuchitsu K (2007) Elicitor-induced cytoskeletal rearrangement relates to vacuolar dynamics and execution of cell death: In vivo imaging of hypersensitive cell death in tobacco BY-2 cells. Plant Cell Physiol **48**: 1414–1425

15. Ashtamker C, Kiss V, Sagi M, Davydov O, Fluhr R (2007) Diverse subcellular locations of cryptogein-induced reactive oxygen species

production in tobacco bright yellow-2 cells. Plant Physiol **143**: 1817–1826

16. Liu P, Luo L, Guo JH, Liu HM, Wang BQ, Deng BX, Long CA, Cheng YJ (2010) Farnesol induces apoptosis and oxidative stress in the fungal pathogen *Penicillium expansum*. Mycologia **102**: 311–318

17. Koh S, André A, Edwards H, Ehrhardt D, Somerville S (2005) *Arabidopsis thaliana* subcellular responses to compatible *Erysiphe cichoracearum* infections. Plant J **44**: 516–529

18. Robatzek S, Chinchilla D, Boller T (2006) Ligand-induced endocytosis of the pattern recognition receptor FLS2 in *Arabidopsis*. Genes Dev **20**: 537–542

19. Wang WM, Wen YQ, Berkey R, Xiao SY (2009) Specific targeting of the *Arabidopsis* resistance protein RPW8.2 to the interfacial membrane encasing the fungal haustorium renders broadspectrum resistance to powdery mildew. Plant Cell **21**: 2898–2913

20. Doehlemann G, van der Linde K, Aßmann D, Schwammbach D, Hof A, Mohanty A, Jackson D, Kahmann R (2009) Pep1, a secreted effector protein of *Ustilago maydis* is required for successful invsion of plant cells. PLoS Pathogens **5**: e1000290

21. Godfrey SAC, Mansfield JW, Corry DS, Lovell HC, Jackson RW, Arnold DL (2010) Confocal imaging of *Pseudomonas syringae* pv. *phaseolicola* colony development in bean reveals reduced multiplication of strains containing the genomic island PPHGI-1. Mol Plant-Microbe Interact **23**: 1294–1302

22. Czymmek KJ, Fogg M, Powell DH, Sweigard J, Park SY, Kang S (2007) *In vivo* time-lapse documentation using confocal and multi-photon microscopy reveals the mechanisms of invasion into the *Arabidopsis* root vascular system by *Fusarium oxysporum*. Fungal Genetics & Biology **44**: 1011–1023

23. Baskin TI, Busby CH, Fowke LC, Sammut M, Gubler F (1992) Improvements in immunostaining samples embedded in methacrylate: localization of microtubules and other antigens throughout developing organs in plants of diverse taxa. Planta **187**: 405–413

24. Gubler F (1989) Immunofluorescence localization of microtubules in plant root tips embedded in butyl-methyl methacrylate. Cell Biol Int Rep **13**: 137–145

25. Kobayashi I, Kobayashi Y, Hardham AR (1994) Dynamic reorganization of microtubules and microfilaments in flax cells during the resistance response to flax rust infection. Planta **195**: 237–247

26. Bracha-Drori K, Shichrur K, Katz A, Oliva M, Angelovici R, Yalovsky S, Ohad N (2004) Detection of protein-protein interactions in plants using bimolecular fluorescence complementation. Plant J **40**: 419–427

27. Bhat RA, Lahaye T, Panstruga R (2006) The visible touch: *in planta* visualization of protein-protein interactions by fluorophore-based methods. Plant Methods **2**: 12

28. Ohad N, Shichrur K, Yalovsky S (2007) The analysis of protein-protein interactions in plants by bimolecular fluorescence complementation. Plant Physiol **145**: 1090–1099

29. Sekar RB, Periasamy A (2003) Fluorescence resonance energy transfer (FRET) microscopy imaging of live cell protein localizations. J Cell Biol **160**: 629–633

30. Hoppe AD, Seveau S, Swanson JA (2009) Live cell fluorescence microscopy to study microbial pathogenesis. Cell Microbiol **11**: 540–550

31. Lalonde S, Ehrhardt DW, Frommer WB (2005) Shining light on signaling and metabolic networks by genetically encoded biosensors. Curr Opin Plant Biol **8**: 574–581

32. Stewart CN (2006) Go with the glow: fluorescent proteins to light trangenic organisms. Curr Opin Plant Biol **24**: 155–162

33. Shaner N, Campbell RE, Steinbach PA, Giepmans BNG, Palmer AE, Tsien RY (2004) Improved monomeric red, orange and yellow fluorescent proteins derived from *Discosoma* sp. red fluorescent protein. Nature Biotech **22**: 1567–1572

34. Wiedenmann J, Ivanchenko S, Oswald F, Schmitt F, Röcker C, Salih A, Spindler K-D, Nienhaus GU (2004) EosFP, a fluorescent marker protein with UV-inducible green-to-red fluorescence conversion. Proc Natl Acad Sci **101**: 15905–15910

35. Lukyanov KA, Chudakov DM, Lukyanov S, Verkhusha VV (2005) Innovation: Photoactivatable fluorescent proteins. Nat Rev Mol Cell Biol **6**: 885–891

36. Subach FV, Patterson GH, Manley S, Gillette JM, Lippincott-Schwartz J, Verkhusha VV (2009) Photoactivatable mCherry for high-resolution two-color fluorescence microscopy. Nature Methods **6**: 153–159

37. Dhonukshe P, Aniento F, Hwang I, Robinson DG, Mravec J, Stierhof YD, Friml J (2007) Clathrin-mediated constitutive endocytosis of PIN auxin efflux carriers in *Arabidopsis*. Curr Biol **17**: 520–527

38. Morris AC, Djordjevic MA (2006) The *Rhizobium leguminosarum* biovar *trifolii* ANU794 induces novel developmental responses on the subterranean clover cultivar Woogenellup. Mol Plant-Microbe Interact **19**: 471–479

39. Ladiges P, Evans B, Saint R, Knox R (2010) Biology. McGraw-Hill, North Ryde, NSW,

40. Hardham AR (2007) Cell biology of plant-oomycete interactions. Cell Microbiol 9: 31–39

41. Rafiqi M, Gan PHP, Ravensdale M, Lawrence GJ, Ellis JG, Jones DA, Hardham AR, Dodds PN (2010) Internalization of flax rust avirulence proteins into flax and tobacco cells can occur in the absence of the pathogen. Plant Cell 22: 2017–2032

42. Murdoch LJ, Kobayashi I, Hardham AR (1998) Production and characterisation of monoclonal antibodies to cell wall components of the flax rust fungus. Eur J Plant Pathol 104: 331–346

Chapter 19

The Use of Open Source Bioinformatics Tools to Dissect Transcriptomic Data

Benjamin M. Nitsche, Arthur F.J. Ram, and Vera Meyer

Abstract

Microarrays are a valuable technology to study fungal physiology on a transcriptomic level. Various microarray platforms are available comprising both single and two channel arrays. Despite different technologies, preprocessing of microarray data generally includes quality control, background correction, normalization, and summarization of probe level data. Subsequently, depending on the experimental design, diverse statistical analysis can be performed, including the identification of differentially expressed genes and the construction of gene coexpression networks.

We describe how Bioconductor, a collection of open source and open development packages for the statistical programming language R, can be used for dissecting microarray data. We provide fundamental details that facilitate the process of getting started with R and Bioconductor. Using two publicly available microarray datasets from *Aspergillus niger*, we give detailed protocols on how to identify differentially expressed genes and how to construct gene coexpression networks.

Key words: Filamentous fungi, Microarray, Transcriptomics, Expression profiling, Bioconductor, Gene coexpression networks

1. Introduction

The open source and open development project Bioconductor (1) (http://bioconductor.org/) is actively developed and maintained by members of the academic community, thus providing scientists with leading edge computational biology tools. Bioconductor is well suited for the academic environment where it can be used for research and education. Rather than providing a graphical user interface, most Bioconductor packages depend on command line input. Therefore, getting started with Bioconductor requires some effort, especially for those having limited computational background.

Melvin D. Bolton and Bart P.H.J. Thomma (eds.), *Plant Fungal Pathogens: Methods and Protocols*,
Methods in Molecular Biology, vol. 835, DOI 10.1007/978-1-61779-501-5_19, © Springer Science+Business Media, LLC 2012

In the following, we give a step-by-step tutorial on how Bioconductor can be used for transcriptomic data analysis and provide the reader with the most important theoretical background on the statistics involved. We use two Affymetrix microarray data-sets that were recently published for *Aspergillus niger* (2, 3). The datasets as well as all Bioconductor packages are publicly available, allowing the reader to repeat each step of the analysis. We start with a brief description on how the statistical programming language R (4) (http://www.r-project.org/), Bioconductor core packages, and additional Bioconductor packages can be installed (see Subheading 3.1). For the identification of differentially expressed genes, we demonstrate how to import CEL files and associate them with phenotypic labels (see Subheading 3.2.1), how to preprocess microarray data and asses its quality (see Subheading 3.2.2), how to perform multiway comparisons (see Subheading 3.2.3), and finally, how to export the data (see Subheading 3.2.4). For the construction of gene coexpression networks, we subsequently import CEL files (see Subheading 3.3.1), assess the data quality and preprocess the CEL files (see Subheading 3.3.2), perform nonspecific filtering (see Subheading 3.3.3), and construct gene coexpression networks (see Subheading 3.3.4).

2. Materials

In the examples described thereafter, the open source and open development packages from the Bioconductor project (1) (http://bioconductor.org/), which use the statistical programming language R (4) (http://www.r-project.org/), are used for the analysis of transcriptomic data. The following packages available from the Bioconductor homepage are required: affy (5), affycoretools (6), affyPLM (7), limma (8), genefilter (9), and makecdfenv (10).

Both transcriptomic datasets used for demonstration purposes were recently published for *A. niger* and have been deposited at public databases. The first dataset (2) used for the identification of differentially expressed genes comprises nine Affymetrix microar-rays corresponding to three different time points during maltose-limited retentostat cultivations and is available via its accession number GSE21752 at the NCBI Gene Expression Omnibus (GEO) database (http://www.ncbi.nlm.nih.gov/geo/). The second dataset (3) used for the construction of gene coexpression net-works comprises 29 Affymetrix microarrays from multiple time points after transfer of mycelial biomass from fructose to a variety of carbon sources: rhamnose, xylose, sorbitol, galacturonic acid, polygalacturonic acid, and sugar beet pectin. The dataset is avail-able at the EMBL ArrayExpress database (http://www.ebi.ac.uk/arrayexpress/) via the accession number E-MEXP-1626.

Both transcriptomic datasets are based on the Affymetrix chip dsmM_ANIGERa_coll511030F. The corresponding chip description file (CDF) can be downloaded from the NCBI GEO database via the GEO accession number GPL6758.

3. Methods

3.1. Installation

Before installing Bioconductor packages, the statistical programming language R has to be installed. R installation packages and platform-specific help files are available for Linux, Mac OS, and Windows at the Comprehensive R Archive Network (CRAN: http://www.r-project.org/). After successful installation, the R command window can be accessed via its application icon or by typing R at the command prompt (see Note 1). To install the Bioconductor core packages, make sure to have an Internet connection and type at the R command window (see Note 2):

```
> source("http://bioconductor.org/biocLite.R")

> biocLite()
```

Further packages have to be installed:

```
> biocLite("affycoretools")

> biocLite("makecdfenv")

> biocLite("limma")
```

For Affymetrix chips, analysis of raw data usually starts with CEL files. They contain the information from intensity calculations of the pixel values from the raw image data (DAT files) obtained with the chip scanner. To be able to import the Affymetrix raw data from CEL files, Bioconductor requires information about the corresponding Affymetrix chip, which has to be provided as a chip-specific package. While those packages can be downloaded for many popular Affymetrix chips, they first have to be built for custom-made arrays such as the *A. niger* chip. The Affymetrix CDF provides all information for building the package.

To build the CDF package, first download the dsmM_ANIGERa_coll511030F CDF archive file via its GEO accession number GPL6758 from the GEO database (http://www.ncbi.nlm.nih.gov/geo/) and decompress it (see Note 3). Create a new directory (from now on referred to as working directory, WD), copy the CDF file to the WD, and rename it into "dsmM_ANIGERa_anColl.CDF". At the R command line, define the WD with the function setwd (see Notes 4 and 5), load the required package makecdfenv with the library command, and build the chip-specific package with the make.cdf.package function:

```
> setwd("absolute path to WD/")

> library("makecdfenv")

> make.cdf.package("dsmM_ANIGERa_anColl.CDF", species = "Aspergillus niger")
```

The newly built package is saved in a new folder in the WD (WD/dsmmanigeraancollcdf/). Next, open a new command prompt window, change to the WD, and install the package (see Note 1):

```
> R CMD INSTALL dsmmanigeraancollcdf
```

3.2. Computation of Differentially Expressed Genes

The identification of differentially expressed genes will exemplarily be shown on a transcriptomic dataset recently published for *A. niger* (2). The data have been deposited at the NCBI GEO database and can be downloaded via its accession number: GSE21752. *A. niger* was cultivated under controlled conditions in carbon-limited retentostat cultures, which is a specific form of continuous cultivation. The biomass was retained in the bioreactor, while old medium was removed and fresh defined medium was fed at a constant rate. With increasing cultivation time, the biomass-specific availability of the sole limiting carbon source maltose decreased and the fungus underwent carbon starvation. The RNA used for microarray hybridizations derived from three different time points: Batch phase referred to as day 0 (d0), day 2, and day 8 of retentostat cultivation (d2 and d8, respectively). With three replicate cultures, the number of microarrays adds up to a total of nine. Below, we describe how to identify sets of genes that are differentially expressed comparing d2 vs. d0, d8 vs. d0, and d8 vs. d2.

3.2.1. Importing CEL Files into R

Create a new WD and download the raw data (CEL files) from the NCBI GEO database (http://www.ncbi.nlm.nih.gov/geo/) via the dataset accession number GSE21752. Next, decompress (see Note 3) the data and copy the CEL files into the WD. Open the R command line, load the packages that will be used for the data analysis, and define the WD using the function setwd (see Notes 4 and 5):

```
> library("affy")

> library("affycoretools")

> library("affyPLM")

> library("limma")

> setwd("absolute path to WD/")
```

Before importing the CEL files into R, generate a file that allocates phenotypic labels to them. It consists of tab-separated columns and should be saved in the WD as a plain text file named "phenotypic_labels.txt". The required sample information can be obtained from the NCBI GEO database via the corresponding dataset accession number (GSE21752). The content of the phenotypic label file should look like:

```
sample  FileName        Target

d0.1    GSM542228.CEL  d0

d0.2    GSM542335.CEL  d0

d0.3    GSM542336.CEL  d0

d2.1    GSM542337.CEL  d2

d2.2    GSM542338.CEL  d2

d2.3    GSM542339.CEL  d2

d8.1    GSM542340.CEL  d8

d8.2    GSM542341.CEL  d8

d8.3    GSM542342.CEL  d8
```

Use the function read.table to import the text file and assign (see Note 6) the phenotypic labels to the variable pData, which represents a data frame with row names equal to the unique file names. Subsequently, call the function ReadAffy to import the raw data from all annotated CEL files:

```
> pData <- read.table("phenotypic_labels.txt", row.names = 2, header = TRUE)

> rawData <- ReadAffy(filenames = rownames(pData), verbose = TRUE)
```

3.2.2. Quality Assessment and Preprocessing

Because RNA is very sensitive to degradation by ribonucleases, we recommend checking the average integrity of RNA transcripts. For this purpose, make a degradation plot showing mean intensities of the probes ordered from 5′ to 3′ end of their target transcripts. The 5′/3′ ratio of mean probe intensities should be comparable between the samples, as shown in Fig. 1 (see Note 7). The following code will create a degradation plot (see Note 8):

```
> plotDeg(rawData, filenames = pData$sample)
```

Next, use the Robust Multiarray Average (RMA) package (11) for preprocessing probe level data. The function rma performs background correction, quantile normalization, and robust median polish to summarize intensities of probe sets to expression values. Background correction aims to differentiate specific from nonspecific hybridization signals. For this purpose, Affymetrix chips are designed to provide pairs of nearly identical probes for hybridization, so-called perfect match (PM) and mismatch (MM) probes. The MM probes are identical to the PM probes, except for their central nucleotide at position 13, which is changed to its complement. The Affymetrix MAS5.0 algorithm for background correction uses the intensities of both probes to correct for nonspecific hybridization signals. However, it has been shown that MM probes in many cases give stronger signals than their corresponding PM probes

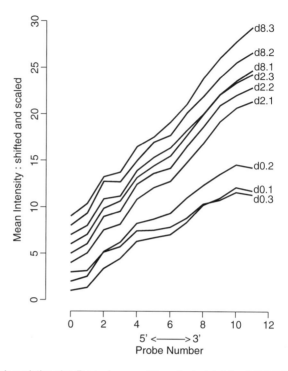

Fig. 1. RNA degradation plot. For each array of the retentostat dataset (GSE21752), mean probe intensities are plotted ordered from 5' to 3' end. For assessment of RNA quality, it's important that the 5'/3' intensity ratios are comparable between samples rather than the actual values. Typical 5'/3' ratios depend on the Affymetrix chip used (1, 6).

indicating that besides nonspecific specific hybridization occurs (12). Therefore, RMA does not use MM probe intensities, but applies a global empirical model for the distribution of PM probe intensities for background correction. Especially for lowly expressed genes, the RMA algorithm for background correction has been shown to be superior to MAS5.0 (11). Before normalization, differences in probe intensities between arrays are due to biological and technical differences. The later is mainly introduced during labeling and hybridization, which normalization aims to compensate for. After normalization, differences in probe intensities between arrays should in theory exclusively be due to biological differences. The quantile normalization performed by RMA makes probe intensity distributions comparable between the set of arrays under investigation. Finally, RMA fits a robust multiarray model to summarize the probe intensities of each probe set into an expression value. Use the rma function to preprocess the probe level data and compute expression values. The results are organized as an object of the class ExpressionSet. To allocate phenotypic labels to the ExpressionSet object, call the function new:

```
> eset <- rma(rawData)

> phenoData(eset) <- new("AnnotatedDataFrame", data = pData)
```

To show the effect of quantile normalization, generate two box plot diagrams for the raw and RMA processed data. First, use the function par to define a new window in which two plots can be placed next to each other (see Note 9). Subsequently, generate two box plot diagrams with the function boxplot. The two box plots nicely illustrate the effect of RMA preprocessing. As a result of the quantile normalization, the arrays have nearly identical median intensities and comparable intensity distributions after RMA pre-processing compared to the raw data (see Fig. 2).

```
> par(mfrow = c(1,2))

> boxplot(rawData, las = 2, cex.axis = 0.5, main = "raw data")

> boxplot(eset, las = 2, cex.axis = 0.5, main = "RMA pre-processed data")
```

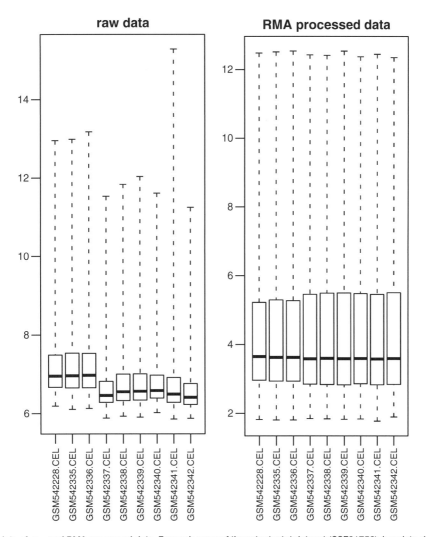

Fig. 2. Boxplots of raw and RMA processed data. For each array of the retentostat dataset (GSE21752), boxplots show the log intensity distributions of raw and RMA processed data. As a result of the quantile normalization, the log intensity distributions are comparable after RMA processing (11).

Next, convert the ExpressionSet object eset, to a matrix with RMA expression values (see Note 10) using the function exprs and assign the resulting matrix to the variable e. The matrix has nine columns (one for each CEL file) and 14,554 rows, each containing probe-specific RMA expression values. The probes still include Affymetrix control probes beginning with "AFFX". To remove them, use the function grep (see Note 11) and obtain an index referring to all probes that are starting with "AFFX". Subsequently create the matrix e.anig containing exclusively *A. niger*-specific probes (see Note 12):

```
> e <- exprs(eset)

> subIndex <- grep("AFFX", featureNames(eset))

> e.anig <- e[-subIndex,]
```

Before starting the comparative analysis, do a simple test to exclude errors during the allocation of phenotypic labels to the CEL files. Using the function plotPCA, perform a principal component analysis (PCA) and plot the first two principal components against each other. Array data deriving from replicate time points are expected to cluster together. The PCA plot clearly shows that replicate array data from d0, d2, and d8 cluster together and that the three clusters are well separated (see Fig. 3). Therefore, errors during the allocation of phenotypic labels to the CEL files can be excluded.

```
> par(mfrow = c(1,1))

> plotPCA(e.anig, groupnames = labels, pch = c("0","0","0","2","2","2","8","8","8"),
    col = rep("black",9), legend = FALSE)
```

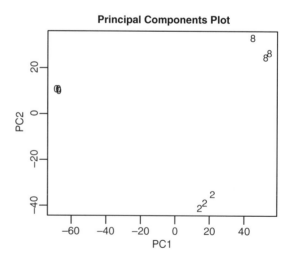

Fig. 3. Principle component analysis (PCA). For each array of the retentostat dataset (GSE21752), the first two principle components are plotted against each other. Ideally, replicate arrays should cluster together and clusters of replicate arrays should be well separated. PCA plots can thus be applied to exclude errors during the allocation of phenotypic labels and to assess the reproducibility.

*3.2.3. Multiway
Comparisons*

When analyzing microarray data, multiple hypothesis testing has to be taken into account (see Note 13). Diverse methods have been suggested to correct *p*-values for multiple testing among which the Benjamini and Hochberg (BH) false discovery rate (FDR) (13) is commonly used. However, multiple testing correction decreases the statistical power (see Note 14), which in many cases is anyway low due to small sample sizes. Different methods have been suggested to compensate for that. A reduction of the number of hypothesis tests to be performed by nonspecific filtering of probes with low information content is one possibility. If the nonspecific filtering is independent from the following hypothesis tests, it has been shown to increase the statistical power (14). The standard deviations or mean values of probe intensities over all arrays (neglecting phenotypic labels) could be used for nonspecific filtering. An alternative approach could be the computation of moderated (15) instead of normal statistics. One such approach is implemented in the `eBayes` method from the Limma package (8), where a global variance estimator borrows information from all genes to infer probe-specific variances. A combination of nonspecific filtering and moderated statistics is obviously an interesting approach. However, it has been shown to potentially decrease the statistical power and therefore either nonspecific filtering with normal *t*-statistics or moderated *t*-statistics with unfiltered data are recommended (14).

It is very convenient to use the `limma` package for multiway comparisons of microarray data. It can even be used for multifactorial experimental designs (see Note 15). Compute moderated *t*-statistics on the unfiltered data by calling multiple `limma` functions. Indicate which experimental conditions should be compared to each other by defining a contrast matrix with the function `makeContrasts`:

```
> f <- factor(pData$Target, levels = levels(pData$Target))

> design <- model.matrix(~0 + f)

> colnames(design) <- levels(pData[,2])

> fit <- lmFit(e.anig, design)

> contrast.matrix <- makeContrasts(d2 - d0, d8 - d2, d8 - d0, levels = design)

> fit2 <- contrasts.fit(fit, contrast.matrix)

> fit2 <- eBayes(fit2)
```

To better understand the concept of the contrast matrix, have

```
> contrast.matrix
```

a look at how R represents it. Type at the R command line:

Next, make Venn diagrams to get an overview of the sets of genes that are differentially expressed comparing d2 to d0, d8 to d2, and d8 to d0. First, use the function `decideTests` to decide whether or not genes are differentially expressed controlling the FDR at 0.005:

```
> results <- decideTests(fit2, adjust.method = "fdr", p.value = 0.005)
```

The results matrix consists of three columns, one for each comparison defined in the contrast matrix, and rows for each probe. The numbers indicate if a gene is not differentially expressed (0), upregulated (+1), or downregulated (-1). To get an impression of the results matrix, take a look at rows 120–130:

```
> results[120:130,]
```

Finally, use the information of the results matrix to generate two Venn diagrams (see Fig. 4) for the up- and downregulated genes:

```
> par(mfrow = c(1,2))

> vennDiagram(results, include = "up", main = "UP, FDR q-value < 0.005", cex = 0.8)

> vennDiagram(results, include = "down", main = "DOWN, FDR q-value < 0.005", cex =
  0.8)
```

Next, obtain some of the results from the moderated *t*-statistics that were computed with the limma package. Extract the *p*-values and assign them to the matrix p.values (see Note 16).

```
> p.values <- fit2$p.value
```

In order to have distinct column names when combining different data later, change the column names of the p.value matrix. Use a for-loop (see Note 17) to access column by column of the

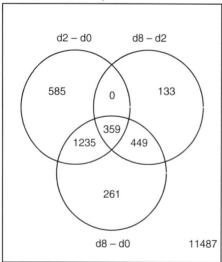
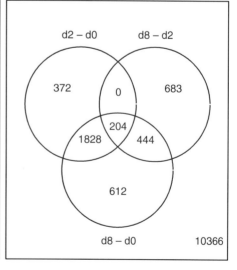

Fig. 4. Venn diagrams. Venn diagrams showing the relations between sets of up- and downregulated genes identified comparing day 0 (d0), day 2 (d2), and day 8 (d8) of retentostat cultivation.

p.value matrix. In each cycle of the for-loop, the function paste (see Note 18) is used to generate a new column name by preceding the corresponding column name from the contrast matrix with "pValue".

```
> for (i in 1:ncol(contrast.matrix)) {
>   colnames(p.values)[i] <- paste("pValue", colnames(contrast.matrix)[i], sep = " ")
> }
```

To obtain FDR q-values, use the function p.adjust and correct the p-values for multiple hypothesis testing. For each column of the p.value matrix, apply the Benjamini and Hochberg method to compute FDR q-values, combine them to a new matrix fdr.values using the function cbind (see Note 19), and change their column names:

```
> fdr.values <- NULL
> for (i in 1:ncol(p.values)) {
>   fdr.values <- cbind(fdr.values, p.adjust(p.values[,i], method = "BH"))
>   colnames(fdr.values)[i] <- paste("qValue", colnames(contrast.matrix)[i], sep = " ")
> }
```

Next, obtain the fold changes applying the limma function topTable (see Note 20). Loop through all comparisons defined in the contrast matrix and call the function topTable to extract the log 2 fold changes for the corresponding comparison. Subsequently transform the log 2 fold changes to normal scale and append them to the matrix FCs. Define the column names of the matrix FCs analogously to the examples given above.

```
> FCs <- NULL
> for (i in 1:ncol(contrast.matrix)) {
>   toptable <- topTable(fit2, coef = i, number = 15000, adjust.method = "BH", sort.by
    = "none", p.value = 1, lfc = 0)
>   FCs <- cbind(FCs,2^toptable$logFC)
>   colnames(FCs)[i] <- paste("FC", colnames(contrast.matrix)[i], sep = " ")
> }
```

Besides the RMA expression data for the nine arrays, it is helpful to provide mean expression data for each of the three time points. Make use of the pData matrix, which allocates phenotypic labels to the CEL files to loop through the three time points. Because the dataset consists of triplicates, use the function unique (see Note 21) to access each time point only once. In each cycle of the loop, call the function which (see Note 22) to obtain an index pointing to the corresponding columns of the expression data matrix. Subsequently, use the function apply (see Note 23) to calculate mean expression values and convert the log 2 expression data to normal scale. At the end of each loop, extend the matrix mean by the computed mean expression data and define the column names accordingly:

```
> means <- NULL

> for(i in 1:length(unique(pData$Target))) {

>   Index <- which(eset$Target == unique(pData$Target)[i])

>   log2.a <- apply(e.anig[,Index],1,mean)

>   a <- 2^(log2.a)

>   means <- cbind(means, a)

>   colnames(means)[i] <- paste("mean", unique(pData$Target)[i], sep = " ")

> }
```

3.2.4. Exporting Data

For sharing data, it is wise to export it in such a way that it can be imported into any spreadsheet program. Use the function cbind to combine the RMA and mean expression data as well as fold changes, p-values, and FDR q-values into the new matrix export:

```
> export <- cbind(e.anig, means, FCs, p.values, fdr.values)
```

Finally, use the function write.table to save the matrix export as a tab-delimited plain text file "results.txt" in the current WD:

```
> write.table(export, quote = FALSE, row.names = TRUE, col.names = NA, sep = "\t",

    file = "results.txt")
```

3.3. Construction of Gene Coexpression Networks

In the following, we give a short demonstration on how gene coexpression networks can be built when starting with raw CEL file data. The transcriptomic dataset used was published in 2006 for *A. niger* (3) and comprises 29 microarrays, which can be downloaded from the EMBL ArrayExpress database via its accession number: E-MEXP-1626. Briefly, the data derive from shakeflask experiments in which fungal biomass from 18 h pregrown cultures was transferred to seven carbon sources: rhamnose, xylose, sorbitol, fructose, galacturonic acid, polygalacturonic acid, and sugar beet pectin. In total, one reference sample was taken from the preculture and subsequently four samples were taken within 24 h after transfer for each of the seven carbon sources. Thus, no biological replicates are available for the transcriptomic data. The carbon sources cover repressing (fructose) and nonrepressing (sorbitol) carbon sources as well as complex carbon sources (polygalacturonic acid and sugar beet pectin) and common monomeric constituents (rhamnose, xylose, and galacturonic acid).

3.3.1. Importing CEL Files

Download the expression dataset from the ArrayExpress database (http://www.ebi.ac.uk/arrayexpress/), decompress (see Note 3), and copy the CEL files into a new WD. On the R command line, load the required libraries and define the WD:

```
> library("affy")

> library("affyPLM")

> library("affycoretools")

> library("genefilter")

> setwd("absolute path to WD/")
```

Next, use the function ReadAffy to import all CEL files present in the WD. Because we are not interested in differential expression between specific conditions, no phenotypic labels are associated with the CEL file names.

```
> rawData <- ReadAffy(verbose = TRUE)
```

3.3.2. Quality Assessment and Preprocessing

The affyPLM package has been suggested for quality assessment of transcriptomic data and detection of outlier arrays (7). First, fit a linear model to the Affymetrix probe level data using the function fitPLM. Based on the linear model, generate two plots that are particularly helpful to detect outlier arrays: The normalized unscaled standard error (NUSE) and the relative log expression (RLE) plots.

```
> rawDataPLM <- fitPLM(rawData)

> par(mfrow = c(1,2))

> boxplot(rawDataPLM, main = "NUSE", ylim = c(0.95, 1.25), las = 2, whisklty = 0,
   staplelty = 0, cex.axis = 0.5)

> abline(h = 1)

> Mbox(rawDataPLM, main = "RLE", whisklty = 0, staplelty = 0, ylim = c(-0.4, 0.4), las
  = 2, cex.axis = 0.5)

> abline(h = 0)
```

Problematic arrays can be recognized from the NUSE plot as having elevated and more spread intensities. In the RLE plot, problematic arrays tend to have a larger spread and a median different from zero (7). Inspecting the NUSE and RLE plots, all seven arrays from the 24 h samples can be considered as outlier arrays (see Fig. 5). This could be due to secondary effects such as carbon and oxygen limitations during shakeflask cultivation that are independent from the carbon sources. Furthermore, the first sorbitol time point appears to be problematic, probably because the fungus requires more time to induce the enzymatic machinery to metabolize it and therefore, during the first time points, suffers from carbon limitation. We recommend removing these eight arrays before continuing with further analysis (see Notes 11 and 12).

```
> problematicArrays <- grep("24|oS04", sampleNames(rawData))

> rawData.sub <- rawData[,-problematicArrays]
```

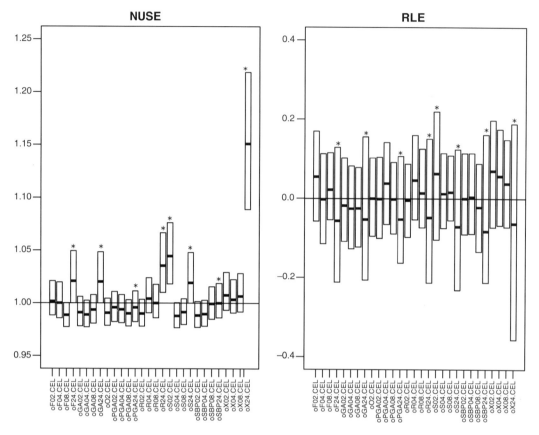

Fig. 5. Quality assessment using probe level models (PLM). Normalized unscaled standard error (NUSE) and relative log expression (RLE) plots for each array of the carbon source transfer experiments (E-MEXP-1626). Outlier arrays (marked with *asterisks*) can be recognized by their deviation from the remaining arrays (7).

Using a spike-in and dilution benchmark dataset, it was shown that the RMA algorithm is superior to the Affymetrix MAS5.0 algorithm for the detection of differentially expressed genes (11). However, as recently reviewed (16), the preprocessing method of choice for gene correlation studies is less clear. With two different approaches (17, 18), it was shown that RMA preprocessing is not suited for gene coexpression studies because of gene correlations introduced as artifacts by the RMA algorithm. For *Escherichia coli*, it was reported that a combination of the MAS5.0 background correction procedure with nonlinear normalization (invariantset method) and probe summarization according to Li-Wong (19) gave the best correlations between cotranscribed operon genes (17).

To combine MAS5.0 background correction with the nonlinear normalization and Li-Wong probe summarization, use the expresso function from the affy package. Assign the results, which are organized as an object of the class ExpressionSet, to the variable eset. Subsequently, convert the ExpressionSet object to the expression matrix e and remove all Affymetrix control probes to finally obtain the matrix e.anig:

```
> eset <- expresso(rawData.sub, bgcorrect.method = "mas", pmcorrect.method = "mas",
    normalize.method = "invariantset", summary.method = "liwong")
> e <- exprs(eset)
> subIndex <- grep("AFFX", featureNames(eset))
> e.anig <- e[-subIndex,]
```

3.3.3. Nonspecific Filtering

For the construction of gene coexpression networks, the expression patterns of every pair-wise combination of genes are checked for significant correlation. The number of probe pairs that can be drawn from n probes can be calculated with the binomial coefficient (n over 2) and is equal to $n! \bullet (n! \bullet (n-2)!)^{-1}$. With the function choose, calculate the number of comparisons required for the e.anig matrix:

```
> choose(nrow(e.anig), 2)
```

The number of comparisons is 105,248,286, which will require a lot of computing time. The number of comparisons can be reduced by not taking those genes into consideration that have a low overall variability across all arrays. Removal of all genes with standard deviation (SD) lower than the mode of the distribution of SDs has been suggested (20) as a good threshold for nonspecific filtering based on variability (see Note 24). First, calculate row-wise SDs with the rowSds function, determine the mode of the distribution of SDs with the function shorth, and select all probes with SD larger than the mode:

```
> sds <- rowSds(e.anig)
> sds.mode <- shorth(sds)
> e.anig.sub <- e.anig[sds > sds.mode,]
```

After this nonspecific filtering step, the number of pair-wise comparisons reduced approximately by half. 10,647 probes are left for which a total of 56,673,981 pair-wise comparisons have to be computed.

3.3.4. Building Gene Coexpression Networks

To compute pair-wise correlation coefficients, use the function cor. For each pair-wise combination of columns from a given matrix, cor computes correlation coefficients as specified by the method parameter. Before the function can be applied, the matrix e.anig.sub has to be transposed using the function t. In the resulting matrix e.anig.sub.t, columns (CEL files) and rows (probes) are interchanged.

```
> e.anig.sub.t <- t(e.anig.sub)
```

The application of different correlation coefficients for the
construction of gene coexpression networks has been discussed in
a recent review (16). The Pearson correlation coefficient is often
used as a measure for gene coexpression. However, it is sensitive to
outliers and high correlation coefficients can only be obtained for
linear relationships. The Spearman rank correlation coefficient has
been supposed as a good alternative. The Spearman correlation
coefficient is less sensitive to outliers because it is not directly

```
> e.anig.sub.t.cor <- cor(e.anig.sub.t, method = "spearman")
```

calculated from the expression values but from ranks of expression
values. Whereby, it can also detect correlations for nonlinear relationships such as the Michaelis–Menten kinetic. To calculate pair-wise
Spearman correlation coefficients for the transposed matrix
e.anig.sub.t, use the function cor (see Note 25):

The correlation matrix is symmetric with respect to its main
diagonal. Row-column combinations define pairs of genes. Probes

```
> cutoff.min <- 0.7
```

```
> cutoff.max <- 0.95
```

```
> cutoff.increment <- 0.05
```

correlated to themselves are lying on the main diagonal and both
halves (above and below the main diagonal) of the matrix contain
equal information. Therefore, only correlation coefficients from
one half of the correlation matrix excluding its main diagonal have
to be checked and compared to a threshold.

Before checking all pair-wise comparisons, set a range of critical
correlation coefficients by defining the minimal and maximal
correlation coefficients cutoff.min and cutoff.max, as well as
the increment cutoff.increment. The code below exemplarily
defines a range of critical correlation coefficients, for which five
different coexpression networks (with absolute correlation coefficients smaller than 0.70, 0.75, 0.80, 0.85, 0.90, and 0.95) will
be constructed within a for-loop (see Note 17).

In order to check one half of the correlation matrix excluding
its main diagonal, basically two for-loops have to be initiated. The
outer for-loop increments its counter i, used as row index, by 1
starting with the first row until it reaches the last but one row. The
inner for-loop increments its counter j, used as column index, by 1
starting at i plus 1. In the inner for-loop, the row and column indices i and j are used to access the corresponding correlation coefficient and compare it to the range of thresholds defined above. For
this, a third for-loop is initiated where probe pairs are linked to
each other by positive or negative correlations if the correlation
coefficient is larger or smaller in case of positive or negative values,
respectively. The results are exported to files in the simple interaction format (SIF) format and saved in the WD (see Note 24).

```
> for(i in seq(1, nrow(e.anig.sub.t.cor) - 1, 1)){

>   print(i)

>   for(j in seq(i + 1, nrow(e.anig.sub.t.cor), 1)){

>   for (cutoff in seq(cutoff.min, cutoff.max, cutoff.increment)) {

>     file.name <- paste("network.", cutoff, ".sif")

>     if (e.anig.sub.t.cor[i,j] > 0 && e.anig.sub.t.cor[i,j] > cutoff) {

>       write(paste(row.names(e.anig.sub)[i], "POScorrelation",
   row.names(e.anig.sub)[j], sep = " "), file = file.name, append = TRUE)

>     }

>     if (e.anig.sub.t.cor[i,j] < 0 && e.anig.sub.t.cor[i,j] < (-1) * cutoff) {

>       write(paste(row.names(e.anig.sub)[i], "NEGcorrelation",
   row.names(e.anig.sub)[j], sep = " "), file = file.name, append = TRUE)

>     }

>   }

> }

> }
```

The resulting SIF files can be imported into Cytoscape (21), an open source platform for network analysis and visualization. Cytoscape provides a graphical user interface and various plug-ins for network analysis exist, such as MCODE (22) for the identification of clusters of highly connected genes or BinGO (23) for enrichment analysis of Gene Ontology terms.

4. Notes

1. If R cannot be called from every directory at the command line, the directory containing its executable has to be added to the PATH environment variable. Details are depending on the operating system.

2. For R code and command line commands, each new line begins with a "larger-than" symbol (>). For copy and pasting the code to the corresponding command prompt, the "larger-than" symbols have to be removed.

3. For Windows systems, the open source program 7-zip (http://www.7-zip.org/) can be used for decompression.

4. In general, most R and Bioconductor functions/packages are well documented. Typing a question mark directly in front of any function at the R command line should open a short documentation (for example ?mean opens the documentation for the function mean). Furthermore, many packages provide additional documentations (so-called vignettes) containing

executable examples. When putting two question marks directly in front of any word, a list can be obtained with functions matching that word (for example `??mean` lists all functions matching mean).

5. The function `setwd` can be used to define the WD. The forward slash (/) is used as path separator. It is most robust, to define the absolute path. While on Mac OS and Linux, `setwd("/Users/someUser/Documents/someWD/")` could for example be used, it could look something like `setwd("C:/Documents and Settings/someWD/")` on a Windows system.

6. In R, values can be assigned to variables using `<-` or `->`, which are assigning a value on their right to a variable on their left, or vice versa.

7. There is no critical slope indicating RNA degradation since different chip architectures result in different chip-specific slopes. Rather than the actual values of the slopes, it is important that the slopes are comparable for different samples.

8. Plots can be exported to PDF files by simply preceding the plot command(s) with `pdf("desired_filename.pdf")`. With `dev.off()`, the PDF file is finalized and saved in the current WD. If multiple plots are generated after initializing the PDF document, they will be appended to each other resulting in a single PDF file.

9. The function `par` can be used to set diverse graphical parameters among which `mfrow` can be used to define the number of rows and columns. Defining `par(mfrow = c(i, j))` before plotting graphs will define i rows and j columns allowing to place i times j plots on one page or window.

10. Unlike most other preprocessing procedures, RMA expression values are in $\log(2)$ scale.

11. The function `grep` can be applied on a vector and returns an index of elements that match a given pattern.

12. To obtain parts of vectors or matrixes, subscripts can be used in R. Subscripts are given in squared brackets directly behind the corresponding variable. With `v[3]`, the third element from the one-dimensional vector v can be obtained. For matrixes, the squared brackets contain two subscripts separated by a comma. The first subscript refers to the row and the second to the column. For the matrix m, the value of the fifth row and the second column can be obtained by `m[5,2]`. Furthermore, the values of the fifth row and columns 2–5 can be obtained by `m[5,2:5]`. With `m[5,c(2,4,5)]`, the values of the fifth row and columns 2, 4, and 5 can be obtained. By using negative subscripts, elements can be dropped from a vector or matrix. In addition, subscripts can consist of logical expressions (for

example > or <) and subscripts can also be used for sorting. Many more details can be found in basic R documentations.

13. If one performs 10,000 independent *t*-tests on a set of 10,000 genes to identify differentially expressed genes applying a critical *p*-value of 0.05, the number of false positives (Type I error α) accounts to 500 (5% of 10,000). Without an adjustment for multiple testing, the Type I error rate is not controlled at the level indicated by the *p*-value, especially because the majority of genes is not expected to be differentially expressed in microarray studies.

14. The Type II error β accounts for false negatives and the statistical power is defined as $1-\beta$.

15. The code can easily be adapted for similar analysis. The only lines that have to be changed are line 4 and line 5. Please consult the documentation of the limma package to understand the different commands in more detail.

16. The $ operator can be used in R to obtain a component from an object.

17. For-loops can be applied to generalize or automate repeating blocks of code. For-loops consist of two parts. The first part, which is enclosed by brackets, initiates the loop by setting the loop counter, comparing it to a defined limit and incrementing it in each cycle. The second part of the loop, the body, is enclosed by curly brackets and contains the code to be executed in each cycle until the loop counter reaches the defined limit. By initializing a loop with (i in 1:ncol(contrast.matrix)), the loop counter i is defined to start at 1, being incremented in each cycle by 1 until it is equal to the number of columns of the contrast matrix. Alternatively, the function seq can be used to define the range of the loop counter i as follows (seq(start value, end value, increment)).

18. The function paste can be used to concatenate strings. The argument sep defines how the strings are separated. Calling paste("one", "two", sep = "-") would results in "one-two".

19. The function cbind is used here to append matrixes or columns from matrixes together. Because cbind does not take the row names into consideration, the number and the order of rows have to be identical. If that is not the case, the function merge can be used.

20. The limma function topTable can be used to list the top-ranked genes from the limma analysis based on fold change, FDR, or other criteria. Besides other information, the function topTable returns the logarithmic fold changes for the corresponding comparisons. In this example, the argument sort.by = "none" represses any sorting, because the data

will be combined with other data based on the original row order using the function `cbind` later. The argument `coef` specifies which comparison of the contrast matrix is of interest.

21. The function `unique` can be used to remove redundant elements from a vector.

22. To check within a logical expression if two arguments are equal, a double equal sign (==) has to be used. If only a single equal sign (=) is used, the argument on the right site gets assigned to the variable on the left.

23. The function `apply` can be used to apply a function to rows or columns of a matrix. While `apply(x, 1, mean)` computes column-wise mean values for all rows, `apply(x, 2, mean)` computes row-wise mean values for all columns of the matrix x.

24. Depending on the available computational resources, one might consider to not perform any nonspecific filtering or to apply a more severe threshold. We recommend applying a range of filter settings and comparing the resulting coexpression networks.

25. With the method parameter for the cor function, different correlation coefficients can easily be calculated. Should R give an error message related to problems with memory allocation, apply a more stringent threshold for nonspecific filtering to reduce the number of pair-wise combinations.

Acknowledgments

This work was supported by a grant of SenterNovem IOP Genomics project IGE07008. Part of this work was carried out within the research program of the Kluyver Centre for Genomics of Industrial Fermentation, which is part of the Netherlands Genomics Initiative/Netherlands Organization for Scientific Research. We thank T.G. Homan for discussions and proof reading of the manuscript.

References

1. Gentleman, R. C., Carey, V. J., Bates, D. M., Bolstad, B., Dettling, M., Dudoit, S., Ellis, B., Gautier, L., Ge, Y., Gentry, J., Hornik, K., Hothorn, T., Huber, W., Iacus, S., Irizarry, R., Leisch, F., Li, C., Maechler, M., Rossini, A. J., Sawitzki, G., Smith, C., Smyth, G., Tierney, L., Yang, J. Y., and Zhang, J. (2004) **Bioconductor: open software development for computational biology and bioinformatics** *Genome Biol* **5**, R80.

2. Jorgensen, T. R., Nitsche, B. M., Lamers, G. E., Arentshorst, M., van den Hondel, C. A., and Ram, A. F. **Transcriptomic insights into the physiology of Aspergillus niger approaching a specific growth rate of zero** *Appl Environ Microbiol* **76**, 5344–55.

3. Martens-Uzunova, E. S., Zandleven, J. S., Benen, J. A., Awad, H., Kools, H. J., Beldman, G., Voragen, A. G., Van den Berg, J. A., and Schaap, P. J. (2006) **A new group of exo-acting family 28 glycoside hydrolases of Aspergillus niger that are involved in pectin degradation** *Biochem J* **400**, 43–52.

4. Team, R. D. C. (2010) **R: A language and environment for statistical computing.**, R Foundation for Statistical Computing, Vienna, Austria.

5. Gautier, L., Cope, L., Bolstad, B. M., and Irizarry, R. A. (2004) **affy--analysis of Affymetrix GeneChip data at the probe level** *Bioinformatics* **20**, 307–15.

6. MacDonald, J. W. (2008) **affycoretools: Functions useful for those doing repetitive analyses with Affymetrix GeneChips**.

7. Bolstad, B. M., Collin, F., Brettschneider, J., Simpson, K., Cope, L., Irizarry, R. A., and Speed, T. P. (2005) **Quality Assessment of Affymetrix GeneChip Data.** *in* "Bioinformatics and Computational Biology Solutions Using R and Bioconductor.", Springer, New York.

8. Smyth, G. K. (2004) **Linear models and empirical bayes methods for assessing differential expression in microarray experiments** *Stat Appl Genet Mol Biol* **3**, Article3.

9. Gentleman, R., Carey, V., Huber, W., and Hahne, F. **genefilter: methods for filtering genes from microarray experiments**.

10. Irizarry, R. A., Gautier, L., Huber, W., and Bolstad, B. (2006) **makecdfenv: CDF Environment Maker**

11. Irizarry, R. A., Bolstad, B. M., Collin, F., Cope, L. M., Hobbs, B., and Speed, T. P. (2003) **Summaries of Affymetrix GeneChip probe level data** *Nucleic Acids Res* **31**, e15.

12. Naef, F., Lim, D. A., Patil, N., and Magnasco, M. (2002) **DNA hybridization to mismatched templates: a chip study** *Phys Rev E Stat Nonlin Soft Matter Phys* **65**, 040902.

13. Benjamini, Y., and Hochberg, Y. (1995) **Controlling the False Discovery Rate: A Practical and Powerful Approach to Multiple Testing** *Journal of the Royal Statistical Society. Series B (Methodological)* **57**, 289–300.

14. Bourgon, R., Gentleman, R., and Huber, W. **Independent filtering increases detection power for high-throughput experiments** *Proc Natl Acad Sci USA* **107**, 9546–51.

15. Baron, R. M., and Kenny, D. A. (1986) **The moderator-mediator variable distinction in social psychological research: conceptual, strategic, and statistical considerations** *J Pers Soc Psychol* **51**, 1173–82.

16. Usadel, B., Obayashi, T., Mutwil, M., Giorgi, F. M., Bassel, G. W., Tanimoto, M., Chow, A., Steinhauser, D., Persson, S., and Provart, N. J. (2009) **Co-expression tools for plant biology: opportunities for hypothesis generation and caveats** *Plant Cell Environ* **32**, 1633–51.

17. Harr, B., and Schlotterer, C. (2006) **Comparison of algorithms for the analysis of Affymetrix microarray data as evaluated by co-expression of genes in known operons** *Nucleic Acids Res* **34**, e8.

18. Lim, W. K., Wang, K., Lefebvre, C., and Califano, A. (2007) **Comparative analysis of microarray normalization procedures: effects on reverse engineering gene networks** *Bioinformatics* **23**, i282–8.

19. Li, C., and Wong, W. H. (2001) **Model-based analysis of oligonucleotide arrays: expression index computation and outlier detection** *Proc Natl Acad Sci USA* **98**, 31–6.

20. Hahne, F., Huber, W., Gentleman, R., and Falcon, S. (2008) **Bioconductor case studies**, Springer Verlag.

21. Shannon, P., Markiel, A., Ozier, O., Baliga, N. S., Wang, J. T., Ramage, D., Amin, N., Schwikowski, B., and Ideker, T. (2003) **Cytoscape: a software environment for integrated models of biomolecular interaction networks** *Genome Res* **13**, 2498–504.

22. Bader, G. D., and Hogue, C. W. (2003) **An automated method for finding molecular complexes in large protein interaction networks** *BMC Bioinformatics* **4**, 2.

23. Maere, S., Heymans, K., and Kuiper, M. (2005) **BiNGO: a Cytoscape plugin to assess over-representation of gene ontology categories in biological networks** *Bioinformatics* **21**, 3448–9.

Chapter 20

Population Biology of Fungal Plant Pathogens

Zahi K. Atallah and Krishna V. Subbarao

Abstract

Studies of the population genetics of fungal and oomycetous phytopathogens are essential to clarifying the disease epidemiology and devising management strategies. Factors commonly associated with higher organisms such as migration, natural selection, or recombination, are critical for the building of a clearer picture of the pathogen in the landscape. In this chapter, we focus on a limited number of experimental and analytical methods that are commonly applied in population genetics. At first, we present different types of qualitative and quantitative traits that could be identified morphologically (phenotype). Subsequently, we describe several molecular methods based on dominant and codominant markers, and we provide our assessment of the advantages and shortfalls of these methods. Third, we discuss various analytical methods, which include phylogenies, summary statistics as well as coalescent-based methods, and we elaborate on the benefits associated with each approach. Last, we develop a case study in which we investigate the population structure of the fungal phytopathogen *Verticillium dahliae* in coastal California, and assess the hypotheses of transcontinental gene flow and recombination in a fungus that is described as asexual.

Key words: Population genetics, Genotype, Phenotype, Dominant marker, Codominant marker, Structure, Phylogenetic tree and network, Linkage equilibrium, Coalescent, Migration

1. Introduction

Fungal and oomycetous pathogens behave as dynamic populations with demographic phenomena similar to those observed in other organisms, including humans. The significance of population biology therefore comes from advancing the knowledge related to epidemiology and population genetics. Here, we shall restrict the discussion to the population genetics of fungal pathogens as influenced by natural selection, genetic drift, recombination, mutation, and gene flow on the gene diversity and structure of populations through time. A plethora of studies have centered on the

Melvin D. Bolton and Bart P.H.J. Thomma (eds.), *Plant Fungal Pathogens: Methods and Protocols*,
Methods in Molecular Biology, vol. 835, DOI 10.1007/978-1-61779-501-5_20, © Springer Science+Business Media, LLC 2012

populations of plant pathogenic fungi and oomycetes using both molecular and phenotypic markers. The common thread in all these studies is improving the understanding of the evolution of target organisms and to apply this knowledge for improved management of plant diseases.

Information gathered from population genetics studies may improve models of disease epidemics and forecasting, enhance the evaluation of risks to established plant cultivars, or assist in targeting control measures. For instance, incorporating knowledge of a pathogen's population structure into breeding for disease resistance may provide insight into the potential long-term and global effectiveness of resistant breeding lines. The alternative approach, which relies on using individual strains, is fraught with a lack of knowledge on the representativeness of such strains, and the possibility of a sudden upswing in strains capable of overcoming such resistance. Such information may also serve as a means of clarifying the viability and longevity of fungicide treatments, and avoiding fungicide resistance and its long-term repercussions. Furthermore, knowledge of the means and magnitude of pathogenic migration is also critical in the formulation of regulations governing the international trade in plant germplasm.

The definition of population is not as clear-cut as it may intuitively sound, and many definitions are available (1). In this chapter, populations are defined *sensu* Futuyama as "a group of conspecific organisms that occupy a more or less well-defined geographic region and exhibit reproductive continuity from generation to generation" (1). Several of the methods described in this chapter may be applied to study processes involved in speciation and the evolutionary history of genera and species. However, the current chapter is solely aimed at analyzing fungal populations at the subspecies level.

2. Methods

2.1. Phenotypes

2.1.1. Morphology

Morphological traits such as growth rate, colony color and architecture, spore numbers, distribution of sexual or asexual structures, etc., were repeatedly used to differentiate strains of fungi and oomycetes, usually in conjunction with other methods. Numerous studies have included aspects of colony morphology in the analysis of the fungal/oomycete population biology (2–5). A commonly used morphological trait is radial growth, which involves measuring the diameter of colonies over time. It is commonly suggested that two measurements at 90° angles be made, and their mean recorded. If these measurements are made in a time series, it becomes crucial to apply a time series statistical analysis, rather than a common analysis of variance (ANOVA). This is due to the fact that the measurements made over time are not independent variables, i.e., the colony diameter at t_{n+1} is not independent of the diameter of the same colony at t_n. The size, quantity, and distribution of resistance structure (e.g., sclerotia, microsclerotia, and oospores) on the culture medium may also be recorded to compare among strains. The microscopic size and number of asexual spores (e.g., conidia, sporangia) produced may also be quantified. These traits are also valuable when comparing the response of studied strains to fungicides or other chemicals, as each strains may respond differently thus providing a means of interspecies differentiation.

2.1.2. Vegetative and Mycelial Compatibility Grouping

The formation of heterokaryons is common among fungi and is governed by a mechanism of self/nonself recognition (6–8). Vegetative compatibility groups (VCG) were successfully applied for the analysis of fungal populations such as *Verticillium dahliae* (8, 9), *V. albo-atrum* (10), and *Fusarium oxysporum* (11, 12). Similarly, mycelial compatibility groups (MCG) were applied to *Sclerotinia sclerotiorum* population analyses (2, 13, 14).

Hyphal fusion occurs between any two strains, regardless of compatibility. Multiple *het* loci control the vegetative compatibility of fungi, whereby hyphae of isolates compatible at all these loci anastomose (fuse), and form a stable heterokaryon. No demarcation between strains is visible in a compatible reaction between two strains. Conversely, if two strains are incompatible at one or more loci, following hyphal fusion, the heterokaryotic cells die and a barrage at the junction between the two strains is observed. In practice, the two strains look distinct and separated by an obvious gap on a Petri dish. In certain fungi such as *Neurospora crassa*, the mating type locus is responsible for compatibility, in addition to sexual mating (6).

In some fungi (e.g., *Ophiostoma* sp. and *Cryphonectria* sp.) incompatibility is readily observed by the formation of asexual fruiting bodies (6). In others, using auxotrophic markers is essential to the identification of VCGs. The most commonly employed auxotrophic marker is the nitrate nonutilizing (*nit*) mutants, usually generated on a chlorate-amended culture medium. *Nit* mutant strains cannot degrade chlorate, hence grow unaffected on chlorate-amended culture media, whereas those that are *nit*+ grow poorly on these media (11). Upon pairing a *nit* strain with a complementary *nit*+ strain on the culture medium amended with a source of nitrate, hyphal anastomosis leading to heterokaryosis will allow for complementation. In this situation, a *nit* strain will grow similar to a wild-type, whereas growth is severely impeded in incompatible pairings. Typically, *nit* strains would be paired with tester strains from each VCG to identify the VCG to which the tested strain belongs to.

Unlike VCGs, MCGs require only the addition of red food coloring to the culture medium (14). Compatible reactions will result in a complete fusion of the paired strains. In contrast, incompatible reactions will result in a marked barrage between paired strains, and a visible accumulation of the red color in hyphae close to the barrage.

Use of compatibility grouping has provided a major understanding of the population structure of a number of fungi (7). However, compatibility grouping suffers from significant shortcomings in fungi that undergo recombination (frequent or intermittent) (15). Recombination tends to upset the linkage of *het* loci and other unlinked makers. Such a phenomenon makes compatibility grouping ineffective in sexual fungi. In addition to the latter deficiencies, the great number of pairings and the length of incubation period required, as well as the significant margin of ambiguity, make compatibility grouping impractical for many studies. However, a combination of VCG typing and other markers may provide an added benefit to asexual fungi (16). This advantage, in asexual organisms, stems from the linkage (high correlation) of phenotypic and genotypic markers, hence a high association of VCG assignment with other trait and molecular data.

2.1.3. Virulence (Races) and Fitness

The gene-for-gene type interaction between a fungal or oomycete pathogen may be used to study their population structure. Examples of fungal or oomycete pathogens with well-defined races abound in the literature including *Puccinia striiformis* f. sp. *tritici* (wheat and barley stripe rust), *Puccinia triticina* (wheat leaf rust), *V. dahliae* (Verticillium wilt), *Bremia lactucae* (lettuce downy mildew), *Aphanomyces euteiches* (Aphanomyces root rot), or *Peronospora farinosa* f. sp. *spinaceae* (spinach downy mildew) (17–22). In such cases, the acquired pathogen population is used to challenge an established array of host differentials varying in the locus/gene providing host resistance. Strains that differ in their ability to cause disease on particular host genotypes (one or combination of multiple genotypes) form specific races. The number of races identified in one pathogen is totally dependent on the number of resistance genes identified in its host(s).

The use of races is very fluid because of the persistent risk from the evolution of new races. However, races are assigned based on the interaction of individual isolates with a set of differential host cultivars. For instance, about 70 new races of *P. triticina* are identified yearly in the US based on their interaction with a set of 20 differential lines of wheat (18). In contrast, because of the current availability of one gene for resistance to Verticillium wilt (the *Ve* gene), *V. dahliae* strains are screened on two cultivars, a Ve^+ and a Ve^-. The latter approach suggests that there are only two possible types of strains of *V. dahliae* in nature, whereas it is conceivable that more races may be present, but the ability to identify these is limited by the unavailability of varying sources of resistance. Additionally, in fungi and oomycetes that are prone to outcrossing sexual recombination (regardless of the frequency of this event), the avirulence genes could be spread in a population more effectively, and hence the ability of a pathogen to generate novel races is greatly enhanced (23).

Here, we define fitness as the ability of organisms to survive and reproduce in a particular environment. A number of studies have tried to identify a possible correlation between the fitness of fungal and oomycetous phytopathogens in agricultural conditions with traits such as resistance to fungicides (24–29). Whether a positive or negative correlation is encountered is solely associated with the particular pathosystem.

Both fitness and virulence may be affected by factors such as selective sweep, induced by agricultural practices. Such dramatic changes may drastically influence the studied population. Interestingly, the widespread release of the spring wheat cultivar Zak in the US Pacific Northwest in 2002, was accompanied by a selection for isolates of *P. striiformis* f. sp. *tritici* harboring avirulence genes exhibiting an incompatible reaction with the host cultivar (30). In the mid-1990s, a marked displacement of widespread *Phytophthora infestans* (potato and tomato late blight) genotypes by the metalaxyl-resistant genotype US8 was likely influenced by the convergence of: an increased fitness, enhanced aggressiveness, and continued application of metalaxyl on affected potato crops (31, 32). All of these factors likely synergistically led to a selective sweep that virtually wiped out the previously cosmopolitan US1 genotype.

2.2. Molecular Markers

2.2.1. Sequences

Acquiring DNA sequences is becoming a more feasible approach to studying populations. Additionally, the significant gains made though the use of newer sequencing methods providing increased sequence length, high genome coverage, decreased cost, etc., make comparisons of entire genomes a likely feasible option in the near future. Large amounts of such data may also pose significant logistic and analytical problems. However, the use of sequences acquired for specific loci has become the mainstay of many research programs, and is likely to remain, for the foreseeable future, an effective tool because of the wealth of analytical tools available for such data.

Ribosomal DNA (rDNA) sequences are among the most commonly used because they are multicopy and provide a variety of fast and slow evolving regions. Due to the large amount of rRNA required to produce ribosomes, multiple copies of rDNA are commonly present in the genome of fungi (33). For instance, 59 copies were identified in *Thanatephorus praticola* (*Rhizoctonia solani* AG4) (34), whereas 38–91 copies, depending on the strain used, were counted in *Aspergillus fumigatus* (35), and close to 150 were identified in *Saccharomyces cerevisiae* (36). However, as few as one copy of the rDNA was identified in the genome of *Pneumocystis carinii* f. sp. *hominis* (37). The significance of the rDNA is that it provides regions that are conserved, as well as others that are rapidly evolving.

Conserved regions are coding regions, which include the three types of rRNA, namely the 18S, 5.8S, and 28S, in addition to the internally transcribed spacer (ITS) region. The ITS is further divided into two areas ITS1, which separates the 18S from the 5.8S, and ITS2 that spans the 5.8S and the 28S. The rDNA tandems are separated by the intergenic spacer regions (IGS), which is also further divided as it encompasses the 5S rDNA (33). Further information on the organization and primer sequences may be found in White et al. (38) and on the website of Prof. Rytas Vilgalys of Duke University (http://www.biology.duke.edu/fungi/myco-lab/primers.htm).

The ITS rDNA is commonly used to distinguish among species within one genus (33), which is likely associated with the higher rate of variability and lower evolutionary pressure compared with the highly conserved 18S, 5.8S, and 28S rRNAs. The ITS may be the most routinely targeted DNA region, particularly for taxonomic purposes or evolutionary studies of related species. As of September 2010, more than 2.64 million fungal ITS sequence entries were available in GenBank, which highlights the preponderant role that this marker plays in population biology. In contrast, the IGS rDNA is a non-transcribed and rapidly evolving region with little evolutionary pressure. This high variability makes the IGS a commonly used tool for subspecies population analyses (39–43). The prevalence of indels and inversions in the IGS may impede proper sequence alignment, hence making this region hard to utilize in certain fungi.

Because of the need to preserve the function they code for, the evolution of coding genes, primarily central function genes, lends itself ideally to ascertaining the identity of fungal species (44, 45). Such coding sequences are typically highly conserved within a species because of the significance of the proteins they code for to the integrity and viability of fungi. However, populations that have remained divergent, with little to no migration or recombination, for many generations may accumulate mutations in such genes (in exons, not just introns), hence in such situations these sequences may be informative at a population level (42, 46, 47). Sequences of the mating type (*MAT*) gene idiomorphs were used in the analyses of fungal populations (48).

Although highly informative for phylogenetic analyses at the species level, sequences of exonic regions of coding genes may not lend themselves readily to subspecies analyses. The propensity to have monomorphic sequences, because of purifying selection, is a guarantee for the fungal population of a continued successful function. However, natural selection forces, such as host or environmental adaptation, in both allopatric (an evolutionary process in which one species divides into two because the original homogenous population has become separated and both groups diverge from each other) and sympatric (an evolutionary process in which a species splits into two without any separation of the ancestral species' geographic range) populations, may provide useful polymorphic sequences.

Introns are stretches of DNA present within the sequence of genes, and are transcribed but through splicing are removed from the mRNA, hence not translated into proteins. Unlike exons, introns are not subject to stringent evolutionary pressure, which likely provides higher chances to observe polymorphisms at the subspecies level. Introns have successfully been used in population analyses of fungi (49, 50). Recombination is thought to be involved in the loss of introns, but that DNA repeat expansion may be required to create new introns (51). Noncoding mutations in introns are unlikely to be associated with specific phenotypic traits.

The frequency of recovery of mating type (MAT) loci in fungal populations is commonly employed to determine the reproductive status (16, 42, 48, 52–54). In an infinite population that is not undergoing selective pressure or is not subject to migration, where there are no impediments to random mating, i.e., at Hardy-Weinberg equilibrium (HWE), the ratio of MAT idiomorphs should not be significantly different from 1:1. Hence, if the frequency of recovery of both MAT idiomorph is at parity, and the fungus is heterothallic (i.e., requiring two different mating types for sexual reproduction to occur), it would be possible to conclude that the population is at equilibrium. Moreover, isolates are called homothallic if they possess both MAT idiomorphs.

In a heterothallic fungus or oomycete, if only one of the two idiomorphs is present in certain locality, then sexual recombination is ostensibly disrupted. Subsequently, the population will likely appear clonal, with a few clones being widely distributed geographically, but also over several generations. Such a phenomenon was observed with *P. infestans* prior to the 1990s (31), and in *Ascochyta rabiei* of chickpea in the US Pacific Northwest (54). In both cases, a founder effect (the reduction in genetic variability resulting from the establishment of a new population by a restricted number of individuals that emigrate from a larger population) had restricted the migration to one of the two mating types. In the case of *P. infestans*, genotype US8 (mating type A2) supplanted the widely dispersed genotype US1 (mating type A1) in the mid-1990s; subsequently, epidemics of a different genotype, US22 (mating type A2) swept potato and tomato production regions of the US in 2009 (http://www.plantpath.cornell.edu/Fry/New_tomato_strain.pdf).

2.2.2. Dominant Markers

These molecular markers are usually universal and fairly painlessly applicable to fungi and oomycetes for which little genetic information is available. However, these types of markers are usually non-specific and issues of non-reproducibility may be common (55). Additionally, these markers only allow the identification of a dominant allele. Thus, in a heterozygous diploid organism, the information obtained masks the heterozygosity, which biases the outcome of the analysis. Despite such deficiencies, these markers have a role to play in population biology, especially if obtained products are sequenced or linked to a given phenotypic marker.

Random amplified polymorphic DNAs (RAPD) is a direct application of the concept of polymerase chain reaction (PCR) (56–58). RAPDs rely on the ability of short (~10 bp) and arbitrary primers to anneal to a given genome and amplify products at random, in highly permissive PCR conditions (55, 59, 60). The outcome is observed in a gel electrophoresis, and is a set of bands from the amplification of DNA regions that are complementary to the primer used. These PCR products may differ in size, as well as identity. In other words, RAPDs may amplify one locus in two isolates, but indels may occur and the size of the ensuing product will be different. Alternatively, RAPDs may amplify different loci that migrate similarly upon gel electrophoresis.

Only one primer is used in a reaction, and such a primer is usually randomly generated, which in turn means that little knowledge of the fungal/oomycete genome is required. And because of the small size of the primer, the likelihood of annealing to the genome is high, which made RAPDs a prime method for the analysis of populations in the 1990s. The concern about the identity of amplified products led to the development of SCAR (sequenced characterized amplified region) makers based on RAPD products. In brief, a RAPD produced fragment is sequenced and a pair of specific primers is generated from this product, which are in turn used as codominant markers for their target locus.

Although highly effective in detecting variability within populations, and extreme simplicity and rapidity, RAPDs suffer from issues pertaining to their reproducibility. Small changes in PCR conditions or mixes may affect amplification, and subsequently the outcome of RAPDs. It is not uncommon for the outcome of analyses performed using RAPDs to differ markedly between labs, or even users or PCR thermocyclers within a lab, let alone from one batch of PCR reagents to the other.

The amplified fragment length polymorphism (AFLP) method (61) is generally regarded as considerably more reproducible than RAPDs, and equally "easy" to implement on fungal/oomycete organisms with little genome information available (55, 60). This method combines aspects of the PCR with those of RFLPs (restriction fragment length polymorphisms). In a nutshell, AFLP relies on the amplification of restriction fragments following digestion with two common enzymes (usually *Eco*RI (6 bp recognition site) and *Mse*I (4 bp recognition site)). Following the digestion, universal adapters are enzymatically ligated to the restriction fragments, which will be used in two PCR amplifications. The adapters include: (a) a core sequence identical to the adaptors; (b) the sequence specific to the used restriction digestion enzyme; and (c) an extension of one to several nucleotides, which is selective (60, 61). The longer the latter extension the more stringent the AFLP becomes, which in return reduces the number of restriction sites that are identified.

The AFLP is inherently capable of generating a custom range of target loci in a short amount of time, based on the requirements of researchers. Unlike RAPDs, AFLP amplifications are conducted under stringent PCR conditions, which make AFLP a highly reproducible method that produces very stable fragments. Similar to RAPD amplified fragments, those produced by AFLP may be sequenced and converted into SCARs (62). The problem of dominance is a significant impediment to the use of AFLP, which may be overcome by converting the anonymous AFLP loci into SCAR markers. However, this method is routinely used when large numbers of qualitative markers are required.

2.2.3. Codominant Markers   Dominant markers provide information about one allele per locus in heterozygous isolates. In contrast, codominant markers allow for the identification of both alleles per locus in heterozygous organisms. Therefore, the use of dominant markers likely leads to a loss of information, which may be critical in diploid organisms.

RFLP is a method that relies primarily on the digestion of the target organism's DNA by restriction digestion enzymes. Isolates are subsequently compared following hybridization with a specific DNA probe (55). Differences in fragment sizes indicate differences in DNA sequence, which is used to identify polymorphisms among individuals within populations. Such polymorphisms include mutations in the restriction sites, indels, or even sequence rearrangements (55).

The choice of single- or multilocus DNA probes used for hybridization is critical. Single locus probes are easy to interpret, whereas multilocus probes produce a fingerprint, which is not very different from the outcome of RAPD or AFLP analyses. But unlike the latter two methods, RFLP results are highly reproducible, although the method itself may be demanding (55, 60, 63). Large amounts of clean DNA are required, and a good proficiency is critical to handle the various intricate steps involved in RFLP, but also the use of radioactively labeled probes or silver staining, and the ensuing generation of biohazards. RFLPs remain a very effective tool that has helped clarify the population structure of plant pathogens such as *Mycosphaerella graminicola* (64–66).

To circumvent the requirement for large amounts of DNA, it is possible to amplify the targeted genomic region by PCR before digesting it with restriction enzymes, and DNA hybridization. This method is referred to as PCR-RFLP.

Simple sequence repeat (SSR), microsatellite, and short tandem repeat (STR) refer to a repeated series of a contiguous motif of 2–10 nucleotides in length. These SSR are randomly distributed in eukaryotic genomes (67), and are commonly flanked by conserved regions, which makes it possible to design PCR primers. Originally described by Hamada et al. (68), the method was rapidly established in various eukaryotic systems as it exhibited great variability useful for genotyping at the subspecies level (69–72). Subsequently, such SSR were identified in fungi (73–75) and oomycetes (76, 77). Use of SSRs for interspecific analyses is commonly considered to be less effective than for subspecies population analyses (78, 79).

In *Neurospora* sp., the mutation rate in microsatellite was estimated to reach ~2,500 higher than the mutation rate in flanking regions; and the average rate was estimated in the range of 2.8×10^{-6} to 2.5×10^{-5} per locus and per generation (78). In *Neurospora*, the stepwise mutation model (SMM) was deemed to describe the evolution of SSR loci better than the two-phase mutation model (TPM) or the infinite-allele model (78). The SMM assumes that variation in the number of repeats is incremental, one step at a time going up or down (80). This type of mutation raises concerns about size homoplasy arising from an "identity by state" of fragment carrying SSR motifs, which may be a concern for di- and trinucleotide repeat motifs. But Estoup et al. (81) consider such homoplasy to be minimal and offset by the high variability at SSR. Nevertheless, it is perhaps advisable to limit the number of di- and trinucleotide repeats in the analyses, to reduce the probability of homoplasy, but also to circumvent occurrence of stutter bands commonly observed when using these motifs.

The identification of SSR motifs is an intricate and lengthy process that may require a significant level of expertise. We recommend consulting Sharma et al. (82). The most straightforward approach requires a sequenced fungal or oomycete genome, from which computer-based searches using algorithms such as Phobos (83) or Tandem Repeats Finder (TRF) (84). In the absence of a genome sequence, it is possible to search for SSR motifs by producing and sequencing a DNA library of the target fungus/oomycete organism, which would then be screened for SSRs. However, casting such a broad net will generate lots of sequence data that may not be helpful, and at a very high expense. For that reason, fishing-out these specific repeat motifs may be achieved by using specially generated probes attached to beads, or used in regular Southern blot hybridization. This method allows the user to identify the specific motifs for which the probes were designed, rather than all types of SSR motifs available in the genome. If SSRs in coding regions are sought out, it is perhaps advisable to generate an expressed sequence tag (EST) library. This approach involves a higher level of expertise for the normalization of the EST library, prior to clone sequencing, or probing for specific SSR motifs.

Few studies have reported on the use of single nucleotide polymorphisms (SNP) in fungi and oomycetes (85–89). SNPs are the most common type of sequence variation in genomes (55), and are estimated at ~90% (90, 91). SNPs are codominant markers.

The mutation rate of SNPs is much lower than that of SSRs, 10^{-8}–10^{-9} per generation and per locus compared with 10^{-4}, respectively (90). SNPs are restricted to only four possible character states (A, C, T or G), whereas SSRs may generate a larger number of alleles per locus. Therefore, individual SNPs may be less informative than SSRs for population analyses, but this may be corrected by increasing the number of targeted SNP loci (90, 91). However, a very large number of targeted SNP loci may increase the chance of linkage among loci.

SNPs may be identified using the same methods as those described above for the identification of SSRs. It is important to note that a certain bias, termed the "ascertainment bias," may arise from the decision-making process associated with the choice of SNP loci, especially when transferring SNPs between species; or when SNPs are identified in a small group of individuals and subsequently used on large cohorts (90–92). This bias is especially pronounced when only highly variable loci are included in data analyses. Furthermore, SSRs may also suffer from a similar ascertainment bias.

In addition to sequencing alleles containing potential SNPs, it is possible to utilize methods relying on the high-resolution melt curve analyses to identify predetermined SNPs (93). These methods permit the differentiation of alleles by the shift in the temperature required to denature the double strands of DNA. Such methods are rapid, reproducible, and relatively inexpensive.

3. Analytical Tools

Population genetics was in great part built on the fixation index (F_{ST}), which was used to study the structure of populations, compare subpopulations, and estimate gene flow (94, 95). F_{ST} is markedly formulated on the basis of "idealized" population models that may be hard to apply in a blanket fashion in natural conditions (96). Since the inception of the coalescent theory (97, 98), questions of a population genetic nature may be addressed using phylogeographic (phylogeography is defined here as the study of historical factors that impact today's population genetics) approaches, which have become powerful tools to investigate growth rate, size, bottlenecks, gene flow, and divergence time, among many other possible scenarios (99). Such analyses may be completed without relying on preset population models, and by utilizing the increasingly powerful computational tools available to researchers.

Pearse and Crandall state that "(…) in the situations where results have the most practical importance—guiding conservation efforts and identifying genetically distinct management units, evolutionarily significant units, or species boundaries—traditional population genetic analytical methods may be at their worst, and are less-than-optimal at best" (96). Therefore, we recommend exploring the population biology using summary statistics such as the F_{ST}, but to include coalescent-based analyses. Several studies of fungal and oomycetes population genetics have effectively utilized such approaches to the origin of organisms, or their migration paths, population status (42, 46, 48, 100–105).

3.1. Phylogenetic Trees and Networks

Exhibiting phylogenetic trees to reflect the relationship among organisms is an elegant method that relies on the genetic drift as means of speciation from a common ancestor. And for this purpose the different mathematical algorithms available rely on specific evolutionary models, to depict in a radiating fashion the putative evolutionary history of a gene or taxa. Such an evolutionary model may be based on the genetic distance separating the individuals studied (e.g., neighbor-joining [NJ], or unweighted pair group method with arithmetic mean [UPGMA]); or conversely, other methods that rely on information gathered from the characters in a sequence (DNA, RNA, or protein) alignment (e.g., maximum parsimony [MP], maximum likelihood [ML]). However, regardless of the algorithm used to build phylogenetic trees, the use of common phylogenetic analyses may not always reflect the true evolutionary history within a species. Smouse (106) states this problem eloquently: "Where subspecific evolution is highly reticulate, our strictly radiating trees are suspect (…). Forcing an evolutionary reticulogram into a strictly radiating tree form involves some distortion; that distortion can occasionally be severe and our inferred trees positively misleading."

Phylogenetic trees have commonly been used in intraspecific studies, but they do not account for basic population genetics processes, such as speciation, recombination, population structure, etc. For this reason it may be fitting to include the use of networks to analyze subspecies populations (106, 107). Such networks are inherently more effective at displaying the reticulate nature of evolution commonly observed within a species. We highly recommend readers to consult Posada and Crandall (107) and Huson and Bryant (108) for further information regarding the various methods that may be used to build phylogenetic networks to explain the population genetics within a fungal or oomycete species.

If the investigator is using molecular markers other than sequences, it is possible to employ distance-based cladograms and networks to analyze the studied population. NJ is a commonly used method that is based on the concept of minimum evolution, which is related to clustering (109). Hence, a distance matrix is required to generate such a tree. An added benefit to NJ trees is that the investigator does not assume a molecular clock, which in turn means that branches may assume different evolutionary rates. However, if a molecular clock may be applied and the root is known, then the

UPGMA model may be applicable. But such an assumption may not be legitimate without adequate testing. A range of different programs provide such analyses: PAUP (110), Phylip (111), NTSYS (112), etc. Networks may be generated in programs such as: SplitsTree4 for a number of different types of networks (108, 113), Arlequin for the minimum spanning tree (114), or TCS for the statistical parsimony (115), etc.

Many of these same programs may be applied for both sequences as well as dominant and codominant markers.

3.2. Population Structure

Structuring of populations occurs when gene flow between subpopulations is impeded over several generations, because of factors such as geographic separation or mating behavior, which leads to diverging evolutions. Measuring the summary statistic F_{ST}, which is based on gene diversity, was commonly used to estimate the differentiation among subpopulations, with values closer to zero indicating a lack of differentiation. However, evidence indicates that analyses based on the descriptive statistic G_{ST} and related statistics, may lead to wrong conclusions especially when using highly variable markers, like SSRs (116, 117). Jost (117) also indicates that F_{ST} and G_{ST} tend to decline when polymorphisms increase. Furthermore, statistics such as F_{ST} may fail to identify gene flow between subpopulations because of variable diversity among loci (118).

While the use of summary statistics may carry a risk of erroneous conclusions, F_{ST} is still widely used especially when applied in the framework of an analysis of molecular variance (AMOVA) (95). The AMOVA uses an analysis of variance (ANOVA) framework to partition the genetic variability within and between subpopulations and groups of subpopulation, and computes a parameter analogous to F_{ST} (Φ) (95). These authors (95) describe Φ_{ST} as "reflecting the correlation of haplotypic diversity at different levels of hierarchical subdivision." The hierarchical subdivision separates the haplotypic variance into its various covariance components hence providing insight into the diversity within a subpopulation, among subpopulations within a population, or among populations. And in order to avoid the assumption of normality associated with the ANOVA, the significance of the variance components of the AMOVA and Φ_{ST} are subjected to a permutation test (95). Two widely used analysis packages for AMOVA tests are: Arlequin and GenAlEx (114, 119).

The population structure may also be tested through a variety of methods that do not rely on summary statistics such as F_{ST}. For example, the principal coordinates analysis (PCO or PCoA), which is a type of multivariate analysis whereby the various correlated parameters are separated into covarying variables that account for various amounts of the variance (119–121). A distance matrix among subpopulations or individuals is computed, which is then used to compute eigenvectors that form principal coordinates. Use of the Microsoft Excel add-in GenAlEx allows for a straightforward application of this method (119).

A second approach for identifying the structure of a population divides individuals into clusters using highly developed models based on the application of the Markov chain Monte Carlo (MCMC) method to test the maximum likelihood of a certain set of parameters. This is based on the proportion of affiliation of each individual to individual k genetic clusters identified (122, 123). Each cluster is identified to minimize the deviation from a HWE and linkage equilibrium (96). Subsequently, no a priori knowledge of the population structure is required. Such analysis can be implemented in the program Structure (123). The program BAPS (124) produces a similar type of analysis where the HWE and linkage equilibrium are assumed. However, unlike Structure, BAPS deals with populations rather individuals and includes geographic information in the analysis. Another major difference between BAPS and Structure, is that the former builds clusters on the basis of allele frequencies, rather than concatenating individuals in clusters that fit the HWE (96). BAPS also allows the number of clusters to vary within an interval (the minimum being one cluster, and the maximum being the number of sampling groups from which the isolates were obtained), while in Structure a new analysis is required for each k studied (k is assigned by the investigator). Structurama (125) is yet another program for the analysis of population structures, which implements the basic method of Structure, but also allows for k to be drawn from a Dirichlet probability distribution (the Dirichlet is the conjugate prior of the multinomial distribution). Subsequently, the value of k would not be fixed or chosen arbitrarily.

Comparing the significance of F_{ST} or Φ_{ST} values is contingent on formulating the proper hypothesis, and methods based on MCMC may require a keen knowledge of the models used and their implications, in addition to requiring significant amounts of time to go through the thousands (if not millions) of iterations required. In contrast, a graphical approach provides a fast and fairly straightforward way to reveal the population structure without engaging a priori models or the need for hierarchically nested subpopulations. Population Graphs (126) offers such a visual approach methodologies that describes the complexity of the studied population without forcing the analysis to compare averages, which may not reflect the true structure.

3.3. Migration

Migration or gene flow into (immigration) or out (emigration) of a subpopulation results in alterations in allele frequencies, especially immigration of new genotypes. The type migration that we are addressing in this section does not pertain to transiting individuals but to those that establish in a new location, survive, possibly interbreed, and whose genes are preserved for multiple generations. Typical examples may include the global dispersal of A2 mating type strains of *P. infestans* (127), the transcontinental gene flow of *V. dahliae* in spinach seed (100), or *Rhynchosporium secalis* in barley seed (128).

Gene flow may be measured using F_{ST} (and its analogs), especially that non-differentiated subpopulations remain as such through migration, and using the $F_{ST} \approx 1/(4Nm+1)$ with Nm being the number of immigrants per generation. However, the restrictions of the model used in the latter equation make it exceedingly unlikely that F_{ST} provide reasonable estimates of migration (129). In contrast, methods based on the coalescent analysis provide an effective set of tools to estimate migration rates among subpopulations without the inconvenience of rigid evolutionary models based on isolation. And these coalescent-based methods provide far more realistic estimates of gene flow compared with F_{ST}, as these methods try to identify the best genealogy through exploration of large number of genealogies, by use of the MCMC, rather than relying on summary statistics (118). Such methods are included in programs such as: Migrate (130, 131); IM, IMa, and IMa2 (132); Lamarc (133); and BayesAss⁺ (134). Except for BayesAss⁺ the remainder of the programs accept the sequence RFLP, SNP, and SSR data. RAPD and AFLP data may not be analyzed using anyone of these programs, especially that no evolutionary model is available. Unlike Migrate, IM and Lamarc, BayesAss⁺ does not require that the studied population be at equilibrium, as it does not require loci to be at HWE, and it focuses on the short-term migration (96). It is also worth noting that only a few evolutionary models are incorporated into the latter programs, but significant effort is expended to increase the scope and simulate more biologically realistic evolutionary models.

We highly recommend consulting the following references for further information regarding migration analyses: Beerli et al. (130, 131, 135), Nielsen and Wakeley (136), Hey and Nielsen (132, 137), Hey (138, 139), Kuhner (118), Nielsen and Beaumont (99), and Pearse and Crandall (96). The primary reason for consulting such a large number of references is to build a deeper understanding of the process of coalescence, and also to fully comprehend the statistical approaches that are associated with the calculations, and which could greatly influence the results. In addition to the use of MCMC analyses, the choice of maximum likelihood based approach with the likelihood ratio test (LRT) compared with a Bayesian approach with the Bayes factor (BF) plays a monumental role in the speed of the analysis, and the potential outcome. For instance, the typical question of how long to run an analysis, may not be answered by a benign guess, but requires knowledge of the basics in order to devise tests that lead to a convergence of the MCMC, which would then provide meaningful information. Nevertheless, simulations have shown that for coalescent-based methodologies large numbers of entries may not be required; however, a larger number of loci to analyze may hold the key to a meaningful understanding of the evolutionary history of the studied organism (140).

3.4. Recombination

Fungi and Oomycetes may undergo outcrossing or selfing recombination, sexual or parasexual recombination, or may be totally clonal with no recombination. For further information on recombination in fungi we refer the readers to Milgroom's review of the impact of recombination on population structure (23).

Recombination is the avenue by which new combinations of genes are produced by exchange of nuclei between parents, and subsequent independent chromosome assortment. This in turn leads to higher genotypic diversity within the population, and to a random association among alleles at various loci. It is crucial to distinguish genotypic diversity from gene diversity. Gene diversity refers to the variety and number of alleles per locus, whereas genotypic diversity refers to the assortment of alleles across all studied loci. In recombining populations, no new alleles are introduced, but alleles at individual loci are reassorted with a variety of different alleles at other loci. This scenario leads to a phenomenon of linkage equilibrium (in haploids it is referred to as gametic equilibrium). Subsequently, recombination makes it difficult to test for natural selection (118). In contrast to recombination, gene flow, natural selection, and divergence coupled with mutations, typically lead to the increase in genetic diversity.

In asexual fungi and oomycetes, distinctive characteristics of clonality may be observed in populations. These features include the broad incidence of identical genotypes and a high correlation of loci. This high correlation among loci (especially distant loci) leads to linkage (or gametic) disequilibrium.

In organisms that are not known to have a sexual stage, or in populations where recombination is suspected but no glaring evidence is observed, evidence for recombination may be inferred by testing for linkage equilibrium. Such testing may be performed in the program Multilocus (141) that computes the index of association (I_A) among the studied loci (142). I_A values not significantly different from zero are expected for randomly mating populations, whereas I_A values are different from zero in cases of linkage disequilibrium (143). It is important to keep in mind that linkage equilibrium is affected by gene flow.

A strictly bifurcating tree may not precisely capture and display the significance of recombination on the evolutionary history of a studied population (144). In contrast a phylogenetic network would be more effective at exhibiting the impact of recombination events, especially that they form a typical webbing that disassociates them from divergence. Recombinants may be identified from networks, such as those produced by SplitsTree4 (113), which provides a variety of models that target recombination. And SplitsTree also includes the Φ_w test (145) for testing and confirmation of the occurrence of recombination. Additional methods are detailed in the chapter on recombination detection by Lemey and Posada (144).

Quantifying the rate of recombination of a fungus or oomycete is critical to clarifying the evolutionary history of the studied organism. The program Lamarc (133) provides such a tool, in addition to quantifying migration and growth rates (in fact growth or shrinkage). Lamarc also assumes that recombinations are homologous (between the same loci), that the rate has been constant for a long period, and that recombination events are selectively neutral. Moreover, it is worth mentioning that at the current time, combining populations with different recombination rates is impractical in Lamarc.

3.5. Population Assignment

Deducing the source of individuals in a population may be critical for conservation biology, forensic plant pathology, or studies of evolutionary biology of fungi and oomycetes. Such analyses may be answered by asking the following question: To which of these known populations does this individual belong? Or conversely by asking this question: Can any of these known populations be excluded as the source of this individual? To answer these pertinent questions, an array of computational methods is available (96). The program GeneClass2 (146) provides a number of these methods that allow the identification of the origin of recent migrants, or the exclusion of populations as potential sources of such immigrants. These methods include a test that excludes populations that are not likely to be the sources of the individual genotypes tested (147). Using the exclusion approach allows for testing the source of migrant genotypes, while acknowledging that not all possible source populations were included in the analysis (96). GeneClass2 also includes a migrant assignment method relying on a Bayesian approach to place genotypes into possible source populations based on allele frequencies (148). And another commonly used method is included in GeneClass2, which relies on a Monte Carlo resampling method (149) that generates hypothetical gametes rather than alleles, to reduce the Type I error rate (150, 151).

A similar assignment test may be performed in Structure (123). This test has the advantage of testing all populations and individuals together, which increases the power of the analysis. However, Structure assumes that all possible populations have been sampled, which may not be practical. This last point favors the models included in GeneClass2 (96).

4. Case Study

In the following example, we will describe how some of the methods presented above may be applied to the study of actual fungal populations. For this purpose, we will use the soilborne fungal phytopathogen *V. dahliae* to detail its population structure in coastal

California (USA), its migration paths with imported seed, and the potential for recombination. We employed both codominant and dominant molecular and phenotypic markers. It is also worth noting that *V. dahliae* is a haploid ascomycete and that significant differences may accompany analyses of diploid fungi and oomycetes.

4.1. Development of SSR Markers

The entire genome sequence of *V. dahliae* was acquired from the *Verticillium* Group Database (http://www.broadinstitute.org/annotation/genome/verticillium_dahliae). Each of 52 supercontigs was probed for the presence of SSR motifs using the program TRF (84). Motifs of 2–10 bp were targeted, and 100 such motifs, with no indels, were identified. Di- and trinucleotide repeats are usually harder to deal with especially that they tend to generate stutter bands upon gel electrophoresis. Therefore, we opted to have a restricted number of such repeats and more of the larger sized motifs. Furthermore, care was taken to request, in TRF, the sequences of the 500 bp flanking the SSR motif on either side. These sequences were used to design PCR primers in Primer3 (152), which would amplify fragments of 300–600 bp.

These 100 primer pairs were first tested on a restricted number of *V. dahliae* strains. Of the 100 primer pairs, 22 primer pairs exhibited consistent PCR amplification and polymorphism. One of these primers was then labeled with a fluorescent dye at the 5′ end, with either one of four fluorophores: 6-FAM, HEX, ROX, and TAMRA. Such a labeling is intended for multiplex automated fragment analysis. However, it is not required and PCR products of these primers may be run on regular polyacrylamide gel electrophoresis for fragment separation and analysis.

4.2. Verticillium dahliae Collection

We included more than 200 strains of *V. dahliae* that were isolated from hosts growing in coastal California, Central California, and Wisconsin in the US, and also from spinach seed that is grown in northern Europe (Netherlands and Denmark) and the US state of Washington (see Table 1). These spinach seed lots were destined for planting in coastal California.

The identity of all strains was confirmed morphologically, microscopically and by sequencing the ITS rDNA. Complete sequence identity was found between the sequences of all strains and those deposited in GenBank as *V. dahliae*.

4.3. Molecular Marker Analysis

4.3.1. IGS rDNA Sequencing

The IGS rDNA was amplified from 141 strains using primers SK34 (5′-GGGTCCTGTAAGCAGTAG-3′) and SK35 (5′-GAGC-CATTCGCAGTTTCG-3′). Sequences were edited in BioEdit (153) and aligned in Clustal W (154, 155), and manually refined in McClade v. 4.06 (156). Subsequently, sites that violated the infinite-site model were removed and indels were recoded, in the program SNAP Workbench (157) using SNAP Map (158).

Table 1
Origin of Verticillium dahliae strains used in population analysis

Sampling group	# Individuals
Coastal California	*123*
Non-lettuce Asteraceae	15
Lettuce	65
Pepper	25
Strawberry	18
Central California	*60*
Tomato	60
Wisconsin	*17*
Acer sp.	7
Rubus sp.	1
Rosa sp.	3
Fraxinus sp.	4
Echinacea sp.	2
Spinach seed	*43*
Washington State	9
Denmark	8
Netherlands	26
Total	242

The general time reversible (GTR) model was determined as the most likely nucleotide substitution model, upon testing in Modeltest (159). A bifurcating phylogeny was completed in MrBayes (160), and implementing 4 runs of 5,000,000 generations each, with trees sampled every 100th generation and a heating temperature of 0.001 to reduce the average standard deviations of the split frequencies below 0.01. Posterior probabilities were determined from a set of 10,000 trees, 1,000 per run, with sampling after 100,000 generations, allowing the log likelihood values to plateau. Midpoint rooting was used to display the unrooted IGS rDNA tree in FigTree v. 1.2.2 (161).

Because the analysis was interspecific, a phylogenetic network analysis was also completed to identify potential recombination or homoplasy events. The neighbor-net network, which is based on a distance method, was generated in SplitsTree4 (108) for AFLP data sets. Furthermore, haplotypes were generated in SNAP Map, which were then used build a 95% connection level statistical parsimony network in the program TCS (115).

The evolutionary history of the *V. dahliae* populations studied was analyzed by coalescent analysis in Genetree version 9.0 (162, 163), which is included in SNAP Workbench. This analysis incorporated information gathered on the migration and population growth rate from separate analyses. A full model migration analysis was performed using Migrate ver. 3.0 (130, 131) using the maximum likelihood method. Short chains were performed with 1,000 samples and 20 sampling increments and 1,000 trees were recorded; whereas long chains were performed with 5,000 samples and a burn-in of 10,000, and 5,000 trees were recorded. And the population growth rate was estimated in Lamarc ver. 2.1.3 (133), using the maximum likelihood method, including four heated chains, ten initial chains of 5,000 sampled genealogies, and five final chains of sampled 10,000 genealogies. The migration and population growth rate data were fed into Genetree and the tree exhibiting the highest likelihood and with the highest root probability was chosen to map the chronological mutation events and to identify the time to the most recent common ancestor (TMRCA). Five independent analyses were completed, each including a different random seed number and 1,000,000 coalescent simulations.

Regardless of the phylogenetic method used to analyze and display the IGS rDNA sequences, one may clearly detect in Fig. 1 spinach strains (labeled So) in the same clades as strains from hosts growing in coastal California. And from the Neighbor-net network (see Fig. 1) distinct netting may be observed along several branches, which suggests recombination events. Furthermore, the Φ_w test included in SplitsTree indicated that there is evidence for recombination among the sequences, which was reflected in a $P = 0.0$ that leads to the failure to accept the null hypothesis of no recombination. Additionally, the coalescent analysis (see Fig. 2) indicated a possible recent expansion and a trend toward divergence in *V. dahliae*, regardless of the host of origin. Strains from the various hosts growing in coastal California seem to be undifferentiated, indicating that the hosts do not seem to select for specific genotypes.

4.3.2. AFLP

DNA was digested with *Eco*RI and *Mse*I restriction enzymes, adapters were then ligated, and *Eco*RI and *Mse*I primers (without additional selective nucleotides) were used for pre-amplification. Subsequently, selective PCR amplifications were carried out with 15 *Eco*RI and *Mse*I primer combinations (labeled with 6-FAM florescent dye at the 5' end), which included one or two selective bases (61). Products were separated by capillary electrophoresis, and fragment analysis was completed in GeneMarker ver. 1.80 (SoftGenetics, State College, PA). Fragments 50–500 bp long and showing a minimum intensity of fluorescence of 250 were retained. The binary data were assembled using the AFLPDAT function (164) in the statistical package R (165, 166). Thereafter, the AFLP results were used to analyze the population structure in an AMOVA

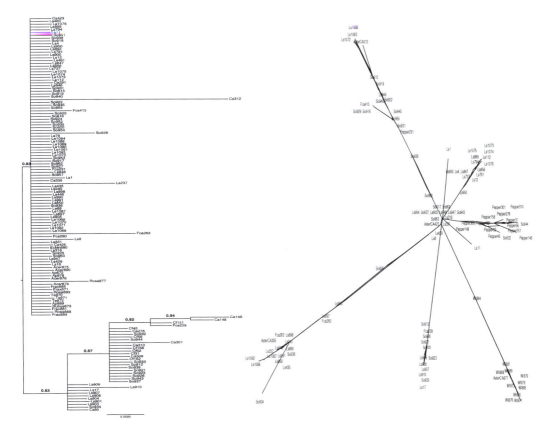

Fig. 1. Unrooted Bayesian (*left*) (reprinted with permission from APS Press, St. Paul, MN, USA) and Neighbor-net (*right*) phylogenies using the IGS rDNA sequences of 142 strains of *Verticillium dahliae*. In the Bayesian tree, the values above branches indicate Bayesian posterior probabilities (Strain abbreviations: non-lettuce Asteraceae = Cs and Te; lettuce = Ls; pepper = Ca and Cf; strawberry = Fca; spinach = So; tomato = Le; and Wisconsin = Acer, Ap, Echin, Frax, Rubus, Rosa, or simply Wl).

framework in Arlequin ver. 3.11 (114). To analyze the hierarchical structure of IGS rDNA sequence data, the sampling groups were nested into three clusters: (a) all strains collected in the coastal California (lettuce, pepper, non-lettuce Asteraceae, and strawberry); (b) strains from Wisconsin; and (c) strains from spinach.

The AMOVA showed that 91% of the genotypic variability was measured within sampling groups (see Table 2). The Φ_{ST} value for the AFLP was 0.087 ($P < 0.0001$). The differentiation among clusters, $\Phi_{CT,}$ was 0.033, not significantly different from 0 ($P = 0.296$), and amounted to 3% of the total variability (see Table 2). The differentiation among sampling groups within clusters, Φ_{SC}, was 0.055 and was significantly different from 0 ($P < 0.0001$). This in turn means that little differentiation exists among the three sampling groups of *V. dahliae*, regardless of the hosts from which the pathogen is isolated. This may also be used to conclude that there is significant gene flow of *V. dahliae* between the various regions and hosts. Because there are no established evolutionary models for AFLPs, it

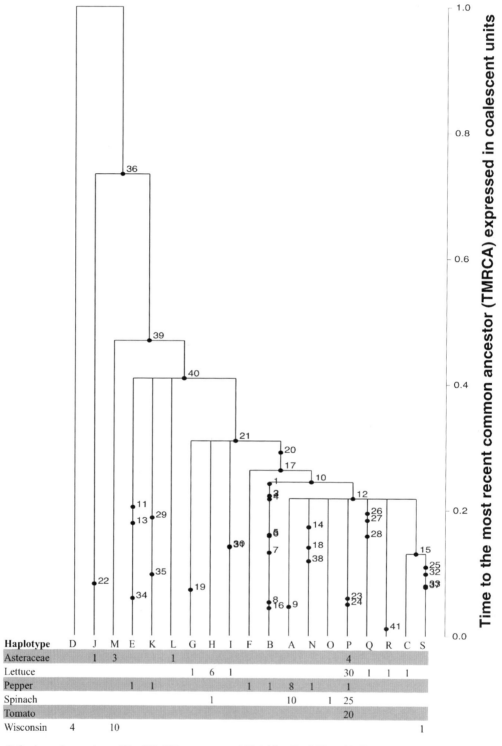

Fig. 2. Coalescent genealogy of the IGS rDNA sequences of *V. dahliae* ($L = 1.96e - 62$; $SD = 6.23e - 64$). *Numbered black circles* indicate mutations positions. All 19 identified haplotypes and the frequencies of recovery from sampling groups are shown at the *bottom*. Reprinted with permission from APS Press, St. Paul, MN, USA.

Table 2
Results of the AMOVA using the AFLP data comparing three sampling groups of *V. dahliae* from: coastal California, Wisconsin, and spinach seed

Source of variation	Percentage	Φ	*P*-value
Among clusters	3.35	0.033	0.296
Among sampling groups within clusters	5.33	0.055	<0.0001
Within sampling groups	91.32	0.087	<0.0001

is the investigator who is bound to building phylogenies, applying AMOVAs and membership analysis in Structure (123), which we will present in the next section.

4.3.3. SSRs

All 22 SSR primer pairs were used on the above-mentioned subpopulations of *V. dahliae*. Fragment analysis was completed in GeneMarker, as with the AFLP analysis. The membership of each strain to individual genetic clusters was tested using Structure v. 2.2 (123). For each genetic cluster K (1–7), defined as genetic clusters, ten runs were performed with a burn-in period of 10,000 generations and 10,000 MCMC simulations. To estimate the most likely value of K, the ad hoc statistic ΔK was computed (167). These results were plotted graphically using the Distruct (168). The possibility of recombination was tested using the Index of Association (I_A) in Multilocus ver. 1.3b (141). Because gametic equilibrium is affected by admixture, I_A values were computed for each genetic cluster obtained in the previous analysis, using 50,000 artificially recombined datasets generated for the entire set of haplotypes. The rate of gene flow was tested in Migrate. Short chains were performed with 1,000 samples and 20 sampling increments and 1,000 trees were recorded; whereas, long chains were performed with 5,000 samples and a burn-in of 10,000, and 5,000 trees were recorded.

Strains of *V. dahliae* isolated from central California tomato belonged to a distinctly different genetic cluster from strains isolated from spinach or coastal California hosts (see Fig. 3). This suggests that there is not enough gene flow between the central California and other regions and hosts, to prevent this population from diverging from the others. This was also evident in the analysis of migration (see Fig. 4), where we found that on average 0.5 migrants per generation immigrate into or emigrate from central California. Conversely, more than one migrant per generation immigrated into coastal California from the spinach seed sources in northern Europe and Washington State. This long-distance migration accompanying seed contributed to the *V. dahliae* subpopulation from coastal California.

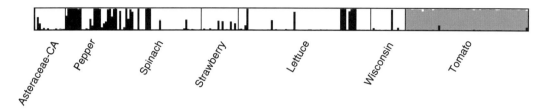

Fig. 3. Membership of individual genotypes of *V. dahliae* to each of three genetic clusters (*K* = 3) measured in Structure. Haplotypes are separated into discrete vertical bars that are organized by sampling groups. Differences in color within a vertical bar indicate a multi-cluster membership, with the height of each color within an individual indicating the proportional membership.

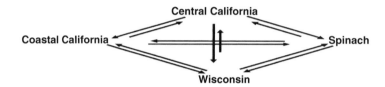

Fig. 4. Bilateral estimates of gene flow among the various subpopulations of *V. dahliae* analyzed in Migrate. The direction of each *arrow* indicates the direction of gene flow, and the arrow length is proportional to the number of migrants per generation that are exchanged between subpopulations. Only Coastal California and Spinach exchange more than one migrant per generation.

Table 3
Index of association (I_A)

Sampling group	I_A
Asteraceae	1.852*
Lettuce	2.097*
Lettuce blue genetic cluster	0.281
Pepper	2.093*
Spinach	3.058*
Strawberry	0.665*
Wisconsin	2.507*
Tomato	1.900*

Analyses were performed using multilocus on each sampling group of *V. dahliae*, including lettuce sampling group strains affiliated with the blue genetic cluster of Fig. 3. I_A of sampling groups in disequilibrium are significantly different from 0 as denoted by the asterisks

Interestingly, evidence of recombination was discovered in the lettuce subpopulation, when admixed genotypes were removed (see Table 3). This finding is novel for *V. dahliae* for which no teleomorph is known, and has long been described as clonal. It is clear though that clonality plays a significant role in the evolution of the fungus.

References

1. Wells, J. V., and Richmond, M. E. (1995) Populations, metapopulations, and species populations: What are they and who should care?, *Wildlife Society Bulletin 23*, 458–462.

2. Atallah, Z. K., Larget, B., Chen, X., and Johnson, D. A. (2004) High genetic diversity, phenotypic uniformity, and evidence of outcrossing in Sclerotinia sclerotiorum in the Columbia basin of Washington state, *Phytopathology 94*, 737–742.

3. Caten, C. E., and Newton, A. C. (2000) Variation in cultural characteristics, pathogenicity, vegetative compatibility and electrophoretic karyotype within field populations of *Stagono-spora nodorum*, *Plant Pathol. 49*, 219–226.

4. Chaijuckam, P., and Davis, R. M. (2010) Characterization of diversity among isolates of *Rhizoctonia oryzae-sativae* from California rice fields, *Plant Dis. 94*, 690–696.

5. Dorrance, A. E., Miller, O. K., and Warren, H. L. (1999) Comparison of *Stenocarpella maydis* isolates for isozyme and cultural characteristics, *Plant Dis. 83*, 675–680.

6. Glass, N. L., and Kuldau, G. A. (1992) Mating type and vegetative incompatibility in filamentous ascomycetes, *Annual Review of Phytopathology 30*, 201–224.

7. Leslie, J. F. (1993) Fungal vegetative compatibility, *Annual Review of Phytopathology 31*, 127–150.

8. Puhalla, J. E., and Hummel, M. (1983) Vegetative compatibility groups within *Verticillium dahliae*, *Phytopathology 73*, 1305–1308.

9. Bhat, R. G., Smith, R. F., Koike, S. T., Wu, B. M., and Subbarao, K. V. (2003) Characterization of *Verticillium dahliae* isolates and wilt epidemics of pepper, *Plant Dis. 87*, 789–797.

10. Correll, J. C., Gordon, T. R., and McCain, A. H. (1988) Vegetative compatibility and pathogenicity of *Verticillium albo-atrum*, *Phytopathology 78*, 1017–1021.

11. Correll, J. C., Klittich, C. J. R., and Leslie, J. F. (1987) Nitrate non-utilizing mutants of *Fusarium oxysporum* and their use in vegetative compatibility tests, *Phytopathology 77*, 1640–1646.

12. Ploetz, R. C., and Correll, J. C. (1988) Vegetative compatibility among races of *Fusarium oxysporum* f. sp. *Cubense*, *Plant Dis. 72*, 325–328.

13. Kohn, L. M., Carbone, I., and Anderson, J. B. (1990) Mycelial interactions in *Sclerotinia sclerotiorum*, *Exp Mycol 14*, 255–267.

14. Kohn, L. M., Stasovski, E., Carbone, I., Royer, J., and Anderson, J. B. (1991) Mycelial incompatibility and molecular markers identify genetic variability in field populations of *Sclerotinia sclerotiorum*, *Phytopathology 81*, 480–485.

15. Taylor, J. W., Jacobson, D. J., and Fisher, M. C. (1999) The evolution of asexual fungi: Reproduction, speciation and classification, *Annual Review of Phytopathology 37*, 197–246.

16. Grubisha, L. C., and Cotty, P. J. (2010) Genetic isolation among sympatric vegetative compatibility groups of the aflatoxin-producing fungus *Aspergillus flavus*, *Mol. Ecol. 19*, 269–280.

17. Michelmore, R., and Wong, J. (2008) Classical and molecular genetics of *Bremia lactucae*, cause of lettuce downy mildew, *Eur. J. Plant Pathol. 122*, 19–30.

18. Bolton, M. D., Kolmer, J. A., and Garvin, D. F. (2008) Wheat leaf rust caused by *Puccinia triticina*, *Molecular Plant Pathology 9*, 563–575.

19. Line, R. F. (2002) Stripe rust of wheat and barley in North America: A retrospective historical review, *Annu. Rev. Phytopathol. 40*, 75–118.

20. Klosterman, S. J., Atallah, Z. K., Vallad, G. E., and Subbarao, K. V. (2009) Diversity, pathogenicity, and management of *Verticillium* species, *Annu. Rev. Phytopathol. 47*, 39–62.

21. Grau, C. R., Muehlchen, A. M., Tofte, J. E., and Smith, J. E. (1991) Variability in virulence of *Aphanomyces euteiches*, *Plant Disease 75*, 1153–1156.

22. Irish, B. M., Correll, J. C., Koike, S. T., and Morelock, T. E. (2007) Three new races of the spinach downy mildew pathogen identified by a modified set of spinach differentials, *Plant Dis. 91*, 1392–1396.

23. Milgroom, M. G. (1996) Recombination and the multilocus structure of fungal populations, *Annual Review of Phytopathology 34*, 457–477.

24. Cafe, A. C., and Ristaino, J. B. (2008) Fitness of isolates of *Phytophthora capsici* resistant to mefenoxam from squash and pepper fields in North Carolina, *Plant Dis. 92*, 1439–1443.

25. Kato, M., Mizubuti, E. S., Goodwin, S. B., and Fry, W. E. (1997) Sensitivity to protectant fungicides and pathogenic fitness of clonal lineages of *Phytophthora infestans* in the United States, *Phytopathology 87*, 973–978.

26. Peever, T. L., and Milgroom, M. G. (1994) Lack of correlation between fitness and resistance to sterol biosynthesis-inhibiting fungi-

cides in *Pyrenophora teres*, *Phytopathology 84*, 515–519.

27. Porter, L. D., Miller, J. S., Nolte, P., and Price, W. J. (2007) In vitro somatic growth and reproduction of phenylamide-resistant and -sensitive isolates of *Phytophthora erythroseptica* from infected potato tubers in Idaho, *Plant Pathol. 56*, 492–499.

28. Raposo, R., Delcan, J., Gomez, V., and Melgarejo, P. (1996) Distribution and fitness of isolates of *Botrytis cinerea* with multiple fungicide resistance in Spanish greenhouses, *Plant Pathol. 45*, 497–505.

29. Suzuki, F., Yamaguchi, J., Koba, A., Nakajima, T., and Arai, M. (2010) Changes in fungicide resistance frequency and population structure of *Pyricularia oryzae* after discontinuance of MBI-D fungicides, *Plant Dis. 94*, 329–334.

30. Sui, X. X., Wang, M. N., and Chen, X. M. (2009) Molecular mapping of a stripe rust resistance gene in Spring wheat cultivar Zak, *Phytopathology 99*, 1209–1215.

31. Goodwin, S. B., Smart, C. D., Sandrock, R. W., Deahl, K. L., Punja, Z. K., and Fry, W. E. (1998) Genetic change within populations of *Phytophthora infestans* in the United States and Canada during 1994 to 1996: Role of migration and recombination, *Phytopathology 88*, 939–949.

32. Miller, J. S., and Johnson, D. A. (2000) Competitive fitness of *Phytophthora infestans* isolates under semiarid field conditions, *Phytopathology 90*, 220–227.

33. Deacon, J. W. (2006) *Fungal biology*, Fourth ed., Blackwell Publishing, Malden, MA.

34. Vilgalys, R., and Gonzalez, D. (1990) Organization of ribosomal DNA in the basidiomycete *Thanatephorus praticola*, *Current Genetics 18*, 277–280.

35. Herrera, M. L., Vallor, A. C., Gelfond, J. A., Patterson, T. F., and Wickes, B. L. (2009) Strain-dependent variation in 18S ribosomal DNA copy Numbers in *Aspergillus fumigatus*, *J. Clin. Microbiol. 47*, 1325–1332.

36. Ide, S., Miyazaki, T., Maki, H., and Kobayashi, T. (2010) Abundance of ribosomal RNA gene copies maintains genome integrity, *Science 327*, 693–696.

37. Tang, X., Bartlett, M. S., Smith, J. W., Lu, J. J., and Lee, C. H. (1998) Determination of copy number of rRNA genes in *Pneumocystis carinii* f. sp. *hominis*, *J. Clin. Microbiol. 36*, 2491–2494.

38. White, T. J., Bruns, T., Lee, S., and Taylor, J. W. (1990) Amplification and direct sequencing of fungal ribosomal RNA Genes for phylogenetics, in *PCR - Protocols and Applications - A Laboratory Manual*, pp 315–322, Academic Press, New York, NY.

39. Qin, Q. M., Vallad, G. E., Wu, B. M., and Subbarao, K. V. (2006) Phylogenetic analyses of phytopathogenic isolates of *Verticillium* spp, *Phytopathology 96*, 582–592.

40. Carbone, I., Anderson, J. B., and Kohn, L. M. (1999) Patterns of descent in clonal lineages and their multilocus fingerprints are resolved with combined gene genealogies, *Evolution 53*, 11–21.

41. Carbone, I., and Kohn, L. M. (2001) Multilocus nested haplotype networks extended with DNA fingerprints show common origin and fine-scale, ongoing genetic divergence in a wild microbial metapopulation, *Mol. Ecol. 10*, 2409–2422.

42. Malvarez, G., Carbone, I., Grunwald, N. J., Subbarao, K. V., Schafer, M., and Kohn, L. M. (2007) New populations of *Sclerotinia sclerotiorum* from lettuce in California and peas and lentils in Washington, *Phytopathology 97*, 470–483.

43. Zhang, N., and Blackwell, M. (2002) Population structure of dogwood anthracnose fungus, *Phytopathology 92*, 1276–1283.

44. Carbone, I., and Kohn, L. M. (1999) A method for designing primer sets for speciation studies in filamentous Ascomycetes, *Mycologia 91*, 553–556.

45. Glass, N. L., and Donaldson, G. C. (1995) Development of primer sets designed for use with the PCR to amplify conserved genes from filamentous Ascomycetes, *Appl. Environ. Microbiol. 61*, 1323–1330.

46. Stukenbrock, E. H., Banke, S., Javan-Nikkhah, M., and McDonald, B. A. (2007) Origin and domestication of the fungal wheat pathogen *Mycosphaerella graminicola* via sympatric speciation, *Mol. Biol. Evol. 24*, 398–411.

47. Ceresini, P. C., Shew, H. D., James, T. Y., Vilgalys, R. J., and Cubeta, M. A. (2007) Phylogeography of the *Solanaceae*-infecting Basidiomycota fungus *Rhizoctonia solani* AG-3 based on sequence analysis of two nuclear DNA loci, *BMC Evol. Biol. 7*.

48. Zaffarano, P. L., McDonald, B. A., and Linde, C. C. (2009) Phylogeographical analyses reveal global migration patterns of the barley scald pathogen *Rhynchosporium secalis*, *Mol. Ecol. 18*, 279–293.

49. Cote, M. J., Prud'homme, M., Meldrum, A. J., and Tardif, M. C. (2004) Variations in sequence and occurrence of SSU rDNA group I introns in *Monilinia fructicola* isolates, *Mycologia 96*, 240–248.

50. Gobbi, E., Firrao, G., Carpanelli, A., Locci, R., and Van Alfen, N. K. (2003) Mapping and characterization of polymorphism in mtDNA of *Cryphonectria parasitica*: evidence of the presence of an optional intron, *Fungal Genet. Biol. 40*, 215–224.

51. Sharpton, T. J., Neafsey, D. E., Galagan, J. E., and Taylor, J. W. (2008) Mechanisms of intron gain and loss in *Cryptococcus*, *Genome Biol. 9*.

52. Sommerhalder, R. J., McDonald, B. A., and Zhan, J. S. (2006) The frequencies and spatial distribution of mating types in *Stagonospora nodorum* are consistent with recurring sexual reproduction, *Phytopathology 96*, 234–239.

53. Linde, C. C., Zala, M., Ceccarelli, S., and McDonald, B. A. (2003) Further evidence for sexual reproduction in *Rhynchosporium secalis* based on distribution and frequency of mating-type alleles, *Fungal Genet. Biol. 40*, 115–125.

54. Peever, T. L., Salimath, S. S., Su, G., Kaiser, W. J., and Muehlbauer, F. J. (2004) Historical and contemporary multilocus population structure of *Ascochyta rabiei* (teleomorph : *Didymella rabiei*) in the Pacific Northwest of the United States, *Mol. Ecol. 13*, 291–309.

55. Avise, J. C. (2004) *Molecular markers, natural history, and evolution*, Second ed., Sinauer Associates, Inc., Sunderland, MA.

56. Mullis, K., Faloona, F., Scharf, S., Saiki, R., Horn, G., and Erlich, H. (1986) Specific enzymatic amplification of DNA in vitro - the polymerase chain reaction, *Cold Spring Harbor Symp. Quant. Biol. 51*, 263–273.

57. Saiki, R. K., Gelfand, D. H., Stoffel, S., Scharf, S. J., Higuchi, R., Horn, G. T., Mullis, K. B., and Erlich, H. A. (1988) Primer directed enzymatic amplification of DNA with a thermostable DNA polymerase, *Science 239*, 487–491.

58. Saiki, R. K., Scharf, S., Faloona, F., Mullis, K. B., Horn, G. T., Erlich, H. A., and Arnheim, N. (1985) Enzymatic amplification of beta-globin genomic sequences and restriction site analysis for diagnosis of sickle cell anemia, *Science 230*, 1350–1354.

59. Williams, J. G. K., Kubelik, A. R., Livak, K. J., Rafalski, J. A., and Tingey, S. V. (1990) DNA polymorphisms amplified by arbitrary primers are useful as genetic markers, *Nucleic Acids Res. 18*, 6531–6535.

60. Xu, J. (2006) Fundamentals of fungal molecular population genetic analyses, *Curr. Issues Mol. Biol. 8*, 75–89.

61. Vos, P., Hogers, R., Bleeker, M., Reijans, M., Van de Lee, T., Hornes, M., Friters, A., Pot, J., Paleman, J., Kuiper, M., and Zabeau, M. (1995) AFLP: a new technique for DNA fingerprinting, *Nucleic Acids Res. 23*, 4407–4414.

62. Radisek, S., Jakse, J., and Javornik, B. (2004) Development of pathotype-specific SCAR markers for detection of *Verticillium albo-atrum* isolates from hop, *Plant Dis. 88*, 1115–1122.

63. McDonald, B. A. (1997) The population genetics of fungi: Tools and techniques, *Phytopathology 87*, 448–453.

64. Linde, C. C., Zhan, J., and McDonald, B. A. (2002) Population structure of *Mycosphaerella graminicola*: From lesions to continents, *Phytopathology 92*, 946–955.

65. Zhan, J., Kema, G. H. J., and McDonald, B. A. (2004) Evidence for natural selection in the mitochondrial genome of *Mycosphaerella graminicola*, *Phytopathology 94*, 261–267.

66. Zhan, J., Pettway, R. E., and McDonald, B. A. (2003) The global genetic structure of the wheat pathogen *Mycosphaerella graminicola* is characterized by high nuclear diversity, low mitochondrial diversity, regular recombination, and gene flow, *Fungal Genet. Biol. 38*, 286–297.

67. Li, Y. C., Korol, A. B., Fahima, T., Beiles, A., and Nevo, E. (2002) Microsatellites: genomic distribution, putative functions and mutational mechanisms: a review, *Mol. Ecol. 11*, 2453–2465.

68. Hamada, H., Petrino, M. G., Kakunaga, T., Seidman, M., and Stollar, B. D. (1984) Characterization of genomic poly(dT-dG). poly(dC-dA) sequence - Structure, organization, and conformation, *Mol. Cell. Biol. 4*, 2610–2621.

69. Litt, M., and Luty, J. A. (1989) A hypervariable microsatellite revealed by invitro amplification of a dinucleotide repeat within the cardiac muscle actin gene, *Am. J. Hum. Genet. 44*, 397–401.

70. Tautz, D. (1989) Hypervariability of simple sequences as a general source for polymorphic DNA markers, *Nucleic Acids Res. 17*, 6463–6471.

71. Tautz, D., Trick, M., and Dover, G. A. (1986) Cryptic simplicity in DNA is a major source of genetic variation, *Nature 322*, 652–656.

72. Weber, J. L., and May, P. E. (1989) Abundant class of human DNA polymorphisms which can be typed using the polymerase chain reaction, *Am. J. Hum. Genet. 44*, 388–396.

73. Bridge, P. D., Pearce, D. A., Rivera, A., and Rutherford, M. A. (1997) VNTR derived oligonucleotides as PCR primers for population studies in filamentous fungi, *Lett. Appl. Microbiol. 24*, 426–430.

74. Geistlinger, J., Weising, K., Kaiser, W. J., and Kahl, G. (1997) Allelic variation at a hypervariable compound microsatellite locus in the

ascomycete *Ascochyta rabiei*, *Mol. Gen. Genet.* *256*, 298–305.

75. Meyer, W., and Mitchell, T. G. (1995) Polymerase chain reaction fingerprinting in fungi using single primers specific to minisatellites and simple repetitive DNA sequences - Strain variation in *Cryptococcus neoformans*, *Electrophoresis 16*, 1648–1656.

76. Dobrowolski, M. P., Tommerup, I. C., and O'Brien, P. A. (1998) Microsatellites in the mitochondrial genome of *Phytophthora cinnamomi* failed to provide highly polymorphic markers for population genetics, *FEMS Microbiol. Lett. 163*, 243–248.

77. Gobbin, D., Pertot, I., and Gessler, C. (2003) Identification of microsatellite markers for *Plasmopara viticola* and establishment of high throughput method for SSR analysis, *Eur. J. Plant Pathol. 109*, 153–164.

78. Dettman, J. R., and Taylor, J. W. (2004) Mutation and evolution of microsatellite loci in Neurospora, *Genetics 168*, 1231–1248.

79. Goldstein, D. B., Linares, A. R., Cavalli-Sforza, L. L., and Feldman, M. W. (1995) An evaluation of genetic distances for use with microsatellite loci, *Genetics 139*, 463–471.

80. Valdes, A. M., Slatkin, M., and Freimer, N. B. (1993) Allele frequencies at microsatellite moci - the stepwise mutation model revisited, *Genetics 133*, 737–749.

81. Estoup, A., Jarne, P., and Cornuet, J. M. (2002) Homoplasy and mutation model at microsatellite loci and their consequences for population genetics analysis, *Mol. Ecol. 11*, 1591–1604.

82. Sharma, P. C., Grover, A., and Kahl, G. (2007) Mining microsatellites in eukaryotic genomes, *Trends Biotechnol. 25*, 490–498.

83. Mayer, C. (2006) Phobos 3.3.11.

84. Benson, G. (1999) Tandem repeats finder: a program to analyze DNA sequences, *Nucl. Acids Res. 27*, 573–580.

85. Akamatsu, H. O., Grunwald, N. J., Chilvers, M. I., Porter, L. D., and Peever, T. L. (2007) Development of codominant simple sequence repeat, single nucleotide polymorphism and sequence characterized amplified region markers for the pea root rot pathogen, *Aphanomyces euteiches*, *J. Microbiol. Methods 71*, 82–86.

86. Amend, A., Garbelotto, M., Fang, Z. D., and Keeley, S. (2010) Isolation by landscape in populations of a prized edible mushroom *Tricholoma matsutake*, *Conserv. Genet. 11*, 795–802.

87. Amend, A., Keeley, S., and Garbelotto, M. (2009) Forest age correlates with fine-scale spatial structure of Matsutake mycorrhizas, *Mycol. Res. 113*, 541–551.

88. Carter, D. A., Taylor, J. W., Dechairo, B., Burt, S., Koenig, G. L., and White, T. J. (2001) Amplified single-nucleotide polymorphisms and a (GA)(n) microsatellite marker reveal genetic differentiation between populations of *Histoplasma capsulatum* from the Americas, *Fungal Genet. Biol. 34*, 37–48.

89. Morgan, J. A. T., Vredenburg, V. T., Rachowicz, L. J., Knapp, R. A., Stice, M. J., Tunstall, T., Bingham, R. E., Parker, J. M., Longcore, J. E., Moritz, C., Briggs, C. J., and Taylor, J. W. (2007) Population genetics of the frog-killing fungus *Batrachochytrium dendrobatidis*, *Proc. Natl. Acad. Sci. 104*, 13845–13850.

90. Brumfield, R. T., Beerli, P., Nickerson, D. A., and Edwards, S. V. (2003) The utility of single nucleotide polymorphisms in inferences of population history, *Trends Ecol. Evol. 18*, 249–256.

91. Morin, P. A., Luikart, G., and Wayne, R. K. (2004) SNPs in ecology, evolution and conservation, *Trends Ecol. Evol. 19*, 208–216.

92. Ramirez-Soriano, A., and Nielsen, R. (2009) Correcting Estimators of theta and Tajima's D for Ascertainment Biases Caused by the Single-Nucleotide Polymorphism Discovery Process, *Genetics 181*, 701–710.

93. Nguyen-Dumont, T., Le Calvez-Kelm, F., Forey, N., McKay-Chopin, S., Garritano, S., Gioia-Patricola, L., De Silva, D., Weigel, R., Sangrajrang, S., Lesueur, F., and Tavtigian, S. V. (2009) Description and validation of high-throughput simultaneous genotyping and mutation scanning by high-resolution melting curve analysis, *Hum. Mutat. 30*, 884–890.

94. Wright, S. (1978) *Variability within and among natural populations*, Vol. 4, University of Chicago Press, Chicago.

95. Excoffier, L., Smouse, P. E., and Quattro, J. M. (1992) Analysis of molecular variance inferred from metric distances among DNA haplotypes - application to human mitochondrial-DNA restriction data, *Genetics 131*, 479–491.

96. Pearse, D. E., and Crandall, K. A. (2004) Beyond FST: Analysis of population genetic data for conservation, *Conserv. Genet. 5*, 585–602.

97. Kingman, J. F. C. (1982) The coalescent, *Stochastic Process. Appl. 13*, 235–248.

98. Kingman, J. F. C. (2000) Origins of the coalescent: 1974-1982, *Genetics 156*, 1461–1463.

99. Nielsen, R., and Beaumont, M. A. (2009) Statistical inferences in phylogeography, *Mol. Ecol. 18*, 1034–1047.

100. Atallah, Z. K., Maruthachalam, K., du Toit, L., Koike, S. T., Davis, R. M., Klosterman, S. J., Hayes, R. J., and Subbarao, K. V. (2010) Population analyses of the vascular plant pathogen *Verticillium dahliae* detect recombination and transcontinental gene flow, *Fungal Genet. Biol. 47*, 416–422.

101. Goss, E. M., Carbone, I., and Grunwald, N. J. (2009) Ancient isolation and independent evolution of the three clonal lineages of the exotic sudden oak death pathogen *Phytophthora ramorum, Mol. Ecol. 18*, 1161–1174.

102. Brunner, P. C., Schurch, S., and McDonald, B. A. (2007) The origin and colonization history of the barley scald pathogen *Rhynchosporium secalis, J. Evol. Biol. 20*, 1311–1321.

103. Stukenbrock, E. H., Banke, S., and McDonald, B. A. (2006) Global migration patterns in the fungal wheat pathogen *Phaeosphaeria nodorum, Mol. Ecol. 15*, 2895–2904.

104. Gladieux, P., Zhang, X.-G., Afoufa-Bastien, D., Valdebenito Sanhueza, R.-M., Sbaghi, M., and Le Cam, B. (2008) On the origin and spread of the scab disease of apple: Out of central Asia, *PLoS ONE 3*, e1455.

105. Munkacsi, A. B., Stoxen, S., and May, G. (2008) Ustilago maydis populations tracked maize through domestication and cultivation in the Americas, *Proceedings of the Royal Society B: Biological Sciences 275*, 1037–1046.

106. Smouse, P. E. (2000) Reticulation inside the species boundary, *J. Classif. 17*, 165–173.

107. Posada, D., and Crandall, K. A. (2001) Intraspecific gene genealogies: trees grafting into networks, *Trends Ecol. Evol. 16*, 37–45.

108. Huson, D. H., and Bryant, D. (2006) Application of phylogenetic networks in evolutionary studies, *Mol. Biol. Evol. 23*, 254–267.

109. Van de Peer, Y. (2009) Phylogenetic inference based on distance methods, in *The phylogenetic handbook* (Lemey, P., Salemi, M., and Vandamme, A. M., Eds.), pp 142–160, Cambridge University Press, Cambridge, UK.

110. Swofford, D. L. (2002) Phylogenetic analysis using parsimony (*and other methods), version 4.0b4a ed., Sinauer Associates, Sunderland, MA.

111. Felsenstein, J. (1989) PHYLIP - Phylogeny Inference Package (Version 3.2), *Cladistics 5*, 164–166.

112. Rohlf, F. J. (2002) NTSYS-PC: numerical taxonomy and multivariate analysis system Exeter Software, Setauket, NY.

113. Huson, D. H. (1998) SplitsTree: A program for analyzing and visualizing evolutionary data, *Bioinformatics 14*, 68–73.

114. Excoffier, L., Laval, G., and Schneider, S. (2005) Arlequin ver. 3.0: An integrated software package for population genetics data analysis, *Evol. Bioinfo. Online 1*, 47–50.

115. Clement, M., Posada, D., and Crandall, K. A. (2000) TCS: a computer program to estimate gene genealogies, *Mol. Ecol. 9*, 1657–1659.

116. Balloux, F., and Lugon-Moulin, N. (2002) The estimation of population differentiation with microsatellite markers, *Mol. Ecol. 11*, 155–165.

117. Jost, L. (2008) G_{ST} and its relatives do not measure differentiation, *Mol. Ecol. 17*, 4015–4026.

118. Kuhner, M. K. (2009) Coalescent genealogy samplers: windows into population history, *Trends Ecol. Evol. 24*, 86–93.

119. Peakall, R., and Smouse, P. E. (2006) Genalex 6: genetic analysis in Excel. Population genetic software for teaching and research, *Mol. Ecol. Notes 6*, 288–295.

120. Engelhardt, B. E., and Stephens, M. (2010) Analysis of population structure: A unifying framework and novel methods based on sparse factor analysis, *PLoS Genet. 6*, e1001117.

121. Reich, D., Price, A. L., and Patterson, N. (2008) Principal component analysis of genetic data, *Nature Genet. 40*, 491–492.

122. Falush, D., Stephens, M., and Pritchard, J. K. (2003) Inference of population structure using multilocus genotype data: Linked loci and correlated allele frequencies, *Genetics 164*, 1567–1587.

123. Pritchard, J. K., Stephens, M., and Donnelly, P. (2000) Inference of population structure using multilocus genotype data, *Genetics 155*, 945–959.

124. Corander, J., Waldmann, P., and Sillanpaa, M. J. (2003) Bayesian analysis of genetic differentiation between populations, *Genetics 163*, 367–374.

125. Huelsenbeck, J. P., and Andolfatto, P. (2007) Inference of population structure under a Dirichlet process model, *Genetics 175*, 1787–1802.

126. Dyer, R. J., and Nason, J. D. (2004) Population Graphs: the graph theoretic shape of genetic structure, *Mol. Ecol. 13*, 1713–1727.

127. Goodwin, S. B., and Drenth, A. (1997) Origin of the A2 mating type of *Phytophthora infestans* outside Mexico, *Phytopathology 87*, 992–999.

128. Linde, C. C., Zala, M., and McDonald, B. A. (2009) Molecular evidence for recent founder populations and human-mediated migration in the barley scald pathogen *Rhynchosporium secalis, Mol. Phylogenet. Evol. 51*, 454–464.

129. Whitlock, M. C., and McCauley, D. E. (1999) Indirect measures of gene flow and migration: F_{ST} not equal $1/(4Nm+1)$, *Heredity 82*, 117–125.

130. Beerli, P., and Felsenstein, J. (1999) Maximum likelihood estimation of migration rates and effective population numbers in two populations using a coalescent approach, *Genetics 152*, 763–773.

131. Beerli, P., and Felsenstein, J. (2001) Maximum likelihood estimation of a migration matrix and effective population sizes in n subpopulations by using a coalescent approach, *Proc. Natl. Acad. Sci. 98*, 4563–4568.

132. Hey, J., and Nielsen, R. (2007) Integration within the Felsenstein equation for improved Markov chain Monte Carlo methods in population genetics, *Proc. Natl. Acad. Sci. 104*, 2785–2790.

133. Kuhner, M. K. (2006) LAMARC 2.0: maximum likelihood and Bayesian estimation of population parameters, *Bioinformatics 22*, 768–770.

134. Wilson, G. A., and Rannala, B. (2003) Bayesian inference of recent migration rates using multilocus genotypes, *Genetics 163*, 1177–1191.

135. Beerli, P., and Palczewski, M. (2010) Unified framework to evaluate panmixia and migration direction among multiple sampling locations, *Genetics 185*, 313–326.

136. Nielsen, R., and Wakeley, J. (2001) Distinguishing migration from isolation: A Markov chain Monte Carlo approach, *Genetics 158*, 885–896.

137. Hey, J., and Nielsen, R. (2004) Multilocus methods for estimating population sizes, migration rates and divergence time, with applications to the divergence of Drosophila pseudoobscura and D-persimilis, *Genetics 167*, 747–760.

138. Hey, J. (2006) Recent advances in assessing gene flow between diverging populations and species, *Curr. Opin. Genet. Dev. 16*, 592-596.

139. Hey, J. (2005) On the number of New World founders: A population genetic portrait of the peopling of the Americas, *PLoS. Biol. 3*, 965–975.

140. Felsenstein, J. (2006) Accuracy of coalescent likelihood estimates: Do we need more sites, more Sequences, or more loci?, *Mol. Biol. Evol. 23*, 691–700.

141. Agapow, P. M., and Burt, A. (2001) Indices of multilocus linkage disequilibrium, *Mol. Ecol. Notes 1*, 101–102.

142. Maynard Smith, J., Smith, N. H., Orourke, M., and Spratt, B. G. (1993) How clonal are bacteria?, *Proc. Natl. Acad. Sci. 90*, 4384–4388.

143. Burt, A., Carter, D. A., Koenig, G. L., White, T. J., and Taylor, J. W. (1996) Molecular markers reveal cryptic sex in the human pathogen *Coccidioides immitis*, *Proc. Natl. Acad. Sci. 93*, 770–773.

144. Lemey, P., and Posada, D. (2009) Introduction to recombination detection, in *The phylogenetic handbook* (Lemey, P., Salemi, M., and Vandamme, A. M., Eds.), pp 493-518, Cambridge University Press, Cambridge, UK.

145. Bruen, T. C., Philippe, H., and Bryant, D. (2006) A simple and robust statistical rest for detecting the presence of recombination, *Genetics 172*, 2665–2681.

146. Piry, S., Alapetite, A., Cornuet, J.-M., Paetkau, D., Baudouin, L., and Estoup, A. (2004) GeneClass2: a software for genetic ssignment and first-generation migrant detection, *J. Hered. 95*, 536–539.

147. Cornuet, J. M., Piry, S., Luikart, G., Estoup, A., and Solignac, M. (1999) New methods employing multilocus genotypes to select or exclude populations as origins of individuals, *Genetics 153*, 1989–2000.

148. Rannala, B., and Mountain, J. L. (1997) Detecting immigration by using multilocus genotypes, *Proc. Natl. Acad. Sci. 94*, 9197–9201.

149. Paetkau, D., Slade, R., Burden, M., and Estoup, A. (2004) Genetic assignment methods for the direct, real-time estimation of migration rate: a simulation-based exploration of accuracy and power, *Mol. Ecol. 13*, 55–65.

150. Bergl, R. A., and Vigilant, L. (2007) Genetic analysis reveals population structure and recent migration within the highly fragmented range of the Cross River gorilla (*Gorilla gorilla* diehli), *Mol. Ecol. 16*, 501–516.

151. Proctor, M. F., McLellan, B. N., Strobeck, C., and Barclay, R. M. R. (2005) Genetic analysis reveals demographic fragmentation of grizzly bears yielding vulnerably small populations, *Proceedings of the Royal Society B: Biological Sciences 272*, 2409–2416.

152. Rozen, S., and Skaletsky, H. (2000) Primer3 on the WWW for general users and for biologist programmers, in *Bioinformatics methods and protocols: Methods in molecular biology* (Krawetz, S., and Misener, S., Eds.), pp 365-386, Humana Press, Totowa, N.J.

153. Hall, T. A. (1999) BioEdit: a user-friendly biological sequence alignment editor and analysis program for Windows 95/98/NT, *Nucl. Acid Symp. Ser. 41*, 95–98.

154. Thompson, J. D., Gibson, T. J., Plewniak, F., Jeanmougin, F., and Higgins, D. G. (1997) The CLUSTAL_X windows interface: flexible strategies for multiple sequence alignment

aided by quality analysis tools, *Nucl. Acids Res. 25*, 4876–4882.

155. Thompson, J. D., Higgins, D. G., and Gibson, T. J. (1994) Clustal-W - Improving the sensitivity of progressive multiple sequence alignment through sequence weighting, position-specific gap penalties and weight matrix choice, *Nucl. Acids Res. 22*, 4673–4680.

156. Maddison, D. R., and Maddison, W. P. (2003) MacClade, 4.06 ed., Sinauer Associates Inc., Sunderland, MA.

157. Price, E. W., and Carbone, I. (2005) SNAP: workbench management tool for evolutionary population genetic analysis, *Bioinformatics 21*, 402–404.

158. Aylor, D. L., Price, E. W., and Carbone, I. (2006) SNAP: Combine and Map modules for multilocus population genetic analysis, *Bioinformatics 22*, 1399–1401.

159. Posada, D., and Crandall, K. A. (1998) MODELTEST: testing the model of DNA substitution, *Bioinformatics 14*, 817–818.

160. Huelsenbeck, J. P., and Ronquist, F. (2001) MrBayes: Bayesian inference of phylogenetic trees, *Bioinformatics 17*, 754–755.

161. Rambaut, A. (2009) FigTree, v. 1.2.2 ed., http://tree.bio.ed.ac.uk/software/figtree/.

162. Bahlo, M., and Griffiths, R. C. (2000) Inference from gene trees in a subdivided population, *Theor. Popul. Biol. 57*, 79–95.

163. Griffiths, R. C., and Tavare, S. (1994) Ancestral inference in population genetics, *Stat. Sci. 9*, 307–319.

164. Ehrich, D. (2006) AFLPDAT: a collection of R functions for convenient handling of AFLP data, *Mol. Ecol. Notes 6*, 603–604.

165. Hornik, K., and Leisch, F. (2005) R version 2.1.0, *Comput. Stat. 20*, 197–202.

166. Ihaka, R., and Gentleman, R. (1996) R: a language for data analysis and graphics, *Journal of Computational and Graphical Statistics 5*, 299–314.

167. Evanno, G., Regnaut, S., and Goudet, J. (2005) Detecting the number of clusters of individuals using the software STRUCTURE: a simulation study, *Mol. Ecol. 14*, 2611–2620.

168. Rosenberg, N. A. (2004) DISTRUCT: a program for the graphical display of population structure, *Mol. Ecol. Notes 4*, 137–138.

Chapter 21

Polyethylene Glycol (PEG)-Mediated Transformation in Filamentous Fungal Pathogens

Zhaohui Liu and Timothy L. Friesen

Abstract

Genetic transformation is an essential tool for the modern study of gene function and the genetic improvement of an organism. The genetic transformation of many fungal species is well established and can be carried out by utilizing different transformation methods including electroporation, Agrobacterium, biolistics, or polyethylene glycol (PEG)-mediated transformation. Due to its technical simplicity and common equipment requirements, PEG-mediated transformation is still the most commonly used method for genetic transformation in filamentous fungi. Here, we describe a PEG-based protocol developed for genetic transformation of *Stagonospora nodorum*, a fungal pathogen of wheat. This protocol is directly applicable to other fungi especially those in the Dothideomycete class of fungi.

Key words: Agrobacterium transformation, Electroporation transformation, Filamentous fungus, Fungal pathogen, Fungal transformation, *Phaeosphaeria nodorum*, Polyethylene glycol, *Stagonospora nodorum*

1. Introduction

Genetic transformation is a biological process involving the delivery and integration of exogenous DNA into the genome of an organism. Transformation is fundamental to many basic studies in the biology of eukaryotic organisms and was first successfully demonstrated in model species including *Saccharomyces cerevisiae* (1, 2), *Neurospora crassa* (3), and *Asepergillus nidulans* (4, 5). These early studies established a basic protocol for transforming fungal protoplasts by the addition of a high concentration of polyethylene glycol (PEG) in the presence of calcium ions following the initial step of exposure to DNA. Later, this basic protocol was successfully adapted to a wide range of fungal and fungal-like species including plant fungal

Melvin D. Bolton and Bart P.H.J. Thomma (eds.), *Plant Fungal Pathogens: Methods and Protocols*,
Methods in Molecular Biology, vol. 835, DOI 10.1007/978-1-61779-501-5_21, © Springer Science+Business Media, LLC 2012

pathogens (6). Additional methods including electroporation (7), biolistics (8), and Agrobacterium transformation (9, 10) were also developed as alternatives to overcome difficulties sometimes associated with PEG-based methods. Potential disadvantages of PEG-mediated transformation include difficulty in obtaining high concentrations of viable protoplasts, low transformation efficiency, high percentages of transient transformants, and frequent multiple loci integration. However, due to its simplicity in technical operation and equipment required, the PEG-mediated method remains the most commonly used method to conduct transformation in filamentous fungi. According to Ruiz-Díez (11), genetic transformation has been successful in all major groups of fungi, including different zygomycetes, ascomycetes, basidiomycetes, and phycomycetes as well as some *fungi imperfecti*.

The basic procedure for PEG-mediated transformation contains three major steps: protoplast preparation, transforming DNA uptake, and regeneration on selective media. Transformable protoplasts can be prepared through digestion of the cell walls of mycelium or germinating spores using cell wall degrading enzymes (CWDEs). The CWDEs commonly used are helicase or glusulase from snail stomach preparations, and cellulase, 1, 3-glucanase, chitinase, and driselase from *Trichoderma* sp. or other fungal species. It has been reported that different batches of lytic enzymes may vary in the effectiveness of cell wall degradation, therefore testing of each batch and testing the combination of different CWDEs for protoplasting efficiency is highly recommended (6, 11). The digestions are usually performed in an osmotic buffer containing sorbitol or high salt in order to stabilize the resulting protoplasts (6).

The uptake of DNA by protoplasts is carried out by incubating protoplasts with highly concentrated DNA followed by the addition of up to 10 volumes of a 40–60% PEG 4000 solution and then another period of incubation (6, 11). PEG has been observed to cause clumping and fusion of protoplasts, which is thought to facilitate the trapping of DNA to the fungal cells (6). However, a recent study (12) suggested PEG is unlikely to induce the interaction between DNA and the cell surface, and the fusion of protoplasts is not the direct cause of DNA uptake; therefore, the role of PEG in DNA uptake remains largely unknown. Following the PEG treatment, protoplasts are washed with an osmotic buffer containing sorbitol and transferred to regeneration media for recovery of the cell wall before being plated onto the selective medium. The selection of transformants can be achieved by using an auxotrophic complement or drug resistance depending on the selectable marker gene used in the vector DNA construct.

The filamentous fungus *Stagonospora nodorum* is the causal agent of Stagonospora nodorum blotch (SNB) which affects both leaves and glumes of wheat. The disease occurs wherever wheat is grown and is economically important in many major wheat production areas (13). Transformation of *S. nodorum* was first reported by Cooley et al. (14) using a plasmid containing the hygromycin phosphotransferase gene (*hph*). The authors established a PEG-mediated protocol for *S. nodorum* by modifying protocols developed for model filamentous fungal species. This transformation protocol has since been successfully used to transform *S. nodorum* for the study of fungal genes involved in signal transduction, metabolism, and pathogenicity (13). Using a modification of this protocol, we also have successfully transformed *S. nodorum* in our lab in order to investigate fungal virulence genes encoding necrotrophic effectors (host selective toxins) (15, 16). Furthermore, the protocol was successfully applied to *Cocholiobus sativus*, the causal agent of wheat and barley spot blotch (17), the barley net blotch pathogen, *Pyrenophora teres*, and sugarbeet leaf spot pathogen, *Cercospora beticola* (unpublished data).

2. Materials

All materials are sterilized by autoclaving at 121°C for 20 min or for liquid reagents by filtering through a 0.45-µm filter before use. Ultrapure water should be used for the preparation of all solutions.

2.1. Protoplasting Components

Buffers or solutions should be made fresh or one day before the transformation.

1. Fries medium: add about 800 mL of water to a 2-L glass beaker. The following chemicals are weighed and dissolved in water: 5 g ammonium tartrate, 1.0 g ammonium nitrate, 0.5 g magnesium sulfate, 1.3 g KH_2PO_4, 2.6 g K_2HPO_4, 30.0 g sucrose, 1.0 g yeast extract, and 2-mL trace element stock solutions (1-L water containing LiCl 167 mg, $CuCl_2 \cdot 2H_2O$ 107 mg, H_2MoO_4 34 mg, $MnCl_2 \cdot 4H_2O$ 72 mg, and $CoCl_2 \cdot 4H_2O$ 80 mg). Increase volume to 1 L with water. No adjustment is required for pH. The medium is poured into two different size flasks: 60 in 250-mL Erlenmeyer flasks and 200 in 600-mL Erlenmeyer flasks (see Note 1). Autoclave and store at room temperature.

2. Mycelial wash solution (MWS): 0.7 M KCl and 10 mM $CaCl_2$. Add about 800 mL water to a 2-L glass beaker. Add 52.1 g KCl and 1.47 g $CaCl_2 \cdot 2H_2O$ and dissolve in water. Increase volume to 1 L with water. Split into 500 mL in two 1 L bottles with screw tops. Autoclave and store at room temperature.

3. MWS-based CWDE solution: The solution should be made fresh immediately before use. For each plate digestion, a 40-mL digestion solution is needed. Measure 40 mL of osmotic buffer (MWS, see above) and pour into a 200-mL beaker. Weigh 400 mg β-1,3 glucanase and 200 mg driselase supplied by

Interspex (see Note 2). Stir for 5–10 min and pour into 50-mL centrifuge tubes. The tubes are centrifuged at maximum speed ($3,700 \times g$) at room temperature for 10 min. The supernatant is sterilized by filtering through a 0.45-μm CA membrane.

4. Sterilized water: autoclave at least 1 L ultrapure water.

5. Beakers covered with Miracloth: Use two layers of Miracloth over a 500-mL beaker for harvesting and washing mycelial tissue and four layers over 200-mL beakers for harvesting protoplasts. All assembled beakers are wrapped with aluminum foil and autoclaved for 20 min.

2.2. Transformation Components

1. Tris–HCl stock solution: 1 M Tris–HCl pH7.5. Add 800 mL water into a 2-L beaker. Add 157.62 g Tris–HCl and adjust pH to 7.5 with NaOH. Increase volume to 1 L with water. Store stock at room temperature.

2. $CaCl_2$ stock solution: 1 M $CaCl_2$ stock solution. Add 80 mL water into a 200-mL beaker. Add 14.70 g $CaCl_2 \cdot 2H_2O$ and dissolve into the water. Fill to 100 mL with water. Store at 4°C.

3. PEG solution: 50% PEG, 10 mM $CaCl_2$, and 10 mM Tris–HCl pH 7.5. Add 50 mL of water to a 200-mL beaker. Weigh 50 g of PEG 3350 (see Note 3) and transfer to the beaker. Turn on low heat to speed dissolving. Add 1 mL of a 1-M $CaCl_2$ stock solution (see above) and 1 mL 1 M Tris–HCl pH7.5 solution (see above). Increase volume to 100 mL with water. After the chemical is completely dissolved, the solution is sterilized by vacuum filtration through a 0.45-μm CA membrane. The sterilized PEG can be stored in a 50-mL sterile centrifuge tube at room temperature.

4. STC solution: 1.2 M Sorbitol, 10 mM pH 7.5 Tris–HCl, 10 mM $CaCl_2$. Add 350 mL of water to a 1-L beaker. Weigh 109.30 g d-Sorbitol and transfer to the beaker. Add 5 mL of a 1-M $CaCl_2$ stock solution (see above) and 5 mL of a 1-M Tris–HCl pH 7.5 solution (see above). Increase volume to 500 mL with water. After the chemical is completely dissolved, the solution is sterilized by vacuum filtration through a 0.45-μm CA membrane. The sterilized STC can be stored in a 50-mL sterile centrifuge tube at 4°C.

5. Plasmid DNA: Plasmid DNA can be prepared through the regular alkaline lysis method (see Note 4) as described by Sambrook and Russell (18) followed by the purification of the plasmid DNA using precipitation with PEG 8000 (18). The purified plasmid DNA should be linearized with the appropriate restriction enzyme and then reprecipitated using sodium acetate/100% ethanol. The DNA pellet is resuspended in a solution of STC/PEG with the volume ratio at 4:1 to obtain the concentration of ~1 μg/μL for transformation.

2.3. Regeneration and Selection

1. Regeneration medium: 1.0 M sucrose, 0.1% yeast extract, and 0.1% tryptone. Add 100 mL water to a 200-mL glass beaker. Add the following chemicals with the indicated amount 68.40 g sucrose, 0.2 g yeast extract, 0.2 g tryptone. Mix and increase the volume to 200 mL with water. The solution is autoclaved and stored at room temperature.

2. Regeneration medium agar: 1 M sucrose, 0.1% yeast extract, 0.1% tryptone, and 1.5% agar (w/v). Add 100 mL of water to a 200-mL glass beaker. Add the following chemicals with the indicated amount: 68.40 g sucrose, 0.2 g yeast extract, 0.2 g tryptone, and 3 g agar. Mix and increase the volume to 200 mL with water. The solution is autoclaved and stored at 65°C in a water bath.

3. Hygromycin B: Hygromycin B in PBS 50 mg/mL sterile (see Note 5).

4. Secondary selection medium: V8-PDA plates plus 100 μg/mL hygromycin B. Add 200 mL water to a 500-mL graduated cylinder. Add 75 mL commercial V8 juice (Campbell's, Camden NJ), 5 g Difco PDA, 1.5 g $CaCO_3$, and 5 g agar to the cylinder. Add water up to 500 mL, pour into glass bottles, and autoclave for 20 min at 121°C. After the media has cooled to 55°C, 500 μL of hygromycin B is added and mixed. The media is then poured into Petri dishes and stored at 4°C.

3. Methods

All operations should be done using sterilized labware and should be performed in a laminar flow hood unless otherwise stated.

3.1. Transformable Protoplast Preparations

1. Take a well-maintained *S. nodorum* culture on a V8-PDA plate with visible spores (see Note 6). Flood the culture surface by adding ~5 mL sterilized water and scrape the culture surface to create a conidial suspension. Inoculate 200 μL of the spore suspension into 60 mL of Fries medium. The culture is grown in an environmental incubator shaker (Innova 44, New Brunswick Scientific, Edison, NJ) for 2 days at 27°C and 100 rpm.

2. Pour the whole culture into a sterilized blender cup (see Note 7). Grind the mycelial tissue for 3 times of 10 s at high speed and 10 s at low speed to produce a fine suspension. Inoculate 25 mL of the ground suspension into 250 mL of Fries medium in a 500-mL Erlenmeyer flask (see Note 8). Grow for 2 more days under the same conditions mentioned above.

3. Filter fungal mycelium through two layers of Miracloth over a beaker (see Note 7). Wash 3 times with sterilized water and 3

more times with MWS (see Note 9). Transfer ~2 g (see Note 8) of the washed mycelial tissue and resuspended in a 100×20-mm petri dish with 40 mL of MWS-based CWDE solution (see Note 9).

4. The mixture is then incubated at 30°C with shaking at 70 rpm for ~3 h (see Note 10).

5. Gently pour the digestion solution through four layers of Miracloth over a beaker and transfer the filtered solution into a 50-mL conical centrifuge tube. Centrifuge the protoplast solution for 5 min at $2,000 \times g$ at room temperature. Remove the supernatant without disturbing cells.

6. Resuspend the obtained protoplasts with MWS and combine all protoplasts into one tube.

7. Centrifuge the protoplast solution again for 5 min at $2,000 \times g$ at room temperature.

8. Resuspend the concentrated protoplasts with 300–500 μL STC/PEG (4:1) solution. Dilute 10 μL of protoplasts to 1 mL in MWS. Count the number and calculate the concentration of the protoplasts under a microscope. Commonly, the procedure yields 1 to 5×10^8 protoplasts/mL. The healthy protoplasts are perfectly round in shape and less than 10 μm in diameter (Fig. 1). The protoplasts are now ready for transformation.

3.2. PEG-Based Transformation

1. Label and place the required number (see Note 11) of 15-mL centrifuge tubes on ice and carefully pipette 100 μL of the above fresh protoplasts into each of the tubes.

2. Pipette 20 μg in 20 μL of linearized plasmid DNA into each protoplast solution and mix by gently stirring the solution (see Note 12). Use 20 μL of the STC/PEG (4:1) solution for control tubes.

3. Incubate the tubes on ice for 20 min (see Note 13).

4. Place the tubes on a rack at room temperature, add 100 μL of PEG solution to each tube, and mix well by tapping the bottoms of the tubes.

5. Add 300 μL PEG solution to each tube and mix well by tapping the bottoms of the tubes.

6. Add 600 μL of PEG solution to each tube and mix well by tapping the bottoms of the tubes (see Note 14).

7. Incubate all the tubes at room temperature for 20 min.

8. Add 1 mL STC to each tube and mix well by tapping the bottoms of the tubes.

9. Add 3 mL STC to each tube and mix well by tapping the bottoms of the tubes.

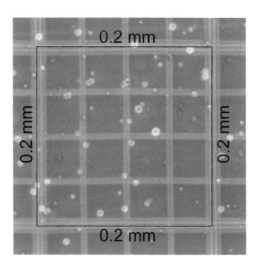

Fig. 1. Cell counts of protoplasts using a hemacytometer under the microscope. In a $0.2 \times 0.2 \times 0.1\text{-mm}^3$ cell on a hemacytometer, there are about 50 round, white protoplasts differing in size but none larger than 10 μm in diameter.

10. Add 4 mL STC to each tube and mix well by gently inverting the tubes several times.

11. Centrifuge the tubes at room temperature for 10 min at $3,000 \times g$ to pellet the treated protoplasts.

12. Carefully pour off the supernatant and resuspend the protoplast pellet in 0.8 mL regeneration media by gently pipetting with 1000 μL tips. Add another 0.8 mL of regeneration media to each tube.

13. Incubate the tubes at 30°C with shaking at 70 rpm for 1–4 h (see Note 15).

14. Mix the protoplast suspension (~1.6 mL) in a 50-mL conical tube with 20 mL of regeneration media agar incubated at 55°C and 40 μL of 50 mg/mL hygromycin B solution. Quickly pour the mixture into a clean 100×15 mm Petri dish. No hygromycin B solution is added for the negative control.

15. Incubate all plates at 30°C without shaking. The regenerated colonies will be visible in 4–6 days (see Note 16).

3.3. Colony Verification and Single-Sporing

1. Incubate the plates at 30°C until the regenerated colonies reach at least 2 mm in diameter.

2. Cut a small piece of mycelium from the edge of the colony using a flame-sterilized scalpel and transfer it to a V8-PDA plate containing 100 μg/mL hygromycin B.

3. Incubate the plates at room temperature under a 24-h photoperiod. True transformants will continue to grow, form new colonies, and sporulate.

4. Isolate single spores for all transformants by picking a single pycnidia with a fine needle tip and transferring the spores to a regular V8-PDA plate (see Note 17).

5. Incubate the plates at room temperature under a 24-h photoperiod. The new colonies are pure lines containing the transformation construct.

4. Notes

1. A small volume of Fries medium (60 mL in a 200-mL flask) is used for the first step of fungal culturing, while a larger volume (200 mL in a 500-mL flask) is used to grow the fungus in large amounts for protoplasting.

2. CWDEs from Interspex were discontinued in 2002. However, Sigma lysing enzymes induce similar digestion.

3. We have experienced variation in batches of PEG that significantly affects transformation efficiency in *S. nodorum*.

4. A large preparation of plasmid DNA is time-consuming and does not consistently yield the high quality DNA necessary for transformation. We often prepare plasmid DNA by combining 10 mini preps using a QIAGEN plasmid min prep kit (Cat# 27106), followed by an ethanol precipitation. The obtained DNA is resuspended in a small amount of sterilized distilled water and linearized by adding the buffer and enzyme only. The digestion solution is added to the STC/PEG (4:1) solution to adjust the concentration to 1 μg/μL and is then used directly for transformation. A single preparation of this procedure can obtain enough high quality DNA for three individual transformations.

5. Hygromycin B is effective for as long as 5 years if stored properly.

6. It usually takes 5–7 days for *S. nodorum* to produce mature pycnidia and conidia under a 24-h photoperiod. A well-maintained culture on V8-PDA is usually full of pycnidia with very limited growth of fungal hyphae.

7. The Fries medium does not contain antibiotics, and therefore it is easily contaminated by bacteria. During the fungal growth, it is recommended to check the culture for contamination before proceeding to the next step.

8. For *S. nodorum*, a 250-mL culture is able to provide enough mycelial tissue for digestion. From a 2-day-old culture, about 2 g can be obtained for setting up one digestion plate. However, for *Pyrenophora teres*, at least four 250-mL cultures are used in setting up the four digestion plates that are needed.

9. We found the following osmotic buffer works better for *Pyrenophora teres*. It contains 1 M NaCl, 10 mM $MgCl_2$,

10 mM potassium phosphate buffer at a pH of 5.8. This buffer is used for making the MWS and the digestion solution.

10. The 3 h digestion is typically enough for *S. nodorum* to produce $>1 \times 10^8$ protoplasts/mL from one plate of digestion solution. For *Pyrenophora teres*, it requires at least 6 h of digestion time, or even overnight digestion at 4°C. The protoplasts usually become swollen after 6 h of digestion, but the buffer mentioned in Note 9 is able to keep the protoplasts viable for a longer period if required.

11. In addition to the transformations for each DNA treatment, two controls are needed including (1) a plate with protoplasts without DNA or antibiotics for testing the viability of the protoplasts and (2) a plate with protoplasts without DNA but with the antibiotic (hygromycin B) which is used to test the effectiveness of the antibiotic. On the former control plate, the protoplasts should germinate and grow, while on the latter control plate, no growth should be visible (see Fig. 2).

12. Protoplasts are very fragile and therefore fast up and down pipetting or vigorous mixing will damage the protoplasts. The mixing can be performed by tapping the bottom of the tube or gentle pipetting up and down. Additionally, the end of the pipette tip can be cut off to reduce the stress on the protoplasts while pipetting.

13. During the incubation, it is recommended that the tube be tapped several times to allow mixing.

14. After adding the 600 μL of PEG solution, the solution sometimes becomes cloudy in the tubes containing DNA and protoplast. This is a result of conjugation of protoplasts. We suspect this phenomenon may be associated with the age of the protoplast since transformation treatment with younger protoplasts is less likely to form this cloudy precipitation. However, based on our observations, this does not typically affect the transformation efficiency.

15. The treated protoplasts need at least 1 h in the regeneration media for sufficient recovery of the cell wall; however, it should not be too long before plating the protoplasts on the selective media, as *S. nodorum* protoplasts will start to germinate around 4 h. For other fungi, it may be less than 4 h.

16. On control plates without antibiotics, protoplasts will germinate and grow quickly, covering the whole plate within 2 days (see Fig. 2). The slow germination and growth of the protoplast can also be visible in 2 days under the microscope if they successfully received and integrated exogenous DNA, but 5–6 days is required for transformed protoplasts to form a 2-mm colony on the plate (see Fig. 2).

17. Efficient single sporing procedures are unique for each fungal species; please refer to the published literature for other fungi.

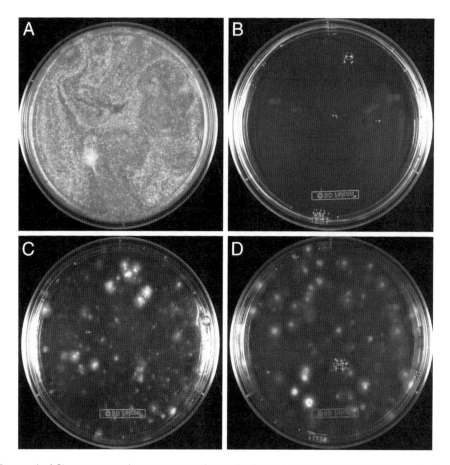

Fig. 2. The growth of *Stagonospora nodorum* on regeneration media plates with or without antibiotics 5 days after transformation. (**a**) Positive control: DNA untreated protoplasts germinate and grow normally on a media plate free of hygromycin B indicating protoplasts are viable after the transformation process; (**b**) negative control: no fungal growth on plates containing 100 μg/mL hygromycin B indicating this concentration of antibiotics is effective in inhibiting the germination and growth of untransformed protoplasts. (**c, d**) Treated protoplasts were grown on plates containing 100 μg/mL hygromycin B. Greater than 100 fungal colonies formed from protoplasts successfully transformed with the hygromycin resistance gene (15).

References

1. Beggs, J.D. (1978) Transformation of yeast by a replicating hybrid plasmid. *Nature* 275, 104–109.

2. Hinnen, A., Hicks, J.B., and Fink, G.R. (1978) Transformation of yeast. *Proc Natl Acad Sci USA* 75, 1929–1933.

3. Case, M.E., Schweizer, M., Kushner, S.R., and Giles, N.H. (1979) Efficient transformation of *Neurospora crassa* by utilizing hybrid plasmid DNA. *Proc Natl Acad Sci USA* 76, 5259–5263.

4. Balance, D.J., Buxton, F.P., and Turner, G. (1983) Transformation of *Aspergillus nidulans* by the orotidine-5′-phosphate decarboxylase gene of *Neurospora crassa*. *Biochem Biophys Res Commun* 112, 284–289.

5. Yelton, M.M., Hamer, J.E., and Timberlake, W.E. (1984) Transformation of *Aspergillus nidulans* by using a trpC plasmid. *Proc Natl Acad Sci USA* 81, 1470–1474.

6. Fincham, J.R.S. (1989) Transformation in fungi. *Microbio Rev* 53, 148–170.

7. Chakraborty, B.N., and Kapoor, M. (1990) Transformation of filamentous fungi by electroporation. *Nucleic Acids Res* 18, 6737.

8. Amarleo, D., Ye, G.-N., Klein, T.M., Shark, K.B., Sanford, J.C., and Johnston, S.A. (1990) Biolistic nuclear transformation of *Sacchormyces cerevisiae* and other fungi. *Curr Genet* 17, 97–103.

9. Bundock, P., den Dulk-Ras, A., Beijersbergen, A., and Hooykaas, P.J. (1995) Trans-kingdom T-DNA transfer from *Agrobacterium tumefaciens* to *Saccharomyces cerevisiae*. *Embo J* 14, 3206–3214.

10. de Groot, M.J., Bundock, P., Hooykaas, P.J., and Beijersbergen, A.G. (1998) *Agrobacterium tumefaciens*-mediated transformation of filamentous fungi. *Nat Biotechnol* 16, 839–842.

11. Ruiz-Díez, B. (2002) Strategies for the transformation of filamentous fungi. *J Appl Microbiol* 92, 189–195.

12. Kuwano, T., Shirataki C., and Itoh, Y. Comparison between polyethylene glycol- and polyethylenimine-mediated transformation of *Aspergillus nidulans*. *Curr Genet* 54, 95–103.

13. Solomon, P.S., Lowe, R.G.T., Tan, K.-C., Waters, O.D.C, and Oliver, R.P. (2006) *Stagonospora nodorum*: cause of stagonospora nodorum blotch of wheat. *Mol Plant Pathol 7*, 147–156.

14. Cooley, R.N., Shaw, R. K., Franklin F.C.H., and Caten, C.E. (1988) Transformation of the phytopathogenic fungus *Septoria nodorum* to hygromycin B resistance, *Curr Genet* 13, 383–389.

15. Friesen, T.L., Stukenbrock, E.H., Liu, Z.H., Meinhardt, S., Ling, W., Faris, J.D., Rasmussen, J.B., Solomon, P.S., McDonald, B.A., and Oliver, R.P. (2006) Emergence of a new disease as a result of interspecific virulence gene transfer. *Nat Genet* 38, 953–956.

16. Liu, Z.H., Faris, J.D., Oliver, R.P., Tan, K.-C., Solomon, P.S., McDonald, M.C., McDonald, B.A., Nunez, A., Lu, S., Rasmussen, J.B., and Friesen, T.L. (2009). SnTox3 acts in effector triggered susceptibility to induce disease on wheat carrying the *Snn3* gene. *PLoS Pathog*, 5(9), e1000581.

17. Leng, Y., Wu, C., Liu, Z.H., Friesen T.L., Rasmussen, J.B., and Zhong, S. (2011). RNA-mediated gene silencing in the cereal fungal pathogen *Cochliobolus sativus. Mol Plant Pathol*, 12, 289–298.

18. Sambrook, J., and Russell, D.W. (2001) Molecular cloning: a laboratory manual. Cold Spring Harbor Laboratory Press.

Chapter 22

In Vitro Induction of Infection-Related Hyphal Structures in Plant Pathogenic Fungi

W.R. Rittenour and S.D. Harris

Abstract

In recent years, a voluminous amount of genomic data has been generated for several plant pathogenic fungi. Multiple studies have utilized these genomic data to advance our knowledge about the molecular mechanisms of plant pathogenesis. However, not all plant pathogenic fungi share the same infection strategies, and several genes have been identified that are crucial for plant pathogenesis in one fungus, but dispensable in others. In order for data on biological relevance to keep pace with accumulating genomic data, new biological assays need to be developed for several pathogenic fungi. Accordingly, we have developed an in vitro assay that allows us to monitor morphological changes in hyphal development as the head blight pathogen *Fusarium graminearum* infects wheat. Using previously frozen detached wheat glumes, we are able to monitor both subcuticular and intercellular hyphal development of *F. graminearum*. The method described takes only 3–5 days from inoculation to microscopic observation (depending on time point) and does not require any elaborate laboratory equipment or supplies. This method could be adapted for different necrotrophic or hemi-biotrophic pathogens, on their host tissue types, in order to characterize their hyphal differentiation in vitro.

Key Words: Fusarium head blight, *Gibberella zeae*, Glume, Infection hyphae, Subcuticular hyphae

1. Introduction

Plant pathogenic fungi differentiate morphologically distinct hyphal structures at various times during infection. For example, the spores of the rice blast fungus, *Magnaporthe oryzae*, germinate and eventually differentiate a dome-shaped appressorium that mediates penetration of the rice leaf epidermis (1). The ability to induce *M. oryzae* appressoria in vitro has advanced the understanding of the physiological and genetic cues required for appressoria formation (2–4). A similar in vitro assay in other plant pathogenic fungi,

Melvin D. Bolton and Bart P.H.J. Thomma (eds.), *Plant Fungal Pathogens: Methods and Protocols*,
Methods in Molecular Biology, vol. 835, DOI 10.1007/978-1-61779-501-5_22, © Springer Science+Business Media, LLC 2012

coupled with the increase of readily available genomic data, will help elucidate the genetic requirements of infection-related development.

Fusarium graminearum can cause a myriad of plant diseases, though it is most notorious for causing fusarium head blight of wheat and barley (5,6). Many of the molecular requirements of pathogenicity and virulence have been elucidated over the past decade; however, a simple in vitro assay for pathogenic hyphal development was lacking. Unlike plant pathogenic fungi such as *M. oryzae*, *F. graminearum* does not differentiate an appressorium before invading plant tissue. However, it does differentiate coral-like subcuticular hyphae and wide dikaryotic hyphae upon wheat and barley invasion (7–9). We have used the following protocol to induce these infection-related structures in vitro and to demonstrate that the $\Delta gpmk1$ MAPK mutant is unable to induce wide bulbous infection hyphae (10).

2. Materials

2.1. Inoculation Material

1. *Y*east *M*alt Extract *A*gar (YMA) plates: 4 g malt extract, 4 g sucrose, 4 g yeast extract, 15 g agar, to 1 L of deionized-distilled H_2O (ddH_2O).

2. *C*arboxy*m*ethyl *c*ellulose (CMC) liquid media: 15 g CMC (low viscosity, sodium salt), 1 g yeast extract, 1 g NH_4NO_3, 1 g KH_2PO_4, 0.25 g $MgSO_4...7H_2O$.

3. Macroconidia of *Fusarium graminearum*: macroconidia can be generated in CMC media or on solid YMA plates (see Note 1).

4. 0.05% Tween 20 solution.

5. Frozen wheat heads: Collect wheat heads at anthesis from wheat variety "Norm." Store the wheat heads in a plastic bag at –20°C (see Note 2).

6. Ultraviolet light for sanitizing wheat glumes.

7. 4% (w/v) water agar.

2.2. Components for Fixing Inoculated Glumes

1. 24-well cell culture plate.

2. Clearing solution A: 75 mL ethanol, 25 mL glacial acetic acid.

3. Clearing solution B: 50 mL ethanol, 10 mL glacial acetic acid, 10 mL glycerol.

4. Lactophenol blue staining solution: 13.5 mL lactic acid, 13.5 mL phenol, 12 mL ddH_2O, 1 mL 0.4% trypan blue solution.

5. Destain solution: 60 mL glycerol, 40 mL ddH_2O.

2.3. Components for Dissecting/Mounting Fixed Glumes

1. Dissecting microscope.
2. Surgical Scalpel.
3. Fine metal forceps.
4. Microscope slides and coverslips.
5. Bright-field compound microscope with camera (camera optional).

---

3. Methods

3.1. Growth of F. graminearum on Solid YMA or Liquid CMC Media to Produce Macroconidia

1. Inoculate YMA plates or 50 mL CMC media with 50 µL of a 1×10^5 macroconidia suspension (see Note 1).
2. If using YMA plates to generate macroconidia, place the inoculated plates in an incubator set at 28°C. Incubate for 4–5 days.
3. If using liquid CMC cultures to generate macroconidia, place the 50 mL cultures in a rotary shaker set at 200 rpm and 25°C for 5 days.

3.2. Harvest the Macroconidia

1. If using YMA, add 1.5 mL of 0.05% tween to the plate and release macroconidia with a sterile-glass rod. Collect macroconidia suspension and filter through miracloth (Calbiochem) to remove hyphal debris.
2. If using CMC, filter the media/spore suspension through miracloth to remove hyphal debris.
3. Assess the macroconidia concentration with a hemacytometer and dilute to 1×10^6 in 0.05% tween.

3.3. Prepare Wheat Glumes for Inoculation

1. Remove wheat heads from −20°C. Carefully remove glumes from the florets and place on 4% water agar with the cavity (adaxial side) facing up (see Note 3; Fig. 1).
2. Place water agar plates, without lids, under an ultraviolet light for 30–60 s to sanitize the glumes.

3.4. Inoculate and Incubate Detached Wheat Glumes

1. Collect 200 0L of macroconidia suspension (concentration between 10^4 and 10^6 macroconidia per mL; see Note 4) in a P200 micropipettor.
2. Aliquot approximately 20–30 µL of the suspension onto the adaxial side of each glume (see Note 3; Fig. 1).
3. Incubate the glumes at room temperature for 1 min, and then remove the macroconidia suspension from the glumes.
4. Incubate the wheat glumes at room temperature for given time points (see Note 5).

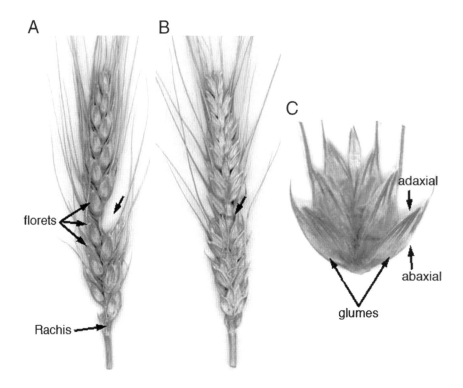

Fig. 1. Wheat heads and florets. (a) Side view of wheat head showing rachis, florets, and area where a floret was detached (*arrow*). (b) Front view of a wheat head showing where floret was detached (*arrow*). (c) Detached floret showing the abaxial and adaxial side of wheat glumes.

3.5. Fixing and Mounting Wheat Glumes for Microscopy

1. After a given time point, remove glumes from water agar plate and place into well of 24-well cell culture plate containing 1 mL of clearing solution A (see Note 6). Incubate overnight (~16 h) at 75 rpm and room temperature.

2. Move the glumes to another well containing 1 mL clearing solution B. Incubate for 4 h at 75 rpm and room temperature.

3. Move the glumes to another well containing 1 mL of lactophenol blue staining solution (see Note 7). Incubate overnight (~16 h) at 75 rpm and room temperature.

4. Move glumes to another well containing 1 mL of 60% glycerol for 5 min to remove most of the residual lactophenol blue. After 5 min incubation, move to fresh 60% glycerol and incubate for at least 3 h.

5. With the aid of a dissecting scope, use the forceps and surgical scalpel to cut washed glumes into ~1 mm² sections for oblique views or 0.25 mm thick strips for cross-section views (see Note 8).

6. Move the sections to a microscope slide containing ~50 μL 60% glycerol. Overlay with coverslip and examine under microscope (see Fig. 2 for some expected results).

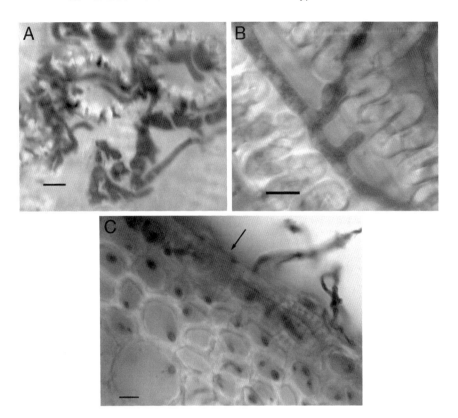

Fig. 2. Some expected results. (**a**) Coral-like subcuticular hyphae after 24 h. (**b**) Bulbous infection hyphae after 48 h. (**c**) Cross-section view of glume inoculated on adaxial side (*arrow*). Note the invasion of lower mesophyll cells by *F. graminearum* hyphae. Hyphae in all panels stained with lactophenol trypan blue. Scale bar = 10 μm in all panels.

4. Notes

1. Many laboratories use CMC to generate *F. graminearum* macroconidia, and it has become an accepted standard for the generation of macroconidia; we have also had good results using this method. However, we obtain 10–100-fold more macroconidia when using YMA, and some of the mutants generated in our laboratory do not sporulate well in CMC. Users are encouraged to try both methods and see which works better for their given scenario.

2. Subcuticular and bulbous invasion hyphae are more readily formed on previously frozen glumes compared to freshly detached wheat glumes (unpublished data).

3. We have inoculated macroconidia and ascospores on both the abaxial and adaxial side of wheat glumes (10). On the abaxial side, the hyphae appear to readily penetrate stomata, but the development of morphologically distinct structures is not as common as when the spores are inoculated onto the adaxial side of the glume.

4. When beginning our experiments, overinoculation often caused heavy invasion and staining that made distinguishing morphological structures difficult. We have since had success with macroconidia concentrations between 5×10^4 mL^{-1} and 1×10^6 mL^{-1}. Before pursuing these experiments, we recommend users empirically test various macroconidia concentrations to see which yield suitable results.

5. Subcuticular hyphae and bulbous invasion hyphae are readily observed at the 24 h incubation time point when *F. graminearum* strain PH-1 and wheat variety "Norm" are used (Fig. 2) (10). However, 48 and 72 h may be necessary for slower-growing mutants. Also, bulbous infection hyphae are often more prevalent after 48 h, as they stain very lightly after only 24 h.

6. In order to save supplies, we often include two glumes of the same treatment type per well when clearing/staining hyphae.

7. While we describe staining with lactophenol blue, we also have stained with calcofluor white and observed infection-related hyphae with fluorescent microscopy (10). Also, we have used a *F. graminearum* strain that constitutively expresses GFP (kindly provided by Dr. Jin-Rong Xu, Purdue University) and electron microscopy to observe these hyphal structures (unpublished data; 10).

8. Certain sections of the wheat glume appear to be more conducive to microscopic observation. Some parts (especially near the base of the glume or near the midrib) are too thick for mounting on a slide and/or background stain heavily with lactophenol blue.

References

1. Howard RJ, Valent B (1996) Breaking and entering: host penetration by the fungal rice blast pathogen *Magnaporthe grisea*. Ann Rev Microbiol 50:491–512

2. Mitchell TK, Dean RA (1995) The cAMP-dependent protein kinase catalytic subunit is required for appressorium formation and pathogenesis by the rice blast pathogen *Magnaporthe grisea*. The Plant Cell 7:1869–1878

3. Xu J-R, Hamer JE (1996) MAP kinase and cAMP signaling regulate infection structure formation and pathogenic growth in the rice blast fungus Magnaporthe grisea. Genes & Devel 10:2696–2706

4. Takano Y, Choi W, Mitchell TK, Okuno T, Dean RA (2003) Large scale parallel analysis of gene expression during infection-related morphogenesis of *Magnaporthe grisea*. Mol Plant Pathol 4:337–346

5. McMullen M, Jones R, and Gallenberg D (1997) Scab of Wheat and Barley: A re-emerging disease of devastating impact. Plant Dis 81:1340–1348

6. Goswami RS, Kistler HC (2004) Heading for disaster: *Fusarium graminearum* on cereal crops. Mol Plant Pathol 5:515–525

7. Pritsch C, Muehlbauer GJ, Bushnell WR, Somers DA, Vance CP (2000) Fungal development and induction of defense response genes

during early infection of wheat spikes by *Fusarium graminearum*. Mol Plant-Microbe Int 13:159–169

8. Jansen C, Wettstein Dv, Shafer W, Kogel KH, Felk A, Maier FJ (2005) Infection patterns in barley and wheat spikes inoculated with wild-type and trichodiene synthase gene disrupted *Fusarium graminearum*. Proc Natl Acad Sci, USA 102:16892–16897

9. Guenther JC, Trail F (2005) The development and differentiation of *Gibberella zeae* (anamorph: *Fusarium graminearum*) during colonization of wheat. Mycologia 97: 229–237

10. Rittenour WR, Harris SD (2010) An *in vitro* method for the analysis of infection-related morphogenesis in *Fusarium graminearum*. Mol Plant Pathol 11:361–369

Chapter 23

Fungicide Resistance Assays for Fungal Plant Pathogens

Gary A. Secor and Viviana V. Rivera

Abstract

Fungicide resistance assays are useful to determine if a fungal pathogen has developed resistance to a fungicide used to manage the disease it causes. Laboratory assays are used to determine loss of sensitivity, or resistance, to a fungicide and can explain fungicide failures and for developing successful fungicide recommendations in the field. Laboratory assays for fungicide resistance are conducted by measuring reductions in growth or spore germination of fungi in the presence of fungicide, or by molecular procedures. This chapter describes two techniques for measuring fungicide resistance, using the sugarbeet leaf spot fungus *Cercospora beticola* as a model for the protocol. Two procedures are described for fungicides from two different classes; growth reduction for triazole (sterol demethylation inhibitor; DMI) fungicides, and inhibition of spore germination for quinone outside inhibitor (QoI) fungicides.

Key words: Triazole, Strobilurin, *Cercospora beticola*

1. Introduction

In vitro assays for fungicide sensitivity have been developed for a number of fungal/fungicide systems and several excellent reviews have been published (1–6). Such assays are useful to monitor changes in sensitivity of the fungal pathogen as fungicides are applied over time, usually years. This information is most often used to make efficacious fungicide recommendations for disease management. For best results, sensitivity assays should be done before fungicides are registered and used commercially in the field in order to establish a baseline that can be used to monitor changes in sensitivity in subsequent years. Many *in vitro* procedures have been developed to measure changes in fungicide sensitivity. Molecular procedures utilize PCR or real time PCR to detect specific nucleic acid base changes that are associated with fungicide resistance (7). In order to use molecular assays, a specific genetic change conferring resistance must be known and specific primers

Melvin D. Bolton and Bart P.H.J. Thomma (eds.), *Plant Fungal Pathogens: Methods and Protocols,*
Methods in Molecular Biology, vol. 835, DOI 10.1007/978-1-61779-501-5_23, © Springer Science+Business Media, LLC 2012

identified. Biological assays for fungicide resistance rely on changes in the growth of the fungus, such as inhibition of radial growth or reduction of spore germination, in response to exposure to the fungicide *in vitro*. Because fungicides have different targets site in the fungus and consequently different modes of action, the assay used may be fungicide dependent.

In this protocol, we describe two methods for monitoring fungicide resistance using the fungus *Cercospora beticola* Sacc., the cause of leaf spot of sugar beet (CLS), as a model for the procedures. This disease is endemic in all sugar beet producing areas and causes a loss in yield and sucrose due to reduced photosynthetic area by diseased leaves. The disease is controlled by crop rotation, resistant varieties and timely fungicide applications. Several fungicides from different classes, including benzimidazoles, dithiocarbamates, tin compounds, triazoles, and strobilurins have been used for managing CLS. During the past 20 years, *C. beticola* has developed resistance to fungicides from all of these classes and fungicide resistance is a real and economically important consideration in sugar beet production. Recent reviews of fungicide resistance in sugar beets in Europe (8) and the US (9) have been recently published. For efficacious disease control and fungicide resistance management, the most widely used fungicides in both the United States and European countries for managing CLS are several fungicides in the triazole and strobilurin classes. The two groups of fungicides have different metabolic targets and modes of action, and therefore require two different procedures for measuring fungicide resistance.

Triazole fungicides are ergosterol biosynthesis inhibiting fungicides that are divided into two groups, demethylation inhibiting fungicides and morpholines (10). This class of fungicides inhibits mycelial growth and the *in vitro* bioassay uses inhibition of colony radial growth to measure resistance to fungicides.

Strobilurin fungicides are QoI inhibitors that block electron transport through the mitochondrial system (11). This class of fungicides inhibits spore germination and the *in vitro* bioassay uses inhibition of spore germination to measure resistance to fungicides.

This chapter will describe two procedures, radial growth, and spore germination, used to assay fungicide resistance in *C. beticola*, to a triazole and strobilurin fungicide, respectively, but the procedures may be applicable to other biological systems that utilize control of a fungal disease by a fungicide.

2. Materials

1. Water agar medium: distilled water 1 L, 15 g agar. Mix agar and water and autoclave 15 min. Pour the molten agar into Petri dishes (see Note 1).

Fig. 1. V8 juice can.

2. Clarified V8 medium (CV8): V8 juice (see Fig. 1, Note 2), 500 mL, 1.5 g calcium carbonate Add the calcium carbonate to the V8 juice and stir to dissolve the $CaCO_3$. Clarify the V8 juice by centrifuging the mixture at $1,700 \times g$ for 10 min in 250-mL bottles using a GSA rotor. The supernatant fluid is used for CV8 medium preparation. Discard the pellet. To prepare the medium combine 100 mL of the clarified V8 supernatant, 900 mL of distilled water and 15 g agar (see Note 3). Sterilize by autoclaving for 20 min (see Note 4).

   Ampillin can be added to CV8 during the initial isolation to reduce bacterial contamination, but is not necessary for subsequent transfers to CV8 medium after a pure culture has been established. Ampicillin is added after the medium has cooled to 50°C at final concentration of 0.2 g/L.

3. Salicylhydroxamic acid (SHAM): Prepare a stock solution of SHAM by adding 400 mg SHAM to 4,000 μL methanol and shake under hot water to dissolve; it dissolves slowly. Add 500 μL to 500 mL water agar cooled to 55°C, which gives a final concentration of 100 mg/L.

3. Methods

All procedures are conducted at room temperature using basic sterile techniques in a clean environment. All testing is conducted using the technical grade active ingredient of each fungicide, not the formulated commercial fungicide in order to eliminate potential interference of fungicide activity or fungal growth due to compounds present in commercial fungicide formulations. The term µg/mL is equivalent to parts per million (ppm).

3.1. Collection of Cercospora Isolates for Testing

Sugar beet leaves with *Cercospora* leaf spot (CLS) are collected and processed immediately to insure viability of spores. From each field sample, *C. beticola* spores are collected from a minimum of five spots per leaf from five leaves per field. The spores are collected by applying 20–30 µL of sterile distilled water to a *Cercospora* spot using a microliter pipette and flushed several times to dislodge spores. The spores from the five spots/leaf are pooled in a 1.5-mL centrifuge tube and a composite of 120 µL of the pooled spore suspension is transferred to a Petri plate containing water agar plus ampicillin (0.2 g/L). This plate is used as a source of germinated single spore subcultures for subsequent testing of triazole resistance by radial growth and strobilurin resistance by spore germination. Spore germination time is approximately 16 h.

3.2. Inhibition of Radial Growth Procedure for Triazole Resistance Testing

For triazole fungicide sensitivity testing, a radial growth procedure for *C. beticola* is used because these fungicides inhibit mycelial growth. Sensitive isolates have reduced growth in the presence of fungicide compared to sensitive isolates in the presence of fungicide on artificial media. Radial growth is easy to measure, since most fungi naturally grow radially in artificial media.

1. Select a single spore subculture from the original non-amended media and transfer to CV8 medium.

2. Incubate the culture at 20°C in a continuous light regime until the colony covering about 60% most of the plate is produced (ca. 15 days for *C. beticola*).

3. Remove an agar plug 4 mm in diameter from the active growth area of the colony and place in the center of a set of Petri dishes containing non-amended CV8 medium and CV8 media amended with serial tenfold dilutions of a technical grade triazole fungicide active ingredient from 0.001 to 1.0 µg/mL (see Note 5). A separate test is conducted for each triazole fungicide.

4. Incubate the dishes for 15 days in the dark at a temperature of 20°C.

5. Evaluate growth by making two perpendicular measurements across the colonies and averaging the diameters.

6. This data is converted to a percent reduction of growth by comparing the average colony diameter data on amended media to the average colony diameter data on non-amended water agar medium. This growth reduction data is used to calculate an EC_{50} value for each isolate; EC_{50} is the effective concentration of fungicide that reduces radial growth by 50%

Table 1
Colony radial growth measurements of a *C. beticola* isolate growing on media amended with five concentrations of tetraconazole (0.0–1.0 µg/mL)

	0	0.001	0.01	0.1	1	EC$_{50}$
Diameter 1	38	38	35	21	7	
Diameter 2	37	35	34	18	9	
Average diameter	37.5	36.5	34.5	19.5	8	
% of growth	100	97.3	92	52	21.3	
% Growth reduction	0	2.7	8	48	78.7	0.159

compared to the growth on non-amended media. Higher EC$_{50}$ values indicate reduced sensitivity, and possible resistance, to the fungicide.

7. Percent growth is calculated for each dilution is using the following formula: (average colony diameter with fungicide)/(average colony diameter without fungicide) times 100 (see Table 1). Percent growth reduction is then calculated by subtracting percent growth from 100 (see Table 1).

8. Plot logarithmic fungicide concentration vs. reduction of colony growth. For mathematical reasons, fungicide concentrations of zero (the non-amended medium) cannot be used to construct a graph using a logarithmic scale. Therefore, it is necessary to assign a value to the non-amended control fungicide concentration of tenfold lower than the lowest fungicide dilution used in the assay. In Fig. 1, this assigned value is the 0.0001 µg/mL.

9. On the resultant curve, find the fungicide concentration point on the curve where growth is reduced by 50% (see Fig. 2), or more precisely by regression curve fitting.

3.3. Reduction of Spore Germination Procedure for Strobilurin Resistance Testing

For strobilurin fungicide sensitivity testing, it is necessary to use a procedure that measures spore germination because these fungicides act by inhibiting spore germination. Resistant isolates have higher spore germination rates compared to sensitive isolates in the presence of fungicide on artificial media.

1. Transfer a subculture from the original non-amended medium to non-amended CV8 medium and incubate in conditions to induce sporulation (see Note 6).

2. Spores are induced from a 15-day-old *C. beticola* culture. Add 2 mL of sterile distilled water to the plate and gently scrape the

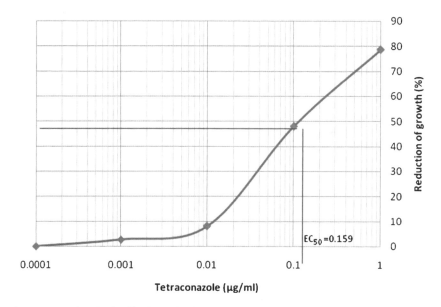

Fig. 2. Plot of percent growth reduction (Y axis) against fungicide concentration (X axis) using data from Table 1 to calculate an EC_{50} value.

mycelium using the edge of a microscope slide. Transfer about 500 µL of the aqueous phase onto CV8 medium and spread the solution using a bent glass rod, avoiding transfer of large fragments of mycelium. Allow the plate to dry in the hood and incubate the plate at room temperature under continuous light using an equal mixture of cool white fluorescent light and plant growth black light bulbs. Spores will be visible after 3 days and spores should be harvested after 5 days for testing.

3. Add 3 mL of sterile distilled water containing to the plate and gently shake the plate to dislodge the spores.

4. Transfer a 120 µL aliquot of the spore suspension to non-amended water agar medium and a water agar series amended with serial tenfold dilutions of technical grade strobilurin fungicide from 0.001 to 1.0 µg/mL and containing salicylhydroxamic acid (SHAM) at a concentration of 100 µg/mL (see Note 7).

5. Incubate the Petri dishes at room temperature.

6. Studies in our lab have demonstrated that *C. beticola* spores reach >80% germination in about 16 h; therefore, view germination of 100 spores at random 16 h after plating (see Note 8).

7. Calculate a percent germination for each fungicide concentration in the series.

8. Calculate percent growth reduction by subtracting percent growth from 100 (see Fig. 1).

9. Plot logarithmic fungicide concentration vs. reduction of colony growth. For mathematical reasons, fungicide concentrations of zero (the non-amended medium) cannot be used to construct a graph using a logarithmic scale. Therefore, it is necessary to assign a value to the non-amended control fungicide concentration of tenfold lower than the lowest fungicide dilution used in the assay. In Fig. 2, this assigned value is the 0.0001 μg/mL.

10. On the resultant curve, find the fungicide concentration point on the curve where growth is reduced by 50%, or more precisely by regression curve fitting.

Higher EC_{50} values indicate reduced sensitivity and possible resistance to the fungicide.

4. Notes

1. Any agar will work for the protocol regardless of purity. Add approximately 15 mL/dish. The volume of agar added is not critical, and with experience the volume can be estimated while pouring. No nutrients are provided by the water agar. Use of disposable plastic dishes facilitates testing.

2. The original V8 is made mainly from tomatoes and the juices from seven additional vegetables, specifically: beets, celery, carrots, lettuce, parsley, watercress, and spinach. A photo is included as many countries outside the US do not recognize V8 juice.

3. Any agar will work for this purpose.

4. Autoclave 10–15 min longer if several liters are prepared simultaneously.

5. Because the technical formulations of the fungicide active ingredients used for resistance testing have limited solubility in water, they must first be dissolved in solvent for dilution and addition to the agar medium. We prepare a stock solution at 10 μg/mL of each technical ingredient to use as a stock solution for subsequent dilutions. For triazole and strobilurin fungicides, acetone is the choice. The only precaution is to add the dissolved fungicide to the media when it is about 55°C, by injecting the fungicide plus solvent into the media using a pipette to prevent evaporation of the acetone.

6. Conditions for inducing sporulation, viz. light quality, light duration, light intensity, temperature, growth medium composition, etc, vary among fungi; consequently it is necessary to know a procedure for spore production in order to conduct this procedure.

7. Salicylhydroxamic acid (SHAM) is added to prevent alternate oxidation pathways in the fungus to overcome resistance and give false resistance results (12).

8. *C. beticola* spores are hyaline (clear) and needle-shaped which makes them difficult to see through a stereoscope; the light must be adjusted to diffract the light to enhance visibility. Spores are considered germinated when the mycelial growth from the spore (conidium) is at least double of the length of the spore. Percent germination is calculated by the number of spores germinated divided by the number of spores viewed multiplied by 100.

References

1. Brent KJ, Holloman, DW (2007) Fungicide resistance in crop pathogens: How can it be managed? FRAC Monograph 1, 2nd Ed. FRAC, Brussels, Belgium

2. Brent, K J, Holloman, D W (2007) Fungicide resistance: The assessment of risk. FRAC Monograph 2, 2nd Ed. FRAC, Brussels, Belgium

3. Delp, CJ (1988) Fungicide resistance in North America. APS Press, St. Paul, MN

4. Russell, PE (2003) Sensitivity baselines in fungicide resistance research and management. FRAC Monograph 3, Crop Life International, Brussels, Belgium

5. Staub, T, Sozzi, D (1984) Fungicide resistance. Plant Dis. 68:1026–1031

6. van den Bosch, F, Gilligan, CA (2008) Models of fungicide resistance dynamics. Ann. Rev. Phytopathol. 46:123–147

7. Pasche, JS *et al* (2004) Shift in sensitivity of *Alternaria solani* response to QoI fungicides. Plant Disease 88:181–187

8. Karaoglandis, GS, Ioannidis PM (2010) Fungicide resistance of *Cercospora beticola* in Europe. Pp 189–211 *in* Cercospora leaf spot of sugar beet and related species. Eds. Lartey *et al.* APS Press. St. Paul, MN

9. Secor, GA *et al* (2010) Monitoring fungicide sensitivity of Cercospora beticola of sugar beet for disease management decisions. Plant Disease 94: 1272–1282

10. Ioannidis, PM, Karaoglanidis, GS (2000) Resistance of *Cercospora beticola* Sacc. to fungicides. In Advances in Sugar beet Research Vol. 2: *Cercospora beticola* Sacc. Biology, agronomic influence and control measures in sugar beet 2000. Pp. 123–145. Inter. Inst. Beet Res. Brussels

11. Bartlett, DW *et al* (2002) The strobilurin fungicides. Pest Mgmt. Sci. 58:649–662

12. Ziogas, BN et al (1997) Alternative Respiration: a Biochemical Mechanism of Resistance to Azoxystrobin (ICIA 5504) in *Septoria tritici* Pestic. Sci. 50:28–34

Chapter 24

Identification of Lipid-Binding Effectors

Shiv D. Kale and Brett M. Tyler

Abstract

In recent years, the functional roles of effectors from a wide variety of fungal and oomycete pathogens have begun to emerge. As a product of this work, the importance of effector-lipid interactions has been made apparent. Phospholipids are not only important signaling molecules, but they also play important roles in the trafficking of endosomes and the localization of proteins. Characterizing effector-lipid interactions can provide novel information regarding the functions of effectors relevant to their cellular and subcellular targeting and their potential effects on host signaling and vesicle trafficking. We present here two techniques that can be used to screen for and validate protein-lipid interactions without the need to access highly specialized machinery. We describe in detail how to perform lipid filter and liposome-binding assays and provide suggestions for troubleshooting potential problems with these assays.

Key words: Phospholipids, Phosphoinositides, Phosphatidylinositol-3-phosphate, Liposomes, Lipid filter assay, Effectors, Oomycetes, Fungi, Pathogenesis, Protein translocation

1. Introduction

Phospholipids play several important roles in cellular structure, development, and signaling. These lipids generically consist of a head group linked to the glycerol backbone via a phosphate group. The glycerol backbone is linked to two fatty acyl chains. Phosphatidylcholine (Ptd-Cho) is the most abundant phospholipid in both plants and animals, and predominantly plays a structural role due to its zwitterionic nature. Phosphatidylethanolamine (Ptd-Eth) is also abundant in all living cells and is the principal phospholipid in bacterial cells. Both Ptd-Cho and Ptd-Eth are the key lipids in the building of lipid bilayers in cellular membranes. Phosphatidic acid (Ptd-Acid) plays diverse roles in the cell. Ptd-Acid accumulation leads to membrane curvature (1), is a precursor for all cell membrane

Melvin D. Bolton and Bart P.H.J. Thomma (eds.), *Plant Fungal Pathogens: Methods and Protocols*,
Methods in Molecular Biology, vol. 835, DOI 10.1007/978-1-61779-501-5_24, © Springer Science+Business Media, LLC 2012

phospholipid biosynthesis (2), can recruit proteins to membranes such as sphingosine kinase I (3), and is an important signaling molecule (4–6). Phosphatidylserine (Ptd-Ser) is normally found on the inner leaflet of membranes and its asymmetric distribution is maintained through the function of flippases (7). Ptd-Ser also functions as a protein-recruiting lipid, for example localizing spectrin to the lipid bilayer (8). Interestingly, Ptd-Ser distribution is altered in mammalian cells undergoing apoptosis (9), where it is important for the recruitment of macrophages and other types of scavenger cells that remove the apoptotic cells without the induction of inflammation (10). These abundant phospholipids play important roles in signaling, structure, and recruitment of proteins.

Another important group of phospholipids, though scarce in abundance, are the mono-, bis-, and tri-phosphorylated phosphatidylinositides (PtdIns). Mono-PtdIns make up less than 1% of the lipid content of cells yet are essential for cellular function (11). PtdIns-3,4,5-P is the product of the class I PI 3-kinases and plays a role in the activation of signaling components such as the protein kinase AKT, which is responsible for downstream signaling in multiple cellular processes including but not limited to proliferation, apoptosis, and glucose metabolism (12, 13). The role of intracellular PtdIns-4-P has been viewed principally as a precursor to PtdIns-4,5-P, an important phospholipid in signaling specifically through phospholipase C (14). Recently however, PtdIns-4-P has been shown to play roles in recruitment and signaling independently of PtdIns-4,5-P (11, 15). Intracellular PtdIns-3-P has been implicated in the formation and differentiation of endosomes, and in growth factor signaling (16, 17). Extracellular PtdIns-3-P has recently been identified to mediate entry of effector proteins into host cells from both fungal and oomycete pathogens (18).

There are several phospholipid-binding domains that have proved useful tools for the study of protein-phospholipid interactions both *in vivo* and *in vitro*. The first phospholipid-binding domain discovered was Protein Kinase C (PKC) that contains two highly conserved domains known as C1 and C2 (19). C1 and C2 bind specifically to diacylglycerol and phosphatidylserine respectively (20, 21). A substrate of PKC is pleckstrin, which binds specifically to PtdIns-3,4,5-P (22). Pleckstrin contained the first phosphoinositide-binding region, which subsequently became known as the Pleckstrin homology (PH) domain (19). With the ability to search through genomes, several putative PH domain-containing proteins were identified and subsequently characterized. These include the PEPP1 and FAPP1 PH domains that are highly specific for PtdIns-3-P and PtdIns-4-P, respectively (23). Broad specificity PH domains also exist such as the LL5α/β PH domain that binds to all mono and poly-phosphorylated phosphoinositides (23). These three proteins provide examples of the lipid-binding diversity and specificity found among PH domains. The Hrs FYVE and VAM7p PX (phox) domains are also well characterized PI-3-P-binding domains (24, 25). The myristoylated, alanine-rich C kinase substrate (MARCKS) family binds specifically to phosphatidylserine (26). The Raf-1 kinase and cyclic-AMP phosphodiesterase 4A1 contain well-characterized phosphatidic acid-binding domains (27, 28). Analysis of these proteins has not only revealed the importance and function of lipid-protein interactions in several cellular processes, but has also created a useful toolbox of techniques to characterize lipid-protein interactions.

Symbionts, including mutualists and pathogens, utilize a variety of mechanisms to sustain their colonization or infection of plant and animal hosts. One mechanism used by eukaryotic plant pathogens is the secretion of small proteins that have the ability to enter host cells without the use of pathogen-encoded machinery (18, 29, 30). In the oomycetes, *Phytophthora infestans* and *Phytophthora sojae*, intracellular effectors Avr3a and Avr1b, respectively, require a pair of N-terminal amino acid motifs, RXLR and dEER, to enter host cells (18, 29, 31). Bioinformatic predictions show that RXLR-dEER-containing proteins make up a large super family of effectors that are rapidly evolving (32). Many of these intracellular RXLR-dEER effectors have the ability to suppress host defense mechanisms (33–35). Putative RXLR-like motifs have been identified in a variety of well-studied fungal effector proteins that can enter host cells in the absence of pathogen-encoded machinery (18, 30). RXLR and RXLR-like effector entry is mediated through binding to PtdIns-3-P on the outer surface of the plasma membrane of both plant and animal cells (18). Lipid-mediated effector and toxin entry has also been co-opted by many bacterial pathogens. For example, many bacterial toxins such as tetanus, cholera, and botulinum toxins enter host cells by endocytosis after binding glycolipids (36), while the CagA effector of *Helicobacter pylori*, enters by binding Ptd-Ser on the outer leaflet (37). The N-terminus of the type III secreted effector (T3SE) YopM has been shown to mediate the entry of human HeLa cells in the absence of pathogen-encoded machinery (38). Lipid binding also plays an important role in the virulence function of effectors upon entry into the cytoplasm. The *Legionella* T4SE's DrrA and SidM bind PtdIns-4-P after entering host cells (39, 40), while the *Legionella* T4SE, LidA, binds PtdIns-3-P, PtdIns-4-P, and PtdIns-5-P (41). Lipids continue to be identified in novel roles such as PtdIns-3-P in the apicoplast of the apicomplexan *Plasmodium falciparum* (42) and PtdIns-3,4,5-P and Ptd-Ser on the outer leaflet of human cells during infection (37, 43). The role of these lipids is presently unknown but may play roles in protein recruitment or cell signaling.

Given the growing importance of lipid-protein interactions in microbe-host interactions, and in particular the function of effector proteins, this chapter is focused on two valuable techniques for identifying and characterizing lipid-protein interactions. The lipid filter assay and liposome-binding assay are described in detail along with several pointers to avoid common difficulties and artifacts associated with these techniques. We have chosen these techniques primarily because they do not require expensive instrumentation, thus researchers who do not focus primarily on lipid-protein interactions can learn them relatively easily. For more detailed studies we recommend the use of surface plasmon resonance (SPR), isothermal titration calorimetry testing, and nuclear magnetic resonance (NMR) (44). Those advanced techniques are costly and time-consuming, but they provide a significant added degree of knowledge regarding the thermodynamics and structural details of an interaction.

2. Materials

2.1. Lipid Filter Assay Hardware and Disposables

1. Rotary Shaker.

2. Petri-dishes (100×15 mm).

3. Hybond C-Extra membrane.

4. Macrovial (National Scientific, Catalog # C4010-LV2, 350-μL Amber Glass, 12×32 mm, Fused Insert, Target Macrovial Screw Thread Vial).

5. Macrovial screw caps (National Scientific, Catalog # C4010-30A).

6. Lipids (Cayman Chemical or Avanti Polar Lipids) (see Note 1).

7. Bovine serum albumin (BSA, fatty acid free; Sigma Aldrich, Catalog # A7030) (see Note 1).

8. Vacuum and containment unit.

9. Saran Wrap.

10. Super Signal West Pico (Thermo Scientific, Prod #34080).

11. Cassette (20×25 cm).

12. X-ray film.

13. X-ray film developer.

14. Primary anti-green fluorescent protein (GFP) antibody from rabbit (Immunology Consultants Laboratory Inc., RGFP-45A-Z).

15. Primary anti-hexahistidine (His-Tag) antibody from rabbit (Immunology Consultants Laboratory Inc., RHIS-45A-Z).

16. Primary anti-glutathione-S-transferase (GST) antibody from goat (GE Healthcare, 27-4577-01).

17. Secondary anti-rabbit IgG antibody from donkey, horseradish peroxidase (HRP) conjugate (GE Healthcare, NA934).

18. Secondary anti-goat IgG antibody from donkey, HRP conjugate (Santa Cruz Biotechnology, SC-2020). For Items 14–18 see Note 1.

2.1.1. Lipid Filter Assay Solutions

Prepare all solutions using ultrapure water (prepared by purifying deionized water to attain a sensitivity of 18 MΩ cm at 25°C) and analytical grade reagents.

1. CM: 6.5 mL chloroform, 3.5 mL methanol.

2. CMW: 6.5 mL chloroform, 3 mL methanol, 0.8 mL water.

3. Phosphate buffered saline (PBS): 10 mM sodium phosphate dibasic, 2 mM potassium phosphate monobasic, 138 mM NaCl, 2.7 mM KCl, pH 7.4.

4. PBS containing 0.1% Tween-20 (PBST).

5. PBST + 3% Albumin Blocking Solution: Add 3 g of BSA per 100 mL of PBST. Each blot requires either 10 or 20 mL of blocking solution. Scale as required (see Note 1).

6. Tris buffered saline (TBS; 10×): 1.5M NaCl, 500 mM Tris–HCl, pH 8.0.

7. TBS: Add 100 mL of 10× TBS to 850 mL of filter-sterilized water. pH to 8.0. Raise volume to 1 L with filter-sterilized water.

8. TBS containing 0.1% Tween-20 (TBST).

9. TBST + 3% Albumin Blocking Solution: Add 3 g of Albumin from BSA per 100 mL of TBST. Each blot requires 10 or 20 mL of blocking solution. Scale as required (see Note 2).

2.2. Liposome-Binding Assay Hardware and Disposables

The hardware to make liposomes comes in a package from Avanti Polar Lipids (cat# 610000) and includes items 1–5 below.

1. Mini-extruder.

2. 100 Polycarbonate membranes (0.2 μm).

3. Filter Supports.

4. Avanti Extruder Heating Adapter.

5. 2 gas tight syringes (1 mL).

6. Macrovial (see above).

7. Macrovial screw caps (see above).

8. Hot plate.

9. Vacuum desiccator.

10. Liquid nitrogen.

11. Sorvall Adapter (Model 00381). Adapter to hold 1.5-mL eppendorf tubes in a SS-34 rotor.

2.2.1. Liposome-Binding Solutions

1. Liposome Buffer: 25 mM Tris buffer pH 6.8, 100 mM NaCl.

2. 5× Loading Buffer: 250 mM Tris–HCl; pH 6.8, 10% SDS, 50% Glycerol, 5% β-Mercaptoethanol, 62.5 mM EDTA, 0.1% Bromophenol Blue.

3. Staining Solution: 0.2% Coomassie Blue, 5.0% Acetic Acid, 50% Ethanol.

4. Destaining Solution: 5.0% Acetic Acid, 50% Ethanol.

3. Methods

3.1. General Information on Preparation and Storage of Lipids

Lipid preparation and storage is a key component to successfully assaying proteins for binding. Prolonged storage (greater than 1 week in CMV or more than 2 weeks in CM) can lead to degradation of lipid head groups, especially multiphosphorylated phospholipids. We strongly recommend using fresh lipids whenever possible.

This section of the chapter describes preparation of your own lipid filters. For initially screening proteins of interest against a diverse array of lipids we recommend first using a commercially made lipid strip (Echelon Biosciences). Since screening a large library of lipids can be time-consuming and expensive, the commercial lipid strips provide a cheap tool to identify potential lipid interactions. However, we do find that these commercial blots are significantly variable in binding specificity and further analysis is required using a gradient of lipid concentrations to check specificity of lipid binding. Dried lipids can be purchased in a variety of quantities. The remainder of this section will describe the preparation of dilutions and aliquots from commercially purchased vials of 50 μg lipid.

1. Label all macrovial tubes with the name of the lipid and the concentration of the lipid before beginning dilutions (see Fig. 1b).

2. Stock lipids should be stored as a dried powder under an inert gas such as nitrogen or helium at –20°C (preferably –80°C) before use. We recommend ordering lipids in small 50–100 μg amounts. 50 μg will provide between 75 and 100 blots when dissolved and diluted appropriately. If the lipids are purchased in larger quantities we recommend dividing the purchase into single use aliquots by dissolving the lipids in CM, dividing the solution into aliquots, and then re-drying them by either vacuum or lyophilization (preferred) (see Note 3).

3. Lipids for the lipid filter assay should be dissolved in either CM, CMW, or other appropriate solvent at a concentration of 200 pmol/μL (see Table 1) (see Note 3). For a 50 μg tube of PtdIns-3-P we dissolve the dry lipid in 270 μL of CM solution directly in the sealed container (see Note 4) (see Fig. 1a).

4. Transfer 250 μL of the lipid from the stock container to a macrovial insert tube. This is the 200 pmol/μL stock solution.

5. To prepare a 100 pmol/μL dilution, 125 μL of the 200 pmol/μL stock solution should be transferred to a new macrovial insert tube. 125 μL of CMV or appropriate solvent should then be added (see Note 5).

6. A series of twofold dilutions should then be carried out to prepare solutions of 50, 25, and 12.5 pmol/μL. Further twofold dilutions may be made if desired (see Note 6).

7. The diluted lipids in the macrovial tubes should be stored at –80°C until required (but no longer than 1 week).

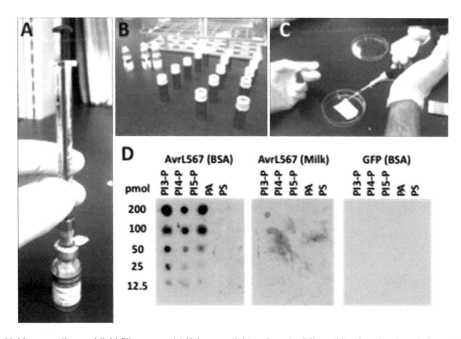

Fig. 1. Lipid preparation and lipid filter assay. (**a**) Using gas tight syringe to deliver chloroform/methanol (CM) solution to commercially purchased dry lipids. (**b**) Prelabeling all macrovial containers and adding CM solution for dilutions. (**c**) Preparation of lipid filters. Note finger covering open macrovial. The top right corner of the lipid blot has been trimmed to keep reference orientation. The Hybond wrapper is still under hybond-C membrane. (**d**) Results of lipid filter assay for full length AvrL567 fused to GFP with BSA and milk blocking agents, and a GFP control. X-ray film exposure 2.5 min. 10 μg protein in 10 mL overnight incubation at 4°C. Note near-complete loss of binding due to inappropriate blocking agent milk. *PI-3-P* phosphatidylinositiol-3-phosphate; *PI-4-P* phosphatidylinositiol-4-phosphate; *PI-5-P* phosphatidylinositiol-5-phosphate.

Table 1
Dilution of lipids

Lipid	Type	MW	Amount (μg)	Solvent	Volume of solvent (μL)
PtdIns-3,4-P	Dipalmitoyl	1036.9	100	CMW	482
PtdIns-4,5-P	Dipalmitoyl	1022.1	50	CMW	245
PtdIns-3,4,5-P	Dipalmitoyl	1138.9	500	CMW	2,195
PtdIns-3-P	Dipalmitoyl	925.1	50	CM	270
PtdIns-4-P	Dipalmitoyl	925.1	50	CM	270
PtdIns-5-P	Dipalmitoyl	925.1	50	CM	270
PtdIns	Dipalmitoyl	828.1	50	CM	302

3.2. General Information on Lipid Filter Assays

The purpose of the lipid filter assay is to provide an initial screen of a protein of interest against a large variety of lipids. For testing a wide assortment of lipids the lipid filter assay is much more economical than the preparation of liposomes or quantitative measurements such as isothermal calorimetry (ITC). Using a gradient of lipid concentrations on a filter can also provide a semiquantitative estimate of the strength of an interaction. Detectable interaction of a protein with only the highest level of lipid (200 pmol) is an indication of weak binding and a possible artifact. Interaction of a protein with a lipid spotted at less than 25 pmol gives confidence in the interaction. Note that commercial lipid strips typically contain 100–200 pmol of each lipid, so that a positive result with such a strip is not a reliable indication of biologically significant binding. The lipid filter assay also suffers from not providing a physiological presentation of the lipids to proteins. For example, Narayan and Lemmon (44) pointed out that a given phospholipid such as PtdIns-4,5-P may constitute only 1% of a biological membrane with the remainder consisting of other lipids such as Ptd-Cho, Ptd-Eth, and Ptd-Ser (44). Thus assaying for a protein interaction with a 100% pure phospholipid preparation may not provide a reliable indication of biologically significant binding. Despite these limitations, the lipid filter assay does provide an easy initial screen for binding of a protein against many phospholipids, and the use of a gradient does provide a semiquantitative indicator of the strength of binding. However, lipid filter results must be validated through the use of additional assays such as liposome binding, ITC, SPR, or NMR.

There are several variations in the protocols for the lipid filter assay. A primary antibody that is directly conjugated to HRP may be used as a single step detection probe. The incubations may be done at varying temperature and times; two standard variations are at room temperature for 1–2 h and at 4°C for overnight incubation. Different detection methods can be used such as alkaline phosphatase conjugated to an antibody or radioactively labeled protein can be generated to directly probe the lipid filters without the use of antibodies (44). Another source of variation in the technique is the choice of blocking buffer. High-grade blocking agents such as albumin from bovine serum work very well and reduce a large amount of background binding. We have found using lower grade blocking agents such as dried milk powder can prevent lipid-protein interactions (see Fig. 1d). Echelon Biosciences also comments that certain protein-lipid interactions are not altered through the use of milk while other interactions are altered or lost (45). We strongly recommend that when performing this assay, include both positive and negative controls to verify that any modifications to the protocol are not impacting known interactions.

3.2.1. Lipid Filter Preparation

1. Handle Hybond-C Extra membranes wearing gloves, and cut membranes to appropriate sizes for the number of lipids used for the experiment (see Fig. 1) (see Note 7).

2. Each membrane should be placed in its own Petri-dish with the protective backing in place and in contact with the Petri-dish (see Note 8).

3. Take the lipids out of the –20 or –80°C storage, gently warm them to room temperature, e.g. with your hand, and then mix using gentle finger tapping.

4. Pipette 1 μL of each lipid onto the membrane using decreasing concentrations from 200 to 12.5 pmol/μL (see Note 9) (see Fig. 1c).

5. Place the lid of the Petri-dish over the dish containing the membrane(s), wrap the Petri-dishes in aluminum foil, then leave at room temperature for a minimum of 1 h and a maximum of 4 h (see Note 10).

3.2.2. Lipid Filter Assay

The methods for assaying the lipid filters in this section assume that each lipid filter is relatively small in size, e.g., 2.5 × 4 cm and thus requires only 10 mL of solution. A larger filter, e.g. 7.5 × 4 cm, would require 20 mL and the volumes used in the protocol should be scaled up.

1. Remove the protective backing from each lipid filter and add 10 mL of the blocking buffer to the Petri-dish. Each Petri-dish should contain a single lipid filter. Replace the Petri-dish lid and cover each dish with a piece of aluminum foil including the edges to keep the blots in the dark.

2. The lipid filters should be shaken at ~85 rpm for 1 h at room temperature and should be kept covered with foil throughout the experiment.

3. Pour off the lipid blocking buffer and add 10 mL of PBST. Add 1–10 μg of the protein of interest to the lipid filters. Shake the lipid filters overnight (14–16 h) at ~85 rpm at 4°C, preferably in a cold room (see Note 11). Do not incubate the filters longer than this, as the more hydrophilic lipids (e.g., PtdInsPs) will begin to leach from the filter.

4. Bring the lipid filters and shaker out of the cold room and back into the lab.

5. Wash the lipid filters 3 times with 10 mL of PBST for 10 min at ~85 rpm at room temperature, thoroughly removing the PBST after each wash.

6. Add 10 mL of fresh PBST to the lipid filters. 1 μL of primary antibody (1 mg/mL) should be added to the lipid filters (see Note 12). Incubate the lipid filters at ~85 rpm for 1 h at room temperature. For this example, the primary antibody should bind GFP which is a C-terminal fusion for AvrL567 (see Fig. 1).

7. Wash the lipid filters 3 times with 10 mL of PBST for 10 min at ~85 rpm at room temperature, thoroughly removing the PBST after each wash.

8. Add 10 mL of fresh PBST to each lipid filter. 1 μL of secondary antibody should be added to the lipid filter (1 mg/mL starting concentration), i.e., a 1:10,000 dilution (see Note 12). The lipid filter should be incubated at ~85 rpm for 1 h at room temperature, but no longer (see Note 13).

9. Wash the lipid filters 2 times with 10 mL of PBST each for 10 min at ~85 rpm at room temperature, thoroughly removing the PBST after each wash.

10. Rinse the lipid filters in 10 mL of PBS (not PBST) for 10 min at ~85 rpm at room temperature and leave the lipid filters in the PBS.

11. Immediately take the lipid filters for development.

12. Take the lipid filters and solutions to the dark room.

13. In the dark, on a large sheet of saran-wrap, add 0.5 mL drops of Super Signal West Pico Stable Peroxide Solution, one for each filter, separated from each other.

14. Add 0.5 mL Super Signal West Pico Luminol/Enhancer Solution to each drop of the peroxide solution.

15. Mix each 1 mL drop with gentle pipetting.

16. Remove each filter from the Petri-dish using forceps and gently dab each one on a dry region of the saran-wrap, before placing the filter face downward in a 1 mL drop of solution for 5 min.

17. Remove the filter from the solution and place the wet filter in a developing cassette on a clear plastic sheet facing upward for exposure to X-ray film.

18. Place clear plastic sheet over the wet lipid filters to avoid contact with X-ray film.

19. Immediately expose each filter to X-ray film repeatedly for times ranging from 30 s to 10 min (see Note 14).

20. Develop X-ray films in a developer.

3.2.3. Evaluating the Results

The results of the lipid filter assay can vary from experiment to experiment. We strongly recommend having a negative and positive control every time they are probed with the same primary and secondary antibody. Figure 1d shows three examples of blots. Positive interactions are observed between AvrL567-GFP and PtdIns-3-P (all the way to 12.5 pmol) along with some binding to PtdIns-4-P and PtdIns-5-P (50 pmol). We can have confidence in this binding reaction since the negative control GFP did not bind any of the phospholipids, while AvrL567-GFP bound all five dilutions quite strongly with dark spots. Note the near-complete loss of binding by AvrL567 when an inappropriate blocking buffer, such as milk, is used. With the use of milk AvrL567 produces faint circles that are not much darker than the background signal on

X ray film. Further confidence in the binding reaction (when BSA is used) is provided by the exposure time required for these blots. The exposure time of 2.5 min was well below the maximum of 10 min. A moderate initial concentration of protein was used, namely 1 μg/mL (10 μg in 10 mL). This is a standard concentration of protein for the experiments. Thus from this experiment we can conclude that AvrL567-GFP likely binds PtdIns-3-P and there is an indication of some affinity for PtdIns-4-P and PtdIns-5-P. To confirm the interaction, another experimental system such as the liposome assay described below should be used.

3.3. Liposome-Binding Assay

The liposome-binding assay provides an experimental system where the protein interacts with the target phospholipid in the context of a membrane-like environment, diluted by the presence of membrane phospholipids. This results in an added level of biological context. The binding assay can be evaluated semiquantitatively, as described here, or it can be quantified, for example by measuring protein fluorescence (18). Liposomes are relatively easy to prepare, require minimal materials, and do not require expensive instrumentation such as required for ITC, SPR, and NMR, nor the very large amount of protein and lipid required for ITC and NMR. Caveats for the liposome assay include the possibility of the protein precipitating in the presence of the liposomes and variability in the quality of liposome preparations. Precipitated protein can give a false appearance of (stronger) binding. Control liposomes lacking the lipid of interest should thus be assayed concurrently to determine if the proteins are precipitating or binding nonspecifically.

3.3.1. Preparing the Mini-Extruder for Preparation of Liposomes

Figure 2 shows the labeled parts, and pictures for each step below. The following section will be clearer if Fig. 2 is referred to while reading through each step.

1. Dip the two filter supports and the polycarbonate membrane in filter-sterilized water and let sit for a few minutes (see Fig. 2a).

2. Place one wet filter support on top of each internal membrane support (see Fig. 2b, d).

3. Add the wet polycarbonate membrane on top of one of the filter supports so that it is sitting evenly on the O-ring. It does not matter which filter support you decide to place the membrane on top of (see Fig. 2c).

4. Take the internal membrane support that has just the filter support and add it on top of the other internal membrane support that is holding the membrane. This creates a stack. Make sure the polycarbonate membrane is sealed by the O-rings (see Note 15). You should have, from top to bottom: internal membrane support, filter support, polycarbonate membrane, filter support, internal membrane support (see Fig. 2e).

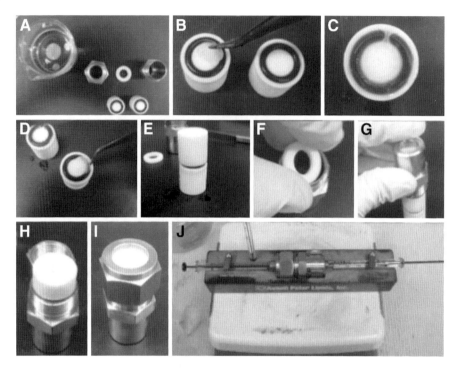

Fig. 2. Assembly of liposome extruder. (**a**) Components of extruder. *Top row left to right*: in the water, 2 filter supports and 1 polycarbonate membrane 0.1 μm, retaining nut, teflon bearing, extruder outer casing. *Bottom row*: two internal membrane supports. (**b**) Placing filter support onto internal membrane support. (**c**) Polycarbonate membrane over filter support and internal membrane support from (**b**). (**d**) Placing filter support onto the other internal membrane support. (**e**) Stacking the internal membrane supports sealing the polycarbonate membrane and filter supports between the two internal membrane supports. (**f**) Placing the teflon bearing inside the retainer nut. (**g**) Placing the extruder outer casing over the stack. (**h**) Flipping over the apparatus. (**i**) Adding the retainer nut over the stack. Tighten securely using only hands. (**j**) Fully assembled extruder in Extruder Heating Adapter on top of hot plate with gas tight syringes containing lipids attached.

5. Place the teflon bearing inside the retaining nut (see Fig. 2f).

6. Place the extruder outer casing over the stack (see Fig. 2g, h).

7. Add the retainer nut to the stack and tighten with your hands (see Fig. 2i).

8. Take both syringes, one filled with Liposome Buffer solution, and attach them to the lipid extruder.

9. Push the liposome buffer solution through the lipid extruder. A slight back pressure should be felt during this process if the there are no leaks (see Note 15).

10. Discard the liposome buffer solution from the syringe.

11. Set the Avanti Extruder Heating Adapter to the desired temperature. Place the assembled extruder in the heating block on top of a hot plate 10–15 min prior to lipid extrusion. For most liposomes a phase transition temperature between 55 and 60°C is suitable (see Notes 16 and 17) (see Fig. 2j).

3.3.2. Liposome
Preparation

The amount of liposomes produced can be varied depending on experiment size. The following protocol is sufficient to make five experimental liposome preparations (50 μL each). Bulk liposome preparations should always be made in larger quantities than desired. For example, 300 μL of liposomes should be made for five experiments (250 μL total required), 600 μL of liposomes should be made for 10 experiments (500 μL total required) (see Table 2). The following instructions are to test binding to PtdIns-3-P, but other target lipids may be substituted for PtdIns-3-P.

1. Add 22 μL of Ptd-Cho (10 μg/μL) to the macrovial tube labeled control and 19.8 μL to a tube labeled PI3P.

2. Add 8 μL of Ptd-Eth (10 μg/μL) to the macrovial tube labeled control and 7.2 μL to the tube labeled PI3P.

3. Add 30 μL of PtdIns-3-P (1 μg/μL) to the tube labeled PI3P.

4. Tighten the lids on the tubes and mix well by finger tapping.

5. Open the tubes and dry off the solvent by either vacuum, under a stream of nitrogen, or lyophilization. This process will take approximately 45–60 min, but can take longer depending on the size of the preparation.

6. Add 300 μL of liposome buffer to the liposomes to bring the final liposome concentration to 1 μg/μL (see Note 18).

7. Thoroughly vortex the liposome mixture for 2 min. Let the mixture sit for 1 h at room temperature (see Note 19).

Table 2
Liposome formula

	Phosphatidylcholine	Phosphatidylethanolamine
Control preparation		
Ratio	5.00	2.00
Percentage	71%	29%
1 μM lipid Mixture	0.71 μM	0.29 μM
766 μg/L	561	205
100 μg liposomes	73.3 μg	26.7 μg
300 μg liposomes	219.9 μg	80.1 μg
μL of lipid for 300 μg liposomes	22 μL	8 μL
Variable preparation		
300 μg (90%)	198 μg	72 μg
μL of lipid (10 μg/μL)	19.8 μL	7.2 μL
μL of PtdIns-3-P (1 μg/μL)	30 μL	

8. Freeze-thaw the liposome preparation 3 times using a liquid nitrogen and a warm water bath. Allow the liposomes to thaw each time and then mix by tapping the tube with your finger.

9. Clean gas tight syringes by rinsing with deionized (DI) water repeatedly (at least 5 times). Let air dry.

10. Using the gas tight syringe, transfer your lipid emulsion to the lipid extruder. Insert the gas tight syringe gently and slowly into the device (see Fig. 2j).

11. Place the connected miniextruder into the prewarmed Extruder Heating Adapter (55–60°C).

12. Push the sample through the lipid extruder slowly so that all of the samples are transferred from one syringe to the other. Repeat this process 20 times (see Fig. 2j).

13. Transfer the liposomes from the gas tight syringe to a 1.5-mL microfuge tube and make 50 μL aliquots into 1.5-mL microfuge tubes.

3.3.3. Conducting the Assay

1. Precentrifuge proteins at $100,000 \times g$ for 15 min, to remove any protein aggregates. If an ultracentrifuge is unavailable, centrifuge for 20 min at maximum speed ($\sim 45,000 \times g$) in large centrifuge (e.g., Sorvall) (see Note 20).

2. Add 10 μL of protein (1 μg/μL) to 50 μL of the control liposome preparation and to 50 μL of the PtdIns-3-P liposomes.

3. Mix the protein liposome mixtures with gentle finger tapping.

4. Incubate at room temperature for 1 h. Cover with aluminum foil.

5. Load 1.5-mL microfuge tubes containing the samples into an ultra-centrifuge and centrifuge at $100,000 \times g$ for 15 min. If an ultra-centrifuge is unavailable, centrifuge for 20 min at maximum speed ($\sim 45,000 \times g$) in large centrifuge (see Note 20).

6. Carefully remove the tubes from centrifuge and gently pipette all of the supernatant into a new microfuge tube without disturbing the pellet (see Note 21).

7. Add 5 μL of 1× loading buffer to the pellet and add 6 μL of 10× loading buffer to the supernatant. Mix gently.

8. Boil samples for 5 min and load samples onto SDS-PAGE gel (see Note 22).

9. Perform SDS-PAGE till dye front has reached the end of the gel.

10. Soak gels in staining solution for 2 h.

11. Soak gels in destaining buffer until coomassie dye is removed.

12. Image gels (see Fig. 3a, b).

3.3.4. Evaluating Results

The results of the liposome assay are straightforward. Binding of protein to the liposomes results in appearance of the protein in the pellet fraction. In many cases, all of the input protein will bind to

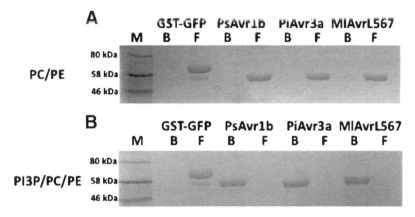

Fig. 3. Liposome-binding assay. Liposome assay of the binding of GTS-GFP, *Phytophthora sojae* Avr1b, *Phytophthora infestans* Avr3a, and *M. lini* AvrL567 to PI-3-P. (**a**) Binding to control liposomes made with Ptd-Eth and Ptd-Cho. (**b**) Binding to liposomes containing PtdIns-PE, PtdIns-PC, and 10% PtdIns-3-P. *B* bound; *F* free; *M* markers.

the liposomes containing the target lipid. Figure 3a shows an example: the control protein (GFP) does not bind to the control liposomes or the liposomes made with PtdIns-3-P. If the GFP or the protein of interest had bound the control liposomes, the experimental conditions should be modified either by raising the pH and/or by increasing the salt concentration so as to decrease the electrostatic interactions between the lipid and protein. Nonspecific electrostatic binding is a more common problem if the liposomes contain an anionic lipid such as Ptd-Ser. If the control GFP had bound to the PtdIns-3-P liposomes but not to the control liposome it might have been that the concentration of PtdIns-3-P was too high and the concentration of PtdIns-3-P should be reduced. If the protein of interest did not bind the control or PtdIns-3-P liposomes or only partially binds the PtdIns-3-P liposomes, but not the control liposomes, try modifying the salt concentrations and pH to create a more physiological environment. In many cases, altering the experimental conditions by pH and salt can convert partial binding into full binding (46), but in such cases attention should be paid to the negative controls to avoid artifactual binding. Greater physiological relevance can be revealed by understanding the role of pH and salt in a protein-lipid interaction.

4. Notes

1. For our experiments we have used synthetic lipids generated from Cayman Chemical and naturally occurring lipids from Avanti polar lipids. When selecting lipids for screening, be mindful of number and location of double bonds, and chain length in relation to physiological relevance. For the lipid filter

assay a specific fraction of BSA (fatty acid free; Sigma Aldrich, Catalog # A7030) should be used. Other types of BSA and blocking agents contain fatty acids and other forms of lipids, which can hinder certain lipid interactions (see Subheading 3.2.3). The primary and secondary antibodies used above have been proven reliable in our hands and give consistent results between replicates. They have strong affinity to their respective ligand and generate minimal background signal during exposure. We recommend these antibodies when probing for His-tag, GFP, and GST-fusion proteins.

2. Lipid filters can be various sizes due to the number of lipids blotted. A lipid filter with three different lipids blotted with their gradient dilutions would be about 2.5×4 cm in size and require only 10 mL of solution to submerge and cover the blot. A lipid filter with 13 different lipids and their gradient dilutions (approx. 7.5×4 cm) would require 20 mL of solution to submerge and cover the blot.

3. Add the CM solution directly to the sealed container of dried lipids using a gas tight syringe. Do not open the vial under negative pressure as it may disperse the dried lipids.

4. Since the lipids are under negative pressure and are lyophilized, adding 270 µL of CM or CMW will result in the lipids only being dissolved in ~250 µL. We recommend adding a little bit more solvent to make sure the final volume is suitable. This can be done after adding the initial CM or CMW and then using a gas tight syringe to measure the volume.

5. This note is very important. CM and CMW evaporate very quickly, especially if the work is being done in a fume hood. The macrovial tubes must have the 350-µL insert in them or else your lipid solutions will evaporate in a matter of minutes, even in a sealed container. The macrovial tubes with the 350-µL insert create a smaller interface between the surface of the liquid and the surrounding air. The decrease in exposed surface area prevents rapid evaporation during use and storage. An unfortunate example is the following early experience. When we first performed the lipid dilutions in macrovial tubes without the inset, we made enough solution for 150 experiments. We did not finish making 20 lipid filters before we ran out of solution due to evaporation. Conversely when using the tubes with the 350-µL inserts we can easily get 75–100 blots from 125 µL of solution. Evaporation also results in uncontrolled amounts of lipid being placed in the filters due to concentration of the lipids in the stock solution.

6. Gloves must be worn when handling the lipids. Prelabel all the macrovial tubes and fill them with 125 µL of CM or CMW, if required, prior to beginning the dilutions. Immediately add

the macrovial screw cap back onto the macrovial once the solvent is in there. Evaporation can be quite a nuisance at this step and can alter lipid concentrations uncontrollably. It is strongly recommended to keep the lids on the macrovials loose and then quickly pipette from one container to the rest during the dilutions. Replace the lids, tighten them, and then mix.

7. After cutting the membranes we suggest to mark very lightly with pencil where you are going to put every 1 µL of lipid solution on the filter directly. Once you have pipetted the 1 µL of lipid solution the spot dries very quickly and is hard to see unless the light catches it from the correct angle. The dried spots become easier to view after you have been doing the experiments regularly for a few months, but initially can be quite difficult to see. The pencil marks also allow you to keep straight rows and columns, and keep track of where the lipids should go.

8. Leave the protective back sheet on the hybond-C Extra membrane while adding the lipids and while they dry. If the protective backing is removed during the drying process, the blots in contact with the Petri-dish or another plastic surface will melt onto the surface due to the nature of the CM mixture.

9. When spotting lipids onto the filters it is always easier to spot all membranes with a given lipid concentration and then proceed to spot all the membranes with the next given lipid solution. This saves a lot of time screwing and unscrewing the macrovial tubes. We use the thumb technique, i.e., when not pipetting solution out of the macrovial tube, cover the opening with your thumb while adding the lipid to the lipid filters. A large volume may be lost to evaporation when not covering the opening with your thumb. Also, certain lipids have a sensitivity to light, so we recommend that you pipette the lipids in a low light environment (see Fig. 1c).

10. Freshly prepared lipid filters should be dried for 1 h to allow for complete drying. After prolonged drying the lipids may begin to oxidize in the air. We do not recommend storing them at 4°C since the lipids may degrade. We have seen increased nonspecific binding when the lipid filters are stored at 4°C overnight or for long period of time. This is also true for commercial lipid strips. Commercial strips should be used within 1–2 days of arrival. For the same reason, some batches of commercial strips have given irreproducible results because they were not prepared freshly prior to shipment.

11. The concentration of protein used for lipid filter-binding assays depends on the affinity of the proteins for the target phospholipids. In very few cases 1 µg of protein in 10 or 20 mL is sufficient to observe binding. In other cases, 10 or 20 µg of protein in 10 or 20 mL respectively is required. Using more than 20 µg or

more than 1 µg/µL of protein can lead to nonspecific background binding. In many cases the effects of mutations in the putative lipid-binding domain can be masked due to the use of an excessive amount of protein. It is strongly recommended that the mutant and wild-type proteins be tested at the concentration at which the wild-type protein produces an even and nonsaturated binding signal.

12. Antibodies have various binding affinities. For the anti-GFP antibody described (see Materials list), best results were obtained with a 1:10,000 dilution. For the anti-HIS tag antibody described, best results were obtained with a dilution of 1:5,000. It is strongly recommended that the user spend time optimizing the dilution of each new batch of primary antibody. Ideally the same antibody preparation should be used for all proteins analyzed across a project. This allows the user to readily compare a range of interactions. For the secondary antibodies described, we find that no more than a 1:10,000 dilution should be used. Effector proteins can be generated in a variety of ways with a variety of tags. We have used both N and C-terminus His tags, and C-terminus GFP tags, and N-terminus GST tags as probes for their respective primary antibodies.

13. Do not let the secondary antibody incubate for even 5 min over an hour. The antibody has a tendency to bind the membrane with overincubation and this results in dirty lipid filter images when exposed to X-ray film.

14. Certain protein-lipid interactions are stronger than others. In the case of PsAvr1b and PiAvr3a, the interaction is so strong that the protein concentration can be lowered to 0.25 µg/mL and a strong interaction can still be seen. We strongly discourage using protein concentrations over 1 µg/mL as nonspecific binding becomes a problem. A negative control such as GST and GFP should always be used. We suggest adjusting experimental conditions such as altering salt concentrations, trying different blocking buffers, and pH ranges to increase or decrease observed binding. If the negative control binds at the same relative affinity as the protein of interest it cannot be concluded that the protein of interest interacts specifically with that lipid under those experimental conditions.

15. The polycarbonate membrane must be evenly lined up over the O-ring. If the membrane is not aligned perfectly over the O-ring to create a seal, the liposomes will not be formed. This can easily be discovered by feeling the back pressure when extruding the liposome buffer with the gas tight syringe. If the membrane is not aligned there will be little or no back pressure when pushing on the syringe. A small leak will prevent the liposomes from being formed correctly and will cause erroneous results during the assay.

16. Certain liposome mixtures require a warmer temperature to pass the phase transition temperature (47). We recommend heating all liposomes as it helps with hydration, keeps the lipids more fluid through the membrane during extrusion, and we see less aggregation of lipids. Temperature can be determined by looking up the composition of your liposomes in (47) or by consulting the same online table at http://Avantilipids.com "Phase Transition Temperatures for Glycerophospholipids."

17. Monitoring the temperature to prevent overheating is very important, not only to prepare the liposomes properly, but to prevent damage to equipment, especially the gas-tight syringes. Temperatures should not fluctuate past 75°C at anytime during the warming process. The hot plate must reach the desired temperature and stabilize at the temperature.

18. pH and salt concentration are two important factors in the liposome assay that can lead to misinterpretation of an interaction. The monovalent cation concentration should not be lowered below 50 mM and we find that concentrations above 200 mM reduce some known interactions, but be aware that these ranges may differ for your protein of interest. Certain proteins such as the Vps27p-FYVE domain require an electrostatic interaction enabled by moderate salt concentrations (50–100 mM KCl) to facilitate binding (48). We therefore recommend starting at 100 mM NaCl and lowering the concentration if necessary for weaker interactions. The strength of an interaction may also be impacted by pH. As the pH is lowered a protein may become more positively charged and have a stronger interaction. Conversely a domain might require a higher pH to fold properly and higher pH values can help remove background binding (10–20%). We do not recommend performing the experiments outside the pH range of 6.8–8.0.

19. The purpose of the 1 h incubation in buffer at room temperature is to allow the lipids to hydrate. If the liposome preparation has a phase transition temperature that is above room temperature then perform the 1 h incubation at the phase transition temperature. If the lipids fail to resuspend for some unknown reason, the solution can be incubated at 37°C for 15–30 min during the 1 h incubation to help with the lipid hydration.

20. In any given protein preparation there is a certain degree of protein that is denatured. Precentrifugation of protein at the same g force used to centrifuge the liposomes, prior to performing the liposome assay, is necessary to remove all precipitated protein. Not performing centrifugation prior to the liposome assay results in precipitated protein in the pellet of the liposomes. This will lead to erroneous results. Do not centrifuge large volumes (1 mL) of protein and then pipette out the entire supernatant. Simply take the desired volume of protein from

the top of the supernatant, to avoid inadvertently recovering any precipitated protein.

21. After centrifugation, the liposome pellet can readily be disturbed. A centrifuged tube should be taken out gently and the supernatant should be very gently pipetted out immediately. Then the next centrifuged liposome preparation should be handled. Do not take out all the tubes then let them sit as you gather materials. Have everything prepared ahead so you can separate the samples immediately after centrifugation. The longer you take to perform this task the more likely the pellet can become disturbed. Do not take 15 min to slowly remove the supernatant from all tubes. Work quickly with precision to separate the supernatant from the pellet. In case accidental mixing does occur, re-centrifuge the sample(s).

22. There may be instances where the volume of the supernatant sample is greater than the volume of the well in the gel. In this case, load as much of the sample as possible. We do not recommend drying down the sample prior to addition of the loading buffer as it may lead to protein degradation. If lack of sufficient well volume becomes a common occurrence prepare SDS-PAGE gels using wider spacer gaps that are greater than the traditional 1 mm.

References

1. Kooijman EE et al (2003) Modulation of membrane curvature by phosphatidic acid and lysophosphatidic acid. Traffic 4: 162–174.

2. Athenstaedt K, Daum G (1999) Phosphatidic acid, a key intermediate in lipid metabolism. Eur J Biochem 266: 1–16.

3. Delon C et al (2004) Sphingosine kinase 1 is an intracellular effector of phosphatidic acid. J Biol Chem 279: 44763–44774.

4. Young BP et al (2010) Phosphatidic acid is a pH biosensor that links membrane biogenesis to metabolism. Science 329: 1085–1088.

5. Anthony RG et al (2006) The *Arabidopsis* proteinkinase PTI1-2 is activated by convergent phosphatidic acid and oxidative stress signaling pathways downstream of PDK1 and OXI1.

6. Park J et al (2004) Phosphatidic acid induces leaf cell death in *Arabidopsis* by activating the rho-related small G protein GTPase-mediated pathway of reactive oxygen species generation. Plant Physiol 134: 129–136.

7. Daleke DL (2003) Regulation of transbilayer plasma membrane phospholipid asymmetry. J Lipid Research 44: 233–242.

8. An X et al (2004) Phosphatidylserine binding sites in erythroid spectrin: location and implications for membrane stability. Biochemistry 43: 310–315.

9. Sahu SK et al (2007) Phospholipid scramblases: an overview. Arch Biochem Biophys 462: 103–114.

10. Zwaal RF et al (2005) Surface exposure of phosphatidylserine in pathological cells. Cell Mol Life Sci 62: 971–988.

11. Vermeet JEM et al (2009) Mapping phosphatidylinositol 4-phosphate dynamics in living plant cells. Plant J 57: 356–372.

12. Garofalo RS et al (2003) Severe diabetes, age-dependent loss of adipose tissue, and mild growth deficiency in mice lacking Akt2/PKB beta. J Clinic Invest 112: 197–208.

13. Franke TF et al (2003) PI3K/Akt and apoptosis: size matters. Oncogene, 22: 8983–8998.

14. Godi A et al (2004) FAPPs control Golgi-to-cell-surface membrane traffic by binding to ARF and PtdIns(4)P. Nat Cell Biol 6: 393–404.

15. D'Angelo et al (2008) The multiple roles of PtdIns(4)P- not just the precursor of PtdIns(4,5) P_2. J Cell Sci 121: 1955–1963.

16. Wurmser AE, Emr SD (1998) Phosphoinositide signaling and turnover. PtdIns(3)P, a regulator of membrane traffic, is transported to the vacuole and degraded by a process that requires luminal vacuolar hydrolase activities. EMBO Journal 17: 4930–4942.

17. Zoncu R et al (2009) A Phosphoinositide switch controls the maturation and signaling properties of APPL endosomes. Cell 136: 1110–1121.

18. Kale SD et al (2010) External lipid PI3P mediates entry of eukaryotic pathogen effectors into plant and animal host cells. Cell 142: 284–295.

19. Stahelin RV (2009) Lipid binding domains: more than simple lipid effectors. J Lipid Res 50: 299–304.

20. Colon-Gonzalez F et al (2006) C1 domains exposed: From diacylglycerol binding to protein-protein interactions. BBA- Mol Cell Biol L 1761: 827–837.

21. Bolsover SR et al (2003) Role of Ca^{2+}/phosphatidylserine binding region of the C2 domain in the translocation of protein kinase Cα to the plasma membrane. J Biol Chem 278: 10282–10290.

22. Craig KL, Harley CB (1996) Phosphorylation of human pleckstrin on Ser-113 and Ser-117 by protein kinase C. Biochem J 314: 937–942.

23. Dowler S et al (2000) Identification of pleckstrin-homology-domain containing proteins with novel phosphoinositide binding specificities. Biochem J 351: 19–31.

24. Gaullier JM et al (2004) FYVE finger bind PtdIns(3)p. Nat Cell Biol 5: 393–404.

25. Lee SA et al (2006) Molecular Mechanism of Membrane Docking by the Vam7p PX Domain J Biol Chem 281: 37091–37101.

26. McLaughlin S et al (2005) Reversible - through calmodulin - electrostatic interactions between basic residues on proteins and acidic lipids in the plasma membrane. Biochem Soc Symp 72: 189–198.

27. Ghosh S et al (1996) Raf-1 kinase possesses distinct binding domains for phosphatidylserine and phosphatidic acid. J Biol Chem 271: 8472–8480.

28. Grange M et al (2000) The cAMP-specific phosphodiesterase PDE4D3 is regulated by phosphatidic acid binding. J Biol Chem: 275, 33379–33387.

29. Dou D et al (2008) RXLR-mediated entry of Phytophthora sojae effector Avr1b into soybean cells does not require pathogen encoded machinery. Plant Cell 20: 1930–1947.

30. Rafiqi M et al (2010) Internalization of flax rust avirulence proteins into flax and tobacco cells can occur in the absence of the pathogen. Plant Cell 22: 2017–2032.

31. Whisson SC et al (2007) A translocation signal for delivery of oomycete effector proteins into host plant cells. Nature 450: 115–118.

32. Jiang RHY et al (2008) RXLR effector reservoir in two Phytophthora species is dominated by a single rapidly evolving superfamily with more than 700 members. Proc Natl Acad Sci USA 12: 4874–4879.

33. Dou D et al (2008) Conserved C-terminal motifs required for avirulence and suppression of cell death by Phytophthora sojae effector Avr1b. Plant Cell 20: 1118–1133.

34. Bos JI et al (2007) The C-terminal half of Phytophthora infestans RXLR effector AVR3a is sufficient to trigger R3a-mediated hypersensitivity and suppress INF1-induced cell death in Nicotiana benthamiana. Plant Journal 48: 165–76.

35. Sohn KH et al (2007) The Downy mildew effector proteins ATR1 and ATR13 promote disease susceptibility in Arabidopsis thaliana. Plant Cell 19: 4077–4090.

36. Lafont F et al (2004) Bacterial subversion of lipid rafts. Curr Opin Microbiol 7: 4–10.

37. Murata-Kamiya N et al (2010) Helicobacter pylori Exploits Host Membrane Phosphatidylserine for Delivery, Localization, and Pathophysiological Action of the CagA Oncoprotein. Cell Host Microbe 7: 338–339.

38. Rüter C et al (2010) A newly identified bacterial cell-penetrating peptide that reduces the transcription of pro-inflammatory cytokines. J Cell Sci 123: 2190–2198.

39. Schoebel, S et al (2010) High-affinity binding of phosphatidylinositol 4-phosphate by Legionella pneumophila DrrA. EMBO reports 11: 598–604.

40. Weber SS et al (2006) Legionella pneumophila exploits PI(4)P to anchor secreted effector proteins to the replicative vacuole. PLoS Pathog 2(5): e46. doi:10.1371/journal.ppat.0020046.

41. Brombacher E et al (2009) Rab1 guanine nucleotide exchange factor SidM is a major PtdIns(4)P-binding effector protein of Legionella pneumophila. J Biol Chem 284: 4846–4856.

42. Tawk L et al (2010) Phosphatidylinositol 3-phosphate, an essential lipid in Plasmodium, localises to the food vacuole membrane and the apicoplast. Eukaryotic Cell doi:10.1128/EC.00124–10.

43. Lee SW et al (2005) PilT is required for PI(3,4,5) P3-mediated crosstalk between *Neisseria gonor-rhoeae* and epithelial cells. Cell Microbiol 7: 1271–1284.

44. Narayan K, Lemmon MA (2006) Determining selectivity of phosphoinositide-binding domains. Methods 39: 122–133.

45. Echelon Biosciences Inc. (2005). Technical Data Sheet. Q&A P-6000 Rev: 1 (08/08/05). http://www.echelon-inc.com/corp/p-6000%-20qa.pdf.

46. Ju H et al (2009) Membrane insertion of the FYVE domain is modulated by pH. Proteins 76: 852–860.

47. Silvius JR (1982) Thermotropic phase transitions of pure lipids in model membranes and their modification by membrane proteins. In: Jost PC, Griffith OH (eds) Lipid-protein interactions, vol 2. Wiley, New York.

48. Diraviyam K et al (2003) Computer modeling of the membrane interaction of FYVE Domains. J Mol Biol 328: 721–736.

Chapter 25

In Silico Identification and Characterization of Effector Catalogs

Ronnie de Jonge

Abstract

Many characterized fungal effector proteins are small secreted proteins. Effectors are defined as those proteins that alter host cell structure and/or function by facilitating pathogen infection. The identification of effectors by molecular and cell biology techniques is a difficult task. However, with the availability of whole-genome sequences, these proteins can now be predicted *in silico*. Here, we describe in detail how to identify and characterize effectors from a defined fungal proteome using *in silico* techniques.

Key words: Secretome, Effector, Pathogen, Host, Interaction, PHI-base, SignalP, InterProScan, GO Terms, WoLF PSORT

1. Introduction

Whole-genome sequencing has become a popular tool for the study of microbe–host interactions. Genome sequences are available for many fungi, including plant pathogens, symbiotic fungi, and saprophytic fungi, but also for opportunistic mammalian fungal pathogens. Moreover, sequencing new species and additional strains of particular species has become much faster and cheaper with the introduction of next generation sequencing (NGS). Current genome sequencing projects focus on high-throughput methods, as they favor speed, accuracy, and low price to base pair ratios. Available NGS techniques have been reviewed recently by Metzker (1). Sequence assembly and subsequent gene model prediction are the next steps in a genome sequencing project. Various tools are available for sequence assembly and gene model prediction, but precise procedures and methods for these tools are

Melvin D. Bolton and Bart P.H.J. Thomma (eds.), *Plant Fungal Pathogens: Methods and Protocols*,
Methods in Molecular Biology, vol. 835, DOI 10.1007/978-1-61779-501-5_25, © Springer Science+Business Media, LLC 2012

not included in this chapter. Genome assembly methods and algorithms have been reviewed extensively by Miller et al. (2). Furthermore; the SEQanswers wiki (http://seqanswers.com/wiki/Software) and SEQanswers forum (http://seqanswers.com/) contain many links to, and tips on, programs for NGS sequence assembly. Prediction of genes in fungal genomes (or any other eukaryote) can be performed using a variety of different approaches which were recently reviewed by Martinez et al. (3).

To characterize effector catalogs, first the genome is annotated by assigning putative functions to as many genes as possible. Subsequently, the set of secreted proteins, or secretome, is defined and ultimately the putative effector catalog is identified and characterized.

2. Methods

2.1. Genome Annotation

Gene annotation describes methods to deduct (putative) functions from gene sequences. Various methods for large-scale annotation exist, including blast analyses against the nonredundant (*nr*), the *Uniprot* or the *Swissprot* sequence database, and the use of Hidden Markov Models (HMMs) such as those which are deposited in the Pfam database (4). At present, various pipelines are available for automated annotation of a large set of protein sequences like that of a fungal proteome. InterProScan (IPS; (5)) and Blast2GO (B2G) are regularly used for whole-genome annotation (6, 7). For IPS the following procedure is used:

2.1.1. IPS Is Installed Locally on a 64-bit Linux Server

(a) Info and the download repository can be found through: http://www.ebi.ac.uk/Tools/pfa/iprscan.

(b) The initial installation requires the/DATA/section, the pre-compiled binaries (32-bit and 64-bit Linux are supported) and IPS itself (Perl architecture). Decompress all files (according to instructions) using "% gunzip –c fileX.tar.gz | tar –xvf"—and follow the installation instructions as in the "Installing_InterProScan.txt" document (present in the IPS Perl package).

(c) IPS has been developed in Perl5, and requires that various Perl modules are installed beforehand. A list of required modules can be found in the installation manual, and installation should be done by CPAN for convenience (manual for Perl CPAN Shell: http://www.troubleshooters.com/codecorn/littperl/perlcpan.htm).

(d) The IPS installation is basically a configuration process. Run the "%/perl Config.pl" from within the iprscan main directory and answer the questions displayed. Options are not permanent; they can later be modified in the configuration files, or by rerunning the Config.pl script.

2.1.2. Testing the IPS Installation

(a) The IPS package comes with a set of test sequences, located in the fasta formatted file. Run a test analysis from the ./iprscan/ bin/using syntax: "%./iprscan -cli -i ../test.seq -iprlookup -goterms." Each run produces an output directory, containing all the individual files and a file summarizing all the data (importable to e.g., Excel).

2.1.3. Running an IPS Analysis

(a) To identify as much information as possible, run the IPS analyses using all available modules. The modules typically used are HMMPfam, HMMPanther, BlastProDom, FPrintScan, HMMSmart, HMMPIR, HMMTigr, ProfileScan, HAMAP, patternScan, SuperFamily, and Gene3D.

(b) Syntax is: "% ./bin/iprscan –cli –i ./inputseqs.fasta." If initialized using the –iprlookup –goterms syntax, IPS tries to retrieve the corresponding InterPro entry and GO term (useful for further analysis). For problems related to computational size (see Note 1).

(c) Data output can be analyzed using Excel. B2G (6, 7) can also be used for automated annotation.

As B2G is written in Java, it can be used on multiple platforms (such as Windows OS, Linux, and Mac OS). The software is user-friendly, owing to its graphical interface and intuitive applications. We typically use it for annotation, GO-term assignment, and GO-term enrichment analyses. We use the following procedure (largely adapted from the B2G tutorial; http://www.blast2go.org/downloads):

2.1.4. Running a B2G Analysis

(a) Run the B2G suite from the web start, available at: http://www.blast2go.org/start_blast2go. You can run the software by determining the proper amount of memory (depends on the amount available in the machine running the analyses) and clicking the relevant link (e.g., 1,500 or 2,048 MB web start) or by manually changing the link setting (see website).

(b) After installation and initialization, protein fasta files can be loaded by {(File), (Load Fasta File)}. Take care to choose the right format (protein fasta formatted) when opening your data file.

(c) First step in the analysis includes blasting your data against a database {(Blast), (Run Blast Step)}. Various databases are possible, including nr , Swissprot and Refseq but also custom databases can be used if available and formatted locally using the Blast package (8). Various options can be changed when running the Blast analyses, including the number of Blast hits that should be recorded (default is 20), the expect-value (default is 1.0E-03, we use 1.0E-06), the blast algorithm (default is BlastP), and the blast mode (depending on whether you are running the analyses locally) (WWW-blast) or over the NCBI web service (QBlast@NCBI). The latter is advantageous since no local database maintenance is required.

(d) A run for approximately 10,000 proteins (typical for most fungal genomes) takes around 24 h using this approach. If preferred IPS results can be imported or alternatively, IPS can be run from within B2G (see Note 2).

(e) Next, GO-terms can be mapped to your data. To this end, go to {(Mapping), (Run GO-Mapping step)}.

(f) Finally merge the data into the annotation by selecting the {(Annotation), (Run Annotation Step)}.

(g) Data can be exported to e.g., Excel. B2G contains useful tools to extract statistic information from the various analysis steps (see Note 3).

2.2. Secretome Prediction

2.2.1. Introduction

Subcellular localization of protein sequences can be determined using various approaches, including detection of targeting signals (such as the signal peptides, ER retention signals, and nuclear localization signals), but also by a comparative approach (derive the most probable site of activity through homology information). The software programs which are commonly used, and which will be described in this section are SignalP 3.0, Phobius, and WoLF PSORT (9–13). SignalP 3.0 contains two different methods capable of detecting N-terminal signal peptides in proteins targeted to the extracellular space or the mitochondria. SignalP 3.0 can use two distinct methods for signal peptide prediction, i.e., neural network (NN) and HMM. Phobius uses the HMM method based SignalP3.0 algorithm in combination with the transmembrane domain predictor TMHMM2 to discriminate between intracellular, plasma-membrane bound, and extracellular proteins. A completely different strategy, based on feature-selection and the k nearest neighbors (kNN) classifier, is used by WoLF PSORT, a recent extension of the well-known and broadly used programs PSORT and PSORT-II. In addition to these programs a number of alternatives are discussed in this section, including Sigcleave, SigPred, Protein Prowler, and SecretomeP. A comprehensive review on the methods available for the computational prediction of subcellular localization has been published in a previous volume of methods in molecular biology (14). In this section, the most common tools are shortly explained, and a method for genome-scale analysis is proposed.

2.2.2. SignalP 3.0

SignalP 3.0 predicts the presence and location of signal peptide cleavage sites in amino acid sequences (10, 13). The SignalP web server (http://www.cbs.dtu.dk/services/SignalP/) comes with the following set of options: Organism group (Eukaryotes, Gram-negative, and Gram-positive bacteria)

– Method (NN, HMM or both).

– Output format (standard (with graphics), full or short).

– Truncation (default setting is a cutoff at 70 amino acids).

To run a signal peptide prediction for the complete proteome, the short, no-graphics option is most easily applied. Both the NN and HMM prediction method can be used; however, for genome-scale analysis the NN method is preferred, as its accuracy has been shown to be higher as compared to the HMM method (13).

With these options, it is possible to load your proteome in subsets of 2000 protein sequences. Using a simple text editor such as Notepad or WordPad the complete proteome can be divided over multiple files, each containing a maximum of 2000 protein sequences.

Alternatively, the SignalP 3.0 package can be installed locally on your personal computer, or computer cluster depending on necessity and availability. A download is available on the SignalP website, which can be obtained only after signing of the academic license agreement (http://www.cbs.dtu.dk/cgi-bin/nph-sw_request?signalp). Installation instructions and a manual page are found on the same page. However, as the average proteome size of fungi is only around 10,000–15,000 proteins, one would need to run only 5–7 individual web server runs to obtain full results, and therefore running these analyses through the web server is favored for smaller laboratories running these analyses once or for only a few fungal genomes.

The nongraphical output of SignalP consists of two defined sets of results: i.e., one table for the neural network (SignalP-NN) predictions and one table for the hidden Markov model (SignalP-HMM) predictions (see Table 1 for example using *Cladosporium fulvum* Ecp6 data, an extracellular fungal protein involved in inhibition of the chitin-induced plant immunity (15)). The NN algorithm uses two features of a typical signal peptide, i.e., the presence/absence of a signal peptide cleavage site (depicted by the C-score) and the likelihood of a certain amino acid to be part of a signal peptide (depicted by the S-score). The Y-score is derived from both the C-score and the S-score and aims to increase the accuracy of the cleavage site prediction. The S-mean score is derived by averaging the S-score over the signal peptide until the Y-score derived signal peptide cleavage site. With the release of SignalP 3.0 the D-score has been introduced which averages the Y-max and S-mean scores. The D-score (minimum 0.5 for secretory proteins)

Table 1
Typical SignalP3.0 nongraphical output

#SignalP-NN euk predictions													
# Name	Cmax	pos	?	Ymax	pos	?	Smax	pos	?	Smean	?	D	?
C. fulvum Ecp6	0.701	19	Y	0.752	19	Y	0.982	3	Y	0.904	Y	0.828	Y

#SignalP-HMM euk predictions						
# Name	!	Cmax	pos	?	Sprob	?
C. fulvum Ecp6	S	0.956	19	Y	0.998	Y

is used to discriminate between secretory and nonsecretory proteins. This parameter can be varied between 0.4 and 0.6 without major effects on both sensitivity and specificity. Emanuelsson et al. (13) reported that within this range (D-score > 0.4 to D-score > 0.6) sensitivity decreases from 98.8 to 95.1% (3.7% difference) and the rate of false positives (Fp) decreased from 1.4 to 0.4% (1% difference). Similar scores are depicted for the SignalP-HMM based predictions, albeit significantly lower than for the SignalP-NN.

2.2.3. Phobius

Phobius is a HMM which combines transmembrane topology, signal peptide, and signal peptide cleavage site predictions. It has been developed by the same authors that built the SignalP (9, 13) and TMHMM (16) programs, in an attempt to address the issue of overlapping predictions with these two programs (11).

The Phobius web server (http://phobius.sbc.su.se/; (17)) contains only one set of options: the output setup, i.e., short, long without graphics or long with graphics. Use the short output options for whole-genome analysis. The web server runs fast and a complete fungal proteome can be uploaded and run at once (no size restrictions are currently in place). A typical output consists of rows describing the sequence ID, the number of transmembrane domains (TMs), the presence or absence of a signal peptide (SP), and the protein topology in tabular format (Table 2). The output can easily be exported to Excel and further analyzed.

2.2.4. WoLF PSORT

WoLF PSORT (12) is a recent extension to the well-established PSORT-II program (18) but it also uses some PSORT (19) and iPSORT (20) features. WoLF PSORT has been specifically built and trained using various eukaryotic protein sets (including fungal sequences, plant sequences, and animal sequences). The program can predict 12 different compartments or destinations for a protein sequence. It uses information regarding signal peptide sequence, amino acid preference, and homology to other proteins with known subcellular localization. The various features are ranked and summed using a kNN nearest neighbor classifier. At the web server (available at http://wolfpsort.org/), only 250 proteins (file size around 200 Kb) can be uploaded. The (only) input option selects for the type of organism from which the sequence was derived, being animal, plant, or fungi. A typical output (shown in Table 3) consist of single lines per protein sequence describing in tabular

Table 2
Typical short Phobius output

Sequence ID	TM	SP	Prediction
C. fulvum Ecp6	0	Y	n5-13c18/19o*

Table 3
Typical short WoLF PSORT output

k used for kNN is: 27

C. fulvum Ecp6	Details	extr:	27.0

format the protein identifier, a "details" link, and the predicted subcellular localization followed by the *k*NN classifier belonging to this predicted localization. In the case that the predictor is rather uncertain, multiple localizations are shown, each with its calculated *k*NN classifier. Besides the web server, a stand-alone package, not restricted in the number of input sequences, can be obtained and installed on a UNIX system.

For genome-scale analysis the web server cannot be used because of the size restriction; therefore, we run the stand-alone program under Linux. Setting up the system is rather straightforward and consists of the following steps (for more detailed information we refer to the readme and installation documentation contained in the installation package):

(a) Download the gunzip tarball from the server web site (http://wolfpsort.org/).

(b) Uncompress the package using e.g., gunzip.

(c) Copy the binaries for the appropriate platform (either sparc or i-386; i-386 is standard for most computers) from the bin directory to the common/bin/directory of your distribution (typically "./bin/"). For this step, administrator rights are required (% sudo mv ./bin/bin/; fill in password upon request).

(d) The installation is now done; however, if the more detailed HTML table output is preferred or required, an additional installation step should be performed, i.e., go to the folder "./bin/psortModifiedForWolfFiles" and run psortModifiedForWoLF with the −t all.seq option ("%... ./psortModifiedForWoLF -t all.seq").

(e) The installation directory can now be copied to any preferred location as long as the subdirectory structure is preserved.

(f) The software can be run using the following two commands, depending on the output format of choice.

 – Run "%... ./bin/runWolfPsortSummaryOnly.pl fungi < ./bin/testQuery.fasta" for a simple text based result.

 – Run "%..../bin/runWolfPsortHtmlTables.pl fungi testOut/queryName < ./bin/testQuery.fasta" for a more elaborate report, containing HTML links to the PSORT-II and iPSORT output.

Typically, we run the simple text based results and export these to Excel, similarly as for the Phobius and SignalP results.

2.2.5. Alternative Methods for the Prediction of Subcellular Localization

Numerous methods exist for the prediction of subcellular localization of protein sequences. The most commonly used programs are described above, yet a lot more useful tools are available. The types of information used are amino acid content, sequence similarity (homology based), signal peptide prediction, domain signatures, and nonsequence based methods. The different predictors that apply these methods have been extensively reviewed by Nakai and Horton (14). A recent paper by Casadio et al. (21) reviewed some of the latest results from comparisons between various predictors. The prime predictors based on their review are TargetP (SignalP) extension for multiple compartments, (13), Protein Prowler (22), LocTree (23), BaCelLo (24), and WoLF PSORT (12).

2.2.6. Removal of False Positives

In order to minimize the number of falsely identified secreted proteins, a number of methods are employed. Plasma membrane bound proteins are removed by both Phobius and WoLF PSORT. Previously, Klee and Sosa (25) demonstrated that WoLF PSORT was the best method for discriminating secreted from plasma membrane bound proteins. Also, WoLF PSORT includes some feature-based methods to identify nucleolar proteins (by nucleolar localization signals) and ER retention motifs.

2.2.7. Defining the Secretome

In order to define the definitive secretome, an overlap approach is used. The data gathered before using Phobius, SignalP 3.0, and WoLF PSORT is combined and only proteins that are predicted to be extracellular by WoLF PSORT, that have a signal peptide, and signal peptide cleavage site according to SignalP 3.0 with a minimal D-Score of 0.4 and which are predicted to have no internal transmembrane helix (TM = 0 by Phobius) are classified as secreted proteins.

This comparative approach has been applied to the *Verticillium dahliae* and *Verticillium albo-atrum* genome Klosterman et al. (26), and similar methods have been used for *Postia placenta* and *Phanerochaete chrysosporium* by van den Wymelenberg et al. (27), and for *Candida albicans* by Lee et al. (28). By the analysis of unpublished datasets we found in general that a high accuracy is obtained when using this comparative approach.

Alternatively, the secretome can be defined by subsequently adding up all proteins that are predicted to be secreted by any program (or by multiple proteins). This method is in part (the programs are run sequentially, but positively scoring proteins are removed before the next step) deployed within the fungal secretome database (FSD, (29)). Typically this method yields high sensitivity but reduced specificity. Similar results were described recently by Lum and Min (30) which describe another database for fungal protein localization predictions based on the same principles as presented in this manuscript.

In this section a number of methods are described for the characterization and categorization of the secretome, in order to define the set of proteins that may act as effector molecules. The first steps include annotation and categorization. Annotation by Blast, IPS, and B2G has been performed for the complete proteome, and these annotation details can be obtained for the secreted proteins. Categorization is performed by analyses of all forms of annotation. Proteins for which neither domains nor informative BlastP hits are observed (thus for proteins for which no function can be obtained) are defined as hypothetical proteins. This group is further subdivided in hypothetical proteins with only noninformative BlastP hits (conserved hypothetical proteins) to other hypothetical proteins (e.g., hypothetical proteins from other fungi) and proteins with no observed homology in the *nr* database (nonconserved hypothetical proteins). Further subdivision can be performed on the conserved hypothetical proteins based on a number of classifications, i.e., the level of homology and the broadness of observed homology along the tree of life. Besides hypothetical proteins, we cluster secreted proteins in multiple enzymatic categories, dictated by the carbohydrate-active enzyme database, or CAZY (http://www.cazy.org; (31)) Further divisions are made based on specific enzymatic groups (noncarbohydrate acting, such as phosphatases and proteases), carbohydrate binding capacity, and the rest of the proteins are (for now) depicted under miscellaneous proteins. For the next step we compare the secreted protein set to the pathogen-host interaction database (PHI-base; http://www.phi-base.org/; (32)) using stand-alone BlastP. To this end, the protein fasta file containing the PHI-base proteins was downloaded and formatted locally using the formatdb algorithm, which is part of the Blast package (8). The formatted PHI-base database can then be used to annotate the secretome using BlastP analyses (P-value $< 1 \cdot 10^{-6}$).

Also, using intrinsic properties of the secretome proteins we can predict and categorize an additional set of potential effector molecules. Generally, it has been observed that effector molecules are small in size (typically less than 300 amino acids) and rich in cysteine residues (33). These features can be used to annotate the secretome and define a set of putative small secreted proteins.

3. Notes

1. Running IPS analyses for a complete proteome (>10,000 proteins) requires a significant amount of memory and processor computing capacity. Data can be chopped in smaller bits and sequentially run using the "&" command in Linux to prevent

overloading (and subsequent crashes). If problems occur with either memory or processor overload, it can be useful to check and alter the settings in the IPS configuration file related to the chunk size. IPS uses a parallelization procedure to effectively cope with bulk requests. This procedure chops the input file in smaller sets which are subsequently analyzed in parallel. The size of these sets, also known as chunks, is defined by the chunk size parameter. Increasing chunk size will limit the amount of parallel jobs and subsequently reduce processor and memory footprint. For a 64-bit server (8 cores, 12 GB of memory) a rather large chunk size of 500–1,000 is advisable.

2. IPS is included in the B2G program, and full genome annotation is performed using web-based service-access to the IPS repository hosted at the European Bioinformatics Institute (EBI). An IPS analysis can be run from {(Annotation), (InterProScan), (Run InterProScan (online))}. It is also possible to import the data from a previous IPS run (e.g., when the analysis was performed on a stand-alone server), by choosing the (Import InterProScan Results (xml)) option. Remember to use the right output format (the default in fact) for the IPS run "%…-format (raw|xml|txt|ebixml|html)."

3. After each analysis (Blast, Mapping, Annotation and InterProScan) step statistics can be generated in B2G and visualized by choosing the appropriate statistics from the drop down menu under {(Statistics)}.

Acknowledgments

This research was supported by a Vidi grant of the Research Council for Earth and Life Sciences (ALW) of the Netherlands Organization for Scientific Research (NWO), by the European Research Area–Network (ERA-NET) Plant Genomics and by the Centre for BioSystems Genomics (CBSG), which is part of the Netherlands Genomics Initiative and NWO.

References

1. Metkzer ML (2010) Sequencing technologies – the next generation. *Nat. Rev. Genet.* 11, 31–46
2. Miller JR, Koren S and Sutton G (2010) Assembly algorithms for next-generation sequencing data. *Genomics* 95, 315–327
3. Martinez D, Grigoriev I and Salamov A (2010) Annotation of protein-coding genes in fungal genomes. *Appl. Comput. Math.* 9, 56–65
4. Finn RD, et al (2010) The Pfam protein families database. *Nucl. Acid. Res.* 38, 211–222
5. Zdobnov EM and Apweiler R (2001) InterProScan - an integration platform for the signature-recognition methods in InterPro. *Bioinform.* 17, 847–848
6. Conesa A, et al (2005) Blast2go: a universal tool for annotation, visualization and analysis in functional genomics research. *Bioinform.* 21, 3674–3676
7. Götz S, et al (2008) High-throughput functional annotation and data mining with the Blast2GO suite. *Nucl. Acid. Res.* 36, 3420–3435

8. Altschul SF, et al (1990) Basic Local Alignment Search Tool. *J. Mol. Biol.* 215, 403–410

9. Nielsen H, et al (1997) Identification of prokaryotic and eukaryotic signal peptides and prediction of their cleavage sites. *Protein Eng.* 10, 1–6

10. Bendtsen JD, et al (2004) Improved prediction of signal peptides: SignalP 3.0. *J. Mol. Biol.* 340, 783–795

11. Käll L, Krogh A and Sonnhammer ELL (2004) A combined transmembrane topology and signal peptide prediction method. *J. Mol. Biol.* 338, 1027–1036

12. Horton P, et al (2007) WoLF PSORT: protein localization predictor. *Nucl. Acid. Res.* 35, 585–587

13. Emanuelsson O, et al (2007) Locating proteins in the cell using TargetP, SignalP and related tools. *Nat. Protocol.* 2, 953–971

14. Nakai K and Horton P (2007) Computational prediction of subcellular localization. *Method. in Mol Biol.* 390, 429–466

15. de Jonge R, et al (2010) Conserved fungal LysM effector Ecp6 prevents chitin-triggered immunity in plants. *Science* 329, 953–955

16. Krogh A et al (2001) Predicting transmembrane protein topology with a hidden markov model: application to complete genomes. *J. Mol. Biol.* 305, 567–580

17. Käll L, Krogh A and Sonnhammer ELL (2007) Advantages of combined transmembrane topology and signal peptide prediction--the Phobius web server. *Nucl. Acid. Res.* 35, 429–432

18. Horton P and Nakai K (1999) Psort: a program for detecting sorting signals in proteins and determining their subcellular localization. *TIBS* 24, 34–xx

19. Nakai K and Kanehisa M (1992) A knowledge base for predicting protein localization sites in eukaryotic cells. *Genomics* 14, 879–911

20. Bannai H, et al (2002) Extensive feature detection of N-terminal protein sorting signals. *Bioinform.* 18, 298–305

21. Casadio R, Martelli PL and Pierleoni A (2008) The prediction of protein subcellular localization from sequence: a shortcut to functional genome annotation. *Brief. Funct. Genom. Proteom.* 7, 63–73

22. Hawkings J and Boden M (2006) Detecting and sorting targeting peptides with neural networks and support vector machines. *J. Bioinform. Comput. Biol.* 4, 1–18

23. Nair R and Rost B (2005) Mimicking cellular sorting improves prediction of subcelluar localization. *J. Mol. Biol.* 348, 85–100

24. Pierleoni A, et al (2006) BaCelLo: a balanced subcellular localization predictor. *Bioinform.* 22, 408–416

25. Klee EW and Sosa CP (2007) Computational classification of classically secreted proteins. *Drug. Discov. Today* 12, 234–240

26. Klosterman S, et al (2011) Comparative genomics yields insights into niche adaptation of plant vascular wilt pathogens. *PLoS Pathog* 7: e1002137

27. van den Wymelenberg A, et al (2006) Computational analysis of the *Phanerochaete chrysosporium* v2.0 genome database and mass spectrometry identification of peptides in ligninolytic cultures reveal complex mixtures of secreted proteins. *Fungal Genet. Biol.* 43, 343–356

28. Lee SA, et al (2003) An analysis of the *Candida albicans* genome database for soluble secreted proteins using computer-based prediction algorithms. *Yeast* 20, 595–610

29. Choi J, et al (2010) Fungal secretome database: Integrated platform for annotation of fungal secretomes. *BMC Genomics* 11, 105–119

30. Lum G and Min XJ (2011) FunSecKB: the fungal secretome knowledgebase. *Databases (Oxford)* 2011, bar001

31. Cantarel BL, et al (2009) The Carbohydrate-Active EnZymes database (CAZy): an expert resource for Glycogenomics. *Nucl. Acid. Res.* 37, 233–238

32. Winnenburg R, et al (2006) PHI-base: a new database for pathogen host interactions. *Nucl. Acid. Res.* 34, 459–464

33. Rep M (2005) Small proteins of plant-pathogenic fungi secreted during host colonization. *FEMS Microbiol. Lett.* 253, 19–27

Chapter 26

Horizontal Transfer of Supernumerary Chromosomes in Fungi

H. Charlotte van der Does and Martijn Rep

Abstract

Several species of filamentous fungi contain so-called dispensable or supernumerary chromosomes. These chromosomes are dispensable for the fungus to survive, but may carry genes required for specialized functions, such as infection of a host plant. It has been shown that at least some dispensable chromosomes are able to transfer horizontally (i.e., in the absence of a sexual cycle) from one fungal strain to another. In this paper, we describe a method by which this can be shown. Horizontal chromosome transfer (HCT) occurs during co-incubation of two strains. To document the actual occurrence of HCT, it is necessary to select for HCT progeny. This is accomplished by transforming two different drug-resistance genes into the two parent strains before their co-incubation. In one of the strains (the "donor"), a drug-resistance gene should be integrated in a chromosome of which the propensity for HCT is under investigation. In the "tester" or "recipient" strain, another drug-resistance gene should be integrated somewhere in the core genome. In this way, after co-incubation, HCT progeny can be selected on plates containing both drugs. HCT can be initiated with equal amounts of asexual spores of both strains, plated on regular growth medium for the particular fungus, followed by incubation until new asexual spores are formed. The new asexual spores are then harvested and plated on plates containing both drugs. Double drug-resistant colonies that appear should carry at least one chromosome from each parental strain. Finally, double drug-resistant strains need to be analysed to assess whether HCT has actually occurred. This can be done by various genome mapping methods, like CHEF-gels, AFLP, RFLP, PCR markers, optical maps, or even complete genome sequencing.

Key words: Supernumerary chromosomes, Plant pathogenicity, Horizontal gene transfer, Horizontal chromosome transfer, Genome evolution, Fungi, Genome plasticity

1. Introduction

In several species of filamentous fungi supernumerary or "conditionally dispensable" chromosomes have been found. In these species, some strains only have the essential chromosomes (the core genome), whereas other strains have one or more additional,

Melvin D. Bolton and Bart P.H.J. Thomma (eds.), *Plant Fungal Pathogens: Methods and Protocols*,
Methods in Molecular Biology, vol. 835, DOI 10.1007/978-1-61779-501-5_26, © Springer Science+Business Media, LLC 2012

nonessential chromosomes. Typically, the latter chromosomes are relatively small (between 0.4 and 2 Mb) compared to essential chromosomes (1, 2), but they may be larger (3). Supernumerary chromosomes are generally mitotically stable but can be lost during meiosis (2, 4), through benomyl-treatment (5) or, sometimes, spontaneously (6).

In some plant pathogenic fungi, supernumerary chromosomes are related to pathogenicity. In such cases, genes that enhance virulence toward a particular host are located on one of these chromosomes. Loss of such a chromosome will thus result in loss of pathogenicity. Pathogenicity-related dispensable chromosomes have been found in isolates of *Nectria haematococca* (anamorph *Fusarium solani*) that are pathogenic toward chickpea (7, 8) or pea (8–11), in isolates of *Alternaria alternata* pathogenic toward tangerine (12, 13), apple (6, 14), strawberry (15), or tomato (16); and in *Fusarium oxysporum* pathogenic toward tomato (3).

Fungi can gain supernumerary chromosomes through horizontal transfer. This can have significant consequences: transfer of a chromosome that carries genes involved in colonization of a particular plant species can result in gain of host-specific pathogenicity in the receiving fungus. Another consequence of HCT is that different parts of a genome can have different evolutionary histories. If a chromosome required for host-specific pathogenicity has been subject to HCT, the phylogeny of the core genome does not correlate with host-specific pathogenicity. Three examples of horizontal chromosome transfer (HCT) are described briefly below.

Manners and coworkers described the first horizontal transfer of a supernumerary chromosome under laboratory conditions (17). In *Colletotrichum gloeosporioides*, a chromosome of 2 Mb was transferred from a biotype A strain (pathogenic on *Stylosanthes scabra*) to a biotype B strain (pathogenic on *S. guianensis*). Gain of the biotype A chromosome by biotype B did not result in gain of pathogenicity toward *S. scabra*. Interestingly, it proved impossible to transfer one of the "normal" chromosomes horizontally.

In *A. alternata*, Akagi et al. (16) proposed that the existence of different clonal lines pathogenic toward tomato can be explained by horizontal transfer of a supernumerary chromosome. This chromosome is 1.0 Mb and carries the *ALT1* polyketide synthase gene, involved in the production of the host-specific AAL toxin. Under laboratory conditions, protoplast fusion between a tomato-pathogenic strain and a strawberry-pathogenic strain produced strains harboring the pathogenicity chromosomes of both parents, and these were pathogenic toward both hosts (18). Interestingly, a naturally occurring dual host-specific strain of *A. alternata* has been described. This strain, pathogenic toward both tangerine and rough lemon, carries host-specific toxin genes for both hosts, and HCT followed by rearrangement of toxin gene clusters was proposed (19).

A consortium of *Fusarium* researchers showed that the tomato pathogen *F. oxysporum* forma specialis *lycopersici* contains four additional chromosomes compared to the closely related *F. verticillioides* (3). These lineage specific (LS) chromosomes are transposon-rich and appear to have an evolutionary origin different from the core genome, suggesting that LS chromosomes had been acquired by *F. oxysporum* through HCT. Under laboratory conditions, two LS chromosomes of ~1.6 and ~2 Mb were transferred horizontally into a nonpathogenic strain, which thereby became pathogenic toward tomato (3). One of these chromosomes contains all known effector genes of this strain, which encode small secreted proteins involved in colonization of tomato. Similar to the situation in *Alternaria*, the different clonal lines in the *F. oxysporum* species complex that cause wilt disease in tomato must have received these "pathogenicity" chromosomes through HCT (20). As in *C. gloeosporioides*, core chromosomes were not found to transfer horizontally.

The mechanism by which HCT occurs is unknown, but probably requires anastomosis (hyphal fusion). To facilitate horizontal transfer of genetic material between two different fungal strains, both strains have to be co-incubated. In order to document the actual occurrence of HCT, which appears to be a rare event, it is necessary to be able to select for HCT progeny in the background of the parent strains. To accomplish this, drug-resistance genes are transformed into the two parent strains before their co-incubation (see Fig. 1). In one of the strains, the "donor," a drug-resistance gene should be integrated in a chromosome of which the propensity for HCT is under investigation. In the "tester" or "recipient" strain, another drug-resistance gene should be integrated somewhere in the core genome. In this way, after co-incubation HCT progeny can be selected on plates containing both drugs (see Fig. 2). Finally, double drug-resistant strains need to be analysed to assess whether HCT has actually occurred.

2. Materials

2.1. Equipment

1. Incubator, shaker, and waterbaths (range 28–42°C) for cloning and transformation.

2. Incubator and/or shaker to maintain the fungus of choice.

3. Table centrifuge (to centrifuge units of 50 mL at $1,000 \times g$).

4. Microcentrifuge (13,000 rpm).

5. Laminar flow cabinet.

6. Trigalski spatulas.

7. Light microscope, 100–400×.

8. Heamocytometer.

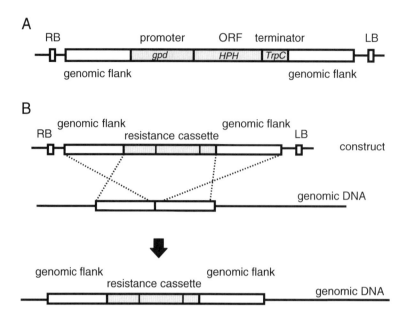

Fig. 1. Creation of parental strains for HCT experiments. (**a**) Example of a construct for transformation of filamentous fungi to drug resistance. This one is composed of a constitutive promoter (from the *gpd* gene of *A. nidulans*), the bacterial *HPH* gene (encodes resistance to hygromycin), a terminator (from the *trpC* gene of *A. nidulans*), genomic flanks that target the cassette to a specific genomic location, and right and left borders (RB, LB) for *A. tumefaciens*-mediated transformation of fungi. (**b**) Transformation of the construct depicted in (**a**) to a fungus can lead to targeted insertion into a predetermined genomic location through homologous recombination.

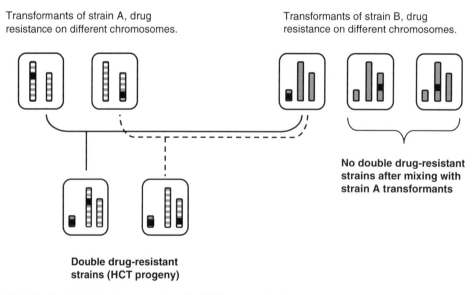

Fig. 2. Selection for horizontal chromosome transfer (HCT) progeny. In this scheme, drug-resistant transformants of strain A are the recipient strains and the left-most transformant of strain B is the donor strain for the double drug-resistant HCT progeny. The strain A transformants and the two right-most transformants of strain B cannot function as donors because in those strains the drug-resistance cassette is inserted on a chromosome that is not subject to HCT.

2.2. Solutions/ Buffers/Media	Media suitable to the fungus under investigation should be used. We here list media that we use for *F. oxysporum*.

1. Liquid medium to harvest asexual spores: 1% KNO_3, 3% sucrose, 0.17% YNB w/o NH_4 or amino acids.

2. Standard growth medium for plates. We commonly use PDA (potato dextrose agar), prepared according to the manufacturer's instructions.

3. Growth medium for selection plates. We commonly use PDA with 100 mM Tris pH 8.0, 100 μg/mL hygromycin, and 100 μg/mL zeocin (see Note 1).

3. Methods

3.1. Preparation of Parent Strains with Drug-Resistance Markers

1. Construct two plasmids, each with a different gene for drug resistance suitable for fungi, such as *HPH* (21) which confers resistance against hygromycin, and *BLE* (21), which confers resistance against phleomycin and zeocin (see Note 2). For random insertion of the construct into the genome, no other sequences are required. For homologous recombination of the construct into a specific site (for instance when insertion into a specific chromosome is desired), flanking regions of DNA homologous to the site of the site of integration have to be present (see Fig. 1, Note 3).

2. Transform each parental strain with a construct harboring a different drug-resistance gene. Depending on the fungal species, transformants can be selected on plates containing 20–200 μg/ mL hygromycin or 20–100 μg/mL phleomycin or zeocin (see Notes 1 and 4).

3. Check the position where the construct integrated into the genome. The position of the resistance gene is important, since it determines which chromosomes will be assessed for the ability to transfer horizontally (see Fig. 2, Notes 5–8).

3.2. Co-Incubation of Parent Strains and Selection of HCT Progeny

1. Harvest microconidia (asexual spores) from the parent strains. For *F. oxysporum*, microconidia can be obtained by growing strains for 3 days (25°C, 175 rpm) in 100 mL minimal medium in a 250–300-mL flask. Mycelium is removed by filtering through one layer of miracloth and conidia are pelleted at 2,000 rpm for 10 min. The conidia can be resuspended in 10 mL water. It is not necessary to wash the conidia.

2. Count the conidia, adjust the conidial density, and mix the parent strains in a 1:1 ratio. 1×10^5 spores of each parent are mixed and plated on a plate without drugs (see Note 9).

3. Incubate the plates with mixed parents until new asexual spores have formed. For *F. oxysporum*, this is after 6–8 days in the dark.

4. Harvest the microconidia from the plates by pipetting 5 mL sterile water on the plates and carefully suspending the conidia with a spatula. Pipette the conidial suspension from the plate and collect it in a tube (see Note 10).

5. Count the conidia and plate several appropriate dilutions on nonselective plates to determine the total number of colony forming units (CFUs). This is done to check whether sufficient CFUs have been plated and to allow calculation of an approximation of the frequency of HCT. For *F. oxysporum*, we found that a series of 20, 100, and 500 μL from a $1:10^6$ dilution directly from the conidial suspension harvested from a co-incubation plate generally works nicely. For the selective plates, plate a million times as much as for the nonselective plates (e.g., 20, 100, and 500 μL directly from the conidial suspension) (see Note 11). Leave the plates for the colonies to emerge. For *F.oxsporum* this is after 2–3 days (dark, 25°C) (see Note 12).

6. Transfer mycelial plugs from the double-resistant colonies to fresh selective plates to confirm their double-resistant nature.

7. Proceed with monosporing. This can be done by pipetting a droplet of water (10 μL) on sporulating mycelium, pipetting it back up, and plating this on a fresh plate. Usually, this renders separate colonies.

8. Confirm the presence of both resistance genes in the double-resistant colonies by PCR, for example with the primers used to assess the fungal transformants (see Note 6), or with primers annealing to the drug-resistance gene itself (see Note 7).

3.3. Analysis of Double Drug-Resistant Strains for HCT

Theoretically, in any double drug-resistant strain retrieved after co-incubation, the tagged chromosome from parent A could have been transferred to parent B, or vice versa, the tagged chromosome from parent B may have been transferred to parent A. Alternatively, double-resistant strains may contain a mix of chromosomes from each parent. In the two studies published so far with this method, only HCT of supernumerary chromosomes has been demonstrated (3, 17). Nevertheless, it remains crucial to determine the origin of the chromosomes of the presumed HCT progeny. Below, we list several methods to analyze the genomic content of the double drug-resistant progeny, to determine the contribution of each parent genome to the genome of the progeny.

(a) Separate the chromosomes by pulsed field gel electrophoresis (PFGE) (see Note 8) (22).

(b) Visualize polymorphisms in restriction fragment or amplified fragment lengths (RFLP or AFLP) (23–25).

(c) Identify and use PCR markers on each of the parental chromosomes that distinguish both parents. Such markers may be based on transposon insertion sites (3).

(d) Full genome sequencing of the parental strains and the progeny.

(e) Produce optical maps. In this method, single intact chromosomes are stretched on a surface and then digested by restriction enzymes (26). On the positions where the enzymes cut, the DNA retracts a little, leaving a small gap that is visible under the microscope. In this way, each chromosome is "barcoded," and thus can be identified.

4. Notes

1. Zeocin and phleomycin are not functional at low pH. Because fungi tend to acidify their environment, plates with zeocin or phleomycin should be buffered. For *F. oxysporum*, PDA (potato dextrose agar plates) buffered with 100 mM Tris pH 8.0 works fine.

2. The drug-resistance genes have to be preceded by a constitutive promoter and followed by a terminator (21). When *Agrobacterium tumefaciens*-mediated transformation (ATMT) is used, a binary vector must be used containing left and right T-DNA borders surrounding the resistance-gene cassette (27, 28). One example of this is pPK2 (29), which contains the *HPH* coding sequence with *A. nidulans gpd* promoter and *A. nidulans TrpC* terminator, all from pAN7.1 (21).

3. For *Fusarium*, flanking regions can be around 1 kb. The efficiency of homologous recombination differs widely among fungi (28).

4. The required amounts of drugs can vary between different fungal species. Here, we list a few examples from literature. *C. gloeosporioides*: 40 μg/mL hygromycin B, 20 μg/mL phleomycin (17). *Fusarium oxyporum*: 100 μg/mL hygromycin B, 100 μg/mL phleomycin, 100 μg/mL zeocine (22), *Magnaporthe grisea*: 200 μg/mL hygromycin B (30). *A. alternata*: 150–400 μg hygromycin B, 10 μg/mL phleomycin (31, 32), *Penicillium roquefortii*: 200–300 μg/mL hygromycin B, 60 μg/mL phleomycin (33).

5. Note that, occasionally, the resistance gene integrates in more than one chromosome (34).

6. In case of homologous recombination, the position of the construct can be checked by PCR, using one primer on the drug-resistance cassette and one primer in the genome, outside the construct. For a fungus with a sequenced genome, the position of

random insertions can be obtained by PCR-amplifying part of the genome flanking the insert, for example by TAIL-PCR, and sequencing the fragment (34). Insertion sites in unsequenced fungi may be determined by separating the chromosomes on a CHEF gel (see Note 8), making a Southern blot and probing with (part of) the sequence of the construct (e.g., ref. (22)).

7. To assess the presence of a gene by PCR, genomic DNA has to be isolated. A fast procedure for *F. oxysporum* is briefly described below (a similar but longer protocol of DNA isolation, which gives cleaner DNA, can for example be found in (35)). Scrape a bit of mycelium (ca. 2 cm²) from a *F. oxysporum* colony growing on plate using a (sterile) scalpel. Collect the mycelium in a 2-mL safelock microcentrifuge tube, which already contains: 300 μL glass beads and 400 μL TE (100 mM Tris pH 8.0, 10 mM EDTA pH 8.0). Add 300 μL phenol-chloroform (1:1) and vortex for 2 min. Spin the tubes at 13,000 rpm for 5 min. Pipette the aqueous phase (top layer) into a new tube. PCR can be done directly on this DNA (do not use more than 0.2 μL template per PCR reaction!). If PCR fails, try using less DNA, or try cleaning the DNA by precipitation (add 0.1 volume 3 M sodium acetate pH 5.5 and three volumes 96–100% ethanol, incubate at −20°C for 1 h, spin 10 min. at 13,000 rpm in a microcentrifuge, discard the supernatant, dry the pellet and resuspend in water or TE) or by an additional chloroform-isoamylalcohol step (add one volume chloroform-isoamylalcohol 24:1, mix well, spin for 10 min at 13,000 rpm in a microcentrifuge and transfer the waterfase to a new tube).

8. Short description of contour-clamped homogeneous electric field (CHEF) electrophoresis to visualize fungal chromosomes.

   (a) Produce protoplasts: The best procedure for this depends on the fungus and can vary significantly. For *F. oxysporum*, germinate microconidia (washed with sterile water) in PDB (Potato dextrose broth, Difco) (5×10^8 conidia in 40 mL) for 12–17 h, at 25°C, 175 rpm. Harvest the germinated conidia by centrifugation (2,000 rpm, 10 min), resuspend in 50 mL 1.2 M $MgSO_4$, 50 mM Sodium citrate pH 5.8. Pellet again, resuspend in 10 mL of the same buffer supplemented with 10 mg/mL Novozym 234 or 100 mg/mL Glucanex. Incubate overnight. Filter the protoplasts over miracloth, keep the flow-through, add four volumes of 1 M Sorbitol, 50 mM $CaCl_2$, 10 mM Tris pH 7.4 and keep on ice. Pellet the protoplasts (2,100 rpm, 15 min, cooled table centrifuge) and resuspend the pellet carefully in 1 mL of 1 M Sorbitol, 50 mM $CaCl_2$, 10 mM Tris pH 7.4.

(b) Make DNA plugs to load the protoplasts on the CHEF gel: Pellet 1 mL of protoplasts (15 min, 2,100 rpm) and resuspend them in 10 mL of 1 M Sorbitol, 0.1 M NaCl, 10 mM Tris, pH 7.4. Pellet $1-2 \times 10^8$ protoplasts (3,000 rpm, 10 min), resuspend in 0.5 mL STE (1 M Sorbitol, 25 mM Tris, 50 mM EDTA, pH 7.5), warm to 37°C (5 min), mix with 0.5 mL 1.2% InCert agarose in STE at 50°C, and pipette into the plug molds (cleaned with 1 M HCl o/n and rinsed well with sterile water). Cool on ice and take the solidified plugs out, incubate the plugs in 2.5 mL NDS (0.5 M EDTA, 0.1 M Tris, 1% lauroylsarcosine, pH 9.5) plus 2 mg/mL pronase E for 24 h at 50°C. Cool the plugs on ice, refresh the NDS+pronase E buffer and incubate 24 h at 50°C. Cool the plugs on ice and wash the plugs three times 2 h with 5 mL, 50 mM EDTA at room temperature. Plugs can be stored at 4°C.

(c) Run a CHEF gel: Cast a gel of 1% molten Seakem Gold agarose in 0.5× TBE (0.5×: 44.5 mM Tris, 44.5 mM Boric acid, 1 mM EDTA, pH 8.0). Cut the plugs a little smaller than the wells and load the plugs. Glue the plugs to the wells with a bit of 1% Seakem Gold agarose in 0.5× TBE. Mount the gel in the electrophoresis equipment with cool (4–14°C) 0.5× TBE buffer. Equilibrate the gel for 30 min. Set the following parameters: voltage: 1.5 V/cm (results in 24–27 mA), switch time: 1,200–4,800 s, run time: approximately 260 h. Run the gel at 4–14°C and refresh the buffer every 2–3 days.

(d) Visualize the chromosomes: Stain the gel in 0.5× TBE plus 1 µg/mL ethidium bromide for 20–30 min. Destain in 0.5× TBE for 30 min. Visualize chromosomes on a UV transilluminator. To make a Southern blot, nick the DNA by exposure to UV for 2–3 min. Denature the DNA two times 30 min in 1.5 M NaCl, 0.5 M NaOH. Transfer the DNA to a Hybond H+membrane by capillary blotting in 0.4 M NaOH. Wash the blot for 5 min in two times SSC (300 mM NaCl, 30 mM trisodiumcitrate, pH 7.0). Air-dry the blot on filter paper, cross-link the DNA on the blot for 5 min with UV. Wrap the blot in plastic and store at 4°C. Visualize chromosomes of interest with a probe using technology of choice.

9. For *F. oxysporum*, PDA (potato dextrose agar, no drugs) can be used. *Colletotrichum spp.* can be co-incubated on oatmeal agar plates (no drugs). A single co-incubation plate per combination is normally enough to obtain HCT progeny, but results may be more robust if more plates per combination are used (e.g., four) and pooling of the microconidia that emerge from these plates.

10. At this stage, pool conidia from several co-incubation plates, if desired.

11. It is advisable to make a series of plates, both for the selective and nonselective media, like the suggested series of 20, 100, and 500 μL (diluted) spore suspensions.

12. In our experience, colonies emerging later are generally false-positives (unstable or partial drug-resistance unrelated to the presence of both antibiotic resistance genes).

References

1. Covert, S. F. (1998) *Curr. Genet.* **33**, 311–9.

2. Wittenberg, A. H. J., van der Lee, T. A. J., Ben M'Barek, S., Ware, S. B., Goodwin, S. B., Kilian, A., Visser, R. G. F., Kema, G. H. J., and Schouten, H. J. (2009) *Plos One* **4**, -.

3. Ma, L. J., van der Does, H. C., Borkovich, K. A., Coleman, J. J., Daboussi, M. J., Di Pietro, A., Dufresne, M., Freitag, M., Grabherr, M., Henrissat, B., Houterman, P. M., Kang, S., Shim, W. B., Woloshuk, C., Xie, X. H., Xu, J. R., Antoniw, J., Baker, S. E., Bluhm, B. H., Breakspear, A., Brown, D. W., Butchko, R. A. E., Chapman, S., Coulson, R., Coutinho, P. M., Danchin, E. G. J., Diener, A., Gale, L. R., Gardiner, D. M., Goff, S., Hammond-Kosack, K. E., Hilburn, K., Hua-Van, A., Jonkers, W., Kazan, K., Kodira, C. D., Koehrsen, M., Kumar, L., Lee, Y. H., Li, L. D., Manners, J. M., Miranda-Saavedra, D., Mukherjee, M., Park, G., Park, J., Park, S. Y., Proctor, R. H., Regev, A., Ruiz-Roldan, M. C., Sain, D., Sakthikumar, S., Sykes, S., Schwartz, D. C., Turgeon, B. G., Wapinski, I., Yoder, O., Young, S., Zeng, Q. D., Zhou, S. G., Galagan, J., Cuomo, C. A., Kistler, H. C., and Rep, M. (2010) *Nature* **464**, 367–73.

4. Funnell, D. L., Matthews, P. S., and VanEtten, H. D. (2002) *Fungal Genet Biol* **37**, 121–33.

5. VanEtten, H., Jorgensen, S., Enkerli, J., and Covert, S. F. (1998) *Current Genetics* **33**, 299–303.

6. Johnson, L. J., Johnson, R. D., Akamatsu, H., Salamiah, A., Otani, H., Kohmoto, K., and Kodama, M. (2001) *Current Genetics* **40**, 65–72.

7. Covert, S. F., Enkerli, J., Miao, V. P., and VanEtten, H. D. (1996) *Mol. Gen. Genet.* **251**, 397–406.

8. Funnell, D. L., and VanEtten, H. D. (2002) *Mol. Plant-Microbe Interact.* **15**, 840–6.

9. Han, Y., Liu, X., Benny, U., Kistler, H. C., and VanEtten, H. D. (2001) *Plant J.* **25**, 305–14.

10. Rodriguez-Carres, A., White, G., Tsuchiya, D., Taga, M., and VanEtten, H. D. (2008) *Applied and Environmental Microbiology* **74**, 3849–56.

11. Coleman, J. J., Rounsley, S. D., Rodriguez-Carres, M., Kuo, A., Wasmann, C. C., Grimwood, J., Schmutz, J., Taga, M., White, G. J., Zhou, S. G., Schwartz, D. C., Freitag, M., Ma, L. J., Danchin, E. G. J., Henrissat, B., Coutinho, P. M., Nelson, D. R., Straney, D., Napoli, C. A., Barker, B. M., Gribskov, M., Rep, M., Kroken, S., Molnar, I., Rensing, C., Kennell, J. C., Zamora, J., Farman, M. L., Selker, E. U., Salamov, A., Shapiro, H., Pangilinan, J., Lindquist, E., Lamers, C., Grigoriev, I. V., Geiser, D. M., Covert, S. F., Temporini, E., and VanEtten, H. D. (2009) *Plos Genetics* **5**, -.

12. Ajiro, N., Miyamoto, Y., Masunaka, A., Tsuge, T., Yamamoto, M., Ohtani, K., Fukumoto, T., Gomi, K., Peever, T. L., Izumi, Y., Tada, Y., and Akimitsu, K. (2010) *Phytopathology* **100**, 120–26.

13. Miyamoto, Y., Masunaka, A., Tsuge, T., Yamamoto, M., Ohtani, K., Fukumoto, T., Gomi, K., Peever, T. L., and Akimitsu, K. (2008) *Molecular Plant-Microbe Interactions* **21**, 1591–99.

14. Harimoto, Y., Tanaka, T., Kodama, M., Yamamoto, M., Otani, H., and Tsuge, T. (2008) *Journal of General Plant Pathology* **74**, 222–29.

15. Hatta, R., Ito, K., Hosaki, Y., Tanaka, T., Tanaka, A., Yamamoto, M., Akimitsu, K., and Tsuge, T. (2002) *Genetics* **161**, 59–70.

16. Akagi, Y., Akamatsu, H., Otani, H., and Kodama, M. (2009) *Eukaryotic Cell* **8**, 1732–38.

17. He, C. Z., Rusu, A. G., Poplawski, A. M., Irwin, J. A. G., and Manners, J. M. (1998) *Genetics* **150**, 1459–66.

18. Akagi, Y., Taga, M., Yamamoto, M., Tsuge, T., Fukumasa-Nakai, Y., Otani, H., and Kodama,

M. (2009) *Journal of General Plant Pathology* **75**, 101–09.

19. Masunaka, A., Ohtani, K., Peever, T. L., Timmer, L. W., Tsuge, T., Yamamoto, M., Yamamoto, H., and Akimitsu, K. (2005) *Phytopathology* **95**, 241–47.

20. van der Does, H. C., Lievens, B., Claes, L., Houterman, P. M., Cornelissen, B. J. C., and Rep, M. (2008) *Environmental Microbiology* **10**, 1475–85.

21. Punt, P. J., and van den Hondel, C. A. (1992) *Methods Enzymol* **216**, 447–57.

22. Teunissen, H. A., Verkooijen, J., Cornelissen, B. J. C., and Haring, M. A. (2002) *Mol. Genet. Genomics* **268**, 298–310.

23. Elias, K. S., Zamir, D., Lichtmanpleban, T., and Katan, T. (1993) *Molecular Plant-Microbe Interactions* **6**, 565–72.

24. Teunissen, H. A., Rep, M., Houterman, P. M., Cornelissen, B. J. C., and Haring, M. A. (2003) *Mol. Genet. Genomics* **269**, 215–26.

25. Staats, M., van Baarlen, P., and van Kan, J. A. L. (2007) *European Journal of Plant Pathology* **117**, 219–35.

26. Samad, A., Huff, E. J., Cai, W. W., and Schwartz, D. C. (1995) *Genome Research* **5**, 1–4.

27. de Groot, M. J. A., Bundock, P., Hooykaas, P. J. J., and Beijersbergen, A. G. M. (1998) *Nature Biotechnology* **16**, 839–42.

28. Michielse, C. B., Arentshorst, M., Ram, A. F. J., and van den Hondel, C. A. M. J. J. (2005) *Fungal Genetics and Biology* **42**, 9–19.

29. Covert, S. F., Kapoor, P., Lee, M. H., Briley, A., and Nairn, C. J. (2001) *Mycological Research* **105**, 259–64.

30. Leung, H., Lehtinen, U., Karjalainen, R., Skinner, D., Tooley, P., Leong, S., and Ellingboe, A. (1990) *Current Genetics* **17**, 409–11.

31. Akamatsu, H., Itoh, Y., Kodama, M., Otani, H., and Kohmoto, K. (1997) *Phytopathology* **87**, 967–72.

32. Velez, H., Glassbrook, N. J., and Daub, M. E. (2007) *Fungal Genetics and Biology* **44**, 258–68.

33. Durand, N., Reymond, P., and Fevre, M. (1991) *Current Genetics* **19**, 149–53.

34. Michielse, C. B., van Wijk, R., Reijnen, L., Cornelissen, B. J. C., and Rep, M. (2009) *Genome Biology* **10**, R4.

35. Lievens, B., Brouwer, M., Vanachter, A. C., Levesque, C. A., Cammue, B. P., and Thomma, B. P. (2003) *FEMS Microbiol. Lett.* **223**, 113–22.

Chapter 27

The Induction of Mycotoxins by Trichothecene Producing *Fusarium* Species

Rohan Lowe, Mélanie Jubault, Gail Canning, Martin Urban, and Kim E. Hammond-Kosack

Abstract

In recent years, many *Fusarium* species have emerged which now threaten the productivity and safety of small grain cereal crops worldwide. During floral infection and post-harvest on stored grains the *Fusarium* hyphae produce various types of harmful mycotoxins which subsequently contaminate food and feed products. This article focuses specifically on the induction and production of the type B sesquiterpenoid trichothecene mycotoxins. Methods are described which permit in liquid culture the small or large scale production and detection of deoxynivalenol (DON) and its various acetylated derivatives. A wheat (*Triticum aestivum* L.) ear inoculation assay is also explained which allows the direct comparison of mycotoxin production by species, chemotypes and strains with different growth rates and/or disease-causing abilities. Each of these methods is robust and can be used for either detailed time-course studies or end-point analyses. Various analytical methods are available to quantify the levels of DON, 3A-DON and 15A-DON. Some criteria to be considered when making selections between the different analytical methods available are briefly discussed.

Key words: *Fusarium graminearum*, *Gibberella zeae*, Head scab disease, *Fusarium culmorum*, Deoxynivalenol, DON, Trichothecene mycotoxin detection, Chemotype, Wheat ear inoculation, Induction of mycotoxin production in liquid culture

1. Introduction

Many species of Ascomycete fungi in the genus *Fusarium* can cause serious disease on cereal and non-cereal crop plants as well as wild plant species (1). In some of these interactions, the production of various phytotoxins and mycotoxins occurs during infection and these can directly contribute to fungal virulence on certain hosts. For example, the cereal-infecting species *Fusarium graminearum* (teleomorph *Gibberella zeae*), *F. culmorum* and *F. pseudograminearum* produce a range of type B trichothecenes,

Melvin D. Bolton and Bart P.H.J. Thomma (eds.), *Plant Fungal Pathogens: Methods and Protocols*,
Methods in Molecular Biology, vol. 835, DOI 10.1007/978-1-61779-501-5_27, © Springer Science+Business Media, LLC 2012

amongst which dexoxynivalenol (DON) and its acetylated derivatives, which inhibit protein translation in eukaryotes, are known virulence factors on wheat (2–5). Other types of mycotoxins produced by *F. graminearum* during infection, such as the potent estrogenic metabolite zearalenone, do not appear to contribute directly to disease formation on plants (6, 7).

Since the mid-1990s there has been a global re-emergence of *Fusarium* Head Scab disease (www.scabusa.org) which is also referred to as *Fusarium* Ear Blight (FEB). The frequent disease epidemics have the potential to devastate wheat, barley, rye, oat or maize crops just weeks before harvest. The increased occurrence and severity of FEB appears to be driven by changes to our climate and agronomic practices (8, 9). At least 17 *Fusarium* species cause the disease (10). Grains harvested from an infected crop are often contaminated with various types of mycotoxins which may make them unsuitable and unsafe for human consumption, animal feed or malting purposes (9, 11). In addition, if grain is stored or shipped incorrectly, i.e. at too high a moisture content, fungal regrowth can commence and mycotoxin levels then rise further. As a consequence, within the last decade strict upper limits are now in place in the US, the EU and elsewhere, as to the levels of DON mycotoxins permitted to enter into different processed and unprocessed food chains via contaminated cereal grains (Food Standards Agency, www.food. gov.uk; ScabUSA, www.scabusa.org (12)).

The global threat to food safety now posed by mycotoxin-producing *Fusarium* species has resulted in the presence of over 100 active research laboratories worldwide. This research aims to understand the biochemistry of mycotoxin synthesis (13, 14), determine the biotic and abiotic factors which trigger or suppress the induction of these pathways and identify the signals and signalling pathways in the fungal cells controlling the final levels and range of mycotoxins produced (see below). Other research groups are defining the genes and pathways controlling the disease-causing abilities of the various *Fusarium* species on the different cereal hosts (summarised in http:// www.phi-base.org (15)), exploring the infection route in detail (16), evaluating fungicide efficacy (17) or identifying sources of resistance in global cereal germplasm collections and then determining the underlying genetic basis of resistance or enhanced susceptibility (reviewed in ref. (18)). It is already known from various epidemiological studies that within a single *Fusarium* species, different mycotoxin chemotypes exist in the same geographic regions and that over time the ratio of chemotypes in a region can change. For example in the UK, *F. graminearum* exists as 3-ADON, 15-ADON and nivalenol (NIV) chemotypes (19) but over the past 15 years the incidence of the 3-ADON chemotype has decreased whilst 15-ADON types have increased (HGCA Crop Monitor, http://www.cropmonitor.co.uk). Over this same time period in Canada where originally 15-ADON chemotypes predominated, highly aggressive 3-ADON chemotypes are now emerging (20).

In this article, we focus on the globally prevalent species *F. graminearum* and DON/ADON mycotoxin production. The biosynthetic pathway for trichothecene mycotoxin production is well defined and the pathway from primary isoprenoid metabolism to deoxynivalenol has been mapped (21, 22). The first committed step is the conversion of farnesylpyrophosphate to trichodiene catalysed by the enzyme trichodiene synthase which is coded for by the *TRI5* gene. The next nine steps to synthesise calonectrin are conserved for the synthesis of deoxynivalenol, T-2 toxin and nivalenol. Another five reactions are required to convert calonectrin to deoxynivalenol, which can exist in various acetylated forms. Very recently, the *TRI8* gene which codes for a trichothecene C-3 esterase has been shown to control the pivotal reaction which converts the diacetylated 3A, 15A DON intermediate into either 3A-DON or 15A-DON (23). The activity of this enzyme therefore defines the chemotype of the strain. In *F. graminearum*, 10 *TRI* genes are located in a highly co-ordinately regulated dense gene cluster in the middle of chromosome 2. The *TRI5* gene is flanked by the *TRI6* and *TRI10* genes which both code for transcription factors regulating trichothecene mycotoxin production (24, 25). A third transcription factor *TRI15* and two other genes, *TRI1* coding for a cytochrome P450 monooxygenase (26, 27) and *TRI101* coding for trichothecene 3-*O*-acetyltransferase (28) are located on three other chromosomes (29). To monitor the induction of mycotoxins, the expression of several *TRI* genes has been studied. The most highly induced *TRI* genes identified using the available Affymetrix Array appear to be *TRI4*, *TRI9* and *TRI14* whilst *TRI5* was found only to be modestly induced during barley ear infection (30). Several groups have prepared various types of reporter strains by fusing the promoter region of the *TRI5* gene to the Green Fluorescent Protein (GFP) reporter protein and then selecting transformants which only express GFP under DON-inducing conditions (31, 32). These have been used to track DON pathway induction during the early stages of detached barley caryopsis infection and in vitro in the presence/absence of specific fungicides. However, some of these early reporter strains are no longer able to produce DON which has somewhat limited their overall usefulness (32). An alternative non-transgenic approach has been to develop a small custom oligonucleotide microarray which contains most of the relevant mycotoxin biosynthesis genes for the production of aflatoxin, fumonisin, ochratoxin, patulin and the trichothecenes (types A and B) (33). These arrays were then used to explore the temporal expression of all the *TRI* genes during the growth of *F. culmorum* on various inductive and non-inductive solid growth media.

In vitro induction of trichothecene mycotoxin biosynthesis for *F. graminearum* was shown in numerous studies to be variable and the levels produced are often considerably lower than the levels produced in planta. Induction on solid media is difficult to explore because the agar must be melted to extract the mycotoxin from the watery supernatant post-centrifugation. Culturing on autoclaved rice or other cereal grain tends to induce only modest amounts of mycotoxin (~8 ppm), which is also difficult to extract and the cultures need to be grown for up to 40 days. In general, the in planta studies are done under controlled environmental conditions which give more consistent DON induction and production. But these assays are particularly problematic when the strains selected for comparative studies grow at different rates or have differing levels of virulence. The later scenarios are often encountered when comparing near isogenic wild-type and single gene deletion strains, to explore whether the genetic alteration either directly or indirectly influences mycotoxin production (for example refs. (34, 35)). Diverse factors which have been shown/suggested to influence mycotoxin production either singly or in combination (36), include nitrogen and carbon nutrient conditions (37), pH and temperature (38–40); oxidative stress (41), water potential (40, 42), other abiotic stresses (43), chemotype (44) and sub-lethal doses of fungicides (17). In an effort to understand the factors controlling mycotoxin induction and the effects of different gene mutations, reliable and precise methods to induce trichothecene mycotoxin production in liquid cultures and in planta are needed. In the literature, a range of methods are described to quantify with either low or high precision the levels of DON and its acetylated derivatives found in grain and liquid culture. These include the use of semi-quantitative dip sticks as well as quantitative competitive ELISA, HPLC and GC-MS methods (45). Therefore, here only the criteria to be considered when making selections between the different analytical methods available are briefly discussed.

In this article, we describe robust and well-tested high-throughput small and large scale methods to induce DON/ADON production in liquid cultures and a method to induce and reliably quantify DON production of *F. graminearum* strains in wheat ears inoculated at anthesis. Each method is suitable for detailed time-course studies or for end-point analyses. By using these three methods, it is possible to easily intercompare any new data sets obtained with those already generated by the expanding global community of *Fusarium* researchers.

2. Materials

2.1. Plant and Fungal Materials

1. Reference stocks of each isolate or strain are stored at −80°C to maintain long-term viability as high titre macroconidia suspensions either in water (see Note 1) or in 15% glycerol.

2. Numerous wild-type trichothecene-producing strains are maintained at fungal stock centres (i.e. ARS Culture Collection,

Peoria, IL, USA; Fungal Genetics Stock Center, Kansas City, MO, USA; CBS, Centraalbureau voor Schimmelcultures, Utrecht, The Netherlands).

3. A fully susceptible spring wheat cultivar, for example, Bobwhite grown in a controlled environment with a temperature of 18°C for 16 h in the light followed by 16°C for 8 h in the dark and constant 50% relative humidity.

4. *Fusarium* conidia suspension at 2×10^5 conidia/mL.

2.2. Growth Media

Prepare all media using distilled water. Sterilise solutions by autoclaving at 121°C for 15 min (see Note 2) or filter sterilise by passage through a 0.45 µm pore size filter (see Note 2). Store reagents at room temperature (see Note 3).

1. Synthetic nutrient-poor agar (SNA): 1 g/L KH_2PO_4, 1 g/L KNO_3, 1 g/L $MgSO_4 \times 7H_2O$, 0.5 g/L KCl, 0.2 g/L glucose, 0.2 g/L sucrose, 20 g/L agar.

2. TB3: 3 g/L Yeast extract, 3 g/L, Bacto-Peptone, 200 g/L Sucrose.

3. Potato dextrose broth (PDB) (Sigma-Aldrich, Gillingham, UK).

4. 2-Stage medium (2SM): 1 g/L $(NH_4)_2HPO_4$, 3 g/L KH_2PO_4, 0.2 g/L $MgSO_4$ $7H_2O$, 5 g/L NaCl, 40 g/L sucrose, 10 g/L glycerol (36, 46, 47).

5. Alternative nitrogen sources for 2SM DON induction medium: putrescine, agmatine, potassium nitrate, glutamine (48).

2.3. Equipment

2.3.1. General

1. A magnetic filter funnel (47 mm) and base unit (Pall Life Sciences, Ann Arbor, MI, USA).

2. Benchtop vacuum pump/supply (e.g. 1 bar operating pressure; 8 mbar vacuum; 10 L/min flow rate).

3. A biological orbital incubator able to maintain a constant temperature and agitation, with a controllable light source (e.g. Innova 44 model, New Brunswick Scientific, Edison, NJ, USA).

4. Sterile conical flasks (250 mL or 1 L) for standard scale cultures or long time courses.

5. Pipettes, 10–1,000 µL and 25 or 50 mL.

2.3.2. In Planta Trichothecene Analysis

6. 15 and 50 mL polypropylene tubes.

7. Pipette (10 µL) and associated tips.

8. Vortex.

9. A tissue homogenizer, e.g. IKA Ultra-turrax T18.

10. Water bath (30°C).

11. Pestle and mortar.

12. A –20°C freezer.

13. A freeze drier.

1. Pestle and mortar.

2. Magnetic filter funnel and base unit, 47 mm size (Pall Life Sciences, Ann Arbor, MI, USA).

3. Vacuum pump (e.g. 1 bar operating pressure; 8 mbar vacuum; 10 L/min flow rate).

4. Bench top centrifuge, e.g. Eppendorf.

5. Spectrophotometer at 450 nm, e.g. DYNEX Technologies MRX.

6. Tubes: 50, 2, 1.5 mL.

7. Pipettes, e.g. Gilson P10–P1000.

1. Sterile 12-well culture plates for small scale cultures (e.g. VWR, Cat-No. 734-0055).

2. DON/15A-DON ELISA kit (e.g. EZ-Quant vomitoxin kit, Thermo Fisher Scientific, Mississauga, Canada, Cat No. 600312) (see Note 4).

3. Miracloth (Calbiochem, La Jolla, CA, USA) cut to approximately 4×4 cm^2 and sterilised at 121°C for 15 min.

4. 50 mM Tris–HCl pH 8.0.

5. Competitive ELISA kit, e.g. EZ-Quant vomitoxin kit, Thermo Fisher Scientific, Mississauga, Canada, Cat No. 600312.

6. Trichothecene standards: DOM-1 standards, 3A-DON, 15A-DON and DON (Biopure, Tullin, Austria) (see Note 5).

7. Various lateral flow dipsticks are commercially available for the semi-quantitative testing of DON in grain samples for example from Envirologix (http://www.envirologix.com) and Charm Sciences (http://www.charm.com) (see Note 6).

3. Methods

1. *Fusarium* inoculum can be produced easily on agar plates using nutrient poor or nutrient-rich conditions at 22°C when irradiated using a mixture of white light/near UV-light. SNA agar plates only allow limited growth of *Fusarium* and are well suited to check for contaminants and formation of the characteristic banana-shaped *Fusarium* spores. Once a sealed SNA plated is covered with fungal growth it can be overlaid with 300 μL TB3 medium. When the plate is left unsealed and incubated for 2 more days abundant conidiation occurs and conidia can be harvested, washed in water and resuspended at the desired concentration.

2. Spore suspensions can be used either fresh or frozen at –80°C as inoculum for diverse experiments (see Note 7).

3. Diluted spore suspensions can be used to inoculate fresh PDB plates where within 3 days abundant conidiation occurs.

3.2. Production of Trichothecenes In Vitro

3.2.1. High-Throughput Production of Trichothecenes

1. Cut Miracloth tissue into small discs with a diameter of 1.8 cm and sterilise.

2. Add 2 mL of PDB medium into each of the wells of a 12-well multi-well plate (see Note 8).

3. Inoculate each well with 50 µL of a *Fusarium* spore stock at a concentration of 2×10^6 spores/mL.

4. Put one Miracloth disc into each of the wells using tweezers (see Note 9).

5. Seal plate with PVC insulation tape.

6. Place plate into shaking incubator at 170 rpm and 28°C and incubate for 48 h to produce fungal biomass (see Fig. 1).

7. Remove PDB medium carefully using a 5 mL pipette and wash the fungal mycelium sticking to the Miracloth membrane two times with 2 mL 2SM mycotoxin induction medium (see Note 10). Then remove washing fluid as much as possible.

8. Add 2 mL 2SM medium, seal plate again using PVC tape and incubate for seven further days in the incubator to induce trichothecene production.

9. For toxin measurements (see Fig. 2) take 40 µL of growth medium from each well and mix with 160 µL TRIS buffer (50 mM, pH 8.0) to adjust the samples to pH 8.0 for subsequent analysis using a DON ELISA kit (see Note 11) or dipstick.

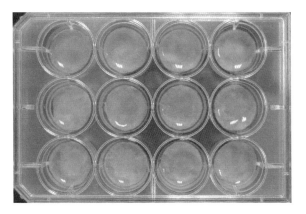

Fig. 1. *Fusarium graminearum* forms even hyphal mats across Miracloth filter discs in 2 mL PDB multi-well cultures after 48 h incubation.

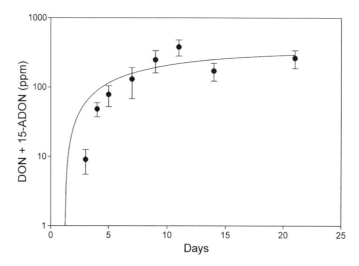

Fig. 2. DON/15A-DON induction time course in 2-mL multi-well cultures of *F. graminearum*. Three biological replicates were analysed per time point. At day 2 the complete medium was replaced with 2SM mycotoxin induction medium. The *bar* indicates the standard deviation.

3.2.2. Large Scale Production of Trichothecenes

This method is based on the trichothecene-inducing media devised by Harris and associates (46) and on the experimental protocol described by Lowe and associates (47).

1. Pipette 50 mL of PDB into a sterile 250 mL conical flask.

2. Inoculate the flask with conidia to a final concentration of 1×10^4 conidia/mL.

3. Transfer the flask to an orbital incubator and incubate at 25°C/150 rpm for 48 h in the dark.

4. Assemble the filtration unit and attach it to the vacuum pump.

5. Place one square of sterile Miracloth on the filtration unit using ethanol-sterilised forceps.

6. Turn on the vacuum pump and pour the culture over the Miracloth. The mycelium will remain on the Miracloth whilst the growth medium will pass through.

7. Wash the mycelium in 100 mL of sterile distilled water three times and filter to remove as much water as possible. The mycelium should collect as a compact mat on the Miracloth.

8. Grasp one edge of the mycelial mat with a set of ethanol-sterilised forceps. Gently lift the mycelium off the Miracloth with a rolling motion to separate the mycelium from the cloth.

9. Transfer the mycelium to a new sterile 250 mL conical flask containing 50 mL of 2SM medium, discard the Miracloth.

10. Gently swirl the medium in the flask to disperse the mycelium in the 2SM.

Table 1
The sequential volumes of culture to be removed from each large flask culture to ensure sufficient fungal biomass at each time point for follow-up technically replicated transcriptomic or metabolomic analyses

Time point (day)	Volume of culture removed (mL)
0	200
1	100
3	25
7	25
14	25

11. Incubate the flask at 28°C/150 rpm for up to 14 days in the dark (see Notes 12 and 13).

12. Assemble the filtration unit as described in steps 4 and 5.

13. Filter the culture through the Miracloth and collect the filtrate. Store filtrate at –20°C until ready to perform the DON analysis.

14. Wash the mycelium as described in step 7.

15. Roll up the mycelium as described in step 8, and transfer it to a polypropylene tube using ethanol-sterilised forceps.

16. Tightly secure the lid of the tube and snap freeze under liquid nitrogen.

17. Store the mycelium at –80°C until required for further analysis (see Note 14).

18. When time course DON experiments are being planned large cultures are established and different volumes are sequentially harvested (see Note 15 and Table 1).

19. It is important to ensure the safe disposal of all liquid cultures (see Note 16).

3.3. Modifications to the Method for the In Vitro Production of Trichothecenes Method

There are several modifications to the growth medium, such as 2SM, that are reported to influence the amount of DON produced. These are either DON-inducing or DON repressing and can be used to enhance production or provide a non-producing control treatment.

3.3.1. Oxidative Stress to Enhance DON Production

One study has reported that if hydrogen peroxide is added to a liquid culture to final concentration of 0.5 mM, DON production will be enhanced (41). As peroxide is rapidly consumed, daily addition to the culture results in a further enhancement, with 100 ppm total trichothecenes reported by 5 dpi instead of 30 ppm without this daily supplement (48). Conversely, addition of catalase enzyme (96 units/50 mL culture) resulted in repression of DON production (17).

3.3.2. Growth on DON-Inducing Nitrogen Sources

When agmatine was used as the sole nitrogen source in a DON-inducing medium such as 2SM, extremely high concentrations of trichothecenes were produced in 7 days (49). Certain other polyamines also have been shown to have a positive effect on toxin production, including arginine, putrescine and ornithine. Polyamines when used at a final concentration of 10 mM in the growth medium can produce approximately 1,000 ppm DON after 7 days growth (39).

3.3.3. Growth on DON-Repressing Nitrogen Sources

There are several nitrogen sources that repress DON production, and these should be avoided if possible in DON-inducing media, or utilised for DON negative control treatments. Growth on nitrate (15 mM) or glutamine (15 mM) will repress DON production (48) (see Note 17).

3.4. Culture Sample Preparation for DON Quantification

1. Decant filtrate from collection flask into 50 mL polypropylene tubes.

2. Centrifuge filtrate at $5,000 \times g$ for 5 min to clarify filtrate.

3. Decant the supernatant into another tube for storage.

4. At this stage the culture pH can be measured in an aliquot of the growth medium.

5. For analysis by DON ELISA, prepare serial dilutions of the sample in Tris–HCl pH 8.0, at 1, 10, 100 and 1,000-fold dilutions. Assay all dilutions according to the manufacturer's instructions. The ELISA quantitative range is limited and serial dilutions extend the range of concentrations from 1 to 1,000 ppm that can be accurately determined.

6. For analysis by GC-MS, ensure that an appropriate internal standard is included (see Note 6).

3.5. In Planta Trichothecenes Analysis

1. Select only full sized wheat ears at early anthesis (see Notes 18–21).

2. Vortex the *Fusarium* spore inoculum slightly to ensure its homogeneity.

3. Pipette 5 μL of conidial suspension (approximately 1,000 conidia) into a tip.

4. Slip the pipette tip between the lemma and the palea of one of the outer floret of the first full size spikelet at the top of the ear and then slowly inject in the conidia suspension.

5. Repeat steps 3 and 4 for the both outer florets of each spikelet and for the subsequent 10–12 spikelets on each ear.

6. Keep the inoculated plants at 100% humidity for the next 3 days. Shade the chamber for the first 24 h (e.g. with a black plastic bin liners) to exclude light.

7. Harvest the entire ears 10 days after inoculation into 50 mL tubes, snap-freeze under liquid nitrogen and store samples at –20°C until required for further analysis (see Note 22).

8. Freeze-dry the material and grind to a fine powder in liquid nitrogen using a mortar and pestle.

9. Resuspend in 15 mL tubes, 1 g of each sample in 5 mL of water and then mix thoroughly using a homogeniser.

10. Incubate the mixture for 30 min at 30°C in a water bath.

11. Centrifuge and collect the supernatant for DON content analysis. The supernatant can be stored at –20°C until required for further analysis.

4. Notes

1. For long-term storage of *Fusarium* strains place at –80°C in water at a concentration of 1×10^6 to 1×10^{10} macroconidia/mL. These stocks will remain fully viable for >1 year.

2. Do not autoclave the media for longer than 15 min otherwise sugars in the media will caramelise.

   For large scale metabolomic (47) experiments it is recommended that each of the components of the media is prepared individually as 10× stock solutions. They should each be filter sterilised by passage through a 0.45 μm pore size filtration unit rather than autoclaved to minimise concentration anomalies due to evaporation and breakdown. For each batch of cultures, fresh media can be prepared from the 10× stock solutions in order to minimise media variation between runs.

3. If growth media are stored at room temperature any microbial contamination is readily identifiable before use.

4. DON ELISA kits to quantify type B trichothecene mycotoxins are available from different sources and provide a convenient method to determine trichothecene mycotoxins. However, due to the cross-reactivity of the antibodies provided, ELISA kits frequently cannot differentiate between DON and 15A-DON or DON and 3A-DON. Care must be taken to select an ELISA kit which matches the chemotype of the *Fusarium* strains used in the experiment. For example, the DON ELISA kit listed in Subheading 2.3.4 can detect DON and 15A-DON but not the presence of 3A-DON.

5. For GC-MS analysis the inclusion of the DOM-1 standard is highly recommended. This standard is added at an early step in the protocol and can be used to monitor the uniformity of sample handling over the entire multi-step method.

6. These dipsticks have not yet been tested for their suitability to assay DON production in liquid cultures where the acetylated forms of DON predominate, in contrast to the grain and ear samples where only DON is detected.

7. Do not repeatedly thaw and refreeze the long-term storage spore stocks. Instead, thaw once, aliquot into smaller volumes, refreeze at −80°C and then use within the next few weeks to prepare inoculum.

8. A micro-method (500 µL) for DON production has recently been published (17).

9. Miracloth has a typical pore size of 22–25 µm. It was chosen as the support matrix for *Fusarium* growth because it is readily available. Alternatively inert Nylon membranes with 10-µm pore size are suitable. Formation of hyphal mats appears to contribute to consistent production of trichothecenes in liquid culture. While hyphal mats can be formed without this support matrix, their variable formation may result in highly variable mycotoxin production. Fungal biomass in each well can be determined by freeze-drying the fungal mycelium together with the filter disc.

10. The use of a manual multi-channel pipette with adjustable spacing will dramatically reduce the amount of pipetting steps required.

11. Trichothecene mycotoxin production in *F. graminearum* liquid cultures rises exponentially before reaching a plateau at about 7 days (see Fig. 2). Mycotoxin end-point concentrations can only be compared when all cultures have reached the plateau phase to reduce variability.

12. DON production may be detected as early as 7 days post-media shift but it is more consistently detected 10–14 days post the media shift. Even if the experiment is done in controlled conditions, a great variation in the amount of produced DON can occur between replicates (e.g. from 2 to 288 ppm at 14 days post-media shift). We recommend anticipating significant culture to culture variation in final DON amounts when designing experiments and to establish "extra" flasks to ensure that those with very similar final DON production levels can be selected for follow-up analyses.

13. A drop in the pH (from 6 to 3 units) occurs during the first 48 h after the transfer into the 2SM medium (see Fig. 3). We recommend monitoring the growth medium pH during this period to determine that the experiment is running as expected.

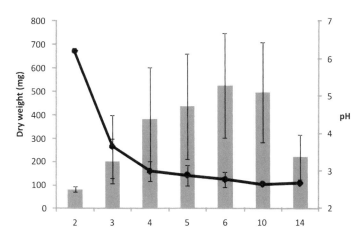

Fig. 3. The changes to pH and biomass typically detected during the in vitro DON/15A-DON induction time course with PH-1 strain. The fungal biomass was monitored by determining the mycelium dry weight contained in a 50 mL aliquot of the harvested liquid culture. Twelve biological replicates from three experiments were monitored. At day 2 the complete medium was replaced with 2SM mycotoxin induction medium. The *bar* indicates the standard deviation.

14. Mycelium harvested from the DON induction experiment may be used for metabolomics, proteomics and transcriptomic analyses. For these purposes, the material must be rapidly frozen in liquid nitrogen, freeze-dried and ground using a pestle and mortar prior to extraction. For metabolomic analysis 15 mg of freeze-dried material is tested in triplicate according to the methodology described in refs. (47, 50). For transcriptomic analysis, approximately 30 mg of freeze-dried material is required for RNA extraction.

15. To perform a time course with serial sampling, the experimental volume must be scaled up in order that sufficient volume is retained for later time points. The modifications to the protocol are as follows:

    Step 1. Aliquot 400 mL of PDB into a sterile 1 L conical flask.

    Step 8. Transfer the mycelium to a new sterile 1 L conical flask containing 600 mL of 2SM medium.

    Step 10. Remove the specified volume of culture from the flask using a serological pipette (25–50 mL) at each time point and filter (see Table 1).

16. The DON mycotoxin produced by several *Fusarium* species inhibits protein synthesis in eukaryotes and prevents polypeptide chain initiation or elongation by binding to the 60S ribosomal subunit (51, 52). When conducting experiments which induce the formation of type B trichothecenes, the researcher needs only to take the normal personal safety precautions. However,

when large scale liquid induction experiments are completed, to protect the environment, the *Fusarium* cultures should be treated with a strong alkali solution (for example, 5M sodium hydroxide) to inactivate DON/ADON prior to autoclaving to kill the fungus. These treated cultures can then be disposed of as autoclaved waste.

17. Alternative nitrogen sources should be prepared to provide an equivalent amount of nitrogen, to help equalise growth rates. A good starting point is 15 mM nitrogen.

18. This method gives quite reproducible results. We recommend inoculating three to five ears per evaluated strain.

19. Only use ears where 40–60% of the ear is covered with visible bright yellow anthers on the day of inoculation. The first and last spikelet inoculated can be identified by placing a small black dot on the outer glume with a waterproof marker pen immediately post-inoculation.

20. Due to the high amount of DON produced with this method, typically 800 ppm for the sequenced *F. graminearum* strain PH-1, DON overproduction by a strain could be hidden by the saturation of the wheat ear tissues with DON. We recommend investigating in parallel the in vitro DON production of each strain as previously described.

21. When selecting the wheat genotype to be used for these assays it is important to determine the exact allele present at the *Rht1* and *Rht2* semi-dwarfing loci present in hexaploid wheat. Both genomic regions are known to affect differentially ear susceptibility to *Fusarium* although the underlying genes and mechanisms involved are currently unclear (53–55).

22. If post-point inoculation the strains tested produce different amounts of macroscopically visible disease, the ear can be divided into symptom and symptomless portions and then these separate tissues assayed for DON levels.

Acknowledgements

Rothamsted Research receives grant-aided support from the Biotechnology and Biological Sciences Research Council (BBSRC) of the UK. MJ was supported by a grant from Bayer CropScience. RL and GC were supported by a BBSRC grant awarded within the special initiative on plant and microbial metabolomics (BB/D007224/1). We thank Neil Brown for critical reading the manuscript.

References

1. Agrios, G. N. (2005) *Plant Pathology*, Academic Press, Inc., London.

2. Cuzick, A., Urban, M., and Hammond-Kosack, K. (2008) *Fusarium graminearum* gene deletion mutants *map1* and *tri5* reveal similarities and differences in the pathogenicity requirements to cause disease on Arabidopsis and wheat floral tissue. *New Phytologist 177*, 990–1000.

3. Mudge, A. M., Dill-Macky, R., Dong, Y., Gardiner, D. M., White, R. G., and Manners, J. M. (2006) A role for the mycotoxin deoxynivalenol in stem colonisation during crown rot disease of wheat caused by *Fusarium graminearum* and *Fusarium pseudograminearum*. *Physiological and Molecular Plant Pathology 69*, 73–85.

4. Proctor, R. H., Hohn, T. M., and McCormick, S. P. (1997) Restoration of wild-type virulence to *Tri5* disruption mutants of *Gibberella zeae* via gene reversion and mutant complementation. *Microbiology 143*, 2583–2591.

5. Proctor, R. H., Hohn, T. M., McCormick, S. P., and Desjardins, A. E. (1995) *Tri6* encodes an unusual zinc-finger protein involved in regulation of trichothecene biosynthesis in *Fusarium sporotrichioides*. *Applied and Environmental Microbiology 61*, 1923–1930.

6. Gaffoor, I., Brown, D. W., Plattner, R., Proctor, R. H., Qi, W., and Trail, F. (2005) Functional analysis of the polyketide synthase genes in the filamentous fungus *Gibberella zeae* (Anamorph *Fusarium graminearum*). *Eukaryot Cell 4*, 1926–1933.

7. Gaffoor, I., and Trail, F. (2006) Characterization of two polyketide synthase genes involved in zearalenone biosynthesis in Gibberella zeae. *Applied and Environmental Microbiology 72*, 1793–1799.

8. Leonard, K. J., and Bushnell, W. R. (2003) *Fusarium head blight of wheat and barley*, The American Phytopathological Society, Minnesota.

9. McMullen, M., Jones, R., and Gallenberg, D. (1997) Scab of wheat and barley: a re-emerging disease of devasting impact. *Plant Disease 81*, 1340–1348.

10. Parry, D. W., Jenkinson, P., and McLeod, L. (1995) Fusarium ear blight (scab) in small grain cereals - a review. *Plant Pathology 44*, 207–238.

11. Wu, F., and Munkvold, G. P. (2008) Mycotoxins in ethanol co-products: modeling economic impacts on the livestock industry and management strategies. *J Agric Food Chem 56*, 3900–3911.

12. Beacham, A., Antoniw, J., and Hammond-Kosack, K. (2009) A genomic fungal foray. *The Biologist 56*, 98–105.

13. Desjardins, A. E. (2006) *Fusarium Mycotoxins - Chemistry, Genetics and Biology*, The American Phytopathological Society, St. Paul, Minnesota USA.

14. Kimura, M., Tokai, T., Takahashi-Ando, N., Ohsato, S., and Fujimura, M. (2007) Molecular and genetic studies of *Fusarium* trichothecene biosynthesis: pathways, genes, and evolution. *Biosci Biotechnol Biochem 71*, 2105–2123.

15. Winnenburg, R., Urban, M., Beacham, A., Baldwin, T. K., Holland, S., Lindeberg, M., Hansen, H., Rawlings, C., Hammond-Kosack, K. E., and Kohler, J. (2008) PHI-base update: additions to the pathogen host interaction database. *Nucleic Acids Res 36*, D572–576.

16. Brown, N. A., Urban, M., Van de Meene, A. M. L., and Hammond-Kosack, K. E. (2010) The infection biology of *Fusarium graminearum*: Defining the pathways of spikelet to spikelet colonisation in wheat ears. *Fungal Biology 114*, 535–571.

17. Audenaert, K., Callewaert, E., Hofte, M., De Saeger, S., and Haesaert, G. (2010) Hydrogen peroxide induced by the fungicide prothioconazole triggers deoxynivalenol (DON) production by *Fusarium graminearum*. *BMC Microbiol 10*, 112.

18. Buerstmayr, H., Ban, T., and Anderson, J. A. (2009) QTL mapping and marker-assisted selection for Fusarium head blight resistance in wheat: a review. *Plant Breeding 128*, 1–26.

19. Polley, R. W., and Turner, J. A. (1995) Surveys of stem base diseases and Fusarium Ear Diseases in winter-wheat in England, Wales and Scotland, 1989-1990. *Annals of Applied Biology 126*, 45–59.

20. Gilbert, J., Clear, R. M., Ward, T. J., Gaba, D., Tekauz, A., Turkington, T. K., Woods, S. M., Nowicki, T., and O'Donnell, K. (2010) Relative aggressiveness and production of 3-or 15-acetyl deoxynivalenol and deoxynivalenol by Fusarium graminearum in spring wheat. *Canadian Journal of Plant Pathology-Revue Canadienne De Phytopathologie 32*, 146–152.

21. Brown, D. W., McCormick, S. P., Alexander, N. J., Proctor, R. H., and Desjardins, A. E. (2002) Inactivation of a cytochrome P-450 is a determinant of trichothecene diversity in *Fusarium* species. *Fungal Genetics & Biology 36*, 224–233.

22. Kimura, M., Tokai, T., O'Donnell, K., Ward, T. J., Fujimura, M., Hamamoto, H., Shibata, T., and Yamaguchi, I. (2003) The trichothecene biosynthesis gene cluster of *Fusarium graminearum* F15 contains a limited number of essential

pathway genes and expressed non-essential genes. *FEBS Letters 539*, 105–110.

23. Alexander, N. J., McCormick, S., Waalwijk, C., and Proctor, R. H. (2010) Genetic basis for the 3ADON and 15 ADON trichothecene chemotypes in *Fusarium graminearum. In: 10th Eur. conf. fungal Genetics, Leiden, The Netherlands.*

24. Tag, A. G., Garifullina, G. F., Peplow, A. W., Ake, C., Phillips, T. D., Hohn, T. M., and Beremand, M. N. (2001) A novel regulatory gene, *Tri10,* controls trichothecene toxin production and gene expression. *Applied and Environmental Microbiology 67*, 5294–5302.

25. Seong, K.-Y., Pasquali, M., Zhou, X., Song, J., Karen, H., McCormick, S., Dong, Y., Xu, J.-R., and Kistler, H. C. (2009) Global gene regulation by *Fusarium* transcription factors Tri6 and Tri10 reveals adaptations for toxin biosynthesis. *Molecular Microbiology 9999.*

26. McCormick, S. P., Harris, L. J., Alexander, N. J., Ouellet, T., Saparno, A., Allard, S., and Desjardins, A. E. (2004) *Tri1* in *Fusarium graminearum* encodes a P450 oxygenase. *Appl Environ Microbiol 70*, 2044–2051.

27. Meek, I. B., Peplow, A. W., Ake, C., Phillips, T. D., and Beremand, M. N. (2003) *Tri1* encodes the cytochrome P450 monooxygenase for C-8 hydroxylation during trichothecene biosynthesis in *Fusarium sporotrichioides* and resides upstream of another new *Tri* gene. *Applied and Environmental Microbiology 69*, 1607–1613.

28. McCormick, S. P., Alexander, N. J., Trapp, S. E., and Hohn, T. M. (1999) Disruption of *TRI101,* the Gene Encoding Trichothecene 3-O-Acetyltransferase, from *Fusarium sporotrichioides. Applied and Environmental Microbiology 65*, 5252–5256.

29. Cuomo, C. A., Guldener, U., Xu, J. R., Trail, F., Turgeon, B. G., Di Pietro, A., Walton, J. D., Ma, L. J., Baker, S. E., Rep, M., Adam, G., Antoniw, J., Baldwin, T., Calvo, S., Chang, Y. L., Decaprio, D., Gale, L. R., Gnerre, S., Goswami, R. S., Hammond-Kosack, K., Harris, L. J., Hilburn, K., Kennell, J. C., Kroken, S., Magnuson, J. K., Mannhaupt, G., Mauceli, E., Mewes, H. W., Mitterbauer, R., Muehlbauer, G., Munsterkotter, M., Nelson, D., O'Donnell, K., Ouellet, T., Qi, W., Quesneville, H., Roncero, M. I., Seong, K. Y., Tetko, I. V., Urban, M., Waalwijk, C., Ward, T. J., Yao, J., Birren, B. W., and Kistler, H. C. (2007) The Fusarium graminearum genome reveals a link between localized polymorphism and pathogen specialization. *Science 317*, 1400–1402.

30. Guldener, U., Seong, K. Y., Boddu, J., Cho, S., Trail, F., Xu, J. R., Adam, G., Mewes, H. W., Muehlbauer, G. J., and Kistler, H. C. (2006) Development of a *Fusarium graminearum* Affymetrix GeneChip for profiling fungal gene expression in vitro and in planta. *Fungal Genetics and Biology 43*, 316–325.

31. Jansen, C., von Wettstein, D., Schafer, W., Kogel, K. H., Felk, A., and Maier, F. J. (2005) Infection patterns in barley and wheat spikes inoculated with wild-type and trichodiene synthase gene disrupted *Fusarium graminearum. Proceedings of the National Academy of Sciences of the United States of America 102*, 16892–16897.

32. Ochiai, N., Tokai, T., Takahashi-Ando, N., Fujimura, M., and Kimura, M. (2007) Genetically engineered Fusarium as a tool to evaluate the effects of environmental factors on initiation of trichothecene biosynthesis. *FEMS Microbiol Lett 275*, 53–61.

33. Schmidt-Heydt, M., and Geisen, R. (2007) A microarray for monitoring the production of mycotoxins in food. *Int J Food Microbiol 117*, 131–140.

34. Baldwin, T. K., Urban, M., Brown, N., and Hammond-Kosack, K. E. (2010) A role for Topoisomerase I in *Fusarium graminearum* and *F. culmorum* pathogenesis and sporulation. *Molecular Plant-Microbe Interactions 23*, 566–577.

35. Urban, M., Mott, E., Farley, T., and Hammond-Kosack, K. (2003) The *Fusarium graminearum MAP1* gene is essential for pathogenicity and development of perithecia. *Molecular Plant Pathology 4*, 347–359.

36. Miller, J. D., Taylor, N., and Greenhalgh, R. (1983) Production of deoxynivalenol and related compounds in liquid culture by Fusarium graminearum. *Canadian Journal of Microbiology 29*, 1171–1178.

37. Jiao, F., Kawakami, A., and Nakajima, T. (2008) Effects of different carbon sources on trichothecene production and *Tri* gene expression by *Fusarium graminearum* in liquid culture. *FEMS Microbiology Letters 285*, 212–219.

38. Gardiner, D. M., Kazan, K., and Manners, J. M. (2009) Nutrient profiling reveals potent inducers of trichothecene biosynthesis in *Fusarium graminearum. Fungal Genet Biol*, 604–613.

39. Gardiner, D. M., Osborne, S., Kazan, K., and Manners, J. M. (2009) Low pH regulates the production of deoxynivalenol by *Fusarium graminearum. Microbiology 155*, 3149–3156.

40. Ramirez, M. L., Chulze, S., and Magan, N. (2006) Temperature and water activity effects on growth and temporal deoxynivalenol production by two Argentinean strains of *Fusarium graminearum* on irradiated wheat grain. *International Journal of Food Microbiology 106*, 291–296.

41. Ponts, N., Pinson-Gadais, L., Verdal-Bonnin, M.-N., Barreau, C., and Richard-Forget, F. (2006) Accumulation of deoxynivalenol and its

15-acetylated form is significantly modulated by oxidative stress In liquid cultures of *Fusarium graminearum*. *FEMS Microbiol Lett 258*, 102–107.

42. Hope, R., Aldred, D., and Magan, N. (2005) Comparison of environmental profiles for growth and deoxynivalenol production by *Fusarium culmorum* and *F. graminearum* on wheat grain. *Letters in Applied Microbiology 40*, 295–300.

43. Schmidt-Heydt, M., Magan, N., and Geisen, R. (2008) Stress induction of mycotoxin biosynthesis genes by abiotic factors. *FEMS Microbiol Lett 284*, 142–149.

44. Ponts, N., Couedelo, L., Pinson-Gadais, L., Verdal-Bonnin, M. N., Barreau, C., and Richard-Forget, F. (2009) Fusarium response to oxidative stress by H_2O_2 is trichothecene chemotype-dependent. *FEMS Microbiol Lett 293*, 255–262.

45. Tacke, B. K., and Casper, H. H. (1996) Determination of deoxynivalenol in wheat, barley and malt by column cleanup and gas chromatography with electron capture detection. *Journal of the Association of Official Analytical Chemists 79*, 472–475.

46. Harris, L. J., Alexander, N. J., Saparno, A., Blackwell, B., McCormick, S. P., Desjardins, A. E., Robert, L. S., Tinker, N., Hattori, J., Piche, C., Schernthaner, J. P., Watson, R., and Ouellet, T. (2007) A novel gene cluster in *Fusarium graminearum* contains a gene that contributes to butenolide synthesis. *Fungal Genet Biol 44*, 293–306.

47. Lowe, R. G., Allwood, J. W., Galster, A. M., Urban, M., Daudi, A., Canning, G. G., Ward, J., Beale, M., and Hammond-Kosack, K. (2010) A combined 1H NMR and ESI-MS analysis to understand the basal metabolism of plant pathogenic *Fusarium* species. *Mol Plant Microbe Interact 23*, 1605–1818.

48. Gardiner, D. M., Kazan, K., and Manners, J. M. (2009) Nutrient profiling reveals potent inducers of trichothecene biosynthesis in *Fusarium graminearum*. *Fungal Genetics and Biology*, 604–613.

49. Gardiner, D. M., Kazan, K., and Manners, J. M. (2009) Novel Genes of *Fusarium graminearum* That Negatively Regulate Deoxynivalenol Production and Virulence. *Molecular Plant-Microbe Interactions 22*, 1588–1600.

50. Ward, J. L., Harris, C., Lewis, J., and Beale, M. H. (2003) Assessment of H-1 NMR spectroscopy and multivariate analysis as a technique for metabolite fingerprinting of *Arabidopsis thaliana*. *Phytochemistry 62*, 949–957.

51. Ehrlich, K. C., and Daigle, K. W. (1987) Protein-synthesis inhibition by 8-Oxo-12,13-epoxytrichothecenes. *Biochimica Et Biophysica Acta 923*, 206–213.

52. Ueno, Y., Nakajima, M., Sakai, K., Ishii, K., Sato, N., and Shimada, N. (1973) Comparative toxicology of trichothene mycotoxins -inhibition of protein-synthesis in animal cells. *J Biochem-Tokyo 74*, 285–296.

53. Hilton, A. J. (1999) Mechanisms of resistance to Fusarium ear blight in winter wheat *(Triticum aestivum L.)*, In *Harper Adams Agricultural College in collaboration with Plant Breeding International*, Open University, Cambridge.

54. Holzapfel, J., Voss, H. H., Miedaner, T., Korzun, V., Haberle, J., Schweizer, G., Mohler, V., Zimmermann, G., and Hartl, L. (2008) Inheritance of resistance to Fusarium head blight in three European winter wheat populations. *Theoretical and Applied Genetics 117*, 1119–1128.

55. Srinivasachary, Gosman, N., Steed, A., Simmonds, J., Leverington-Waite, M., Wang, Y., Snape, J., and Nicholson, P. (2008) Susceptibility to Fusarium head blight is associated with the Rht-D1b semi-dwarfing allele in wheat. *Theoretical and Applied Genetics 116*, 1145–1153.

Chapter 28

Identifying Genes in *Fusarium verticillioides* Through Forward and Reverse Genetics

J.B. Ridenour, R.L. Hirsch, and B.H. Bluhm

Abstract

The increasing availability of sequenced genomes for plant pathogenic fungi has revolutionized molecular plant pathology in recent years. However, the genetic regulatory networks underlying many important components of pathogenesis remain poorly defined. Although the protocols outlined in this chapter can be utilized to identify genes regulating a wide range of biological processes in many filamentous fungi, we focus on describing how to identify genes through forward and reverse genetics, using the plant pathogenic fungus *Fusarium verticillioides* as a model for the protocol. Specifically, this chapter explains how to create a collection of insertional mutants via Restriction Enzyme Mediated Integration (REMI) and how to screen mutants with a high-throughput method to visualize defects in amylolysis. Next, techniques are described to define the genomic lesions in REMI mutants with genome-walker PCR in order to identify candidate genes. Finally, protocols are presented describing a reverse-genetic approach to disrupt candidate genes in the wild-type strain with a split-marker strategy to confirm the phenotype observed in the REMI mutant.

Key words: *Fusarium verticillioides*, Transformation, Protoplasts, Amylolysis, Forward genetics, Reverse genetics

1. Introduction

In filamentous fungi, screening large collections of tagged mutants is an effective approach to identify novel genes involved in a specific biological process (1–3). In this chapter, we describe how a collection of tagged mutants is created in *Fusarium verticillioides*, how mutants are analyzed with a forward genetic screen to identify phenotypes related to amylolysis, and how the underlying genomic lesions are identified (see Fig. 1). Finally, we describe how to confirm the genotype-phenotype relationship through reverse genetics

Melvin D. Bolton and Bart P.H.J. Thomma (eds.), *Plant Fungal Pathogens: Methods and Protocols*,
Methods in Molecular Biology, vol. 835, DOI 10.1007/978-1-61779-501-5_28, © Springer Science+Business Media, LLC 2012

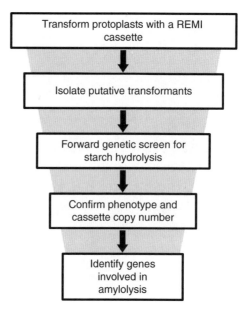

Fig. 1. Conceptual flowchart of the forward genetic screen to identify impaired starch hydrolysis.

by targeting the candidate gene for disruption in the wild-type strain and comparing the resulting phenotype to that of the corresponding Restriction Enzyme Mediated Integration (REMI) mutant.

This approach utilizes a high-throughput, cost-effective, plate-based screen to assay tagged mutants for changes in starch metabolism. Hundreds of mutants can be created from a single transformation event. Therefore, this protocol can be used to produce thousands of REMI mutants if desired. When the protocol is fully optimized, genes can be identified and characterized in approximately 1 month. A REMI mutant collection is an important laboratory resource for functional genomics because the mutants can be repeatedly screened for a wide range of phenotypes.

2. Materials

Equipment:

1. Spectrophotometer (for quantification of DNA).
2. Centrifuge with swinging bucket rotor for 15 and 50 mL conical centrifuge tubes.
3. Incubator shaker.
4. Rocking shaker.
5. Thermocycler.
6. Ultrasonic bath.

Solutions and reagents:

1. Sterile deionized H_2O.

2. 100% ethanol, store at $-20°C$.

3. 95% ethanol, store at $-20°C$.

4. 70% ethanol, store at $-20°C$.

5. V8 agar medium (1 L): 180 mL V8 juice, 2 g $CaCO_3$, 20 g agar. Autoclave at 121°C for 40 min to sterilize.

6. Potato dextrose agar (PDA) medium, 0.2× (1 L): 7.8 g PDA, 10 g agar. Autoclave at 121°C for 40 min to sterilize.

7. Yeast extract peptone dextrose (YEPD) liquid medium: 0.5% yeast extract, 1.0% peptone, 3% dextrose. Autoclave at 121°C for 40 min to sterilize.

8. Enzyme buffer: 1.2M $MgSO_4$, 10 mM K_2HPO_4 pH 5.8. Autoclave at 121°C for 40 min to sterilize, and store at room temperature.

9. Digestion buffer: 50 mL enzyme buffer, 500 mg lysing enzyme from *Trichoderma harzianum* (Sigma-Aldrich, St. Louis, MO), 100 μL beta-glucuronidase (see Note 1).

10. Separation buffer A: 0.6M sorbitol, 100 mM Tris–HCl pH 7.0. Autoclave at 121°C for 40 min to sterilize, and store at room temperature.

11. Separation buffer B: 1.2M sorbitol, 10 mM Tris–HCl pH 7.5. Autoclave at 121°C for 40 min to sterilize, and store at room temperature.

12. STC: 1.2M sorbitol, 10 mM Tris–HCl pH 7.5, 50 mM $CaCl_2 \cdot 2H_2O$. Autoclave at 121°C for 40 min to sterilize, and store at room temperature.

13. Protoplast resuspension buffer: 93% STC, 7% DMSO (see Note 2).

14. 40% polyethylene glycol (PEG) STC: 60% STC, 40% PEG 4000 (see Note 3).

15. Regeneration liquid medium: 1M sucrose, 0.02% yeast extract. Autoclave at 121°C for 40 min to sterilize, and store at room temperature.

16. Regeneration agar medium: 1M sucrose, 0.02% yeast extract, 1% agar. Autoclave at 121°C for 40 min to sterilize (see Note 4).

17. Hygromycin B (see Note 5).

18. PCR buffer (5×): 100 mM Tris–HCl (pH 8.5), 50 mM NH_4SO_4, 50 mM KCl, in 80% glycerol. Autoclave at 121°C for 40 min, and store in 1 mL aliquots at $-20°C$.

19. Glycerol (30%): Autoclave at 121°C for 40 min to sterilize, and store at room temperature.

20. Corn starch minimal (CSM) agar medium: 0.3M $NaNO_3$, 6 mM K_2HPO_4, 4 mM $MgSO_4$, 7 mM KCl, 66 µM $FeSO_4$, 2% agar, and 1% corn starch. Autoclave at 121°C for 40 min to sterilize.

21. Iodine staining solution: 0.25% iodine, 2.5% KI.

22. Extraction kit for genomic DNA isolation from fungal samples.

23. GenomeWalker Universal kit (Clontech, Palo Alto, CA).

24. Gel extraction kit for nucleic acid purification from agarose gels.

25. TA cloning kit for direct cloning of PCR products.

26. Tris EDTA (TE) buffer (1×, pH 8.0): 10 mM Tris–HCl pH 8.0, 1 mM EDTA pH 8.0. Autoclave at 121°C for 40 min to sterilize, and store at room temperature.

27. Plasmid pCB1003.

3. Methods

3.1. REMI Cassette Preparation

REMI has been used to create collections of tagged, insertional mutants in many filamentous fungi. Such collections have proven useful for identifying novel genes through forward genetic screens (1, 3, 4). In this protocol, a REMI cassette is amplified by PCR from a plasmid clone, and the purified PCR product is used for transformation. The 2.1 kb REMI cassette, containing a hygromycin B resistance gene and a promoterless GFP coding region, is amplified from pBLPT01 (see Fig. 2a). Integration of the REMI

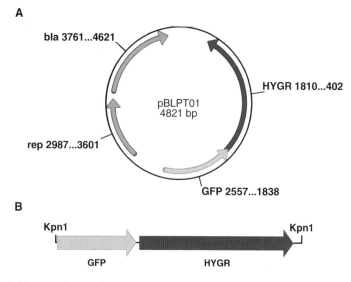

Fig. 2. The promoter-trapping REMI cassette used in this study. (**a**) A GFP ORF lacking a promoter sequence was ligated into pBLPT01 containing a hygromycin resistance (HYG^R) gene. (**b**) The cassette was amplified from plasmid DNA with primers containing *Kpnl* sequences integrated into the 5′ tail.

cassette into the genome can impair gene function by disrupting open reading frames (ORFs), regulatory untranslated regions of genes, or distal regulatory elements. Additionally, the inclusion of an ORF encoding GFP allows the REMI cassette to be used for promoter trapping, in which the activation of endogenous promoters can be monitored in response to selected environmental conditions by visually monitoring the expression of GFP.

1. Primers KpnIPTf and KpnIPTr are designed to amplify the 2.1 kb REMI cassette from pBLPT01 and incorporate *Kpn*I recognition sites on the ends of the PCR product (see Note 6; Fig. 2b).

2. Prepare a PCR reaction to amplify the REMI cassette as follows: combine 10 μL of 5× PCR buffer, 5 μL of 25 mM MgCl$_2$, 1 μL of 10 mM dNTPs, 1 μL of 10 μM primer KpnIPTf, 1 μL of 10 μM primer KpnIPTr, 1 μL of template DNA (pBLPT01 at a concentration of 10 ng/μL), 0.5 μL of *Taq* polymerase, and 30.5 μL of sterile distilled H$_2$O (see Note 7).

3. Amplify the REMI cassette with the following PCR conditions: perform one cycle of denaturation at 94°C for 1 min; 40 cycles with the following parameters: 94°C for 30 s, 58°C for 30 s, and 72°C for 1.5 min; followed by a final polymerization step of 72°C for 5 min.

4. Check the quality of the PCR product using agarose gel electrophoresis (see Note 8).

5. Precipitate the REMI cassette by transferring the PCR product to a 1.5 mL microcentrifuge tube and adding 2 volumes of ice-cold 100% ethanol. Mix thoroughly by inversion and incubate at −20°C for 30 min.

6. Centrifuge the 1.5 mL tube at room temperature for 5 min at 13,000 rpm to pellet DNA. Discard the supernatant.

7. Wash the DNA pellet by adding 750 μL of ice-cold 70% ethanol and inverting the tube 4 times. Centrifuge as described in Subheading 3.1, step 6, and discard the supernatant.

8. Wash the DNA pellet once more by adding 750 μL of ice-cold 95% ethanol and inverting the tube 4 times. Centrifuge as described in Subheading 3.1, step 6, and discard the supernatant.

9. Invert tube on a clean paper towel. Dry the DNA pellet for 10 min (see Note 9).

10. Resuspend the DNA by dissolving the pellet in 50 μL of sterile, deionized H$_2$O.

11. Quantify the amount of the REMI cassette obtained with a spectrophotometer and check the quality using agarose gel electrophoresis. For information regarding expected yield see Note 7.

3.2. Culture Preparation and Protoplast Generation

F. verticillioides produces microconidia that are only slightly larger than protoplasts, thus making protoplast purification by size selection (e.g., filtration) difficult. However, protoplasts can be separated from microconidia based on density. In the following protocol, a highly effective technique based on density gradient centrifugation is described to separate protoplasts from hyphae, microconidia, and cellular debris.

1. Inoculate V8 plates with an agar plug colonized by *F. verticillioides*. Incubate cultures at 25°C for 3–5 days under a 12:12-h light-dark cycle. Under these conditions, the wild-type strain produces large amounts of microconidia (see Note 10).

2. Collect microconidia from the cultures prepared in Subheading 3.2, step 1. Add 1 mL of sterile deionized H_2O to each plate and scrape the surface of the culture repeatedly with a sterile cell spreader to dislodge microconidia. Transfer harvested microconidia to a 15 mL conical centrifuge tube and adjust the concentration to 10^6 conidia/mL with sterile deionized H_2O.

3. Inoculate 50 mL of YEPD medium in a 250 mL flask with 10^6 microconidia. Incubate cultures at 28°C with rotary shaking at 180 rpm for 36–40 h (see Note 11).

4. Collect the tissue by decanting the culture into a 50 mL conical centrifuge tube. Centrifuge at 4°C for 5 min at 5,000 rpm to pellet tissue. Use a transfer pipette to aspirate and discard the supernatant, taking care not to disturb the pelleted tissue (see Note 12).

5. Wash the tissue by gently resuspending in 20 mL of enzyme buffer. Pellet the tissue and discard the supernatant as described in Subheading 3.2, step 4.

6. Gently resuspend the tissue in 25 mL digestion buffer, and decant into a sterile 125 mL Erlenmeyer flask. Incubate at 28°C with rotary shaking at 80 rpm for 2–7 h. After 2 h, check hourly for protoplast release. Once the concentration of protoplasts reaches ~10^6/mL, proceed to the next step (see Fig. 3).

7. Gently decant an equal volume of the tissue suspension prepared in Subheading 3.2, step 6 into three 15 mL conical centrifuge tubes.

8. Gently overlay the tissue suspension with 4 mL of separation buffer A. Separation buffer A will form a distinct interface with the tissue suspension (see Fig. 4a). Centrifuge at 4°C for 5 min at 5,000 rpm. The protoplasts will accumulate at the interface of separation buffer A and the tissue suspension (see Note 13; Fig. 4b).

9. Collect the protoplast fraction from each of the three 15 mL tubes. With a wide-bore transfer pipette, create a gentle

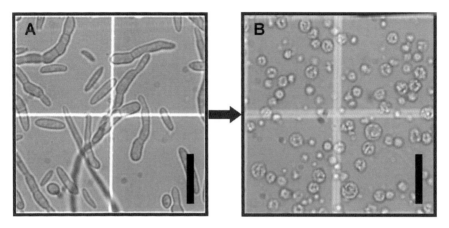

Fig. 3. Digestion of fungal tissue and release of protoplasts. (**a**) Fungal tissue in YEPD medium inoculated with 10^6 micro-conidia incubated for 36 h. (**b**) Protoplasts visible after 7 h digestion with an enzyme solution. *Bars* represent 2.5 μm.

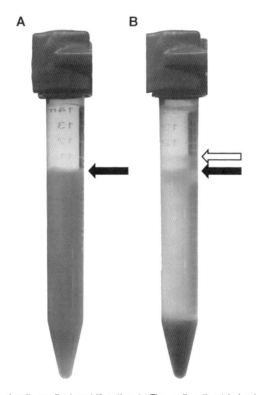

Fig. 4. Protoplast separation by density gradient centrifugation. (**a**) Tissue digestion (*darker layer*) is overlaid with separation buffer A before centrifugation (*black arrow* indicates interface). (**b**) After centrifugation, undigested tissue is pelleted in the bottom of the tube, although some may accumulate near the phase interface (*black arrow*). Protoplasts aggregate in a fine layer above the undigested tissue (*white arrow*).

suction approximately 2 mm above the gradient interface. Combine the protoplast fractions in a new 15 mL conical centrifuge tube (see Note 14).

10. Wash the protoplasts by adjusting the volume to 15 mL with the gentle addition of separation buffer B. Centrifuge at 4°C for 10 min at 5,000 rpm to pellet the protoplasts. Discard the supernatant.

11. Wash the protoplasts once more by gently resuspending in 15 mL of separation buffer B. Pellet the protoplasts and discard the supernatant as described in Subheading 3.2, step 10 (see Note 15).

12. Resuspend the protoplasts in 500 μL of freshly prepared protoplast resuspension buffer. Adjust the concentration to ~10^8 protoplasts/mL with protoplast resuspension buffer. Dispense 100 μL aliquots into 2 mL screw-cap microcentrifuge tubes, and store at –80°C (see Note 16).

3.3. Fungal Transformation

PEG-mediated transformation of protoplasts is a common method for transformation of filamentous fungi (5). In this process, exogenous DNA enters fungal protoplasts due to membrane permeation facilitated by high concentrations of PEG and calcium ions (6). Following the protocol below, several hundred transformants can be obtained per transformation of *F. verticillioides*, and thus the protocol can be used to rapidly create large REMI mutant collections.

1. Prepare protoplasts for transformation by thawing tubes of protoplasts on ice, each tube containing 100 μL of protoplasts at a concentration of ~10^8/mL (see Note 17).

2. Prepare DNA for transformation by co-incubating 10 μg of the REMI cassette (preparation described in Subheading 3.1) with 50 U of the restriction enzyme *Kpn*I. Include a negative control by mixing 10 μL of sterile deionized H_2O and 50 U of *Kpn*I. Gently mix the solution by pipetting, and then incubate at room temperature for 10 min (see Note 18).

3. Add each co-incubation mixture to a tube of protoplasts. Mix by gently agitating the protoplasts, and then incubate on ice for 30 min.

4. Add 100 μL of 40% PEG STC to each tube (including the negative control). Mix by gently rolling the tube. Visually confirm homogenization, and incubate at room temperature for 30 min (see Note 19).

5. Add 1.5 mL of regeneration liquid medium to each tube (including the negative control). Gently mix by inversion, and incubate at room temperature overnight with gentle rocking on a rocking shaker (see Note 20).

6. Following the overnight incubation, separately examine 10 μL of each transformation and the negative control prepared in Subheading 3.3, step 5 with a microscope to ensure the protoplasts are regenerating (see Note 21).

7. Transfer each transformation (including the negative control) to an individual 15 mL conical centrifuge tube using a wide-bore transfer pipette, and adjust the total volume to 15 mL with regeneration liquid medium. Thoroughly mix by inversion, and record the concentration of regenerated protoplasts. The concentration of regenerated protoplasts should be approximately 10^7/mL.

8. Dispense 1 mL of each transformation with the REMI cassette as prepared in Subheading 3.3, step 7, to 15 sterile 100×15 mm Petri dishes. Dispense 1 mL of the negative control transformation prepared in Subheading 3.3, step 7, to six sterile 100×15 mm Petri dishes (see Note 22).

9. It is important to evaluate positive and negative controls by regenerating protoplasts in the presence and absence of antibiotic selection. In each of the 15 transformation plates (see Subheading 3.3, step 8) and three of the negative control plates, pour 7.5 mL of molten regeneration agar medium (~50°C) containing hygromycin B at a concentration of 75 μg/mL. In the remaining three negative control plates, pour 7.5 mL of molten regeneration agar medium (~50°C) containing no hygromycin B (no selection). Immediately after pouring, thoroughly mix each plate with a gentle swirling motion. Ensure protoplasts are evenly dispersed within the molten medium and allow the medium to solidify (see Note 23).

10. Incubate all regeneration plates prepared in Subheading 3.3, step 9, at 25°C for 3–5 days under a 12:12-h light-dark cycle. Check plates daily for regenerating transformants (see Note 24).

11. Isolate putative transformants by transferring 2×2 mm agar blocks containing individual colonies from transformation plates to individual wells of 24-well plates containing 0.2× PDA amended with hygromycin B at a final concentration of 75 μg/mL (see Note 25, Fig. 5).

12. Incubate the 24-well plates at 25°C for 3–5 days under a 12:12-h light-dark cycle to promote fungal growth. Once sufficient biomass has developed, the REMI collection will be ready for long-term storage and high-throughput phenotyping (see Note 26).

13. For long-term storage of the REMI collection, transfer 8–10 agar blocks (2×2 mm) of each transformant to individual 2 mL screw-cap microcentrifuge tubes containing 1 mL of sterile 30% glycerol. Store at −80°C (see Note 27).

A   B   C

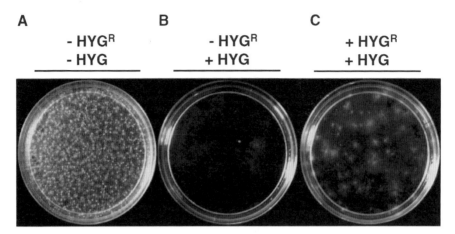

Fig. 5. Protoplast regeneration in the presence and absence of hygromycin selection. (**a**) Untransformed protoplasts incubated in regeneration medium without hygromycin are able to regenerate. (**b**) Untransformed protoplasts incubated with hygromycin do not regenerate. (**c**) A comparatively small percentage of protoplasts transformed with a hygromycin resistance cassette regenerate when incubated with hygromycin B.

3.4. High-Throughput Screen for Amylolysis-Related Phenotypes

Evaluating tagged mutants for phenotypes related to starch hydrolysis is a powerful way to identify previously unknown genes involved in amylolysis. Starch is especially suitable for large-scale enzymatic assays because it is inexpensive and its hydrolysis can be visually assessed by staining with iodine. The following protocol describes a high-throughput technique that can be utilized to assay large collections of mutants for phenotypes related to starch metabolism and identify candidate genes for further characterization.

1. Grow each mutant on CSM agar medium in 100×15 mm Petri dishes. Use individual sterile toothpicks or a sterile needle to point-inoculate each transformant to the center of an individual plate. To provide a reference for the assay, transfer the wild-type strain of *F. verticillioides* to four CSM plates as described above. Incubate at 25°C for 7 days under a 12:12-h light-dark cycle (see Note 28).

2. Prepare the cultures described in Subheading 3.4, step 1, for iodine staining. Gently flood the surface of each plate with H_2O, using a gentle, swirling motion to ensure the culture is fully hydrated. Discard the excess H_2O (see Note 29).

3. Stain the plates prepared in Subheading 3.4, step 2 by applying 5 mL of iodine solution to the surface of each plate. Gently swirl the solution to ensure the medium is evenly stained. After color begins to develop (~5 s), pour off the excess iodine solution. Rinse each plate immediately with H_2O and place it upside down to dry. The unstained halo indicates the region of starch hydrolysis (see Note 30; Fig. 6a).

4. Evaluate starch hydrolysis by recording both quantitative and qualitative observations. For example, the diameter of the

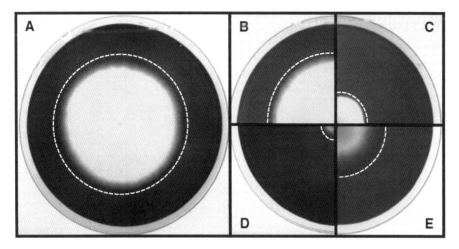

Fig. 6. Examples of phenotypic variability observed among REMI mutants in the amylolysis assay. Amylolysis was measured by staining starch plates with iodine after 7 days of growth. *Unstained areas* indicate starch saccharification. For clarity, *dotted lines* delineate the colony margin. (**a**) A representative wild-type strain stained to visualize starch hydrolysis. Major categories of phenotypes observed include: indistinguishable from wild type (**b**), reduced growth but increased substrate hydrolysis (**c**), severe, pleiotropic growth defects (**d**), and robust growth but reduced starch hydrolysis (**e**).

mycelium and the halo of saccharification should be recorded (see Note 31).

5. Based on data collected in Subheading 3.4, step 4, select a manageable number (2–5% of the collection) of mutants with interesting phenotypes (see Fig. 6b–e).

6. Perform single-spore isolations for each mutant. Collect microconidia from each culture (e.g., as described in Subheading 3.3, step 12) by scraping a small amount of aerial hyphae from the medium surface with a sterile toothpick or sterile needle (see Note 32). In *F. verticillioides*, microconidia are typically formed abundantly in chains on the surface of the mycelium.

7. Transfer the tissue to 1 mL of sterile deionized H_2O in a 15 mL conical centrifuge tube, and examine under a microscope. Adjust the concentration of microconidia to $10^3/mL$ with sterile deionized H_2O.

8. Dispense 200 μL of the suspension described in Subheading 3.4, step 7, onto 100×15 mm plates containing $0.2\times$ PDA amended with hygromycin B at a final concentration of 75 μg/mL. Spread the spore suspension evenly across the surface of the medium with a sterile cell spreader. Incubate at 25°C for 1–2 days under a 12:12-h light-dark cycle. Observe periodically with a stereomicroscope to ensure that emerging colonies originated from a single spore; mark plates with ink where confirmed single-spore colonies are observed.

9. Isolate colonies derived from a single spore by transferring 2×2 mm agar blocks from colonies identified in Subheading 3.4,

step 8, to 0.2× PDA amended with hygromycin B at a final concentration of 75 μg/mL.

10. Repeat the CSM/iodine assay as described in Subheading 3.4, steps 1–4, with four biological replications of each single-spore isolate. For each single-spore isolate, the phenotype should closely match the phenotype of the REMI mutant from which it was derived.

3.5. Characterization of Gene Disruptions in Selected Transformants

The genomic lesion underlying the phenotype of a REMI mutant can be identified with genome walking PCR. With this approach, DNA flanking the REMI cassette is amplified via PCR, and thus the site of insertion can be defined thoroughly.

1. Prepare DNA from the mutants identified in Subheading 3.4 with a DNA extraction kit designed to isolate DNA from fungi. Additionally, extract DNA from *F. verticillioides* wild-type strain 7600 to be used as a control in subsequent experiments.

2. Primers HYG-PF and HYG-PR are designed to amplify a 353 bp sequence from the hygromycin resistance construct contained in the REMI cassette amplified from pBLPT01. Amplify following standard protocols; the product will be used as a probe for a Southern blot (see Subheading 3.5, step 3).

3. Perform a Southern blot to verify that mutants of interest contain a single copy of the REMI cassette. Perform Southern blot analyses with standard protocols (see Note 33; Fig. 7).

4. Amplify genomic DNA flanking the REMI integration sites in each mutant with the Genome Walker Universal kit following the instructions of the manufacturer (see Note 34).

5. Separate the products from the final genome walker PCR using agarose gel electrophoresis.

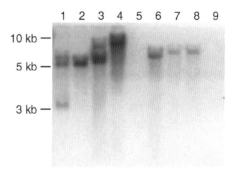

Fig. 7. Representative Southern analysis to determine copy number of the REMI cassette in selected mutants. Nine mutants with amylolysis-related phenotypes were analyzed with a probe derived from the HYG sequence of the REMI cassette. Six of the nine mutants appeared to contain a single copy of the cassette. One mutant appeared to contain multiple copies of the cassette (lane 1) and two mutants were false positives (lanes 5 and 9).

6. For each mutant, extract and purify the major product from the agarose gel as prepared in Subheading 3.5, step 5, with a gel extraction kit designed to purify nucleic acids from an agarose gel following the instructions of the manufacturer.

7. Ligate the purified PCR product for each mutant into a TA-cloning vector with a TA cloning kit following the instructions of the manufacturer.

8. Sequence clones containing putative flanking sequences.

9. To determine the site of integration, use BLAST to query the sequences obtained in Subheading 3.5, step 8 against the *F. verticillioides* genomic database available at www.broadinstitute.org. Once identified, the specific locus disrupted by the REMI cassette can be confirmed as exclusively responsible for the phenotype of interest via targeted deletion (see Note 35).

3.6. Preparation of Split-Marker Deletion Construct

Once a candidate gene has been identified, it should be deleted by targeted mutagenesis in the wild-type strain for phenotypic confirmation. In the approach described below, two PCR products comprising 5′ and 3′ flanking regions of a gene are fused to portions of a hygromycin B resistance gene via PCR to create a split-marker deletion construct (7–9). When transformed into the wild-type strain, the construct integrates into the targeted locus via triple-homologous recombination, and replaces the targeted gene with the hygromycin resistance cassette (see Fig. 8). This protocol can generate multiple, independent deletion strains quickly (within 1 week).

1. Primers F_0, F_2, F_3, and F_5 are designed to amplify the 5′ and 3′ regions flanking the gene of interest. Primer F_1 is nested inside F_0. Primer F_4 is nested inside F_5. Primers F_2 and F_3 incorporate a M13 sequence to facilitate hybridization of each flanking region to the hygromycin resistance cassette (see Table 1; Note 36).

2. Amplify the 5′ and 3′ regions flanking the gene of interest by preparing the following two PCR reactions (A and B) using gene-specific primer combinations (A) F_0/F_2 and (B) F_3/F_5: for each of the two reactions mix 10 μL of 5× PCR buffer, 5 μL of 25 mM $MgCl_2$, 1 μL of 10 mM dNTPs, 1 μL of 10 μM forward primer, 1 μL of 10 μM reverse primer, 1 μL of template DNA (wild-type genomic DNA at a concentration of 100 ng/μL), 0.5 μL of *Taq* polymerase, and 30.5 μL of sterile distilled H_2O (see Note 37).

3. Amplify the 5′ and 3′ fragments of the HYG^R selectable marker by preparing the following two PCR reactions (C and D) using marker-specific primer combinations (C) pUC/M13/HY and (D) YG/pUC/M13 reverse: for each of the two reactions mix 10 μL of 5× PCR buffer, 5 μL of 25 mM $MgCl_2$, 1 μL of 10 mM dNTPs, 1 μL of 10 μM forward primer, 1 μL of 10 μM

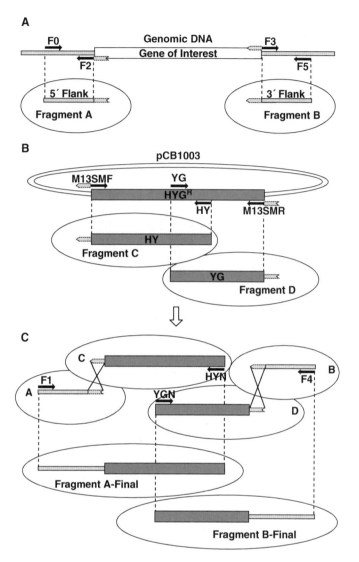

Fig. 8. Schematic representation of the PCR-based split-marker gene disruption strategy. (**a**) 5′ and 3′ flanks of the targeted gene are amplified with primer sets containing M13 tails. (**b**) Partial overlapping regions of the HYG[R] ORF are amplified with primers containing M13 tails. (**c**) Flanking regions of the targeted gene are combined with their respective HYG[R] split-marker cassettes to form the final split-marker gene disruption constructs.

reverse primer, 1 μL of template DNA (pCB1003 plasmid DNA at a concentration of 10 ng/μL), 0.5 μL of *Taq* polymerase, and 30.5 μL of sterile distilled H$_2$O.

4. For PCR reactions A–D, perform an initial denaturation at 94°C for 1 min, followed by 40 cycles of 94°C for 30 s, 56–58°C for 30 s, and 72°C for 45 s, and a final polymerization step of 72°C for 5 min.

5. Check the quality of the PCR products by agarose gel electrophoresis.

Table 1
Primers used in this protocol

Primer [a]	Sequence (5'–3')[b]	T_m (°C)	Function
pUC/M13	GTAAAACGACGGCCAGT	52.6	Forward primer for amplification of REMI cassette (generic)/ amplification of 5' marker fragment
pUC/M13 reverse	CAGGAAACAGCTATGAC	47.0	Reverse primer for amplification of REMI cassette (generic)/ amplification of 5' marker fragment
KpnIPTf	<u>CGGGGTACC</u>GTAAACGAC GGCCAGT	52.1 w/o tail	REMI cassette amplification forward primer
KpnIPTr	<u>CGGGGTACC</u>CAGGAAACAG CTATGAC	47.0 w/o tail	REMI cassette amplification reverse primer
F_0	22–24 nucleotide	58	Forward primer for amplification of gene-specific 5' flank
F_1	22–24 nucleotide	58	Nested forward primer for amplification of gene-specific 5' flank
F_2	<u>AATTCACTGGCCGTCGTTTTAC</u> +22–24 nucleotide	58 w/o tail	Reverse primer for amplification of gene-specific 5' flank
F_3	<u>CGTAATCATGGTCATAGCTGTT TCCTG</u> + 22–24 nucleotide	58 w/o tail	Forward primer for amplification of gene-specific 3' flank
F_4	22–24 nucleotide	58	Nested reverse primer for amplification of gene-specific 3' flank
F_5	22–24 nucleotide	58	Reverse primer for amplification of gene-specific 3' flank
PF	22–24 nucleotide	58	Forward primer for PCR screening putative transformants
PR	22–24 nucleotide	58	Reverse primer for PCR screening putative transformants
HY	GGATGCCTCCGCTCGAAGTA	58	Reverse primer for amplification of 5' marker fragment
HYG-PF	CCAGTGATACACAT GGGGATCAGC	59	Forward primer for Southern blot probe sequence
HYG-PR	GGATATGTCCTGCGGGT AAATAGCTG	59	Reverse primer for Southern blot probe sequence
HY-N	TAGCGCGTCTGCTGCT CCATACAAG	60	Nested reverse primer for amplification of 5' marker fragment
YG	CGTTGCAAGACCTGCCTGAA	58	Forward primer for amplification of 3' marker fragment
YG-N	ACCGAACTGCCCGCTGTTCTC	60	Nested forward primer for amplification of 3' marker fragment

[a]Primers F_0, F_1, F_2, F_3, F_4, F_5, A_1, A_2, PR, and PF are gene specific
[b]Oligonucleotide tails not designed to anneal to the template DNA strand are underlined

6. Precipitate the PCR products individually as described in Subheading 3.1, with the following exception: after the final wash, resuspend the DNA by dissolving the pellet in 100 µL sterile deionized H_2O.

7. Quantify the amount of each fragment with a spectrophotometer and check the quality by agarose gel electrophoresis (for information regarding expected yield see Note 7). Adjust the concentration of each product to 100 ng/µL. The fragments A, B, C, and D will provide template for PCR as described in Subheading 3.6, step 8.

8. Fuse and amplify the split-marker construct fragments by preparing the following two PCR reactions: (A-Final) using the primer combination F_1/HY_N and the template combination A/C and (B-Final) using the primer combination YG_N/F_4 and the template combination B/D.

9. For each of the two PCR reactions, mix 10 µL of 5× PCR buffer, 5 µL of 25 mM $MgCl_2$, 1 µL of 10 mM dNTPs, 1 µL of 10 µM forward primer, 1 µL of 10 µM reverse primer, 1 µL of each template DNA (each at a concentration of 100 ng/µL), 0.5 µL of *Taq* polymerase, and 29.5 µL of sterile distilled H_2O.

10. For fusion PCR, perform an initial denaturation at 94°C for 1 min followed by 40 cycles of 94°C for 30 s, 56°C for 30 s, and 72°C for 1.5 min, followed by a final polymerization step of 72°C for 5 min.

11. Check the quality of the PCR products by agarose gel electrophoresis.

12. Precipitate the PCR products separately as described in Subheading 3.1, with the following exception: After the final wash, resuspend the DNA by dissolving the pellet in 25 µL of sterile, deionized H_2O.

13. Quantify the amount of the split-marker fragments obtained with a spectrophotometer and check the quality by standard agarose gel electrophoresis. Adjust the concentration of both fragments to approximately 1 µg/µL.

3.7. Transformation and Validation of Knockout Transformants

Split-marker gene disruption is facilitated by endogenous fungal recombination mechanisms. In this technique, two split-marker constructs created in Subheading 3.6 are co-transformed into wild-type protoplasts. The triple homologous recombination of the split-marker constructs into the target locus is a dependable methodology to disrupt or delete targeted genes in many fungi (see Fig. 9).

1. Thaw protoplasts (generated as described in Subheading 3.2) on ice.

2. Initiate transformation by adding 5 µg of each split-marker deletion fragment, A-Final and B-Final, to one tube of protoplasts.

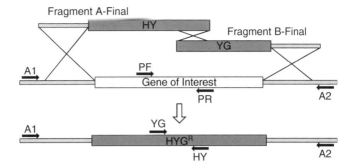

Fig. 9. Transformation strategy of split-marker gene deletion constructs. Final constructs are co-transformed and the targeted gene is disrupted via homologous recombination.

Sterile H_2O (10 μL) is added to the tube of protoplasts to be used as the negative control. Gently mix the split-marker deletion fragments with the protoplasts. Incubate on ice for 30 min.

3. Complete and plate transformation as described in Subheading 3.3, steps 4–10.

4. Isolate 100 regenerating transformants as described in Subheading 3.3, step 11. Incubate cultures at 25°C for 3–5 days under a 12:12-h light-dark cycle (see Note 38).

5. To isolate DNA for PCR, collect tissue from each transformant by harvesting a 5×10 mm area of the mycelium using a sterile toothpick or a sterile needle. Transfer the tissue to 30 μL of 1× TE buffer pH 8.0 in a 1.5 mL microcentrifuge tube (see Note 39).

6. Heat the tissue and buffer mixture for 3 min in a microwave oven.

7. Sonicate the tissue and buffer mixture at room temperature for 3 min in an ultrasonic bath.

8. Pellet the tissue by centrifugation at room temperature for 5 min at 14,000 rpm, and transfer the supernatant to a fresh 1.5 mL microcentrifuge tube. Store at 4°C until needed. The supernatant will serve as template for PCR as described in Subheading 3.7, step 10 (see Note 40).

9. When validating putative disruption mutants, it is important to confirm that both split-marker cassette flanks integrated in the correct location and deleted the targeted sequence. To identify successful gene deletion mutants, prepare four PCR reactions using the following primer combinations for each putative deletion mutant: (A) A_1/PR; (B) PF/A_2; (C) A_1/HY; (D) YG/A_2. Amplicons generated with primer combinations A and B should only be present in the wild-type strain, and C and D should only be present in successful gene disruption mutants (see Fig. 10a).

10. Amplify fragments for screening by preparing four PCR reactions. For each of the four reactions, mix 5 μL of 5× PCR buffer, 2.5 μL of 25 mM $MgCl_2$, 1 μL of 10 mM dNTPs, 1 μL of 10 μM forward primer, 1 μL of 10 μM reverse primer, 2 μL of template DNA (as prepared in Subheading 3.7, step 9), 0.25 μL of *Taq* polymerase, and 12.25 μL of sterile distilled H_2O.

11. PCR conditions for reactions A–D: perform an initial denaturation at 94°C for 3 min followed by 40 cycles of 94°C for 30 s, 56–58°C for 30 s, and 72°C for 1 min, and include a final polymerization step of 72°C for 5 min.

12. Analyze the products of the PCR prepared in Subheading 3.7, step 11 by agarose gel electrophoresis (see Fig. 10a).

13. Purify putative deletion mutants via single-spore isolation as described in Subheading 3.4, steps 6–9, and reevaluate the resulting single-spore isolates by PCR, as described in Subheading 3.7, steps 10–12.

14. Screening putative split-marker mutants by PCR is a powerful and rapid method for identifying deletion mutants. However, screening by PCR does not rule out the possibility of additional ectopic insertions. It is therefore necessary that putative deletion mutants be subjected to Southern blot analysis to verify gene deletion and construct copy number using a gene-specific probe and marker-specific probe respectively (see Fig. 10b).

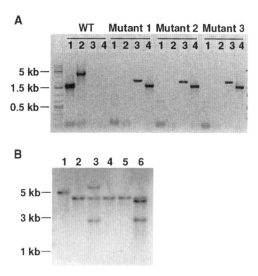

Fig. 10. Validation of putative gene deletion mutants. (**a**) Wild-type and mutant gDNA amplified by PCR with the primers A_1/PR (lane 1); PF/A_2 (lane 2); A_1/HY (lane 3); YG/A_2 (lane 4). Amplicons 1–2 are present only in the wild-type strain. Amplicons 3–4 are present only in mutant strains. (**b**) Southern blots confirming the deletion of the target gene resulting from targeted insertion of the split-marker deletion cassette. Mutants in lane 2, 4, and 5 contain single bands of the expected size for this restriction enzyme digest and are likely successful deletion mutants.

15. Evaluate the phenotype of the knockout strains with the amylolysis screen described in Subheading 3.4. All independent deletion mutants should express a phenotype similar to each other and to the original REMI mutant (see Note 41).

4. Notes

1. Prepare fresh digestion buffer for each experiment. Combine the reagents at room temperature and ensure the solution is well mixed. Sterilize the solution by passage through a 0.45 μm filter and store at 4°C.

2. In a chemical fume hood, add 0.7 mL of DMSO to 9.3 mL of STC. Thoroughly mix the solution, and sterilize by passage through a 0.45 μm filter. Store at room temperature.

3. Dissolve 8 g of PEG 4000 in 10 mL of STC. Heat the solution to ~65°C and stir with a magnetic stir-bar. Adjust the final volume to 20 mL with STC. Sterilize by passage through a 0.45 μm filter and store at 4°C.

4. Subheadings 3.3 and 3.7 require molten regeneration agar medium. The medium can be freshly prepared and autoclaved prior to use. Alternatively, the medium can be prepared, autoclaved, and stored at room temperature until needed. In the latter case, the medium can be melted in a microwave oven prior to use. In either case, it is important that the medium cools to ~50°C before the addition of hygromycin B.

5. The activity of commercially produced hygromycin B is variable, and the effective concentration for fungal selection may vary by supplier and by lot. If the efficacy of the antibiotic is unknown, establish a "kill curve" by testing a range of concentrations on both transformed and untransformed protoplasts. An effective concentration will prohibit the growth of untransformed protoplasts while still allowing the growth of protoplasts transformed with the HYG^R selectable marker.

6. The vector pBLPT01 is derived from the pUC plasmid lineage. For this reason the primers pUC/M13 and pUC/M13 reverse will amplify the 2.1 kb fragment serving as the REMI cassette (see Table 1). To facilitate cassette integration, restriction enzyme recognition sites are added to the 5′-end of these primers. A number of different primer sets can be designed utilizing this technique, with each primer set incorporating recognition sites for different restriction enzymes. Therefore, a single REMI cassette can be designed to integrate into different restriction enzyme sites, greatly increasing the diversity of a collection. However, it is important to avoid enzymes that will

cleave internally within the cassette. In addition, most restriction enzymes require 2–3 nucleotides flanking a recognition site for efficient cleavage.

7. The PCR mixtures and conditions provided in Subheading 3.1, steps 2 and 3 should be tested and adjusted empirically to amplify a specific product. This is true for all PCR performed in this chapter. The PCR mixture described is for a 50 μL reaction, which will yield approximately 7.5 μg of purified PCR product. A larger amount of the REMI cassette will be required to create an adequate mutant collection. Therefore, the volume should be increased ten-fold to yield approximately 75 μg of purified PCR product.

8. Before the DNA is precipitated, it is important to confirm the presence of a single amplicon of desired size with agarose gel electrophoresis. Alternatively, if gel electrophoresis reveals the presence of multiple products, a gel extraction kit designed to purify nucleic acids from agarose gels may be used to extract and purify the fragment of interest. However, the use of a gel extraction kit will reduce the amount of final product.

9. Because residual ethanol can inhibit subsequent reactions, ensure that ethanol has completely evaporated before continuing. It is possible that more than 10 min is needed to ensure ethanol evaporation. If this is the case, place tubes on their sides to prevent loss of the DNA pellet.

10. All steps in Subheadings 3.2 and 3.3 should be carried out using sterile technique.

11. One 50 mL flask will yield approximately 1.5–2 mL of protoplasts at a concentration of $\sim 10^8/\text{mL}$. Protocol may be scaled up to increase protoplast yield.

12. A swinging bucket rotor should be used for all centrifugation steps in Subheading 3.2.

13. A distinct interface between separation buffer A and the tissue suspension must be maintained before and after centrifugation. If the two layers become mixed, centrifugation will not adequately separate the protoplasts from hyphae, microconidia, and cellular debris.

14. When pipetting protoplasts, it is important to use a wide-bore (2 mm) transfer pipette. As an alternative, approximately 5 mm of a 1,000 μL pipette tip can be cut off to increase bore diameter. However, tips modified in this manner need to be autoclaved.

15. Harvested protoplasts must be thoroughly washed to remove digestive enzymes. Residual enzymes could negatively affect protoplast regeneration, particularly during the regeneration of cell walls after transformation.

16. Protoplasts stored at –80°C in resuspension buffer will remain viable for several months.

17. Each transformation reaction will require at least two tubes of protoplasts—one for transformation with the REMI cassette, and one for the negative control transformation. The negative control serves two purposes: confirming that these protoplasts cannot regenerate in the presence of hygromycin indicates the viability of the antibiotic and thus reduces false positives, whereas confirming that protoplasts regenerate in the absence of hygromycin indicates the general viability of the protoplasts and thus reduces false negatives.

18. The REMI cassette and restriction enzyme are coincubated in sterile, deionized H_2O. A restriction enzyme buffer is not utilized.

19. The 40% PEG STC should be less than 1 month old.

20. A 1.5 mL volume of the regeneration medium will create an air bubble that will travel the length of the 2 mL tube during rocking. This will both aerate and mix the regenerating protoplasts during the overnight incubation. Tubes should be gently inverted several times to ensure an adequate air bubble is present before placing on the shaker.

21. Regenerated protoplasts can be identified by the emergence of hyphae. It is important to identify a small number of regenerated protoplasts to ensure protoplast viability.

22. This is performed to standardize the number of regenerated protoplasts in each plate. Volumes may need to be increased or reduced depending on transformation efficiency.

23. Vigorous mixing can splash molten agar onto the sides and lid of the plates. This increases the risk of contamination and should be avoided.

24. Regenerating protoplasts appear as small, white colonies that grow radially. It is important to avoid harvesting mycelium from multiple regenerating protoplasts, so avoid coalescing colonies.

25. If selection is efficient, there should be no growth in the plates containing the negative control transformation. However, the activity of hygromycin B decreases slowly over time. Examine the negative control transformation before isolating mutants and cease harvesting if growth is observed. This is critical to avoid selected nontransformed isolates (false positives).

26. After 5 days of growth, transfer the 24-well plates to 4°C for short-term storage.

27. Document sectoring or potential contamination which may be indicative of impure cultures.

28. Do not perform phenotypic analysis of mutants under antibiotic selection. Antibiotics often cause phenotypic abnormalities such as increased pigmentation that are unrelated to the underlying mutation.

29. Carry out steps 2 and 3 in a large sink. Preflooding each plate with H_2O is critical. The medium will not stain evenly if it is not fully hydrated, which makes scoring phenotypes difficult.

30. To accurately visualize starch hydrolysis, avoid overstaining the plates.

31. Other phenotypes like intensity of hydrolysis (intense/diffuse saccharification), hyphal characteristics (flocculence/circadian growth), and pigment biosynthesis should be documented for future reference. Additionally, the stain intensity will decrease with time. It is important to consistently make assessments within an hour of staining.

32. The mutants may only produce a small number of conidia or perhaps none at all. It would therefore be difficult or impossible to purify such mutants by single-spore isolation. In this case, transformants can be purified by isolating a single hyphal tip.

33. It is important to determine the number of REMI cassettes present in the genome of each mutant. The ideal number of integrations is one, as it is difficult to dissect phenotype-genotype relationships in mutants with multiple copies of the cassette.

34. Blunt cutting enzymes from third party vendors can be utilized to increase the number of libraries for each mutant. If initial PCR is unsuccessful, adjust PCR conditions or design different primers.

35. During insertion, the REMI cassette may delete endogenous DNA during integration. Such deletions can range from a few nucleotides to large regions of a chromosome. DNA flanking each side of the cassette insertion must be identified to rule out the effect of large genomic deletions.

36. The primers used in reactions A, B, C, and D are designed for specific genes (see Fig. 8; for genomic sequences see http://www.broadinstitute.org). Primers should be designed with similar melting temperatures and lengths (58°C and 22–24 bp, respectively; see Table 1). This will allow all four reactions to run simultaneously under the same PCR conditions.

37. Increase the volume of the PCR mixtures described in Subheading 3.6, steps 2, 3, and 9, to 10× to provide 10 50 µL reactions, which will yield approximately 75 µg of purified PCR product.

38. A transformation efficiency of <2% has been published for *F. verticillioides* (10), although with the incorporation of the

split-marker strategy, it is possible to attain a transformation efficiency of 10% or greater (Ridenour and Bluhm, unpublished). Therefore, in an initial screen in order to obtain an adequate number of independent mutants, 100 transformants should be analyzed.

39. Avoid collecting agar medium as it may inhibit the subsequent PCR.

40. We have routinely used PCR template prepared in this manner to amplify specific fragments from various single-copy genes. This method significantly reduces preparation time, which is particularly important when screening a large number of transformants. Alternative methods for template preparation, such as phenol:chloroform extraction or commercial kits, may be used.

41. Mutants may also contain unrelated genomic disruptions that convey additional phenotypes. It is therefore important to visually confirm that the phenotype of each independent deletion strain is highly consistent with the phenotype of the original REMI mutant.

References

1. Thon MR, Nuckles EM, Vaillancourt LJ (2000) Restriction enzyme-mediated integration used to produce pathogenicity mutants of *Colletotrichum graminicola*. Mol Plant Microbe Interact 13:1356–1365.

2. Gold SE, Duick JW, Redman RS, Rodriguez RJ (2000) Molecular transformation, gene cloning and gene expression systems for filamentous fungi, p. 199–238. In G. G. Khachatourians, and D. P. Arora (ed.), Applied mycology and biotechnology, vol. 1. Elsevier Science, Oxford, United Kingdom.

3. Seong K, Hou Z, Tracy M, Kistler HC, Xu JR (2005) Random insertional mutagenesis identifies genes associated with virulence in the wheat scab fungus *Fusarium graminearum*. Phytopathology 95:744–750.

4. Kahmann R, and Basse C (1999) REMI (restriction enzyme mediated integration) and its impact on the isolation of pathogenicity genes in fungi attacking plants. Eur. J. Plant Pathol. 105:221–229.

5. Weld RJ, Plummer KM, Carpenter MA, Ridgway HJ (2006) Approaches to functional genomics in filamentous fungi. Cell Res 16:31–44.

6. Kuwano T, Shirataki C, Itoh Y (2008) Comparison between polyethylene glycol- and polyethylenimine – mediated transformation of *Aspergillus nidulans*. Cur Gen 54:95–103.

7. Fairhead C, Llorente B, Denis F, Soler M, Dujon B (1996) New vectors for combinatorial deletions in yeast chromosomes and for gap-repair cloning using 'split-marker' recombination. Yeast 12:1439–1457.

8. Catlett NL, Lee B, Yoder OC, Turgeon BG (2003) Split-marker recombination for efficient targeted deletion of fungal genes. Fungal Genet Newsl 50:9–11.

9. Kück U, Hoff B (2010) New tools for the genetic manipulation of filamentous fungi. Appl Microbiol Biotechnol 86:51–62.

10. Choi Y-E, and Shim W-B (2008) Enhanced homologous recombination in *Fusarium verticillioides* by disruption of *FvKU70*, a gene required for a non-homologous end joining mechanism. Plant Pathol J 24:1–7.

Chapter 29

Assessment of Autophagosome Formation by Transmission Electron Microscopy

Marina Nadal and Scott E. Gold

Abstract

Autophagy is a complex degradative process by which cytosolic material, including organelles, is randomly sequestered within double-membrane bound vesicles termed autophagosomes and targeted for degradation. Initially described as a nutrient stress adaptation response, the process of autophagy is now recognized as a central mechanism involved in many developmental processes. In this chapter, we provide guidelines to assess the initial steps of autophagy by monitoring autophagic body vacuolar accumulation. We employed a standard electron microscopy approach to observe the vacuoles of nutrient stressed fungal cells.

Key words: Autophagy, Autophagosome, Transmission electron microscopy, Vacuole, Uranyl acetate and lead citrate, Spurr's resin, En bloc

1. Introduction

Autophagy is an intracellular degradative process that allows eukaryotic cells to recycle cytoplasmic material and eliminate obsolete organelles (1). Upon cell exposure to particular stimuli, cytosolic material is sequestered in double-membrane vesicles called autophagosomes. Autophagosomes initiate as cup-shaped double membranes termed preautophagosomes or isolation membranes, which appear in the cytoplasm upon autophagy induction (2). As a preautophagosome grows and expands, it engulfs cytoplasmatic material and ultimately becomes a mature autophagosome. Once fully formed, the autophagosome is docked to the vacuole (plants and fungi) or lysosomes (animals) where its cargo is degraded and recycled (3, 4).

Autophagy has been well conserved among eukaryotes, where it participates in many physiological and developmental processes (5). Among fungi, autophagy has been associated with nutrient

Melvin D. Bolton and Bart P.H.J. Thomma (eds.), *Plant Fungal Pathogens: Methods and Protocols*,
Methods in Molecular Biology, vol. 835, DOI 10.1007/978-1-61779-501-5_29, © Springer Science+Business Media, LLC 2012

stress responses, sporulation, apressorium development, and conidial germination (6–9). The process has been extensively studied in *Saccharomyces cerevisiae*, where 33 autophagy-related genes (*ATG* genes) have thus far been identified (10). Owing to its complexity, a better understanding can be achieved if the entire autophagy process is considered as a series of sequential events. Four recognized stages include: (a) induction, (b) autophagosome assembly, (c) autophagosome docking and fusion, and finally, (d) autophagic body degradation.

In this chapter, we describe a standard method to assess the earlier, initial stages of autophagy induction and autophagosome assembly (11). The methodology employs transmission electron microscopy (TEM) to evaluate accumulation of autophagic bodies within the vacuolar lumen upon exposing cells to autophagy-inducing conditions. The study of the later stages of autophagy, autophagic body degradation and recycling of material, require different approaches (12).

2. Materials (see Note 1)

2.1. Media

1. Non-inducing medium: this medium should represent conditions in which autophagy is not induced (see Note 2).

2. Inducing medium: this medium should represent conditions in which autophagy is induced (see Note 3).

2.2. Buffers and Solutions

1. Protease inhibitor solution: Phenylmethylsulphonyl fluoride (PMSF) should be added to the culture media immediately before inoculation to a final concentration of 1 mM. PMSF is not water soluble. To make the stock solution (50 or 100×), dissolve the appropriate amount in ethanol, methanol, or isopropanyl alcohol. Store at 4°C (see Note 4).

2. Working buffer: 0.1 M potassium phosphate, pH 6.8.

3. Rinsing buffer: 0.05 M potassium phosphate, pH 6.8.

4. 2.5% glutaraldehyde fixing solution. This can be prepared by mixing equal volumes of 5% glutaraldehyde and working buffer.

5. 1% osmium tetroxide (see Note 5).

6. Reynold's lead citrate solution:

   Mix following reagents in the indicated order:

   (a) 30 mL of distilled water.

   (b) 1.33 g of Lead citrate.

   (c) 1.76 g of Sodium citrate, dihydrate. Shake continuously for 1 min, and then intermittently for 30 min. Solution should turn cloudy.

    (d) 8 mL of freshly prepared 1 N NaOH: this should turn into clear solution again.

    (e) Bring to a final volume of 50 mL with distilled water.
       Filter sterilize solution and store in a syringe fitted with a 0.22-μm syringe filter at 4°C.

7. 4% aqueous uranyl acetate: Dissolve 2 g of uranyl acetate in 50 mL distilled water. Adjust pH to 5.0 with a few drops of glacial acetic acid. Filter sterilize solution and store in a syringe fitted with a 0.22-μm syringe filter (This allows easy dispensing of particle free material as needed). Store at 4°C for up to 6 months (see Note 6).

8. 0.5% aqueous uranyl acetate. Dissolve 2 g of uranyl acetate in 400 mL of distilled water. Store at 4°C (see Note 6).

9. Ethanol series: 25, 50, 75, 95, and 100%. Prepare these solutions by mixing appropriate volumes of pure ethanol with distilled water.

10. 100% Acetone.

11. Spurr's resin (firm). Mix the following ingredients in the indicated order:

    E.R.L. 4206 (vinylcyclohexene dioxide): 10 g.

    D.E.R (DER 736 epoxy resin): 6 g.

    N.S.A. (nonenyl succinic anhydride): 26 g.

    Mix well (2–3 min manually) and only then add:

    D.M.A.E. (dimethylaminoethanol): 0.4 g.

    This recipe makes around 45 mL of the resin (see Note 7).

12. 1% Toluidine blue and 2% sodium borate in water. Dissolve the sodium borate in the water first and then add the toluidine blue powder. Mix well until completely dissolved. Filter sterilize solution and store in a syringe fitted with a 0.2-μm syringe filter.

3. Methods

3.1. Culture Growth

1. Inoculate non-inducing medium with your fungal strains to be analyzed. Grow strains overnight or as required until culture enters early exponential growth phase (see Note 8).

2. Harvest cells by filtration or by centrifugation. Wash cells two or three times with distilled water. In the last wash, divide samples in two and resuspend cells separately in a small amount of inducing medium and non-inducing medium.

3 Add inhibitor solution (PMSF) to non-inducing and inducing media to be inoculated to a final concentration of 1 mM. PMSF

is a protein inhibitor and will block the degradation of autophagic bodies and thus allow evaluation of their accumulation in the vacuolar lumen (see Note 9).

4 Transfer the washed cells to the corresponding PMSF amended medium and allow cultures to grow for 4–6 h, depending on the fungal species. This time period should be enough to induce autophagy activity in most fungi. However, some fungi may require shorter or longer times (see Note 10).

5 Pellet cells and precede with the fixation steps. To pellet cells, a centrifuge can be used but avoid high speeds as these might cause cell deformation.

6 An easy way to work through the following steps is to suspend cells in 1% agarose and cut the solidified block into small cubes. Simply pellet the cells, remove supernatant, add 1% cool to the touch (40–50°C) agarose, mix briefly and allow to solidify. Cut the agarose block into smaller portions and transfer to a small glass vial.

3.2. Fixation of Samples for Transmission Electron Microscopy (TEM Fixation)

1. Fix samples in 2.5% glutaraldehyde fixing solution. Add sufficient fixing solution to completely cover the samples and keep overnight at 4°C.

2. Remove fixing solution with a plastic transfer pipette and discard in an appropriate waste container. Add rinsing buffer (0.05 M) and allow to stand for 15°min at 4°C. Remove rinsing buffer. Repeat rinses twice more.

3. Postfix in 1% osmium tetroxide. Mix equal volumes of 2% osmium tetroxide and working buffer (0.1 M). Add enough solution to the vial to cover samples and incubate for 2 h at 4°C. Rinse with water four times for 15 min each.

4. *En bloc* staining. Add enough 0.5% aqueous uranyl acetate to cover the samples. Incubate overnight at 4°C. Rinse with water three times, 15 min each time.

5. Sample dehydration in graded ethanol series. Starting from the lowest concentration, incubate samples in ascending alcohol concentration solutions for 15 min each at room temperature.

6. Incubate samples three times for 10 min each in 100% acetone.

7. Cover samples with a solution of 30% plastic resin and 70% acetone, and incubate overnight at room temperature (see Note 11).

8. Remove resin/acetone mixture from samples and dispose in appropriate waste container. Cover samples with solution made of 70% plastic resin and 30% acetone. Incubate overnight at room temperature.

9. Replace acetone/plastic resin mixture with 100% plastic resin and incubate overnight at room temperature.

10. Replace plastic resin with freshly prepared plastic resin. Place samples in small plastic dishes or similar container that will produce a 0.5 cm thick block. After completely covering the samples with fresh resin, allow polymerization by baking for 24 h at the temperature indicated by the manufacturer (usually between 60 and 70°C) (see Note 12).

11. Working in a chemical fume hood, cut small resin sections containing some of the fixed cells (Fixed material is easily visualized in the resin because of its dark coloration.). This can be done by cramping the hard resin section with a vise and cutting very slowly with a thin jewler's saw. Use gloves all the time and avoid contact with or inhalation of the resin dust which is extremely toxic.

12. Glue the small section onto an appropriate sampler holder (usually small glass cylindrical blocks) designed to fit in the ultramicrotome chuck (the sampler holder). Let the glue dry well (several hours or overnight).

13. Using a low light magnification microscope, trim the edges of the resin block to produce a trapezoidal block (see Note 13).

14. This step should indicate if the resin section selected for sectioning contains fixed material. Place the sample in the microtome equipped with a diamond knife with the longest trapezoid side facing down. Cut thick sections (around 0.5–1 μm). Allow a string of sections to float in the microtome container before trying to transfer them. Very carefully scoop a few sections using a metal loop and place them on a glass slide. Let the water evaporate (can be accelerated by placing slide on a heating block at around 40°C) and then stain with one drop of toluidine blue for 30 s. Rinse extra stain with distilled water. Observe under the light microscope and verify that sample tissue is present in the sections.

15. Once a section containing substantial fixed tissue has been identified using the toluidine blue staining, cut thin sections (80 nm) as described in the previous step. When several sections have been produced, very carefully collect some using a slot grid. Allow grids to dry on a filter paper and store in a grid carrying box (see Note 14).

3.3. Post-Sectioning Staining with Uranyl Acetate and Lead Citrate

1. A strip of parafilm is placed within each of two petri dishes on the bench. Place a few (4–7) NaOH pellets under the second petri dish covered parafilm (NaOH prevents the lead citrate stain from forming a precipitate). Place a drop of uranyl acetate under the first petri dish and a drop of lead citrate under the second. Also prepare two small beakers containing double distilled water.

2. Very carefully take the grid using a pair of needle nose forceps and place gently on top of the uranyl acetate with the sample

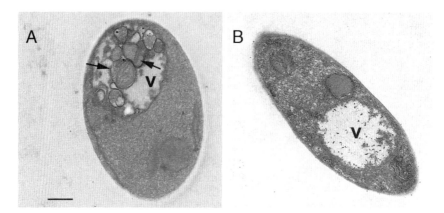

Fig. 1. Autophagosome accumulation in wild type and autophagy-deficient cells of the corn smut fungus, *U. maydis*. Haploid cells from the wild type and the autophagy-deficient (Δ*atg8*) strains were exposed to 5 h of carbon starvation (MM-C). TEM images show the accumulation of autophagic bodies (*arrows*) within the vacuole (v) of wild-type cells (**a**). Note the absence of these structures from vacuoles of Δ*atg8* cells (**b**). Scale bars, 1 μm. Reproduced from Nadal and Gold (16).

side facing the drop. Let float for 3 min. Remove and rinse in water by dipping ten times carefully in the first beaker and then repeat in the second one. Remove the excess water by very careful absorption with a wedge of Whatman paper.

3. With the sample side facing down, place grid on top of lead citrate drop and let float for 3 min. Rinse with water three times for 15 s each and let grids dry on a clean filter paper. Store at room temperature until observing with the TEM (see Note 15).

4. Observe under the TEM. Autophagic bodies should appear in the vacuoles as single membrane vesicles containing different cytoplasmatic material (see Fig. 1). No large single membrane vesicles should be observed inside the vacuole of either autophagy-deficient mutants or cells in which autophagy has not been induced. In the case of some mutants in which autophagosome formation is not completely abolished, it is possible that some vesicles do occur in the vacuoles. In this case, the characterization of such mutants would require the quantification of the autophagic bodies. To achieve this, several cells should be picked randomly and the number of the autophagy bodies in their vesicles compared to an equal number of wild-type cells.

4. Notes

1. Many of the solutions and chemicals used for fixing and staining TEM samples are highly toxic. Therefore, it is strongly recommended to carry out all the procedures wearing gloves and in

a fume hood. For the same reason, waste material should be disposed of in appropriate waste containers according to institutional regulations. Any leftover liquid resin or solid material containing it (beaker, pipettes, etc) waste should be baked at 68°C to allow polymerization before disposal.

2. This methodology to evaluate autophagy activity relies on observation of autophagic body accumulation in the vacuole. Therefore, a good reference condition should be one in which autophagy does not operate. A good assessment is to use media that does not impose a nitrogen or carbon starvation stress. The precise composition will depend on the particular fungal species.

3. In most fungi, nitrogen or carbon stress is known to trigger autophagy, therefore minimal media lacking either of these components could be employed.

4. PMSF solution can crystallize during storage. If so, let the solution warm up at room temperature before use and vortex. PMSF is very toxic; wear gloves while handling and do not exceed 1 mM final concentration.

5. Osmium tetroxide is sold in prescored, sealed 2% osmium tetroxide ampoules. Working with gloves in a fume hood, break open the 2% osmium tetroxide and add an equal volume of working buffer. Solution will turn pale yellow and should be stored at 4°C until used. With time, the solution might turn gray. Only use when color is yellow.

6. Uranyl acetate is slightly radioactive. *Avoid contact.*

7. The Spurr's resin should be prepared in a disposable plastic beaker. The ingredients should be mixed in the indicated order and the best way is to weigh them out on a top loading balance in the same container one after the other. Simply zero the balance after adding each reagent. Before adding the D.M.A.E., stir the materials thoroughly but avoid whipping as this creates small air bubbles. Once mixed, add D.M.A.E. and gently mix again. The container can be kept at the back of the fume hood covered with a plastic petri dish top until required.

8. It is very important that cells are collected during the exponential phase and not once they have reached the stationary phase. It has been shown that autophagy is triggered when *S. cerevisiae* cells enter stationary phase. This is most likely due to the nutrient exhaustion during culture growth (13).

9. PMSF is not water soluble, and should be dissolved in methanol, ethanol, or iso-propanol. Make a 50 mM stock solution and keep at 4°C.

10. For many fungi, 4–5 h of nitrogen or carbon starvation should be sufficient to trigger the autophagy-stress response. An indirect way to establish the precise condition of autophagy induction

would be to use the induction of specific autophagy genes as an indicator. In *S. cerevisiae* and other fungi such as *Ustilago maydis*, autophagy induction correlates with increased levels of several *ATG* gene transcripts (14–16). This could be tested by qRT-PCR or simply RT-PCR on transcripts corresponding to different time points during carbon or nitrogen starvation.

11. Once plastic resin has been prepared following the instructions given in the Subheading 2, it can be stored at room temperature to be used next day. It is recommended that it should stay in the fume hood at all times.

12. When placing samples in the oven, make sure the dishes are perfectly level. If not level, the samples will drift to one side and concentrate as the resin hardens.

13. It is crucial that the sample section is a trapezoid as this will reduce the compression of the resin as the knife cuts through it.

14. Section handling is a critical step and should be performed very carefully. Let the microtome produce a ribbon of sections. Using needle nose forceps grab a grid and submerge it into the knife boat close to the section ribbon that is floating. With the aid of a small eyelash (glued to a thin wooden stick), try to scoop up some of the section as you lift the grid. Ultimately, you should be able to collect a few sections onto the grid. Be very careful and work far away in the boat from the knife to avoid damaging it.

15. For this step, it is critical that the side of the grid facing the stains is the one carrying the sections. The sides of the copper grid can be distinguished from each other as one is shinier than the other. As a rule, choose one side and always use that side when placing the sections.

References

1. Yang, Z. and D.J. Klionsky, *Eaten alive: a history of macroautophagy.* Nat Cell Biol, 2010. **12**(9): p. 814–22.

2. Suzuki, K., et al., *Hierarchy of Atg proteins in pre-autophagosomal structure organization.* Genes Cells, 2007. **12**(2): p. 209–18.

3. Takeshige, K., et al., *Autophagy in yeast demonstrated with proteinase-deficient mutants and conditions for its induction.* J Cell Biol, 1992. **119**(2): p. 301–11.

4. Ishihara, N., et al., *Autophagosome requires specific early Sec proteins for its formation and NSF/SNARE for vacuolar fusion.* Mol Biol Cell, 2001. **12**(11): p. 3690–702.

5. Reggiori, F. and D.J. Klionsky, *Autophagy in the eukaryotic cell.* Eukaryot Cell, 2002. **1**(1): p. 11–21.

6. Kikuma, T., et al., *Functional analysis of the ATG8 homologue Aoatg8 and role of autophagy in differentiation and germination in Aspergillus oryzae.* Eukaryot Cell, 2006. **5**(8): p. 1328–36.

7. Liu, X.H., et al., *Involvement of a Magnaporthe grisea serine/threonine kinase gene, MgATG1, in appressorium turgor and pathogenesis.* Eukaryot Cell, 2007. **6**(6): p. 997–1005.

8. Veneault-Fourrey, C., et al., *Autophagic fungal cell death is necessary for infection by the rice blast fungus.* Science, 2006. **312**(5773): p. 580–3.

9. Hu, G., et al., *PI3K signaling of autophagy is required for starvation tolerance and virulenceof Cryptococcus neoformans.* J Clin Invest, 2008. **118**(3): p. 1186–97.

10. Klionsky, D.J., et al., *A unified nomenclature for yeast autophagy related genes*. Dev Cell, 2003. **5**(4): p. 539–45.

11. Mizushima, N., *Autophagy: process and function*. Genes Dev, 2007. **21**(22): p. 2861–73.

12. Klionsky, D.J., et al., *Guidelines for the use and interpretation of assays for monitoring autophagy in higher eukaryotes*. Autophagy, 2008. **4**(2): p. 151–75.

13. Wang, Z., et al., *Antagonistic controls of autophagy and glycogen accumulation by Snf1p, the yeast homolog of AMP-activated protein kinase, and the cyclin-dependent kinase Pho85p*. Mol Cell Biol, 2001. **21**(17): p. 5742–52.

14. Kirisako, T., et al., *Formation process of autophagosome is traced with Apg8/Aut7p in yeast*. J Cell Biol, 1999. **147**(2): p. 435–46.

15. Rose, T.L., et al., *Starvation-induced expression of autophagy-related genes in Arabidopsis*. Biol Cell, 2006. **98**(1): p. 53–67.

16. Nadal, M. and S.E. Gold, *The autophagy genes ATG8 and ATG1 affect morphogenesis and pathogenicity in Ustilago maydis*. Mol Plant Pathol. **11**(4): p. 463–78.

Chapter 30

Fungal Plant Pathogen Detection in Plant and Soil Samples Using DNA Macroarrays

B. Lievens, A. Justé, and K.A. Willems

Abstract

PCR-based DNA array technology is one of the most suitable techniques to detect and identify multiple pathogens in a single assay. Out of the different array platforms that currently exist, membrane-based DNA macroarrays are the most convenient for plant disease diagnosis because of low costs, great sensitivity, and modest equipment requirements. Here we describe a protocol for routine detection of plant pathogens using DNA macroarrays, i.e., from sampling to analysis of hybridization results. Diagnosis can be completed within 36 h after sample collection.

Key words: Chemiluminescence, Detection, Diagnosis, DNA array, Hybridization, Identification, Membrane, Multiplex, PCR, Reverse dot blot

1. Introduction

Currently, DNA array technology is one of the most suitable techniques for rapid detection and accurate identification of several pathogens in a single assay (1, 2), even when the different target species can only be discriminated by a single-nucleotide polymorphism (SNP) in the gene targeted (3). Arrays typically consist of discrete spots of pathogen-specific oligonucleotide sequences immobilized onto a solid support to which labeled PCR products are hybridized. In general, target sequences are amplified and simultaneously labeled using universal primers spanning a genomic region harboring variable, target-specific sequences such as the ribosomal internal transcribed spacers (ITS). In this way, numerous targets can be amplified with a single primer pair, while target discrimination is performed afterward on the array (1). As the sample DNA is PCR-amplified, sensitivity of DNA arrays is typically

Melvin D. Bolton and Bart P.H.J. Thomma (eds.), *Plant Fungal Pathogens: Methods and Protocols*,
Methods in Molecular Biology, vol. 835, DOI 10.1007/978-1-61779-501-5_30, © Springer Science+Business Media, LLC 2012

high. In addition, the number of pathogens that can be detected simultaneously is in theory unlimited (1). This technology was originally developed for mutation analysis in the screening for human genetic disorders (4, 5), but has also been successfully applied to detect and/or identify plant pathogens of diverse nature (2), including oomycetes (6, 7), fungi (8–13), nematodes (14), and bacteria (10, 15). In addition, hybridization signal intensity enables biomass semi-quantification (9, 10), making DNA arrays even more appealing for plant pathogen diagnosis.

A number of different array platforms have been described so far, ranging from microscopic bead arrays to solid support arrays on nylon membranes (DNA macroarrays) or glass-slides (DNA microarrays) (16). Compared to bead array and microarray platforms, membrane-based macroarrays are more useful in plant disease diagnosis because of lower costs, greater sensitivity, and freedom from the need for highly specialized equipment (2, 17). Here we describe the use of DNA macroarrays for the rapid and efficient detection and identification of several plant pathogenic fungi from environmental samples such as plants and soil. Similar detection tools have been developed by the authors for sugar thick juice bacterial contaminants (18), a panel of 53 *Legionella* species (unpublished results) and several fish pathogens (19, 20). In this chapter, protocols for DNA array production, sampling and DNA extraction, DNA amplification and labeling, DNA array hybridization and image acquisition and analysis are described. For the design and development of target-specific oligonucleotides, we refer to other manuscripts (e.g., (3, 21, 22)).

2. Materials

Always wear a lab coat and disposable powder-free gloves when working in the laboratory. Prepare all solutions using ultrapure water and analytical grade reagents (see Note 1). Prepare and store all reagents at room temperature (unless indicated otherwise). Diligently follow all waste disposal regulations when disposing waste materials.

2.1. DNA Array Production Components

1. Membranes suited for reverse dot blot hybridization such as 0.45 μm Immunodyne ABC membranes (PALL Europe Limited, Portsmouth, UK). Store vacuum sealed packages at room temperature (see Note 2).

2. 5′-C6-amino-labeled oligonucleotides (see Note 3). Dilute (in 10 mM Tris buffer, pH 8.0) to a final concentration of 200 μM. Store at −20°C.

3. Printing buffer: 0.5M $NaHCO_3$, pH 8.4, 0.004% bromophenol blue (BPB) (see Note 4). For 100 mL: Dissolve 4.2 g $NaHCO_3$ in 80 mL water. Adjust pH to 8.4 with 0.2N NaOH. Adjust

volume to 99 mL with water and add 1 mL 0.4% BPB. Filter through 0.2 μm bottle top filter. Store at 4°C.

4. 20× Saline Sodium Citrate (SSC): 3M NaCl, 0.3M Na Citrate, pH 7.0. For 800 mL: Dissolve 140.3 g NaCl and 70.6 g sodium citrate in 650 mL water. Adjust the pH to 7.0 with 1M HCl, and the volume to 800 mL with water. Filter trough 0.2 μm bottle top filter.

5. 2× SSC. For 1 L: Mix 100 mL 20× SSC with 900 mL water.

6. Blocking buffer: 2× SSC, 0.5% casein, 0.05% Tween20 (see Note 5). Add casein to 2× SSC and heat on a stir plate to 80°C or until dissolved. Do not boil the solution. Cool to 20–40°C before adding Tween20 (see Note 6) and use.

2.2. DNA Extraction Components

1. DNA extraction kit: PowerSoil DNA isolation kit (MoBio Laboratories, Solana Beach, CA, USA) for isolating high quality DNA from soils (and most other environmental samples) or PowerPlant DNA isolation kit (MoBio Laboratories, Solana Beach, CA, USA) for isolating high-quality DNA from plants including those high in phenolics and polysaccharides (see Note 7).

2.3. DNA Amplification and Labeling Components

1. 10× DIG DNA labeling mix: 2.0 mM dATP, 2.0 mM dCTP, 2.0 mM dGTP, 1.9 mM dTTP, and 0.1 mM alkali-labile digoxigenin 11-dUTP (DIG-11-dUTP; Roche Diagnostics GmbH, Mannheim, Germany) (see Notes 8 and 9). For a 100 μL labeling mix: Mix 2 μL 100 mM dATP, 2 μL 100 mM dCTP, 2 μL 100 mM dGTP, 1.9 μL 100 mM dTTP, 10 μL 1 mM DIG-11-dUTP, and 81.1 μL water. Vortex briefly. Store at –20°C.

2. PCR primers. Dilute (in 10 mM Tris buffer, pH 8.0) to a working concentration of 20 μM. Store at –20°C.

3. Titanium *Taq* DNA Polymerase (Clontech Laboratories, Palo Alto, CA, USA), including 50× Titanium *Taq* DNA polymerase and 10× Titanium *Taq* PCR buffer (see Note 10). Store at –20°C.

2.4. DNA Array Hybridization Components

1. Anti-digoxigenin alkaline phosphatase (anti-DIG AP) conjugate, Fab fragments (Roche Diagnostics GmbH, Mannheim, Germany). Store at 4°C.

2. CDP-Star (Roche Diagnostics GmbH, Mannheim, Germany). Store protected from light at 4°C (see Note 11).

3. 20× SSC: 3M NaCl, 0.3M Na Citrate, pH 7.0.

4. 5× Buffer I: 0.5M Maleic Acid, 0.75M NaCl, pH 7.5. For 800 mL: Dissolve 46.4 g maleic acid and 35.1 g NaCl in 650 mL water. Adjust to pH 7.5 with NaOH pellets (see Note 12). Adjust volume to 800 mL with water. Filter trough 0.2 μm bottle top filter.

5. 5× Buffer II: 0.5M Trizma Base, 0.5M NaCl, pH 9.5. For 800 mL: Dissolve 48.5 g Trizma base and 23.4 g NaCl in 650 mL water. Adjust to pH 9.5 with 1N NaOH. Adjust volume to 800 mL with water. Filter through 0.2 μm bottle top filter.

6. 10% N-Lauroyl Sarcosine (NLS). For 50 mL: Dissolve 5 g NLS in 40 mL. Adjust volume to 50 mL with water. Filter through 0.2 μm bottle top filter.

7. 10% Sodium Lauryl Sulfate (SLS; synonym Sodium Dodecyl Sulfate (SDS)). For 50 mL: Dissolve 5 g SLS in 40 mL. Adjust volume to 50 mL with water. Filter through 0.2 μm bottle top filter.

8. 6× SSC. For 1 L: Mix 333 mL 20× SSC with 667 mL water.

9. Hybridization buffer: 6× SSC, 0.1% NLS, 0.02% SLS. For 1 L: Add 10 mL 10% NLS and 2 mL 10% SLS to 1 L 6× SSC and mix.

10. Prehybridization buffer: 6× SSC, 0.1% NLS, 0.02% SLS, 1% casein (see Note 5). Add casein to hybridization buffer and heat on a stir plate to 80°C or until dissolved. Do not boil the solution.

11. Hybridization wash buffer I (stringency buffer): 6× SSC, 1% SLS. For 1 L: Add 10 mL 10% SLS to 1 L 6× SSC and mix (see Note 13).

12. 1× Buffer I. For 1 L: Mix 200 mL 5× Buffer I with 800 mL water.

13. Blocking buffer: 1× Buffer I, 1% casein (see Note 5). Add casein to 1× Buffer I and heat on a stir plate to 80°C or until dissolved. Do not boil the solution.

14. Hybridization wash buffer II: 1× Buffer I, 0.3% Tween 20.

15. Detection buffer: 1× Buffer II. For 1 L: Mix 200 mL 5× Buffer II with 800 mL water.

3. Methods

Carry out all procedures at room temperature unless otherwise specified. In order to prevent sample contamination, proper knowledge and application of sample handling technique is essential. A key component in this is the use of separate working areas for the different steps involved. These working areas consist, preferentially, of four different rooms with separate sets of pipettes, plastics, disposables, etc., including:

• Sample preparation area:

  Use a clean set of analysis tools per sample. Use new gloves after handling each sample. Thoroughly clean and disinfect nondisposable analysis tools between samples.

- PCR area I: preparation of PCR reagent mix and DNA array production:

  Avoid DNA cross-contamination. Keep only reagents and primers in this area. UV-sterilize all materials and plastics before bringing them into this area. Prepare the PCR mastermix in a PCR cabinet with UV sterilization. Dispense the mix into PCR tubes and close the tubes firmly. UV-sterilize the cabinet before and after use.

- PCR area II: DNA extraction and PCR preparation:

  Perform DNA extraction. Add extracted DNA to the prepared PCR tubes in a second PCR cabinet. UV-sterilize the PCR cabinet before and after use.

- PCR area III: PCR run and hybridization:

  Run PCR and perform all involved post-PCR activities. Opening tubes may be a major source of contamination as amplicons may be spread by aerosols. Consider all materials in this area as "highly DNA-contaminated." Never bring materials from this area to one of the other areas.

The working direction is from PCR area I to PCR area III (Sample preparation > PCR reagents > Extraction and addition > PCR and hybridization). Working in the reverse direction should be strictly avoided and materials, equipment, glassware, etc. may not be moved from one area to another.

3.1. DNA Array Production

1. Prepare a 40 μM 5′-C6-amino-labeled oligonucleotide solution in printing buffer.

2. Fill a 384-well microtiter plate with the oligonucleotides according to the desired DNA array template (see Notes 14 and 15). Store the plate at −20 or −80°C.

3. Cut Immunodyne ABC membranes to fit the size of the array. Handle membranes with caution using clean forceps. The membrane is best cut through the membrane protection sheet with a paper cutter, leaving ~5 mm border for handling/labeling.

4. Put the V&P Library Copier (V&P Scientific, San Diego, CA, USA) over the microtiter plate.

5. Stick the V&P Multi-Print registration device (using double-sided sticky tape) on a plastified positioning mat representing all possible locations where the oligonucleotides can be deposited (see Note 16).

6. Position the membrane. Stick membrane borders using double-sided tape. Make sure that the membrane is lying flat against the mat.

7. Print oligonucleotides using a 384 pin V&P Multi-Blot replicator (see Note 17). Using the four alignment holes in the Multi-Print, a 1536-spot DNA array can be produced.

8. After printing, seal the plate with sealing tape and store at −20°C until next printing.

9. Assess printing quality of membranes, based on intensity and distribution of spots (see Note 18). Discard low-quality blots.

10. Air-dry membranes for at least 30 min.

11. Label membranes with a unique serial number using a permanent marker to allow tracking of the array during the course of the analysis. Put numbers consistently on the same position on the membrane, e.g., in the lower right corner.

12. Block membranes for 30 min with gentle agitation (~3.5 rpm) in blocking buffer at room temperature in a sterile glass tray (e.g., 15×25 cm) (see Note 18).

13. Wash membranes in 2× SSC for 30 min at room temperature (see Note 19) and membranes are ready to use. Maintain in 2× SSC at 4°C (e.g., in a 50 mL centrifuge tube) for short-term storage or air-dry overnight and keep at room temperature in an envelope for long-term storage. In this case, storing membranes vacuum packaged (plastic sealed) or in a vacuum desiccator is recommended.

3.2. Sampling and DNA Extraction

1. Take samples according to a standardized sampling protocol (see Note 20).

2. Prepare samples for DNA extraction (see Note 21).

3. Extract DNA using the MoBio PowerPlant or PowerSoil DNA isolation kits according to the manufacturer's guidelines (see Note 22). Use a standard amount of sample material (see Note 23).

4. Measure DNA yield and quality spectrophotometrically. A A_{260}:A_{280} ratio ~1.8 is generally accepted as "pure" for DNA A_{260}:A_{230} values are expected in the range of 2.0–2.2 for high-quality DNA extracts.

5. DNA extracts may have to be diluted (5 or 10 times) to avoid detrimental concentrations of potential PCR inhibitors. On the other hand, if too little DNA yield is obtained, DNA samples may have to be precipitated and resuspended (see Note 24).

6. DNA samples are now ready for PCR amplification. Store samples at −20 or −80°C.

3.3. PCR Amplification and Labeling

1. Prepare a PCR mastermix for $n + c + 1$ samples, with n the total number of studied samples and c the number of controls. The extra one is to preempt pipetting volume variation. For each reaction tube, mix 14.1 μL water, 2 μL 10× Titanium *Taq* PCR buffer (final concentration 1×), 1.5 μL 10× DIG DNA labeling mix (final concentration 0.15 mM of each nucleotide), 0.5 μL of each primer (20 μM; final concentration 0.5 μM), and 0.4 μL 50× Titanium *Taq* DNA polymerase (final concentration 1×).

Vortex briefly to mix. Include one or more negative controls that receive water instead of template DNA and one positive control that contains DNA of a strain known to react positively (see Note 25).

2. Divide the mastermix into aliquots of 19 μL in 0.2 mL PCR reaction tubes.

3. Add 1 μL template DNA, vortex briefly, and spin down (2,000 rpm; few seconds). Avoid air bubbles in the tubes.

4. Put tubes in thermocycler. PCR conditions, especially annealing temperature (T_a), will vary with the primers used (see Note 26). For sensitive detection, use 35–40 cycles of PCR. For quantification purposes, run 30 cycles to avoid analysis of samples which were no longer in the exponential phase of the reaction (9).

5. Digoxigenin-labeled amplicons are now ready to be used for hybridization (see Note 27) or can be stored at –20°C until further use.

3.4. Hybridization

Never touch membranes. Always handle membranes with clean forceps. Make sure membranes do not dry out during the process.

3.4.1. Prehybridization

Start prehybridization during the last steps of PCR amplification and labeling.

1. Place each individual membrane in a plastic conical hybridization tube (see Note 28). Make sure the oligonucleotide side of the membrane is facing the inside of the tube. Label each tube to couple with the unique membrane number.

2. Warm the prehybridization buffer to 54°C (warm water bath) and invert the bottle several times to allow complete dissolution of the buffer components.

3. Add prewarmed prehybridization buffer to each membrane (see Note 29). Be sure the cap of the tube is screwed on hand-tight (see Note 30).

4. Balance tubes in hybridization oven. Prehybridize at 54°C (see Note 31) for at least 1.5 h with continuous agitation at about 5 rpm (see Note 32).

3.4.2. Hybridization

1. Warm hybridization buffer to 65°C (warm water bath) and invert the bottle several times to allow complete dissolution of the buffer components.

2. Denature labeled PCR amplicons in 1 mL hybridization buffer (in microcentrifuge tube) by heating at 95°C for 10 min. Then spin briefly to collect condensation.

3. Discard the prehybridization buffer from the hybridization tube and replace by prewarmed hybridization buffer (5 mL in 15 mL tubes or 11 mL in 50 mL tubes) (see Note 29).

4. Add denatured amplicons to the tube (see Note 33). Mix well and put the tube back in the hybridization oven.

5. Hybridize overnight at 54°C with continuous agitation at about 5 rpm.

3.4.3. Washing

1. Heat hybridization wash buffer I to 65°C (warm water bath) and invert the bottle several times to allow complete dissolution of the buffer components.

2. Discard the hybridization buffer. In case you may need to hybridize the same mixture to another array, pour the hybridization buffer from the hybridization tube into a clean tube, label and store at –20°C until reuse (see Note 34).

3. Add prewarmed hybridization wash buffer I to the hybridization tube (see Note 29) and then place the tube back into the hybridization oven. Wash the membrane at 54°C for 45 min with 5–10 rpm agitation (see Note 35).

4. Discard the wash solution and repeat with new warm hybridization wash buffer I.

5. Using forceps, carefully remove each membrane from its hybridization tube and transfer to a baked, sterile glass tray (see Note 36) with hybridization wash buffer II. Wash membranes at room temperature for 5 min with gentle agitation (~3.5 rpm).

3.4.4. Chemiluminescent Detection

All of the following detection steps are performed at room temperature (unless stated otherwise).

1. Block membranes in blocking buffer in a baked, sterile glass tray (see Note 36). Incubate for 30 min with gentle agitation (~3.5 rpm).

2. Dilute anti-DIG-AP 1:25,000 in new blocking buffer and mix well (see Note 37).

3. Incubate membranes for exactly 30 min in a baked, sterile glass tray containing the anti-DIG-AP supplemented buffer with continuous gentle agitation (~3.5 rpm) (see Note 38).

4. Wash the membranes three times in a glass tray with hybridization wash buffer II for 20 min with gentle agitation (see Note 39).

5. Equilibrate membranes for 15 min in a baked glass tray with detection buffer with gentle agitation (see Note 36).

6. Prepare working CDP-Star substrate in a 1.5 mL microcentrifuge tube by diluting stock substrate 100 times in detection buffer. Briefly vortex. CDP-Star substrate is light-sensitive, hence use dark-tinted tubes or aluminum foil to protect the solution from light.

7. Place membranes with the oligonucleotide side facing up on a plastic sheet (see Note 40). Pipette mixed working substrate solution onto each membrane (see Note 41) and overlay the membranes with a second plastic sheet as quick as possible. Ensure that the substrate solution is evenly distributed over the membranes and smooth out any air bubbles. Squeeze out excess liquid.

8. Incubate wrapped membranes for 5 min at 37°C in darkness.

3.5. Image and Data Acquisition and Analysis

After CDP-Star incubation, the membranes are ready for image acquisition.

1. Acquire the images using either X-ray film or a chemiluminescence imaging system (see Note 42). With the AutoChemi System (also known as BioSpectrum AC Imaging System) (UVP, Upland, CA, USA), dynamic integration of 10 exposures of 30 s at 2×2 binning results in an optimal image in a minimum of time (5 min). For more information and further instructions on the use of the system, refer to the user guide.

2. Save the final image, preferably as TIFF format (see Fig. 1).

3. In general, a signal display range of 300 is used for evaluation of signals (see Note 43). Measure the density of the dots (integrated optical density (IOD)) using applicable software such as Labworks 4.6 Image Acquisition and Analysis Software (UVP, Upland, CA, USA) (see Note 44). Make sure background

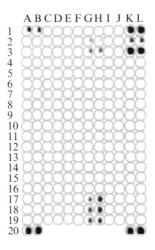

Fig. 1. Example of a DNA macroarray. Each detector oligonucleotide is spotted in duplicate. Specificity of the analysis is enhanced by using multiple oligonucleotides for each target species. In addition to the immobilized pathogen-specific oligonucleotides, the array contains control oligonucleotides for the hybridization (1A & B, 1K & L, 20A & B, 20K & L) and a reference for detection and calibration (2K & L, 3K & L). PCR-labeled amplicons hybridize to genus-specific oligonucleotides for *Pythium* (2G & H, 3G & H) and species-specific oligonucleotides for *P. ultimum* (17G & H, 18G & H, 19G & H).

correction is selected. For more information on the use of the software, we refer to the user guide.

4. Save data in an Excel spreadsheet for further calculations and analysis (e.g., relative IOD (IOD relative to the IOD for the digoxigenin-labeled reference oligonucleotide (8))).

5. Check all controls before drawing conclusions, including a positive and a negative control, PCR control, hybridization control, and the detection and calibration control (see Note 45).

6. If needed, confirm identification with an additional identification technique or by posthybridization amplicon sequencing (20) (see Note 46).

4. Notes

1. Having water at the bottom of the beaker or flask helps to dissolve most buffer components. In addition, heating may help to dissolve solids faster. However, care should be taken to bring the solution to room temperature before adjusting pH.

2. Once the package vacuum seal is opened, the quality of the membrane will diminish quickly as the surface chemistry of an Immunodyne ABC membrane is moisture-sensitive and will react with the moisture in the atmosphere. Typically, at 20°C and 50% relative humidity the active groups responsible for covalent binding will have a half life of 2 days. In addition, following hybridization, increased amounts of dirty background are observed on such membranes.

3. Generally, universal (aspecific) oligonucleotides are included as a control for hybridization. A 3′-digoxigenin-labeled control oligonucleotide with no homology to a known sequence is used as a detection and calibration check. Such sequence can be designed randomly and checked for homology against GenBank. As 3′-digoxigenin-labeled oligonucleotides cannot be synthesized with a 5′-C6-amino linker, these reference oligonucleotides are generally labeled with a 5′-C6-thiol linker.

4. BPB is used as a color marker to check oligonucleotide printing. In addition, as an acid-base indicator, discoloration of the oligonucleotide solution may indicate a pH drop, affecting the binding efficiency. The colored dots will fade by the end of the hybridization process, and the membrane will then appear blank.

5. All buffers containing casein should be freshly made and excess should be discarded.

6. Cut ~5 mm from the of 1,000 μL pipette tips to aspirate Tween20 easily.

7. PCR efficiency can be substantially reduced or even inhibited due to a variety of naturally occurring compounds in difficult samples (e.g., compost, strawberry, tree bark, etc.) that are co-extracted with the nucleic acids, such as phenolic compounds, polysaccharides, humic acids, fulvic acids, and heavy metals. A comparison of different methods and commercially available kits by the authors revealed that these MoBio kits consistently provide the most high-yield, high-quality genomic DNA.

8. In general, a ratio of DIG-11-dUTP to dTTP of 1:19 is sufficient to obtain high sensitivity of the amplicons. Nevertheless, for efficient labeling templates with high GC content, higher concentrations of DIG-11-dUTP should be used. However, as DIG-labeled oligonucleotides in the template may slow down the DNA polymerase, PCR conditions may have to be readjusted when using higher DIG-11-dUTP concentrations.

9. The use of alkali-labile DIG-11-dUTP allows stripping and reprobing, provided that the membranes did not dry out to completion. Nevertheless, stripping and reprobing is not recommended. It may be attempted for troubleshooting purposes, but the results may vary in signal strength and consistency.

10. Other DNA polymerases may be used. However, based on a test with different DNA polymerases, we observed that the yield of the PCR products from almost any template was highest when using Titanium *Taq* DNA polymerase (8).

11. CDP-Star is light-sensitive and susceptible to degradation by bacterial contamination. To minimize this risk, the stock solution may be split in different working solutions in dark-tinted tubes. In addition, use sterile pipette tips when dispensing.

12. Dissolution of NaOH is highly exothermic and may result in boiling when too many pellets are added at once.

13. Color will change from clear to white turbid, which will change again to transparent when heated.

14. The ITS regions are the most frequently used targets for fungal identification. To ensure accuracy of the assay, multiple oligonucleotides should be used per target. Based on our experience we estimate that for each target about three or four specific oligonucleotides should be included in the array. In addition, a second molecular marker such as the β-tubulin gene or the elongation factor-1α gene can be used to support specificity of the assay. At least one oligonucleotide designed as universal target and one oligonucleotide designed as a detection check should be included as controls (see also Note 3).

15. Optimal spots of uniform size and shape are obtained when starting from 35 μL in the wells. From this microplate, oligonucleotides can be blotted approximately 150 or 300 times using the VP384 Multi-Blot pin replicator (1.58 mm diameter

pins; delivering ~0.2 µL hanging drop) or the VP386 Multi-Blot pin replicator (1.19 mm diameter pins; delivering ~0.1 µL hanging drop).

16. The use of a soft underlay such as a mouse mat facilitates array printing.

17. When using the VP384 Multi-Blot replicator, it is not recommended to use more than two alignment holes to avoid overlapping of the blots. Best results are obtained when duplicates of the same oligonucleotides are spotted on the diagonal by repeated printing. When using the VP386 Multi-Blot replicator, higher density DNA arrays can be produced without having overlapping blots.

18. When poor-quality spots (not uniform, too large or too small) are produced, thorough cleaning of the pins (e.g., using VP110 pin cleaning solution) may help.

19. Be generous with buffer to allow the membranes to move freely in the solution. Use at least 250 mL for 20 membranes of 9.5×5 cm or 40 membranes of 9.5×3.5 cm.

20. Sampling has always been a crucial factor in detecting and identifying plant pathogens from environmental samples. As the amount of material that can be processed in a single DNA extraction is generally limited to 1 g or less and since only 1 or 2 µL of the extract is used for PCR amplification, sampling and sample preparation has become even more challenging than it already was.

21. Great care has to be taken in assuring that the extract is prepared from carefully selected representative tissue samples that are most likely to be colonized by the pathogen(s) of concern. Pooling multiple small samples taken from infected plant parts into one extraction may be a good method. To screen for a variety of microorganisms, the authors separately analyze (at least) roots, the lower stem part as well as the foliar part of the plant. With regard to soil analysis, the authors prefer analysis of a homogenized soil sample out of different core samples. Although DNA arrays are suitable for the simultaneous detection and identification of diverse pathogens, different sampling strategies could have to be used as different pathogens may have different spreading capabilities. Consequently, multiple DNA extractions may have to be performed from a single soil sample. For example, pathogen inoculum from resting spore producing fungi such as *Rhizoctonia solani* or *Verticillium dahliae* is preferentially collected by soil sieving. For other pathogens such as *Pythium* sp. or *Phytophthora* sp. direct soil extractions, an intermediate step with baiting techniques or extractions from soil suspensions may be more effective.

22. For thorough sample homogenization and disruption of resistant materials (such as sclerotia, woody materials, etc.), the use of a cell disruptor such as a FastPrep System (Thermo Savant, NY, USA) is highly recommended.

23. To minimize variation between different analyses, a standard sample size should be used (e.g., 0.75 g). If a soil sample is high in water content remove contents from the bead tube (beads and solution) and set aside. Add the soil sample to an empty tube and centrifuge for 30 s at 10,000 rpm. Remove as much liquid as possible with a pipette tip. Add beads and bead solution back to the bead tube and re-start the extraction procedure. If the sample is low in water content absorbing the lysis buffer, prewet samples with sterile water prior to DNA-extraction.

24. To this end, add 2 μL of 5M NaCl to the eluted DNA and mix. Subsequently, add 100 μL of 100% cold EtOH and mix. Centrifuge at 10,000 rpm for 5 min. Decant all liquid. Dry residual ethanol in a speed vacuum dessicator or air-dry and resuspend the precipitated DNA in 10 mM Tris, pH 8.

25. The use of a PCR cooling block is recommended to keep reagents and DNA samples cool.

26. Currently, for fungal pathogens, primers ITS1-F (5′-CTTGGT CATTTAGAGGAAGTAA-3′ (23)) and ITS4 (5′-TCCTCCG CTTATTGATATGC-3′ (24)) are generally used (T_a=59°C (1, 8)). For oomycetes, primers OOMUP18Sc (5′-TGCGGAAG GATCATTACCACAC-3′) and ITS4 have been used (T_a=59°C (9)). Alternatively, UN-up18S42 (5′-CGTAACAAGGTTT CCGTAGGTGAAC-3′) and Oom-lo28S-345H (5′-ACTTGTT CGCTATCGGTCTCGCA-3′) are used which have been validated against a huge collection of oomycete isolates (T_a=68°C (7)). In addition to these universal primers, target-specific primers targeting pathogen-specific sequences may be used (25).

27. Amplification can be checked by agarose gel electrophoresis. On the other hand, as DNA array-based detection is at least 10 times more sensitive, negative results on gel do not necessarily imply DNA has not been amplified.

28. For membranes up to a size of 9.5 × 3.5 cm, use 15 mL disposable conical tubes, for larger membranes use 50 mL tubes. 50 mL tubes can be hold in the disposable tube rotisserie of the hybridization oven. The 15 mL tubes can be placed inside standard hybridization cylinders. Four 15 mL tubes fit perfectly in a standard hybridization cylinder (ID × L = 35 × 300 mm).

29. Use 6 mL for 15 mL tubes, 12 mL for 50 mL tubes. Add buffer using a bottle top dispenser for convenience and consistency when processing multiple samples.

30. Tubes must be tight enough to prevent leakage, but not so tight as to build up pressure.

31. Check the actual temperature inside the hybridization oven with a thermometer regularly. The temperature reading on a hybridization oven could be several degrees off calibration, especially for digital readouts. Temperature is of utmost importance, long exposure to a temperature lower than the intended hybridization temperature leads to unspecific binding to the oligonucleotides.

32. If an analysis has to be stopped after/during pre-hybridization, discard the pre-hybridization buffer from the tube, fill with 2× SSC, and store at 4°C till further use. To continue, repeat pre-hybridization with fresh prewarmed prehybridization buffer.

33. For the analysis of environmental samples, use the entire PCR product (20 μL). For specificity testing using reference cultures, use 10 ng labeled amplicon per mL hybridization buffer to avoid cross-reactions that are not relevant in practice (3, 7, 8, 15). In general, this concentration corresponds to 1–2 or 2–4 μL PCR product per 6 or 12 mL hybridization buffer, respectively.

34. Be sure to pre-heat (until DNA denaturation) the mixture before reuse.

35. Pay close attention to incubation times. Excessive washing of the membrane with hybridization wash buffer I may result in loss of signal.

36. In general, we use trays of 15 × 25 cm. The amount of buffer is dependent on the number and size of the used membranes. Use enough buffer to allow the membranes to move freely in the solution, e.g., 250 mL for 20 membranes of 9.5 × 5 cm or 40 membranes of 9.5 × 3.5 cm.

37. Before adding to the buffer, spin the vial with anti-DIG-AP for 1.5 min at 10,000 rpm to remove aggregated Fab fragments to avoid spotty background.

38. We recommend to use 250 mL for 20 membranes of 9.5 × 5 cm or 40 membranes of 9.5 × 3.5 cm.

39. Use enough washing buffer to remove unbound antibody conjugate, e.g., 350 mL for 20 9.5 × 5 cm membranes or 40 9.5 × 3.5 cm membranes. Do not reuse washing buffer.

40. For example, use a plastic sheet protector. Do not use thin plastic cling wraps as these wrinkle and lead to uneven detection across the membranes. Make sure all arrays are correctly oriented on the plastic sheet (membrane number on the right).

41. 200 cm² of total membrane surface requires approximately 1 mL of diluted substrate. Do not let the membranes dry out.

42. Images can be captured by a number of means, including a variety of cooled CCD cameras or X-ray films. Imaging stations with a highly sensitive CCD camera such as those used by the authors (AutoChemi System, UVP) work best. Advantages of these systems include lower image backgrounds, good dynamic ranges, and rapid and sensitive signal detection. In the absence of a CCD camera image station, certain types of phosphor imagers may also work for detection of chemiluminescence (e.g., Typhoon by GE Healthcare).

43. Low-intensity signals can be visualized by adjusting the actual visible signal display. Nevertheless, because of the system's sensitivity, cross-reactions may be visualized as well, leading to false positive results.

44. When using X-ray films, create an electronic image of the blot (e.g., using a flatbed desktop scanner), then analyze the density of each spot using software with this ability.

45. If signals are not satisfying, this can be due to several factors, including

    (a) Inefficient PCR amplification (check signals for universal oligonucleotides): repeat PCR amplification using diluted DNA and repeat DNA array hybridization. If DNA will still not amplify then PCR optimization may be needed, e.g., by adding additives such as bovine serum albumin (BSA), dimethyl sulfoxide (DMSO), or formamide into the PCR mixture (26). Moreover, amplification efficiency can be monitored by an exogenous control (9).

    (b) Inefficient DNA array hybridization (check signals for universal oligonucleotides and hybridization control): repeat DNA array hybridization.

    (c) Unequal spread of CDP-Star (check signals for the reference control oligonucleotide): Wash membranes with hybridization wash buffer II for 5 min with gentle agitation and repeat from Subheading 3.4.4, step 2.

46. Repeat exposures following overnight storage at –20°C will slightly increase hybridization signals. Incubate membranes for 5 min in darkness at room temperature and repeat membrane exposure from Subheading 3.5.

Acknowledgments

This work was financially supported by the "Vlaams Instituut voor de bevordering van het Wetenschappelijk-Technologisch Onderzoek-Vlaanderen" (IWT) (IWT-010121 and IWT-040169) and De Ceuster Corp, Sint-Katelijne-Waver, Belgium. Additional grants for our DNA array research were obtained through the K.U.Leuven University (IOF-HB/07/014, IOF-HB/09/012, IOF-HB/10/012) and IWT (IWT-070113). We are grateful to A. Lévesque for originally providing us with his protocols. In addition, we would like to thank S. Van Kerckhove, C. Heusdens, and M. Waud for critically reading this manuscript.

References

1. Lievens B, Grauwet TJMA, Cammue BPA et al (2005) Recent developments in diagnostics of plant pathogens: a review. Recent Res Devel Microbiol 9: 57–79

2. Lievens B, Thomma BPHJ (2005) Recent developments in pathogen detection arrays: implications for fungal plant pathogens and use in practice. Phytopath 95: 1374–1380

3. Lievens B, Claes L, Vanachter ACRC et al (2006) Detecting single nucleotide polymorphisms using DNA arrays for plant pathogen diagnosis. FEMS Microbiol Lett 255: 129–139

4. Saiki RK, Walsh PS, Levenson CH et al (1989) Genetic analysis of amplified DNA with immobilized sequence-specific oligonucleotide probes. Proc Nat Acad Sci USA 86: 6230–6234

5. Kawasaki ES, Chehab FF (1994) Analysis of gene sequences by hybridization of PCR-amplified DNA to covalently bound oligonucleotide probes. The reverse dot blot method. Methods Mol Biol 28: 225–236

6. Lévesque CA, Harlton CE, de Cock AWAM (1998) Identification of some oomycetes by reverse dot blot hybridization. Phytopathology 88: 213–222

7. Tambong JT, de Cock AWAM, Tinker NA et al (2006) Oligonucleotide array for identification and detection of pythium species. Appl Environm Microbiol 72: 2691–2706

8. Lievens B, Brouwer M, Vanachter ACRC et al (2003) Design and development of a DNA array for rapid detection and identification of multiple tomato vascular wilt pathogens. FEMS Microbiol Lett 223: 113–122

9. Lievens B, Brouwer M, Vanachter ACRC et al (2005) Quantitative assessment of phytopathogenic fungi in various substrates using a DNA macroarray. Environ Microbiol 7: 1698–1710

10. Sholberg P, O'Gorman D, Bedford K et al (2005) Development of a DNA Macroarray for Detection and Monitoring of Economically Important Apple Diseases. Plant Dis 89: 1143–1150

11. Zhang N, Geiser DM, Smart CD (2007) Macroarray detection of solanaceous plant pathogens in the *Fusarium solani* species complex. Plant Dis 91: 1612–1620

12. Zhang N, McCarthy M, Smart CD (2008) A macroarray system for the detection of fungal and oomycete pathogens of solanaceous crops. Plant Dis 92: 953–960

13. Gilbert CA, Zhang N, Hutmacher RB et al (2008) Development of a DNA-based macroarray for the detection and identification of *Fusarium oxysporum* f. sp. *vasinfectum* in cotton tissue. J Cotton Sci 12: 165–170

14. Uehara T, Kushida A, Momota Y (1999) Rapid and sensitive identification of *Pratylenchus* spp. using reverse dot blot hybridization. Nematology 1: 549–555

15. Fessehaie A, De Boer SH, Lévesque CA (2003) An oligonucleotide array for the identification and differentiation of bacteria pathogenic on potato. Phytopathology 93: 262–269

16. Miller MB, Tang Y-W (2009) Basic Concepts of Microarrays and Potential Applications in Clinical Microbiology. Clinical Microbiol Rev 22: 611–633

17. Cho J-C, Tiedje JM (2002) Quantitative detection of microbial genes by using DNA microarrays. Appl Environ Microbiol 68: 1425–1430

18. Justé A, Lievens B, Frans I et al (2011) Development of a DNA Array for the simultaneous detection and identification of sugar thick juice bacterial contaminants. Food Anal Meth 4: 173–185

19. Frans I, Lievens B, Heusdens C et al (2008) Detection and identification of fish pathogens: what is the future? Isr J Aquac 60: 213–229

20. Lievens B, van Kerckhove S, Justé A et al (2010) From extensive clone libraries to comprehensive DNA arrays for the efficient and simultaneous detection and identification of orchid mycorrhizal fungi. J Microbiol Meth 80: 76–85

21. Chen W, Seifert KA, Lévesque CA (2009) A high density *COX1* barcode oligonucleotide array for identification and detection of species of *Penicillium* subgenus *Penicillium*. Mol Ecol Res 9: 114–129

22. Zahariev M, Dahl V, Chen W et al (2009) Efficient algorithms for the discovery of DNA oligonucleotide barcodes from sequence databases. Mol Ecol Res 9: 58–64

23. Gardes M, Bruns TD (1993) ITS primers with enhanced specificity for basidiomycetes: application to the identification of mycorrhizae and rusts. Mol Ecol 2: 113–118

24. White TJ, Bruns T, Lee S et al (1990) Amplification and direct sequencing of fungal ribosomal RNA genes for phylogenetics. In: Innis MA, Gelfand DH, Sninsky JJ et al (eds.) PCR protocols: A Guide to Methods and Applications. Academic Press, San Diego, CA, USA

25. Lievens B, Claes L, Vakalounakis DJ et al (2007) A robust identification and detection assay to discriminate the cucumber pathogens *Fusarium oxysporum* f. sp. *cucumerinum* and f. sp. *radicis-cucumerinum*. Environ microbiol 9: 2145–2161

26. Palumbi SR (1996) Nucleic Acids II: The Polymerase Chain Reaction. In: Hillis DM, Moritz C, Mable BK (eds) Molecular Systematics, 2nd edn. Sinauer Associates, Inc Sunderland

Chapter 31

Random Insertional Mutagenesis in Fungal Genomes to Identify Virulence Factors

Parthasarathy Santhanam

Abstract

Agrobacterium tumefaciens-mediated transformation (ATMT) has become an important tool for functional genomics in fungi. ATMT-based approaches such as random insertional mutagenesis and targeted knockout are widely used for gene functional analysis in plant-pathogen interactions. Here, we describe a protocol for the identification of pathogenicity and virulence genes through random insertional mutagenesis using the fungal wilt pathogen *Verticillium dahliae* as an example for the protocol.

Key words: ATMT, Insertional mutation, Random mutagenesis, Targeted knockout, *Verticillium dahliae*, Hygromycin

1. Introduction

Agrobacterium tumefaciens has been widely used for the genetic manipulation of plants. However, *A. tumefaciens*-mediated transformation (ATMT) was also shown to be effective in a wide range of fungi (1, 2). ATMT can produce larger numbers of stable transformants and more single copy T-DNA insertions than conventional transformation methods (3). As the transformants are tagged with T-DNA, it is easy to identify the insertion site. Moreover, ATMT is simple, and suitable for both random insertional mutagenesis and targeted mutagenesis. *Verticillium dahliae* is the causal agent of *Verticillium* wilt in over 200 dicotyledonous plant species (4). The molecular mechanisms that govern *Verticillium*-host interactions are of primary importance to combat *Verticillium* wilt diseases. In this protocol, we will demonstrate ATMT as a tool to generate

Melvin D. Bolton and Bart P.H.J. Thomma (eds.), *Plant Fungal Pathogens: Methods and Protocols*,
Methods in Molecular Biology, vol. 835, DOI 10.1007/978-1-61779-501-5_31, © Springer Science+Business Media, LLC 2012

random mutant library in *V. dahliae*, high-throughput pathogenicity screening on susceptible tomato plants and identification of T-DNA insertion site.

2. Materials

2.1. Media

1. Minimal medium: Dissolve $K_2HPO_4 \times 3H_2O$ (3.44 g), KH_2PO_4 (1.45 g), NaCl (0.15 g), $MgSO_4 \times 7H_2O$ (0.5 g), $(NH_4)_2SO_4$ (0.5 g) in 800 mL of water. Adjust the pH to 5.5, the volume to 900 mL with water and autoclave. In parallel, dissolve $CaCl_2 \times 2H_2O$ (0.067 g), $FeSO_4 \times 7H_2O$ (0.0025 g), glucose (2 g) in 100 mL water and filter through a 0.2 μm Whatman filter. Once the autoclaved medium reaches ~50°C add filter sterilized medium and mix well (see Notes 1 and 2).

2. Induction medium: Dissolve $K_2HPO_4 \times 3H_2O$ (3.44 g), KH_2PO_4 (1.45 g), NaCl (0.15 g), $MgSO_4 \times 7H_2O$ (0.5 g), $(NH_4)_2SO_4$ (0.5 g) in 800 mL of water. Adjust the pH to 5.5, the volume to 900 mL with water and autoclave. In parallel, dissolve $CaCl_2 \times 2H_2O$ (0.067 g), $FeSO_4 \times 7H_2O$ (0.0025 g), glucose (1.8 g) MES (7.8 g) and glycerol (5 mL) in 100 mL water and filter through a 0.2 μm Whatman filter. Once the autoclaved medium reaches ~50°C add filter sterilized medium and mix well. Add 1.5% of technical agar for plates.

3. Selection medium: Autoclave PDA (39 g/L) and add hygromycin (50 μg/mL) and cefotaxim (200 μg/mL) immediately before pouring the plates.

2.2. Materials

1. Cefotaxim (200 mg/mL): dissolve required amount in demineralized water and filter sterilize through a 0.2 μm Whatman filter (see Notes 3 and 4).

2. Hygromycin (50 mg/mL): dissolve required amount in demineralized water and filter sterilize through a 0.2 μm Whatman filter (see Notes 5 and 6).

3. Kanamycin (50 mg/mL): dissolve required amount in demineralized water and filter sterilize through a 0.2 μm Whatman filter (see Note 7).

4. Acetosyringone (200 mM): dissolve required amount in DMSO (see Note 8).

5. Buffer: TES (100 mM Tris, pH 8.0, 10 mM EDTA, 2% SDS).

6. Reagents: 5M NaCl, 10% CTAB, 5M NH_4Ac, Chloroform: isoamylalcohol (24:1), Isopropanol and 70% ethanol.

7. Enzymes: *Nco*I (six cutter) and *Msp*I (four cutter), T4 DNA ligase and GO Taq DNA polymerase.

8. Fungal species; *V. dahliae* was used as recipient species for transformation.

9. Bacterial strain: *A. tumefaciens* strain SK1044 (kindly provided by S. Kang, The Pennsylvania State University) with plasmid pBHt2 (5) was used as T-DNA donor strain.

10. Host plant: *Solanum lycopersicum* Mill. cv. Moneymaker is susceptible for *V. dahliae*.

11. Nitrocellulose membrane: Hybond-N+ (GE Health Care).

12. Heat sterilized microcentrifuge tube (1.5 mL).

13. Sterilized glassware: 300 mL conical flask.

14. Glass beads.

3. Methods

3.1. Fungal Transformation (see Fig. 1)

Carry out the following steps in a clean laminar whenever possible.

1. Day 1—Preparation of fungal inoculum. Inoculate *V. dahliae* on PDA plates and allow the fungus to grow at 22°C for 10 days.

2. Day 6—Preparation of *Agrobacterium* culture. Prepare a fresh culture of *A. tumefaciens* from glycerol stock by streaking on LB plates supplemented with kanamycin. Incubate the plates at 28°C for 48 h.

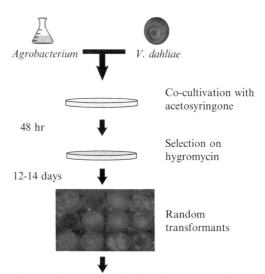

Fig. 1. Flow chart of *Agrobacterium tumefaciens*-mediated transformation of *Verticillium dahliae*.

3. Day 8—Inoculate a fresh colony of *A. tumefaciens* into 50 mL of minimal medium supplemented with 25 μg/mL of kanamycin. Incubate the culture at 28°C for >36 h at 200 rpm.

4. Day 10—Transformation. Centrifuge the *A. tumefaciens* culture at 5,000 rpm for 5 min. Resuspend the bacteria in induction medium to an OD_{600} of 0.15 (see Note 9). Add acetosyringone to a final concentration of 200 μM and incubate the culture at 28°C with 200 rpm for at least 6 h. Harvest and wash *V. dahliae* spores with sterile water by centrifugation at 4,000 rpm for 10 min. Resuspend the pellet with sterile water at spore concentration of 1×10^6 spores/mL. Mix an equal volume of *A. tumefaciens* and *V. dahliae* spore suspension in a microcentrifuge tube. Plate 200 μL of this mixture onto a Hybond-N$^+$ filter placed on induction medium supplemented with acetosyringone. Incubate the plate(s) in dark at room temperature for 48 h.

5. Day 12—Transfer the filter from induction medium to selection medium in upright position. Incubate the plates at room temperature for 10–14 days.

6. Day 22–26—Once colonies appear, transfer one colony per well to a 24-well culture plate with selection medium. Incubate at 22°C for 5–6 days (see Note 10).

7. Day 27–31—Subculture each transformant in 4 cm petriplate with hygromycin, grow for 6–8 days, and harvest conidiospores with sterile tap water. Transfer 500 μL of spore suspension into the tube containing 500 μL of sterile glycerol (25–30%). Store the glycerol stocks at –80°C.

3.2. Pathogenicity Assay

3.2.1. Preparation of Conidiospore Suspension

1. For the pathogenicity assay, subculture each transformant per well in six-well culture plates containing selection medium and for wild-type control subculture on PDA plate. Incubate the plate at 22°C for 7–10 days.

2. Add ~10 glass beads and 3 mL of tap water to each well (see Note 11).

3. Seal the mouth of the culture plate with a PCR tape (see Note 12). Prepare conidiospore suspensions by shaking the sealed culture plate for 10–15 min at 200 rpm.

3.2.2. Root Dip Inoculation

1. Uproot 10-day-old tomato seedlings and rinse the roots in water to remove remaining soil.

2. Dip the roots into the spore suspension (see Note 13) for 5 min and subsequently replant the seedlings in soil.

3.2.3. Scoring

Score the seedlings for up to 14 days postinoculation for typical symptoms of *Verticillium* wilt disease, such as stunting, wilting, chlorosis, and necrosis.

Seedlings exhibiting reduced symptoms compared to wild-type *V. dahliae* inoculated seedlings are retained for rescreening.

3.2.4. Rescreening
(see Note 14)

1. Follow the above mentioned Subheading 3.2.1, step 1.

2. Harvest the conidiospores with 5 mL of tap water and adjust to 1×10^6 spores/mL (see Note 15).

3. Uproot 10-day-old tomato seedlings and remove remaining soil by rinsing in water.

4. Dip the roots into the spore suspension for 5 min and replant the seedlings in soil.

5. Score the seedlings up to 14 days post inoculation and retain transformants that cause reduced symptoms (as described in Subheading 3.2.3).

3.3. T-DNA Insertion Site Identification

3.3.1. Genomic DNA Isolation and Digestion (6)

1. Subculture the selected transformants and incubate at 22°C for 10 days.

2. Harvest spores and transfer to 2 mL tube.

3. Add 500 μL of TES and incubate at 65°C for 60 min (see Note 16) with occasional mixing.

4. Add 140 μL of 5M NaCl, 65 μL of 10% CTAB and mix gently. Incubate again at 65°C for 10 min.

5. Add 800 μL of chloroform: isoamylalcohol (24:1) mix gently and incubate on ice for 20–30 min.

6. Centrifuge at maximum speed in a microcentrifuge, at 4°C for 10 min.

7. Transfer supernatant to a 1.5 mL microcentrifuge tube and add 225 μL of 5M NH_4Ac. Mix gently and incubate on ice for 30 min.

8. Add 10 μL of RNaseA and incubate at 37°C for 20–30 min.

9. Centrifuge at max speed, at 4°C for 10 min.

10. Transfer supernatant to a 1.5 mL microcentrifuge tube, add 500 μL of isopropanol, and mix gently.

11. Centrifuge at maximum speed in a microcentrifuge at room temperature for 5 min.

12. Wash pellets with 250 μL of 70% ethanol.

13. Air-dry and dissolve pellets in 50 μL of demineralized water.

14. Quantify the concentration of genomic DNA.

15. Digest 500 ng of genomic DNA with 5 units of restriction endonuclease (*Nco*I or *Msp*I) in a 50 μL reaction containing 1× buffer and 1× BSA at 37°C overnight (7).

16. Heat-inactivate the enzyme at 65°C for 20 min.

3.3.2. Self-Ligation

1. Add 10 μL of 10× T4 DNA ligase buffer, 5 units of T4 DNA ligase, and 38 μL of H_2O to the digested genomic DNA and mix gently.

2. Allow overnight ligation by incubating the mix at 15°C.

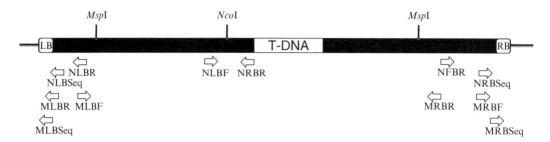

Fig. 2. Priming sites used to amplify the flanking genomic sequences. Primers used for Inverse PCR and for sequencing are shown in *open arrow*. *NcoI* digested genomic DNA are self-ligated and the genomic region flanking the left border is amplified with the primers NRBF/NRBR and sequenced with the primer NRBSeq (see Note 17).

3. Add 10 µL of 3M NaOAc and 250 µL of 95% ethanol to the overnight ligation mixture and centrifuge for 10 min at max speed.

4. Wash the pellets with 250 µL of 70% ethanol and centrifuge for 5 min at maximum speed.

5. Pellets were air-dried and redissolved with 50 µL of demineralized water.

3.3.3. Inverse PCR

1. To amplify the genomic region flanking the right border (see Note 17): 2 µL of ligation product, 1 µL of each primer (NRBF and NRBR, see Fig. 2 and Table 1), 1× PCR buffer, 0.6 µL dNTP (10 mM), 0.8 units GO Taq DNA polymerase and 35.25 µL H$_2$O.

2. Cycling conditions: 1 cycle (2 min 94°C), 35 cycles (30 s 94°C, 30 s 55°C 3 min 72°C), 1 cycle (10 min 72°C) and cool at 15°C.

3. Load 2–3 µL of the amplicon on 1.2% agarose gel and analyze for presence of the product.

3.3.4. Sequencing

1. Purify the remaining PCR products (illustra GFX™, GE Healthcare) and elute with 20 µL of demineralized water (see Note 18).

2. Sequence the purified PCR sample with the primer NRBSeq (see Table 1 and Note 19).

4. Notes

1. While preparing minimal medium, adjust the pH, volume to 900 mL, divide into 45 mL aliquots and autoclave. These aliquots can be stored at room temperature for at least 3 months. In a laminar flow, add the remaining components (prepared freshly each time) and adjust to 5 mL and filter through a 0.2 µm

Table 1
Sequence of primers used for inverse PCR

Primer	Sequence 5′–3′
NLBF	AGTGTATTGACCGATTCCTTGC
NLBR	AGGGTTCCTATAGGGTTTCGCTCATG
NLBSeq	GAATTAATTCGGCGTTAATTCAGT
NRBF	CGTTATGTTTATCGGCACTTTG
NRBR	GGCACTGGCCGTCGTTTTACAAC
NRBSeq	CCCTTCCCAACAGTTGCGCA
MLBF	GGATTTTGGTTTTAGGAATTAGA
MLBR	AATTCGGCGTTAATTCAGTACA
MLBSeq	TCAGTACATTAAAAACGTCCGCAA
MRBF	CAACTGTTGGGAAGGGCGATC
MRBR	CAGCCTGAATGGCGAATGCTA
MRBSeq	GAATGCTAGAGCAGCTTGAGCT

Whatman filter. Mix the 5 mL with the 45 mL to prepare 50 mL of minimal medium, which is enough for one transformation.

2. Minimal medium can be stored at 4°C only for few days. We find precipitation after a week.

3. Cefotaxim is used to kill *Agrobacterium* on selection plates.

4. Hygromycin is used to select fungal transformants containing the T-DNA. Hygromycin is very toxic and carcinogenic, so use proper gloves and work in a laminar flow while preparing stock solutions.

5. Depending on the fungus, the active concentration of hygromycin can vary from 25 to 150 μg/mL.

6. Kanamycin is used to select *Agrobacterium* transformed with the binary vector pBHt2.

7. Store all antibiotics at −20°C except hygromycin (4°C).

8. Acetosyringone enhances the rate of transfer of T-DNA by *Agrobacterium*. Different concentrations of acetosyringone and duration of cocultivation may lead to different numbers of transformants.

9. Based on OD_{600} calculate how much volume of culture needs to be centrifuged to get a final volume at OD_{600} 0.15 (e.g., if the OD_{600} of 36 h culture is 0.220, centrifuge 35 mL and dilute the pellet with 50 mL of induction medium to reach OD_{600} 0.15).

10. Use toothpicks to pick fungal transformants from the selection plates and transfer them to 24-welled culture plates.

11. Glass beads facilitate the release of spores during shaking.

12. Sealing the mouth of culture plate with a PCR tape, avoid cross contamination while shaking.

13. For high-throughput T-DNA mutant screening it is convenient to skip the spore normalization step at this point because it saves a lot of time and labor. As a consequence, it increases the chance of picking up false positive phenotypes (resulting due to lack of spores). Eventually these mutants will be eliminated in the rescreening step where the spore concentrations are normalized.

14. In this step we calibrate the conidiospores to a working concentration of 1×10^6 spores/mL. Hence transformants, which produce fewer spores, are normalized and retested to avoid phenotypes that are resulting from lack of spores.

15. Spore concentration is calculated by counting the spores in a hemocytometer. Dilute the harvested spores 10 times with water and place a drop of diluted spore suspension at the edge of the cover glass mounted on hemocytometer. By capillary action spore suspension will fill the gap between cover glass and the chamber. Count the number of spores in four different squares and calculate the average. Calculate spore concentration per mL based on the type of hemocytometer you use.

16. In this step you can also add proteinase K ($100~\mu g$) to inhibit nuclease activity and to avoid protein contamination.

17. To successfully identify the genomic region flaking the T-DNA use at least two enzymes (*Nco*I and *Msp*I). Amplify both flanking regions (Left and Right) using respective left and right border primers (see Fig. 2 and Table 1).

18. Alternatively you can precipitate the PCR product with $12~\mu L$ of 10M NH_4Ac and $150~\mu L$ of 95% ethanol and mix by inverting. After 10 min of incubation at room temperature, centrifuge at max for 20 min at 4°C. Wash the pellet with 70% ethanol, air-dry, and redissolve the pellets with $20~\mu L$ of demineralized water and sequence the product with respective primer.

19. If the direct cloning of PCR fails, you can clone and sequence the product in a pGEMt-easy vector or similar product following the manufacturer's instructions.

Acknowledgment

This research is supported by a Vidi grant of the Research Council for Earth and Life sciences (ALW) of the Netherlands Organization for Scientific Research (NWO).

References

1. De Groot, M. J. A., P. Bundock, et al. (1998). Agrobacterium tumefaciens-mediated transformation of filamentous fungi. Nature Biotechnology 16(9): 839–842.

2. Michielse, C. B., P. J. J. Hooykaas, et al. (2005). *Agrobacterium*-mediated transformation as a tool for functional genomics in fungi. Current Genetics 48(1): 1–17.

3. Weld, R. J., K. M. Plummer, et al. (2006). Approaches to functional genomics in filamentous fungi. Cell Research 16(1): 31–44.

4. Fradin, E. F. and B. P. H. J. Thomma (2006). Physiology and molecular aspects of Verticillium wilt diseases caused by *V. dahliae* and *V. albo-atrum*. Molecular Plant Pathology 7(2): 71–86.

5. Mullins, E. D., X. Chen, et al. (2001). Agrobacterium-Mediated Transformation of *Fusarium oxysporum*: an efficient tool for insertional mutagenesis and gene transfer. Phytopathology 91(2): 173–180.

6. Möller, E. M., G. Bahnweg, et al. (1992). A simple and efficient protocol for isolation of high molecular weight DNA from filamentous fungi, fruit bodies, and infected plant tissues. Nucleic Acids Research 20(22): 6115–6116.

7. Meng, Y., G. Patel, et al. (2007). A systematic analysis of T-DNA insertion events in *Magnaporthe oryzae*. Fungal Genet Biol 44(10): 1050–1064.

Chapter 32

A Yeast Secretion Trap Assay for Identification of Secreted Proteins from Eukaryotic Phytopathogens and Their Plant Hosts

Sang-Jik Lee and Jocelyn K.C. Rose

Abstract

Secreted proteins from plants and phytopathogens play important roles in their interactions and contribute to elaborate mechanisms of attack, defense, and counter-defense, as well as surveillance and signaling. There is therefore considerable interest in developing techniques to characterize "secretomes." Here, we describe the use of the yeast secretion trap (YST) functional screen to isolate and identify secreted proteins that are accumulated and detected in the extracellular matrix of eukaryotes. This method involves fusing cDNAs generated or derived from plants, pathogens, or infected tissue to a yeast (*Saccharomyces cerevisiae*) invertase (*suc2*) reporter gene lacking its signal peptide, transforming the resulting fusion library into an invertase-deficient yeast strain, and plating the transformants on a sucrose selection medium. A yeast transformant containing a cDNA that encodes a secreted protein can rescue the mutant and the plasmid DNA can then be sequenced to identify the secreted protein. The YST screen can be a very powerful tool in the study of dynamics of plant host-pathogen interactions.

Key words: Apoplast, Extracellular matrix, Plant cell wall, Secreted protein, Secretome, Yeast secretion trap

1. Introduction

The plant cell wall and apoplastic environment represent the front line of defense against microbial pathogens. Structural matrices of polysaccharides, phenylpropanoids in lignified cells and the structural lipids of the specialized cuticular cell wall of epidermal cells provide resilient physical barriers. In addition, the apoplast contains dynamic populations of secreted plant proteins with a wide range of activities, which function to detect and inhibit potential invaders (1). Pathogens also secrete suites of proteins and peptides

Melvin D. Bolton and Bart P.H.J. Thomma (eds.), *Plant Fungal Pathogens: Methods and Protocols*,
Methods in Molecular Biology, vol. 835, DOI 10.1007/978-1-61779-501-5_32, © Springer Science+Business Media, LLC 2012

into the plant wall during invasion, many of which can be termed "effectors," to degrade the host cell walls, subvert defense mechanisms, and co-opt plant metabolism to facilitate successful colonization (2–5). Plant and pathogen secreted proteins have thus co-evolved to provide mechanisms for attack, defense, and counter-defense (6, 7), and there has been a recent surge of interest in characterizing the composition and dynamics of these "secretomes," by profiling secreted protein populations, or by examining the location, structure, and function of specific secreted proteins.

Several approaches have been used to identify and characterize eukaryotic secretomes (8), the most common being to isolate, fractionate, and identify secreted proteins by a combination of electrophoreses and liquid chromatography, followed by mass spectrometry. This has been effectively applied to study secreted proteins from several fungal species, including a number of phytopathogens (e.g., see refs. (9–20)). However, this approach presents several conceptual and technical challenges, such as contamination with intracellular proteins, the difficulties of detecting lower abundance proteins, resistance to extraction and separation, or low stability (21). A popular alternative, which avoids all these issues, is to use computational tools to screen large sequence data sets for eukaryotic proteins that possess a predicted N-terminal signal peptide (SP), and that thereby enter the classical secretory pathway. This is perhaps a less biased approach, although false positives/ negatives should be expected and the effectiveness depends largely on the quality of the primary sequence data. For example, PexFinder, an algorithm based on SignalP v2.0 has been used to identify extracellular proteins from *Phytophthora* that contain predicted SPs based on *Phytophthora infestans* expressed sequence tags (ESTs) (22). However, the presence of an SP only suggests entry in the ER, and it is important to note that a substantial proportion of these proteins may be retained in the secretory pathway or redirected to some other intracellular compartment, rather than being secreted (21). It is therefore advisable to confirm computationally predicted localization with an appropriate experimental analysis, such as immunolocalization or monitoring the accumulation of target proteins fused to fluorescent tags.

In contrast to proteomic studies and in silico predictions, functional screens have also been developed to identify secreted proteins based on the final destination of a protein of interest, utilizing a protein marker whose activities confirm the secretion of a specific protein. One of these, the yeast secretion trap (YST) system, can be employed to screen a cDNA library of interest (23). The YST screen uses invertase, an enzyme that hydrolyzes sucrose to generate glucose and fructose, as a reporter protein, and yeast (*Saccharomyces cerevisiae*) as a host. Complementary DNAs generated from mRNA using random hexamers are fused to the 5′ end of a yeast invertase (*suc2*) gene lacking its SP. The resulting *suc2* fusion library is then transformed into a *suc2* yeast mutant and plated on a sucrose selection medium. All yeast transformants containing a cDNA that encodes a secreted protein should be able to rescue the mutant. The plasmid can be recovered from the yeast transformants and sequenced, allowing identification of the secreted protein. The YST has been very effectively adopted to identify secreted proteins from plant pathogens including obligate biotrophic *Uromyces fabae* (24) and hemibiotrophs such as *P. infestans* (1) and *Colletotrichum graminicola* (25).

In this chapter, we describe the use of the YST functional screen, which can be used to characterize the secretomes of eukaryotic phytopathogens, as well as for the simultaneous analysis of both pathogen and host secretomes from infected tissues. Such a functional approach can be valuable complements to proteomic/bioinformatic studies of cell wall/apoplastic proteins and provide new insights into the molecular mechanisms that contribute to resistance or that modulate infection and disease development.

2. Materials

2.1. Reagents

1. Absolute ethanol.
2. cDNA Synthesis Kit (Stratagene) (see Note 1).
3. Cloning vectors (pYST-0, -1, and -2) (23).
4. Dimethyl sulfoxide (DMSO; Fisher Scientific).
5. DNA ladder molecular weight marker (1 kb Plus DNA Ladder; Invitrogen).
6. DNA sequencing primer (see Note 2).
7. Glass beads 425–600 μm, acid-washed (Sigma).
8. Herring testes carrier DNA (BD Biosciences).
9. Oligotex mRNA Mini Kit (Qiagen) (see Note 1).
10. Perfectprep Plasmid Midi Kit (Eppendorf) (see Note 1).
11. QIAquick Gel Extraction Kit (Qiagen) (see Note 1).
12. QIAquick PCR Purification Kit (Qiagen) (see Note 1).
13. QIAprep Spin Miniprep Kit (Qiagen) (see Note 1).
14. Random hexamer for cDNA synthesis (see Note 3).
15. Restriction enzymes (Promega).
16. T4 DNA ligase (Promega).
17. YEASTMAKER™ Yeast Transformation System 2 (BD Biosciences) (see Note 4).

2.2. Equipment

1. Acrodisc PF syringe filter with supor membrane (Pall Life Science).
2. Agarose gel electrophoresis system.

3. Conical flasks (250 mL).

4. Conical tubes (50 mL).

5. Incubators (30 and 37°C).

6. Laminar flow hood.

7. Microcentrifuge tubes (1.5 and 2.0 mL).

8. MicroPulser Electroporator (Bio-Rad).

9. Petri dishes: 100×15 (VWR Scientific).

10. Petri dishes: 150×15 mm (BD Falcon).

11. Shaking incubators (30 and 37°C).

12. Spectrophotometer (Shimadzu, UV–vis Spectrophotometer UV mini-1240) (see Note 1).

13. Water baths (16 and 42°C).

2.3. Reagent Setup

1. 1% agarose gel: Weigh out 0.5 g of agarose into a 250-mL conical flask. Add 50 mL of 1× TAE. Microwave for about 2–3 min to dissolve the agarose. Leave to cool on the bench for 5 min down to about 55–60°C. Add 1 µL of ethidium bromide (10 mg/mL) and swirl to mix. Pour the gel slowly into the mold with the comb.

2. LB medium: 10 g of tryptone, 5 g of yeast extract, 10 g of NaCl, 20 g of agar (for plates only) per liter. Adjust pH to 7.0 with NaOH solution. Add deionized H_2O to a final volume of 1 L. Autoclave.

3. LB-ampicillin medium: 10 g/L tryptone, 5 g/L yeast extract, 10 g/L NaCl, 20 g/L agar. Adjust pH to 7.0 with NaOH solution. Add deionized H_2O to a final volume of 1 L. Autoclave. Cool to 55–60°C. Add 1 mL of ampicillin (100 mg/mL; filter-sterilized with 0.2-µm membrane filter).

4. 10× LiAc buffer stock solution: 1M LiAc. Adjust pH to 7.5 with dilute acetic acid. Autoclave.

5. 50% (wt/vol) PEG 3350: 50 g of PEG 3350/100 mL. Autoclave or filter-sterilize.

6. Phenol/chloroform/isoamyl alcohol (25:24:1, vol/vol/vol): Mix 25 mL of phenol (Tris-buffered, pH 8.0) with 24 mL of chloroform and 1 mL of isoamyl alcohol.

7. 3M sodium acetate: Dissolve 40.8 g of sodium acetate trihydrate in 80 mL of deionized H_2O. Adjust the pH to 5.2 with glacial acetic acid. Add deionized H_2O to a final volume of 100 mL. Filter to sterilize.

8. Sucrose selection medium: 20 g of Difco peptone, 10 g of yeast extract, 20 g of agar (for plates only) per liter. Adjust the pH to 6.5 with HCl solution. Add deionized H_2O to a final volume of 950 mL. Autoclave. Cool to 55–60°C. Add sucrose

(filter-sterilized with 0.2-μm membrane filter) to 2% (vol/vol) by adding 50 mL of a 40% stock solution.

9. 50× TAE: 242 g of Tris base, 57.1 mL of glacial acetic acid, 100 mL of 0.5M EDTA per liter. Adjust pH to 8.3 with NaOH. Add deionized H_2O to a final volume of 1 L.

10. 10× TE stock solution: 100 mM Tris–HCl (pH 7.5), 10 mM EDTA (pH 8.0). Autoclave and store at room temperature.

11. TE-LiAc solution: Prepare immediately prior to use by combining 0.2 mL of 10× TE with 0.2 mL of 10× LiAc. Bring the total volume to 2 mL using sterile H_2O.

12. TE-LiAc-PEG solution: Prepare immediately prior to use by adding 1 mL of 10× TE, 1 mL of 10× LiAc, and 8 mL of 50% (wt/vol) PEG 3350 to make the final volume of 10 mL.

13. Yeast lysis buffer: 2% (v/v) Triton X-100, 1% (w/v) SDS, 100 mM NaCl, 10 mM Tris–HCl (pH 8.0), 1 mM EDTA.

14. YPD medium: Dissolve 20 g of peptone and 10 g of yeast extract in 900 mL deionized H_2O. Adjust the pH to 6.5 with HCl solution. Add deionized H_2O to a final volume of 950 mL. Autoclave and cool to 55–60°C. Add glucose (filter-sterilized with 0.2-μm membrane filter) to 2% (vol/vol) by adding 50 mL of a 40% stock solution.

2.4. Bacteria and Yeast Cells

1. One Shot® TOP10 Electrocomp™ *Escherichia coli* cells (Invitrogen) (see Note 5).

2. Yeast (*S. cerevisiae*) strain: DBYα2445 (Matα, *suc2Δ9*, *lys2-801*, *ura3-52*, *ade2-101*) (see Note 6) (23).

3. Methods

3.1. Construction of the YST cDNA Library

1. Isolate intact mRNA from total RNA derived from the target eukaryotic tissue(s), using an Oligotex mRNA Mini Kit, according to the manufacturer's instructions (see Note 7).

2. Synthesize the first- and second-strand cDNAs from the mRNA (5 μg) using a random hexamer primer containing a *Not*I restriction site for a directional cloning using a cDNA Synthesis Kit (Stratagene) according to the manufacturer's instructions.

3. Ligate the *Eco*RI adapters (provided with the Stratagene cDNA Synthesis Kit) to the blunted cDNAs and phosphorylate the *Eco*RI ends according to the manufacturer's instructions.

4. Purify cDNA using a QIAquick PCR Purification Kit (Qiagen) (see Note 8).

5. Add 5 μL of 10× *Not*I buffer and 5 μL of *Not*I (10 U/μL) to digest 40 μL of the cDNAs. Transfer the reaction tube to 37°C and incubate for 2 h.

6. Run a 1% agarose gel to fractionate the cDNA fragments and excise a gel slice corresponding to the region from approximately 0.3–1.0 kb, as judged using DNA ladder molecular weight markers (see Note 9).

7. Purify the excised cDNAs with a QIAquick Gel Extraction Kit (Qiagen).

8. Ligate the cDNAs into the *Eco*RI and *Not*I sites of the YST vector(s) by adding the eluted cDNA (20 ng), 1 μL of 10× ligase buffer, 1 μL of T4 DNA ligase, and the pYST vectors (pYST-0, -1 and -2) digested with *Eco*RI and *Not*I (a total of 200 ng) to a final volume of 10 μL. Spin down the reaction in a tube and incubate overnight at 16°C (see Note 10).

9. Apply 2 μL of the ligation mix into 50 μL of a TOP10 electro-competent *E. coli* strain (Invitrogen) for electroporation using a MicroPulser (Bio-Rad) according to the manufacturer's instructions. Add 1 mL of SOC medium (provided with TOP10 electrocompetent cells) to the tube and incubate for 1 h at 37°C. Plate 1 μL of the transformation mixture onto an agar plate supplemented with ampicillin (100 mg/L). Incubate the plate overnight at 37°C. Count the colonies (see Note 11).

10. Transform 8 μL of the remaining ligation mixture in more electroporation reactions (2 μL each per transformation) if a large number of CFU (2×10^5 CFU/mL) and a high percentage of clones with cDNA inserts (>70% recombinants) result from step 9.

11. Plate all the transformed cells by spreading 250 μL of the transformed *E. coli* cells onto each LB-ampicillin agar plate (150×15 mm) at a density of roughly 5×10^4 CFU/plate, using as many plates as necessary to accommodate all the transformed cells, and incubate the plates overnight at 37°C.

12. Add 20 mL of LB liquid medium per plate to harvest the transformants from each plate and incubate at 37°C for 3 h with gentle shaking.

13. Harvest the overlaying LB cultures in a sterile 250-mL flask.

14. Isolate the YST cDNA library DNA from 25 to 50 mL of the bacterial LB cultures according to the manufacturer's instructions as described for the Perfectprep Plasmid Midi Kit.

15. Quantify the amount of the YST cDNA library DNA using a spectrophotometer and store at –20°C until ready for use.

3.2. Preparation of Competent Yeast

1. Streak the yeast glycerol stock from –80°C onto a YPD medium. Incubate the plate for 2–3 days at 30°C.

2. Use one or two fresh yeast colonies (2–3 mm in diameter) to prepare 50 mL of YPD liquid medium in a 250-mL flask.

3. Incubate the culture for 16–18 h at 30°C with shaking at 250 rpm until OD at 600 nm reads 1.2.

4. Transfer the 50-mL culture into 300 mL of YPD in a 1-L flask to produce 0.2–0.3 OD at 600 nm.

5. Incubate the culture for 3 h at 30°C with shaking at 250 rpm until OD at 600 nm reaches 0.4–0.6.

6. Pellet the cells by centrifugation at $4,500 \times g$ for 5 min at room temperature.

7. Resuspend the cell pellet in 50 mL of sterile H_2O by vortexing.

8. Pellet the cells by centrifugation at $4,500 \times g$ for 5 min at room temperature.

9. Discard the supernatant and resuspend cells in 1.5 mL of freshly prepared TE-LiAc solution (see Note 12).

3.3. Yeast Transformation

1. In a 50-mL tube, add 50 µg of YST cDNA library DNA and 2 mg of carrier DNA (herring testes DNA) and mix well.

2. In 1.5-mL tubes, add 100 ng of control DNA and 0.1 mg of carrier DNA (herring testes DNA) and mix well (see Note 13).

3. Add 1 mL of freshly prepared yeast competent cells to the tube containing the YST cDNA library DNA (or 0.1 mL for controls) and vortex to mix well.

4. Add 6 mL of TE-LiAc-PEG solution to the tube containing the YST cDNA library DNA (or 0.6 mL for controls) and vortex to mix well.

5. Incubate at 30°C for 30 min with shaking at 200 rpm.

6. Add 700 µL of DMSO to the tube containing the YST cDNA library DNA (or 70 µL for controls) and mix well gently.

7. Heat the samples by incubating for 15 min in a 42°C water bath and swirl gently every 5 min.

8. Place the tubes on ice for 5 min.

9. Pellet the cells by centrifugation at $4,500 \times g$ for 5 min (or 15 s for controls) at room temperature.

10. Resuspend the cells in 5 mL (or 0.5 mL for controls) of 1× TE buffer.

11. Spread 200 µL of the transformation mixture onto a 150×15 mm plate containing the sucrose selection medium (or 100 µL onto a 100×15 mm plate for controls).

12. Incubate the plates at 30°C for 6–10 days and clear colonies should be apparent, indicating successful transformation with a gene encoding a secreted protein (see Fig. 1 and Note 14).

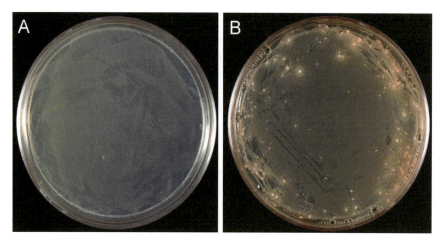

Fig. 1. An invertase-deficient yeast mutant strain is unable to grow on a sucrose selection medium because the absence of secreted invertase results in a lack of sugars to support growth (**a**), but the mutant can be rescued following transformation with a cDNA encoding a secreted protein that is formulated as a fusion protein with invertase (**b**).

3.4. Identification of cDNA Encoding a Secreted Protein

1. Streak out yeast transformants on the sucrose selection medium and incubate at 30°C for 2 days.

2. Inoculate yeast cells from the streaked plate with a sterile spatula into 2 mL of sucrose selection medium and incubate at 30°C for 16–18 h.

3. Transfer the culture into a 2.0-mL microcentrifuge tube and pellet the cells by centrifugation at $14,000 \times g$ for 2 min.

4. Resuspend the pellet in 0.2 mL of yeast lysis buffer with 0.2 g of glass beads by vortexing vigorously for 2 min.

5. Add 0.2 mL of phenol/chloroform/isoamyl alcohol [25:24:1 (v/v/v)] and vortex vigorously for 2 min (see Note 15).

6. Centrifuge at $14,000 \times g$ for 5 min and transfer the upper aqueous layer into a new tube.

7. Precipitate the DNA by adding 20 μL of 3M sodium acetate and 500 μL of absolute ethanol and vortex to mix well. Precipitate the DNA overnight at –20°C.

8. Centrifuge at $14,000 \times g$ for 10 min to pellet the DNA.

9. Wash the DNA pellet by adding 0.5 mL of 70% (v/v) ethanol and spin at $14,000 \times g$ for 2 min at room temperature.

10. Resuspend the DNA pellet in 25–50 μL of sterile H_2O or TE buffer (see Note 16).

11. Transform 1 μL of DNA into electrocompetent *E. coli* cells by electroporation and select transformants on LB plates supplemented with the ampicillin (100 mg/L).

12. Prepare miniprep DNA from the colonies selected on the LB-ampicillin agar plates with a QIAprep Spin Miniprep Kit (Qiagen) and confirm the sequences of cDNA inserts by sequencing using a specific DNA sequencing primer (5′-TCCTCGTCATTGTTCTCGTTCC-3′) derived from YST vector.

13. N-terminal signal peptides of the translated amino acid sequences can be predicted computationally using publicly available signal peptide prediction programs, such as SignalP version 3.0 (http://www.cbs.dtu.dk/services/SignalP) (26) and SecreTary (http://solgenomics.net/secretom/secretary) (see Note 17).

14. To determine the species of origin of the clones, the cDNA sequences can be used for searching sequence databases and probing genomic DNA blots containing total genomic DNA isolated from the target species (3).

15. The full-length cDNA clone should be obtained for further studies, such as biochemical analyses or localization assays.

4. Notes

1. Other similar commercial products are likely to be adequate and could be used as an alternative.

2. DNA sequencing primer (5′-TCCTCGTCATTGTTCTCG TTCC-3′) corresponding to the ADH1 promoter of the YST vector(s) can be used to confirm in-frame fusion between the cDNA and the *suc2* invertase gene (23).

3. A random hexamer primer (5′-GAGAGAGAGAGAGAGAA CCGC<u>GCGGCCGC</u>CNNNNNN-3′) contains a *Not*I (under-lined) restriction site for directional cloning of cDNAs. Other rare-cutting restriction enzyme sites could be used in the random hexamer primer for cDNA synthesis.

4. YEASTMAKER™ Yeast Transformation System 2 (BD Biosciences) contains the following components: 10× LiAc, 10× TE, 50% PEG, and Herring testes carrier DNA, denatured (10 mg/mL).

5. Electrocompetent cells can be prepared manually or obtained from other commercial sources.

6. The yeast strain should carry a mutation in the *SUC2* gene, which has been shown to be critical for growth on a sucrose medium.

7. RNA can be isolated by many other methods. It is important to note that one major factor for successful cDNA synthesis is the isolation of intact RNA.

8. Add 5 volumes of buffer PB (provided in the QIAquick PCR Purification Kit) to the reaction and mix well. Ensure that the final elution volume should be 40 µL for the subsequent reactions.

9. We select the cDNA fragments from approximately 0.3–1.0 kb, which provides a bias toward the 5′ region of the gene, to avoid stop codons found in the shorter (<0.3 kb) or longer (>1.0 kb) cDNA fragments, most likely corresponding to the 5′-UTR or 3′-UTR.

10. Ligation buffer contains ATP, which degrades rapidly during temperature fluctuations. Make aliquots of the buffer and store at –80°C to avoid repeated freeze-thaw cycles.

11. We recommend *E. coli* competent cells with very high transformation efficiency of >1 × 10^9 colony forming units (CFU)/µg. Ideally, there should be approximately 200 colonies on a plate with 1 µL of the transformation mixture. The percentage of recombinant clones and the average insert size of the cDNAs should be determined using randomly selected 50–100 clones from the *E. coli* library. The isolated plasmid DNAs can be digested with *Eco*RI and *Not*I, then fractionated, and visualized on a 1% agarose gel.

12. Yeast competent cells should be used for transformation immediately following preparation.

13. Set up positive (a well-known secreted protein gene) and negative (a gene lacking a signal peptide) control reactions to check the yeast transformation efficiency and efficacy of the YST vector system.

14. Rapidly growing colonies should appear in approximately 4–5 days. Yeast transformants expressing the positive control should grow well on the sucrose selection medium, while yeast transformants expressing the negative control should not grow.

15. This should be done only under a fume hood. Wear appropriate gloves, safety glasses, and lab coat.

16. Dry the DNA pellet to remove residual ethanol before adding sterile H$_2$O or TE buffer.

17. It should be noted that there may be a consistent error rate among publicly available prediction programs.

Acknowledgments

Research in this area was supported by grants from the NSF Plant Genome Program (DBI-0606595), the New York State Office of Science, Technology and Academic Research (NYSTAR), and Cornell Center for a Sustainable Future (CCSF).

References

1. Lee, S.-J., Kelley, B., Damasceno, C.M.B., St. John, B., Kim, B.-S. Kim, B.-D., and Rose, J.K.C. (2006) A functional screen to characterize the secretomes of eukaryotic phytopathogens and their hosts in planta. *Mol. Plant Microbe Interact.* **19**, 1368–1377.

2. Kamoun, S. (2006) A catalogue of the effector secretome of plant pathogenic oomycetes. *Annu. Rev. Phytopathol.* **44**, 41–60.

3. Hardham, A.R. and Cahill, D.M. (2010) The role of oomycete effectors in plant–pathogen interactions. *Funct. Plant Biol.* **37**, 919–925.

4. Hok, S., Attard, A., Keller, H. (2010) Getting the most from the host: how pathogens force plants to cooperate in disease. *Mol. Plant-Microbe Interact.* **23**, 1253–1259.

5. Kelley, B.S., Lee, S.-J., Damasceno, C.M.B., Chakravarthy, S., Kim, B.-D., Martin, G.B. and Rose, J.K.C. (2010) A secreted effector protein (SNE1) from *Phytophthora infestans* is a broadly acting suppressor of programmed cell death. *Plant J.* **62**, 357–366.

6. Misas-Villamil, J.C. and van der Hoorn, R.A.L. (2008) Enzyme-inhibitor interactions at the plant-pathogen interface. *Curr. Opin. Plant Biol.* **11**, 380–388.

7. Oliva, R., Win, J., Raffaele, S., Boutemy, L., Bozkurt, T.O., Chaparro-Garcia, A., Segretin., M.E., Stam, R., Schornack, S., Cano, L.M., Van Damme, M., Huitema, E., Thines, M., Banfield, M.J. and Kamoun, S. (2010) Recent developments in effector biology of filamentous plant pathogens. *Cell. Microbiol.*, **12**, 705–715.

8. Lee, S.-J., Saravanan, R.S., Damasceno, C.M.B., Yamane, H., Kim, B.-D. And Rose, J.K.C. (2004) Digging deeper into the plant cell wall proteome. *Plant Physiol. Biochem.* **42**, 979–988.

9. Bouws, H., Wattenberg, A., and Zorn, H. (2008) Fungal secretomes-nature's toolbox for white biotechnology. *Appl. Microbiol. Biotechnol.* **80**, 381–388.

10. Espino, J.J., Gutiérrez-Sánchez, G., Brito, N., Shah, P., Orlando, R., and González, C. (2010) The *Botrytis cinerea* early secretome. *Proteomics* **10**, 3020–3034.

11. Houterman, P.M., Speijer, D., Dekker, H.L., De Koster, C.G., Cornelissen, B.J.C., and Rep, M. (2007) The mixed xylem sap proteome of *Fusarium oxysporum*-infected tomato plants. *Mol. Plant Pathol.* **8**, 215–221.

12. Medina, M.L., Kiernan, U.A., and Francisco, W.A. (2004) Proteomic analysis of rutin-induced secreted proteins from *Aspergillus flavus. Fungal Genet. Biol.* **41**, 327–335.

13. Medina, M.L., Haynes, P.A., Breci, L., and Francisco, W.A. (2005) Analysis of secreted proteins from *Aspergillus flavus. Proteomics* **5**, 3153–3161.

14. Oda, K., Kakizono, D., Yamada, O., Iefuji, H., Akita, O., and Iwashita, K. (2006) Proteomic analysis of extracellular proteins from *Aspergillus oryzae* grown under submerged and solid-state culture conditions. *Appl. Environ. Microbiol.* **72**, 3448–3457.

15. Paper, J.M., Scott-Craig, J.S., Adhikari, N.D., Cuomo, C.A., and Walton, J.D. (2007) Comparative proteomics of extracellular proteins *in vitro* and *in planta* from the pathogenic fungus *Fusarium graminearum. Proteomics* **7**, 3171–3183.

16. Shah, P., Atwood, J.A., Orlando, R., El, M.H., Podila, G.K., and Davis, M.R. (2009) Comparative proteomic analysis of *Botrytis cinerea* secretome. *J. Proteome Res.* **8**, 1123–1130.

17. Shah, P., Gutiérrez-Sánchez, G., Orlando, R., and Bergmann, C. (2009) A proteomic study of pectin-degrading enzymes secreted by *Botrytis cinerea* grown in liquid culture. *Proteomics* **9**, 3126–3135.

18. Suarez, M.B., Sanz, L., Chamorro, M.I., Rey, M., González, F.J., Llobell, A., and Monte, E. (2005) Proteomic analysis of secreted proteins from *Trichoderma harzianum-* identification of a fungal cell wall induced aspartic protease. *Fungal Genet. Biol.* **42**, 924–934.

19. Tsang, A., Butler, G., Powlowski, J., Panisko, E.A., and Baker, S.E. (2009) Analytical and computational approaches to define the *Aspergillus niger* secretome. *Fungal Genet. Biol.* **46**, S153–S160.

20. Yajima, W. and Kav, N.N. (2006) The proteome of the phytopathogenic fungus *Sclerotinia sclerotiorum. Proteomics* **6**, 5995–6007.

21. Rose, J.K.C. and Lee, S.-J. (2010) Straying off the highway: trafficking of secreted plant proteins and complexity in the plant cell wall proteome. *Plant Physiol.* **153**, 433–436.

22. Torto, T.A., Li, S., Styer, A., Huitema, E., Testa, A., Gow, N.A.R., van West, P., and Kamoun, S. (2003) EST mining and functional expression assays identify extracellular effector proteins from *Phytophthora. Genome Res.* **13**, 1675–1685.

23. Lee, S.-J., Kim, B.-D., and Rose, J.K.C. (2006) Identification of eukaryotic secreted and cell surface proteins using the yeast secretion trap (YST) screen. *Nat. Protoc.* **1**, 2439–2447.

24. Link, T.I. and Voegele, R.T. (2008) Secreted proteins of *Uromyces fabae*: similarities and stage specificity. *Mol. Plant Pathol.* **9,** 59–66.

25. Krijger, J.-J., Horbach, R., Behr, M., Schweizer, P., Deising, H.B., and Wirsel, S.G.R. (2008) The yeast signal sequence trap identifies secreted proteins of the hemibiotrophic corn pathogen *Colletotrichum graminicola*. *Mol. Plant Microbe Interact.* **21,** 1325–1336.

26. Bendtsen, J.D., Nielsen, H., von Heijne, G., and Brunak, S. (2004) Improved prediction of signal peptides: SignalP 3.0. *J. Mol. Biol.* **340,** 783–795.

Chapter 33

Comparing Fungal Genomes: Insight into Functional and Evolutionary Processes

Eva H. Stukenbrock and Julien Y. Dutheil

Abstract

Large amount of genome data are being generated by second- and now also third-generation sequencing technologies. The challenge no longer lies in the generation of the data, but in the analyses of it. We present an overview of approaches and methods to compare complete sequences of related fungal genomes. We focus on evolutionary analyses of genome alignments to describe species divergence and to identify footprints of demography and natural selection within and between species.

Key words: Next generation sequencing, Bioinformatics, Genome comparison, Synteny analyses, Population genomics, Speciation

1. Introduction

Understanding functional and evolutionary differences between species is of fundamental importance in biological research. With the availability of complete genome sequences for a large number of fungal species, comparative analyses become an invaluable tool for understanding and investigating the diversity of fungal species.

In comparative genomics, homologous sequences are compared in the form of an alignment. For individuals of the same species or closely related species, it will in most cases not be difficult to identify homologous sequences based on a high sequence identity between genomes. However, for more distantly related species, the identification of homology and the alignment of sequences can become more challenging. In addition, fungal genomes tend to show a high extent of plasticity, meaning that substantial chromosomal rearrangements additionally can differentiate genomes. These include loss, duplications, inversions, and

Melvin D. Bolton and Bart P.H.J. Thomma (eds.), *Plant Fungal Pathogens: Methods and Protocols*,
Methods in Molecular Biology, vol. 835, DOI 10.1007/978-1-61779-501-5_33, © Springer Science+Business Media, LLC 2012

translocations of whole chromosomes or fragments and even transfer of DNA between species (1–5). This leaves the initial steps of a comparative genome analysis—in which homologous sequences are identified and aligned and where the structural differences of the genomes are described—very critical for the further analyses.

In this chapter, we describe a set of different approaches for comparative genome analyses that can be applied to fungal genomes. We distinguish between genome comparisons at the population level from comparisons at the species level and present different approaches that can be applied to the two different types of genome datasets.

Figure 1 illustrates a workflow in combination with relevant questions to identify and describe differences and similarities between two aligned genomes. We focus on (1) methods used to infer genome correspondences as the synteny between genome sequences, (2) approaches for genome alignments, (3) methods for the identification of protein-coding sequences based on comparative genomics, and lastly we describe (4) approaches to study genome evolution through the comparison of nucleotide and protein sequences.

2. Genome Synteny

Genome synteny describes the extent of colinearity between two genomes and thereby refers to the degree of structural correspondence between chromosomes. In comparative genome studies that include next generation sequence data, the characterization of genome synteny relies entirely on the quality of assemblies. Different measures of assembly quality include the lengths and distributions (N50 values) of contigs as well as assembly coverage and coverage distribution. It is essential to perform such measures of assembly quality and identify those parameter settings of the assembly software which render the best results. Different assembly approaches may be better suited for particular types of data and ideally the first analyses should compare outputs from different assembly methods to choose the appropriate and computationally feasible approach.

The correspondence or mapping of regions between two genomes can be of three types: one-to-one if the genomes have not undergone structural changes, one-to-many if structural changes have occurred in one genome and, many-to-many if both genomes have undergone a high extent of structural rearrangements. It is important to keep in mind that for genomes where many structural changes have occurred, the length of homologous fragments is shorter and thereby leaves less statistical power to infer the correct alignment.

The problem of rearranged fragments is also relevant in reference assemblies of next generation sequence reads onto reference genomes. The mapping of reads to the reference defines corresponding sequences, but does not provide any evidence of colinearity between the genomes. On the other hand, by assembling reads de novo and

Comparative genomics workflow

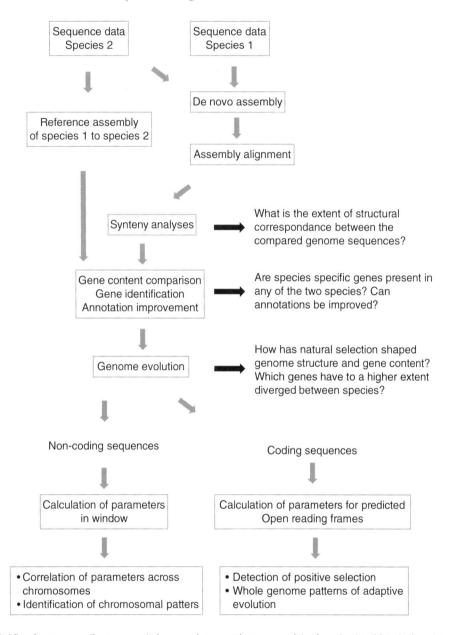

Fig. 1. Workflow for a comparative genome study comparing genomic sequence data of species 1 and 2 and relevant questions for different types of output.

thereafter aligning contigs onto a reference sequence if available, it is possible to additionally assess the extent of colinearity in the alignment of the independently assembled sequences.

The assessment of chromosomal correspondence can be assisted by experimental approaches. The small size of fungal chromosomes allows their electrophoretic separation on agarose gels by pulse field gel electrophoresis. Further homology of chromosomes that show the same size in two individuals can be confirmed by chromosome-specific probes. In particular for fungal species where chromosome numbers and sizes are known to vary, it is relevant to integrate molecular characterization of genome correspondences into comparative genome study.

Macrosynteny describes broadscale chromosomal rearrangements. The Mummer software (http://mummer.sourceforge.net/) provides a good start for a comparison of genomes where the correspondence is not known (6). Alignments of assemblies are initially conducted by the program nucmer and subsequently plotted into a dot plot called a "mummerplot." In the two-dimensional dot plot where the *x*-axis and *y*-axis each represents sequences (chromosomes, scaffolds, or contigs from each of two genomes), the plot shows a comparison of the two sequences by a calculated score for each compared region (see Fig. 2). A dot is drawn for each comparison

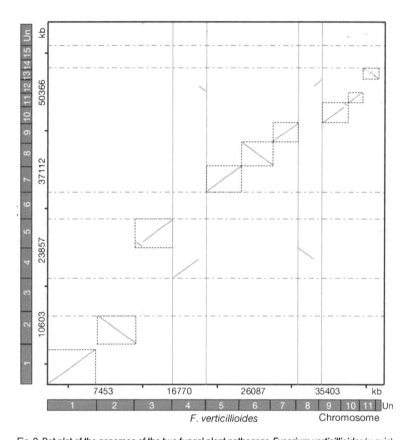

Fig. 2. Dot plot of the genomes of the two fungal plant pathogens *Fusarium verticillioides* (*x*-axis) and *Fusarium oxysporum* (*y*-axis) (3). The dot plot illustrates the alignment of each chromosome and shows conserved (for example, chr 1) or nonconserved (for example, chromosome 2 and 3) synteny. Inversions are shown as *inverted lines* in the dot plot and chromosomal translocations are indicated by *vertical lines* in chromosomes 4 and 8 of *F. verticillioides*. Figure kindly provided by C. Kistler with permission from Nature Publishing Group.

with a score higher than a certain threshold, therefore revealing the global similarity structure by exhibiting repeated regions, inversions, and translocations. Varying the size of the regions initially compared furthermore allows pinpointing macro changes (large regions) to micro changes (smaller regions). An option with the mummerplot is to assign the degree of sequence similarities by color-coding plotted dots. Thus, the mummerplot can illustrate not only the synteny of sequences, but also the extent of sequence similarities in the aligned sequences. The similarity may vary between chromosomes (7). Several genome tools provide options for plotting alignments into dot plots and thereby visualize synteny of two compared genomes.

Additional newly developed applications to assess macrosynteny are the SOAPindel and SOAPsv programs (see Table 1). These two programs are particularly designed for next generation sequence data and allow the quantification of structural differences between a reference genome and an assembly generated by mapping reads to the reference genome. The SOAP software package consists of a complete set of tools for the handling of next generation sequence data from initial assembly of reads to the subsequent analyses of assemblies. The tools can be used in combination and thereby facilitate the workflow of sequence analyses.

Microsynteny is described as the extent of colinearity at a more "local" chromosomal scale. For example, breaks in the microsynteny define gaps, duplications, and translocations of smaller fragments in the genome alignment. In the comparative genome study of the fungal pathogen *Mycosphaerella graminicola*, we analyzed microsynteny between de novo assembled genomes generated from Illumina paired-end reads (8). We used output from a nucmer alignment (6) to assess the extent that contigs where broken up when aligned to a reference genome. The colinearity of contigs and chromosomal fragments in the reference sequence of *M. graminicola* (9) was in our study defined as microsynteny (7, 8). The mummer software generated a table listing coordinates of all aligned fragments from contigs and fragments, and with this list, we ranked aligned fragments into three size categories >100.000 bp, 50.000–99.999, and <49.999 bp. Next, we assessed how many times a contig aligned to one, two, or more chromosomes of the reference genome used for our comparison. While the majority of contigs aligned to only one chromosome as expected, we also saw a number of smaller contigs that aligned to either two or more chromosomes in the reference genome. Such "break-up" of contigs is either due to breakage of synteny, misassembly of contigs, or reference chromosomes. The alignment positions of the "broken" contigs point to regions that merit extra attention in the overall genome comparison, either as being wrongly assigned, or being genuine evidences of structural rearrangements.

Table 1
Suggested software for use in comparative genome studies

Application	Software	Link	Software license	References
Reference assembly	SOAP	http://soap.genomics.org.cn/	Free	(13)
	CLC	http://www.clcbio.com/	Commercial	
	Bowtie	http://bowtie-bio.sourceforge.net/index.shtml	License	(14)
De novo assembly	CLC	http://www.clcbio.com/	Commercial	(13)
	SOAP	http://soap.genomics.org.cn/	Free	
Synteny	Mumer	http://mummer.sourceforge.net/	Open source	(6)
	SOAPindel	http://soap.genomics.org.cn/	Free	
	SOAPsv	http://soap.genomics.org.cn/	Free	
Alignment	Nucmer	http://mummer.sourceforge.net/	Part of Mummer	(6)
	LastZ	http://www.bx.psu.edu/miller_lab/	Open source	(12)
	MultiZ	http://www.bx.psu.edu/miller_lab/	Open source	(11)
	MAUVE	http://asap.ahabs.wisc.edu/mauve/	Open source	(15)
	SOAPaligner		Open source	
	MGA	http://bibiserv.techfak.uni-bielefeld.de/mga/	Free for academic research	(16)
Blast	Blast	http://blast.ncbi.nlm.nih.gov/Blast.cgi	Open source	(20)
	BLAT	http://www.blat.net/	Open source	(21)
Repeat identification	FragmentGluer	http://nbcr.sdsc.edu/euler	Open source	(23)
	LTRharvest	http://www.zbh.uni-hamburg.de/LTRharvest/index.php	Free	(24)
Sequence evolution	MK	http://tree.bio.ed.ac.uk/software/mktest/	Open source	(31)
	CODEML	http://abacus.gene.ucl.ac.uk/software/paml.html	Open source	(32)

3. Genome Alignment

Compared to traditional software for sequence alignment, as for example ClustalW (10), aligning genomes require more complex algorithms that can account for synteny break. Together with the larger size of datasets inherent to the genomic area, this results in considerably more computationally demanding software. The typical output format contains blocks of aligned sequences with coordinates of each aligned fragment. An example of such output format is the Multiple Alignment File (MAF) format, produced by

MultiZ and TBA (11) or LastZ (12). The genome coordinates are fundamental for any further bioinformatic processing of the alignment, as they are required to cross alignment-based analyses with existing annotations like gene tracks.

The choice of a genome alignment program will depend first on the degree of divergence. While the direct mapping of short sequence reads onto a reference genome is problematic for diverged species, it may be the appropriate choice for the rapid and less computationally demanding generation of alignments of highly similar genome sequences. Several applications are available for reference assemblies, e.g., SOAP (13) or Bowtie (14), or as the read mapping procedure incorporated in the CLC Genomic Workbench (see Table 1).

For pairwise comparisons of de novo assembled genome sequences, the programs nucmer in Mummer (6) and LastZ (12) align syntenic sequences. Other applications allow the alignment of multiple genome sequences, for example, MAUVE (15), MultiZ (11), and the MGA multiple genome aligner (16). A possible limitation to the use of these multiple genome aligners is the access to computer facilities that can handle the alignment of fungal genomes of 20–50 Mb.

The underlying algorithm of the LastZ aligner is a stepwise building of local alignments through the search of matching sequences with high scoring values. The highest scoring sets of syntenic alignments are used as anchors to extend local alignments which can contain gaps. The LastZ aligner is a software package that aligns and processes the alignment through the generation of different output formats with different information. For example: LAV files containing information about the alignment "groups" such as contigs, scaffolds, or chromosomes and MAF files (see above) providing information about all coordinates and define minus and plus strands of the sequences. It is valuable for the analyses of particular aligned fragments, but requires specific consideration in the parsing for issues of forward and reverse strand directions and in particular for fragments with rearrangements and indel mutations. R-dotplot output from LastZ can, as the Mummer software, allow the plotting of syntenic blocks and lastly outputs can be fasta alignment files that can serve as input files for several other programs.

4. Proteome Alignment

Comparative genome studies can also rely only on proteome alignments as in the comparison of the maize smut pathogen *Ustilago maydis* and a related smut fungus also on maize *Sporizorium relianum* (17). All predicted open reading frames were translated and aligned and divergences at the amino acid level were compared. A number of algorithms are available for the alignment of protein sequences and for the search of homologous sequences either in local or public databases. Such procedures involve a clustering of the sequences into families of homologous genes, using a distance measure such as the blast score (see for example ref. (18)). Creating such gene families for newly generated genome data can be highly computer demanding. Therefore, the use of existing gene families as "kernels" can be a good starting option for new datasets.

5. Identification of Genes and Regulatory Sequences Using Genome Comparisons

Comparative genomics empowers the annotation of genomes by confirming existing annotations, identifying new open reading frames, and potentially also discovering regulatory sequences (19). Nucleotide conservation in coding or regulatory sequences can be recognized as regions with a lower extent of divergence and thereby be distinguished from noncoding sequences which have evolved under less selective constraints. The potential to identify complete gene structures from their similarity with genes in related species decreases with divergence time. On the other hand, the false-discovery rate increases since the divergence in noncoding sequences will differ very little from the divergence in coding sequences for closely related species. For searching homologous sequences between genomes, the BLAST package from the NCBI homepage can perform searches of related nucleotide sequences (blastn), translated nucleotide sequences (tblastn), and searches of related protein sequences (tblastx) (20). Furthermore, the BLAST-like software BLAT uses an improved algorithm to similarly identify homologous sequences from genome sequences (21). BLAT is developed particularly for efficient and fast searches in large datasets such as full genomes.

Kellis et al. used a comparative genome approach in yeast to revisit the genome annotation in *Saccharomyces cerevisiae* and to identify regulatory sequences (19). Their approach included a computational test to evaluate reading frame conservation. The test provides a score for each predicted reading frame according to its conservation when aligned to an orthologous sequence in another yeast genome. The score is based on the proportion of the open reading frame over which the reading frame is locally conserved in each of the species. Each species with an orthologous alignment cast a vote for accepting or rejecting the open reading frame. The comparative approach was also used to refine gene structure prediction including boundaries of exons and translation start and stop and intergenic regulatory sequences.

C. Genome Evolution

One of the major goals of evolutionary comparative genomics is to understand how and why genomes have diverged. Analyzing the patterns of sequence divergence along a genome alignment is the key to unravel the underlying demography and selection scenarios in extant and ancestral species. This is achieved by looking at the variation of sequence patterns along genome alignments, either by cutting the global alignment into smaller parts, or by sliding a window along it. The size of the segments or windows has to be determined according to the phenomenon of interest and the supposed scale at which its effect is visible and putative technical constraints of the method used to infer it. For a fine scale detection of "hot spots" of a given parameter, small window sizes of 1 kb can be appropriate while other parameters such as frequencies of repetitiveness can be described in windows of 10–25 kb or more depending on the size of the chromosome.

In some cases, it might be necessary to restrain the analysis to a certain subset of the genome. For highly diverged species, for instance, it may be impossible to align nucleotide sequences and instead protein sequences can, as mentioned above, be compared. With a proteome dataset, the evolutionary analysis can focus on changes of the amino acid sequence, the genomic position of the gene encoding the protein, and the associated structural changes.

We here focus on the evolutionary analysis of fully aligned nucleotide sequences and we distinguish between overall descriptions of genome evolution and more specific detection of natural selection in protein-coding genes.

7. Chromosomal Patterns

In our comparison of 13 *Mycosphaerella* genomes, we conducted several analyses for looking at various parameters along genomes (7) (see Fig. 3). They can be sorted in three categories: (1) single genome analyses, (2) pairwise genome analyses, and (3) multiple species comparisons.

Parameters assessed from single genome analyses:

- Repetitive DNA based on repeat identification and repeat masking of genome sequences. Available and user-friendly softwares for the identification of repetitive sequences include RepeatScout (22), FragmentGluer (23), and LTRharvest (24), and for the masking of genome sequences, the program RepeatMasker (see Table 1). Annotating such low-complexity sequences is important because they are likely to exhibit distinct

Chromosome 1

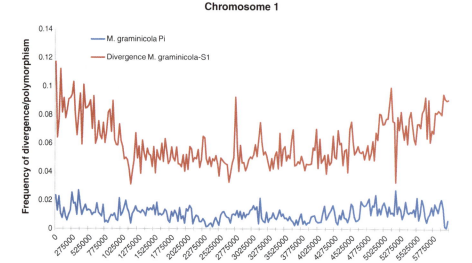

Fig. 3. Overview of mutation frequencies on chromosome 1 in the alignment of three *Mycosphaerella graminicola* sequences and one sequence from a closely related species termed S1. The sliding window illustrates the frequencies of polymorphisms between aligned *M. graminicola* chromosomes (*below line*) and frequencies of substitutions between *M. graminicola* and S1 (*above line*).

mutation patterns and lead to assembly and alignment artifacts, and therefore may be filtered out for particular analyses that rely on nonambiguous homology between the compared sequences.

- Observed GC content calculated in given windows as the proportion of G and C nucleotides. The GC content is an important feature of genomes. It is notably linked to recombination, via the so-called Biased Gene Conversion (BGC) phenomenon, which was demonstrated in yeast and detected in many genomes (25). A related quantity of interest is the equilibrium GC content, noted GC*, that is the GC proportion to which genomes evolve. The computation of GC*, however, requires the use of a substitution model and a maximum likelihood approach. Recombination rate has been demonstrated to be strongly correlated with GC* (26), a result that we also find in our analysis of the fungal genomes (7).

- Exon density counted as the frequency of exonic sites in a given window. For each gene predicted in the reference sequence of *M. graminicola*, we additionally registered the number of aligned codons. The aligned genes were filtered according to the presence of internal stop codons, mutations in start or stop codons, and the percentage of aligned sequence compared to length of the predicted open reading frame.

Parameters assessed from pairwise genomic comparisons of two species:

- Substitutions defined as fixed mutations between species. Substitutions can further be divided into transitions (A↔G or C↔T) or transversion (A↔T, A↔C, C↔G or G↔T). A quantity of interest is the ratio transitions/transversions.

Parameters assessed from multiple genome comparisons:

- For datasets with population sampling, we can infer the amount of polymorphisms defined as variable sites within species. Polymorphisms, shared polymorphisms, and substitutions (see below) can further be distinguished as residing in a coding or noncoding sequence. Coding sequences further allow distinguishing nonsynonymous and synonymous polymorphisms.

- Shared polymorphisms defined as sites that are polymorphic in two or more species.

- Substitutions between species (see above).

- Rates of nonsynonymous and synonymous mutations defined as Ka and Ks (between species) and Pa and Ps (within species) (see below) (27). The rates were calculated per window according to the number of nonsynonymous and synonymous sites (NS and SS) and nonsynonymous and synonymous mutations per window containing coding regions.

- Estimations of ancestral demographic parameters such as effective population size, speciation times, and broadscale ancestral recombination rate, using the CoalHMM software (28) on 100 kb segments of the genome alignment of four species.

By locating the windows and segments on the chromosomes of a given species, one can infer spatial patterns. For calculations based on an alignment involving more than one genome, such localization only makes sense if the synteny is globally conserved between the compared genomes. We found that substitution rates were highly correlated with the physical position on the chromosomes, more specifically with the distance from the telomeres. Windows closer to the telomeres have higher rates of substitutions and more repetitive content (8). Intragenic synonymous and nonsynonymous substitution rates were similarly correlated with the distance to chromosomes ends, and also to exon density, which we interpreted in terms of selective forces.

8. Gene Evolution

For the comparison of aligned predicted open reading frames, we generated tables for each predicted gene listing a number of parameters. The tables served as template for a number of different evolutionary analyses and as an overview of information for the aligned sequences. We used Protein ID as the main identification number. The tables generated from the *Mycosphaerella* genome alignment included:

- Gene/protein ID.
- Chromosome number/contig information.
- Coordinate information.
- Predicted number of codons in the gene.
- Number of aligned codons.
- Number of nonsynonymous and synonymous sites (NA and NS) (27).
- Number of nonsynonymous and synonymous mutations (between species and/or within species if more individuals from the same species are sequenced) (29–31) (see Table 2).
- Rates of nonsynonymous and synonymous mutations defined as Ka and Ks and the ratio Ka/Ks (between species) and Pa and Ps and the ratio Pa/Ps (within species) (see references above) and finally, if applicable, other relevant information (for example annotation, GO/KOG term, signal peptide prediction, etc.).

8.1. Grouping Genes into Categories

Specific categorizations of genes can additionally be included. For pathogenic or symbiotic fungi, proteins potentially involved in fungus–host interactions can be predicted as secreted proteins. Depending on the question of interest, genes can also be divided into Gene Ontology (GO) classes that allow the comparison of particular groups such as cell wall degrading enzymes or sugar metabolism. Being interested in proteins potentially interacting with host molecules, we focused on the groups of genes encoding signal peptides. We could show that this group of genes indeed evolves faster through a higher accumulation of nonsynonymous mutations or a reduced fixing of synonymous mutations measured as a higher Ka/Ks ratio (7, 8).

8.2. Detection of Positive Selection

The genes which have evolved under balancing selection or directional selection in different species can be detected by analysis of ratios of nonsynonymous to synonymous mutations. Of the nonsynonymous mutations, neutral and deleterious mutations predominate in frequency, while the advantageous mutations only occur at low frequencies and therefore are more difficult to detect. We estimate that in the *M. graminicola* genomes, 35% of amino acid changes are fixed by adaptive evolution and positive selection (7, 31). To specifically identify those genes where positive selection has acted,

Table 2
Example of the type of counts and estimates which are required for analyses of adaptive evolution and positive selection

ORF

Within-species comparison

Protein Chr id	Feature	Pos. from	Pos. to	Direction	Aligned codons M. graminicola	KaKs M. graminicola	Ka M. graminicola	Ka M. graminicola	nd M. graminicola	ns M. graminicola	N M. graminicola	S M. graminicola
1 33309	Gene	130943	131722	–1	260	0.0411	0.0056	0.1360	3.3333	25.0000	596.1666	183.8333
1 88592	Gene	132334	134113	–1	316	1.4094	0.0152	0.0108	10.6667	2.6667	701.0002	246.9999
1 29736	Gene	134457	134624	–1	56	0.1661	0.0053	0.0318	0.6667	1.3333	126.1111	41.3889

Between-species comparison

Aligned codons M. graminicola-S1	KaKs M. graminicola-S1	Ka M. graminicola-S1	Ks M. graminicola-S2	nd M. graminicola-S1	ns M. graminicola-S1	N M. graminicola-S1	S M. graminicola-S1
260	0.0566	0.0173	0.3062	10.3333	56.3333	595.9999	183.9999
231	1.1786	0.1355	0.1150	69.5333	20.6667	513.2221	179.7777
56	0.0138	0.0053	0.3820	0.6667	16.0000	126.1111	41.8389

see also ref. (19)

nd counts of non-synonymous mutations; sd counts of synonymous mutations; N Non-synonymous sites; S Synonymous sites

we applied the Nei Gojobori method where the rates Ka and Ks are compared by a Z-test (27):

$$Z \text{ - test} : (Ka - Ks) * \sqrt{(\text{Var } Ka + \text{Var } Ks)},$$

where

$$\text{Var } Ka : Ka * (1 - Ka) / \text{NA},$$

$$\text{Var } Ks : Ks * (1 - Ks) / \text{NS}.$$

Test values, where $Z > 1.65$, are significant with $P < 0.05$. The Z-test is highly conservative and only detects genes with a very strong evidence of positive selection and a much higher average rate of nonsynonymous mutations. Genes where only few codons are positively selected may not be significant by the test of Nei and Gojobori. Therefore, a codon-based model as implemented in the CODEML software from the PAML package can be used (32). CODEML requires fully aligned genes in phylip-like format. Parsing of the aligned gene dataset is therefore required to acquire a filtered input file in this particular sequence format. Such approaches provide more detailed measures of evolutionary parameters. The input dataset can additionally be applied to other models in CODEML such as branch models which compare evolutionary rates in the different species (33, 34) (see below).

9. Comparing Within-Species and Between-Species Variation

For population genomic datasets, the identification of positive selection can be further extended by comparing within-species variation to between-species variation. The McDonald-Kreitman test uses counts of nonsynonymous and synonymous polymorphisms and substitutions to contrast ratios within and between species (35). The test builds on contingency tables of each gene listing the number of nonsynonymous and synonymous polymorphisms vs. substitutions (see Table 2). A G-test can be used to assess the statistical difference in distribution of mutations (36). Under neutral evolution, the ratio of nonsynonymous and synonymous polymorphisms and substitutions should be the same while genes positively selected between species will show an excess of nonsynonymous substitution. For this kind of test that involves the comparison of thousands of genes, it is important to correct for multiple testing. Indeed, if 1,000 genes are analyzed, 50 of them on average are expected to be significant with a p-value threshold of 0.05, just by chance. This means that if the lowest p-value in all tests is, let's say, 0.04, the probability that it is a false positive is actually higher than 4%. The False-Discovery Rate method of (37), as it states, adjust the individual p-values according to a specified, global, false-discovery rate.

9.1. Quantification of Adaptive Evolution

Nonsynonymous mutations in coding genes are deleterious, neutral, or beneficial. Deleterious mutations will be selected against by purifying selection, while the proportion of beneficial mutations will be fixed by selection. The proportion of nonsynonymous divergence attributable to positive selection can be quantified by the parameter α. α contrasts the ratio of nonsynonymous to synonymous polymorphisms (pN/pS) to the ratio of nonsynonymous to synonymous substitutions between species (Ka/Ks) and is calculated as (31):

$$\alpha = 1 - (Ks * pN) / (Ka * pS).$$

The α parameter can be estimated manually for each gene, but can also be computed using the program MK-test (http://tree.bio.ed.ac.uk/software/mktest/) which also allows for the estimation of other evolutionary parameters. For example, it can model the proportion of deleterious mutations as a measure of selective constraints in the genome.

9.2. Estimation of Evolutionary Rates

In a previous study, *Mycosphaerella* pathogens were isolated from agricultural crop hosts and from wild grasses. A key question in our analyses was whether evolution in an agricultural system had increased the evolutionary rate in the wheat adapted *M. graminicola* compared to its "wild" relatives. The dataset included four species forming four branches on an unrooted tree (see Fig. 4). To compare the evolutionary rate among the four branches, we applied a branch model incorporated in the CODEML software (32). To increase the power of the analyses, we generated a dataset of concatenated genes on each chromosome in 100 kb windows. Depending on the number of aligned genes, different window sizes can be chosen and can also be increased to whole chromosomes. The concatenated dataset consists of "supergenes" with nonsynonymous and

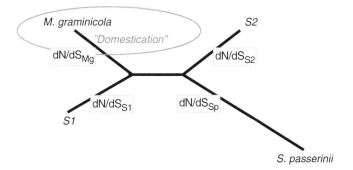

Fig. 4. Unrooted tree of *M. graminicola* and three closely related species: S1, S2, and *Septoria passerinii*. The tree was used to estimate evolutionary rates on each branch in the tree by the program CODEML using a branch model. The evolutionary rate is estimates as the ratio of *d*N/*d*S (nonsynonymous to synonymous mutation rates).

synonymous mutations to be analyzed with the branch model. With CODEML, we first conducted a likelihood ratio test where we compared two models: one with a constant evolutionary rate in all four branches and one where the evolutionary rate is allowed to differ between branches (33, 34). A significantly higher likelihood score for the latter model shows that the four species have evolved at different rates. Next we estimated dN/dS ratios in the four species to identify the potentially significant differences in evolutionary rates of different levels: between species, between chromosomes, and between chunks within chromosomes.

10. Integrating Results

Putting parameters from different analyses together might become cumbersome for such large datasets. The statistical software and programming language R (38) offers a large set of tools for doing so, yet requiring some programming skills, particularly for visualization of the data and output. When full genomes are compared, browsing the data in order to extract features from different sources becomes a computer-expensive task. Databases engines with proper indexing therefore become necessary. A number of different "genome browsers" have been devolved to facilitate the analyses and overview of genome data for biologists with little or no bioinformatics experience. The genome browser at UCSC (http://genome.ucsc.edu/index.html) is one of those, which relies on a MySQL server for database handling. All table data can be browsed remotely or downloaded for local usage. Graphical solutions also exist as standalone software, for example, the previously mentioned CLC Genomics Workbench or the Geneious software (see Table 1). For our analyses, we set up a local MySQL server and developed a dedicated genome browser that allowed the integration and correlation of multiple parameters. Such "global" analyses of parameter correlations are essential to decipher the molecular basis of genome evolution in the fungal species of interest.

The generation of genome data for all types of fungi will in the future provide an invaluable resource to increase our understanding of species biology, diversity, and evolution. Comparative genome approaches will be the tool to extract information from the data and further direct experimental, molecular, and evolutionary studies of fungi.

References

1. Dujon B, Sherman D, Fischer G, Durrens P, Casaregola S, et al. (2004) Genome evolution in yeasts. Nature 430: 35–44.

2. Friesen TL, Stukenbrock EH, Liu Z, Meinhardt S, Ling H, et al. (2006) Emergence of a new disease as a result of interspecific virulence gene transfer. Nat Genet 38: 953–956.

3. Ma L-J, van der Does HC, Borkovich KA, Coleman JJ, Daboussi M-J, et al. Comparative genomics reveals mobile pathogenicity chromosomes in Fusarium. Nature 464: 367–373.

4. Wittenberg AHJ, van der Lee TAJ, Ben M'Barek S, Ware SB, Goodwin SB, et al. (2009) Meiosis Drives Extraordinary Genome Plasticity in the Haploid Fungal Plant Pathogen Mycosphaerella graminicola. PLoS ONE 4: e5863.

5. Poláková S, Blume C, Zárate J, Mentel M, Jarck-Ramberg D, et al. (2009) Formation of new chromosomes as a virulence mechanism in yeast Candida glabrata. Proceedings of the National Academy of Sciences 106: 2688–2693.

6. Delcher AL, Phillippy A, Carlton J, Salzberg SL (2002) Fast algorithms for large-scale genome alignment and comparison. pp. 2478–2483.

7. Stukenbrock EH, Jørgensen FG, Zala M, Hansen TT, McDonald BA, Schierup MH (2010). Whole genome and chromosome evolution associated with host adaptation and speciation of the wheat pathogen Mycosphaerella graminicola. PLoS Genet 6(12): e1001189. doi:10.1371/journal.pgen.1001189.

8. Stukenbrock EH, Bataillon T, Dutheil JY, Hansen TT, Li R, Zala M, McDonald BA, Wang J, Schierup MH (2011). The making of a new pathogen: Insights from comparative population genomics of the domesticated wheat pathogen Mycosphaerella graminicola and its wild sister species. Genome Research. doi: 10.1101/gr.118851.110.

9. Goodwin SB, Ben M'Barek S, Dhillon B, Wittenberg A, Crane CF, et al. (2010). Finished genome of Mycosphaerella graminicola reveals stealth pathogenesis and extreme plasticity. PLoS Genet 7(6): e1002070. doi:10.1371/journal.pgen.1002070.

10. Larkin MA, Blackshields G, Brown NP, Chenna R, McGettigan PA, et al. (2007) Clustal W and Clustal X version 2.0. Bioinformatics 23: 2947–2948.

11. Blanchette M, Kent WJ, Riemer C, Elnitski L, Smit AFA, et al. (2004) Aligning Multiple Genomic Sequences With the Threaded Blockset Aligner. Genome Research 14: 708–715.

12. Schwartz S, Kent WJ, Smit A, Zhang Z, Baertsch R, et al. (2003) Human–Mouse Alignments with BLASTZ. Genome Research 13: 103–107.

13. Li R, Li Y, Kristiansen K, Wang J (2008) SOAP: short oligonucleotide alignment program. Bioinformatics 24: 713–714.

14. Langmead B, Trapnell C, Pop M, Salzberg S (2009) Ultrafast and memory-efficient alignment of short DNA sequences to the human genome. Genome Biology 10: R25.

15. Darling AE, Mau B, Perna NT progressive-Mauve: Multiple Genome Alignment with Gene Gain, Loss and Rearrangement. PLoS ONE 5: e11147.

16. Höhl M, Kurtz S, Ohlebusch E (2002) Efficient multiple genome alignment. Bioinformatics 18: S312–S320.

17. Schirawski J, Mannhaupt G, Karin M, Brefort T, Schipper K, et al. (2010), Pathogenicity determinants in smut fungi revealed by genome comparison. Science 10;330(6010): 1546–1548.

18. Penel S, Arigon A-M, Dufayard J-F, Sertier A-S, Daubin V, et al. (2009) Databases of homologous gene families for comparative genomics. BMC Bioinformatics 10: S3.

19. Kellis M, Patterson N, Endrizzi M, Birren B, Lander ES (2003) Sequencing and comparison of yeast species to identify genes and regulatory elements. Nature 423: 241–254.

20. Altschul S, Madden T, Schaffer A, Zhang J, Zhang Z, et al. (1997) Gapped BLAST and PSI-BLAST: A new generation of protein database search programs. Nucleic Acids Res 25: 3389–3402.

21. Kent WJ (2002) BLAT—The BLAST-Like Alignment Tool. Genome Research 12: 656–664.

22. Price AL, Jones NC, Pevzner PA (2005) De novo identification of repeat families in large genomes. Bioinformatics 21: 351–358.

23. Pevzner PA, Tang H, Tesler G (2004) De Novo Repeat Classification and Fragment Assembly. Genome Research 14: 1786–1796.

24. Ellinghaus D, Kurtz S, Willhoeft U (2008) LTRharvest, an efficient and flexible software for de novo detection of LTR retrotransposons. BMC Bioinformatics 9: 18.

25. Galtier N, Duret L (2007) Adaptation or biased gene conversion? Extending the null hypothesis of molecular evolution. Trends in Genetics 23: 273–277.

26. Duret L, Arndt PF (2008) The Impact of Recombination on Nucleotide Substitutions in

the Human Genome. PLoS Genet 4: e1000071.

27. Nei M, Kumar S (2000) Molecular Evolution and Phylogenetics. Oxford: Oxford University Press.

28. Dutheil JY, Ganapathy G, Hobolth A, Mailund T, Uyenoyama MK, et al. (2009) Ancestral Population Genomics: The Coalescent Hidden Markov Model Approach. Genetics: genetics.109.103010.

29. Eyre-Walker A (2006) The genomic rate of adaptive evolution. Trends in Ecology & Evolution 21: 569–575.

30. Smith NGC, Eyre-Walker A (2002) Adaptive protein evolution in Drosophila. Nature 415: 1022–1024.

31. Welch JJ (2006) Estimating the Genomewide Rate of Adaptive Protein Evolution in Drosophila. Genetics 173: 821–837.

32. Yang Z (2007) PAML 4: Phylogenetic Analysis by Maximum Likelihood. Mol Biol Evol 24: 1586–1591.

33. Yang Z, Nielsen R (2000) Estimating Synonymous and Nonsynonymous Substitution Rates Under Realistic Evolutionary Models. Mol Biol Evol 17: 32–43.

34. Yang Z, Nielsen R (2002) Codon-Substitution Models for Detecting Molecular Adaptation at Individual Sites Along Specific Lineages. Molecular Biology and Evolution 19: 908–917.

35. McDonald JH, Kreitman M (1991) Adaptive protein evolution at the Adh locus in Drosophila. Nature 351: 652–654.

36. Woolf B (1957) The log likelihood ratio test (the G-test) Annals of Human Genetics 21: 397–409.

37. Hochberg YBaY (1995) Controlling the False Discovery Rate: A Practical and Powerful Approach to Multiple Testing. Journal of the Royal Statistical Society Series B 57: 289–300.

38. Team RDC (2008) R: A language and environment for statistical computing. R Foundation for Statistical Computing. Vienna, Austria.

Chapter 34

Multigene Phylogenetic Analyses to Delimit New Species in Fungal Plant Pathogens

Tara L. Rintoul, Quinn A. Eggertson, and C. André Lévesque

Abstract

Supporting the identification of unknown strains or specimens by sequencing a genetic marker commonly used for phylogenetics or DNA barcoding is now standard practice for mycologists and plant pathologists. Does one have a new species when a strain differs by a few base pairs when compared to reference sequences from taxonomically well-characterized species that do not differ morphologically from this new strain? If variation at the intra- and interspecific levels for the locus used for identification is already understood for all the closely related species, it is possible to make a reliable prediction of a new species status, but ultimately this question can only be properly addressed by determining the presence or absence of gene flow among a group of strains of the putative new species and strains of previously delimited species. The Phylogenetic Species Concept (PSC) and its assessment using multigene phylogeny and Genealogical Concordance Phylogenetic Species Recognition (GCPSR) are the basis for this chapter. The theoretical framework and a variety of tools to apply these concepts are explained, to assist in the assessment of whether a species is distinct or new when confronted with some sequence divergence from reference data.

Key words: Species trees, Genealogical concordance, Phylogenetic species concept, Fungal speciation, DNA barcode

1. Introduction

Precise application of refined species concepts combined with accurate species identification affects the academic world of systematics and evolutionary biology, but has tangible consequences for society (1). Many fungal plant pathogens, i.e., true fungi [kingdom Eumycota] and fungal analogs such as oomycetes [kingdom Straminipila], significantly impact the global economy, food supply, and human health. Closely related species can differ greatly in their virulence and the level of threat they pose to agriculture, forestry,

Melvin D. Bolton and Bart P.H.J. Thomma (eds.), *Plant Fungal Pathogens: Methods and Protocols*,
Methods in Molecular Biology, vol. 835, DOI 10.1007/978-1-61779-501-5_34, © Springer Science+Business Media, LLC 2012

or human and animal health. Specific and precise species level identification of these pathogens is essential for effective and appropriate management of disease control, quarantine regulations, free trade between nations, and the conservation of ecosystems (2).

Filamentous fungal plant pathogens can be characterized by their growth pattern of hyphal networks, which break down plant tissue and absorb nutrients by osmotrophy (3). Included in this group are Fungi, i.e., true fungi, and representatives of the oomycetes, sharing consistent morphological and ecological characteristics, although they represent evolutionarily distinct groups (4). There are also other economically important fungal plant pathogens that do not produce a filamentous network and that obtain their nutrients from a close association with the host cells (e.g., Erysiphales or *Synchytrium* species). Determining species limits in these organisms is often difficult because they lack distinct morphological characters and easily monitored or consistent mating behaviors. This chapter gives a comprehensive review of current literature regarding species recognition in fungal plant pathogens and outlines techniques and resources that are available for this purpose.

Determining the limits of species is a two-step process. One must first have a theoretical definition of what separates one species from another, a *species concept*. Secondly, one must have an operational method of recognizing a species under this concept, *species recognition*. Currently, the most widely accepted species concept is the Evolutionary Species Concept (ESC), which asserts that species should represent "… a single lineage of ancestor descendent populations which maintains its identity from other such lineages and which has its own evolutionary tendencies and historical fate" (5). The ESC may sound clear-cut, but unfortunately there is no way to test it directly. Instead, the ESC encompasses three other species concepts that are operationally testable: the Morphological Species Concept (MSC), the Biological Species Concept (BSC), and the Phylogenetic Species Concept (PSC).

The MSC is the traditional concept used for species recognition in fungal plant pathogens, functioning by operational Morphological Species Recognition (MSR). Using this method, morphological and functional characteristics are examined to determine species delineation. Morphological characters include gross phenotypic characters, ultrastructure, growth habit, and optimal growth conditions. Functional characters, which are assessed with physiological and biochemical techniques, include secondary metabolite profiles, ubiquinone systems, fatty acid composition, cell wall composition, protein composition, and pathogenicity. Although the MSC has been used extensively for species diagnosis of fungal plant pathogens, its usefulness is limited because many of these organisms are morphologically cryptic, with relatively few diagnosable characters. Additionally, many fungal plant pathogens have morphologically distinct life stages and the presence of diagnostic characters can vary among strains. As a result, accurate identification of species defined by the MSC requires taxonomists with significant expertise who are usually specialized to identify only a small subset of all fungal plant pathogens. As a consequence of its historical dominance in taxonomy, the MSC is still a valuable tool, especially if one wants to relate a putative new species to older literature and to herbaria specimens. However, it may be less helpful for the identification of economically relevant species that are difficult to differentiate by morphology (6).

The BSC is useful for detecting genetic isolation between two groups by assessing the ability to produce sexually viable offspring in the Biological Species Recognition (BSR). Unfortunately, fungal plant pathogens display a level of reproductive flexibility not seen in other eukaryotes. Some species have been observed to grow solely vegetatively or only reproduce asexually, while many of the sexual species have the ability to self-fertilize, making the identification of outcross progeny arduous. Additionally, sexual reproduction in fungal plant pathogens is often a response to a stress or to certain environmental conditions which may be impossible to recreate in the lab. Furthermore, successful crossing in the laboratory may not indicate actual outcrossing in nature, for example if the compatible populations are geographically isolated. Although the BSC can be very useful in groups that are obligate outcrossers, it cannot be used to define species in many economically important fungi.

The PSC has long been debated, with many variations and definitions being discussed in the literature. Competing definitions and criteria have been hotly debated with no consensus being reached (7–10). For our purposes, we will utilize the definition of the PSC that arises from the standpoint of evolutionary phylogenetics. Under this perspective, species are defined as the smallest monophyletic division of organisms, generally defined as a group of organisms that includes the common ancestor and all of its descendants (1, 11). An addition to the PSC, which can provide clearer methods for recognizing patterns of descent, is the genealogical concordance concept which posits that "extant haplotypes (DNA sequences) present in any 'species' represent the gene lineages that have survived through an organismal pedigree" (12). Using this definition, we can say that we are identifying "genealogical species" under the assumption that these are the basal taxa that occur at the boundary between reticulate and divergent genealogies (11). Advocates of this definition suggest that this interpretation of the PSC is the most sensitive to changes described by the ESC and that genetic changes can be observed long before morphological and behavioral changes evolve (6). These combined criteria provide a method to avoid the bias of determining the limits of a species under the PSC by assessing the concordance of multiple gene genealogies, described as Genealogical Concordance Phylogenetic Species Recognition (GCPSR) (11).

The GCPSR uses multiple gene genealogies to determine phylogenetic topography. These gene topographies are then compared and the concordance of branches supports the presence of separate species, whereas the intersection of conflicting topographies indicates gene flow among individuals below the level of species (6). By using the GCPSR, the subjectivity of recognizing the limits of species groups can be reduced, because one must only identify where the switch from concordance to conflict of topologies exists to define the species limit (13). Similarly, the species limits serve as explicit taxonomic hypotheses that can be further tested with additional data. This results in the GCPSR having the capacity to analyze taxonomic relationships at a higher resolution than other species recognition techniques (6, 14). Applying the GCPSR is the main focus of this chapter. It has been used successfully to identify cryptic species in both plant pathogenic Fungi and the oomycetes (15–22). Such phylogenetic analysis would become part of the new species description that would need to follow the International Code of Botanical Nomenclature (ICBN). Seifert and Rossman (23) recently reviewed the formal requirements and the currently recommended practices to describe a new species. Once species are well established under a robust phylogenetic framework, it becomes possible to identify individual strains or specimens with a more simple approach such as DNA barcoding, whereby only one or a few markers are sequenced for identification (24, 25).

2. Selection of Loci for Analysis

The first step in implementing the GCPSR is the identification and/or selection of neutral, unlinked, single copy loci that are able to accurately represent the evolutionary history of the taxon being examined. A major criticism of identifying species with phylogenetic methods is that single gene trees are only a proxy for species trees, and that the evolution of a specific gene or locus may not always represent the organism as a whole. Because of this, the use of a single gene or locus for species delimitation has been replaced by the GCPSR, which uses multiple, independently assorting or unlinked loci to more accurately depict a species' evolutionary history. This minimizes the bias caused by concerted selection pressure on a single region of the genome. Before selecting loci for inclusion in the GCPSR, characteristics of the gene regions must be considered to avoid potential errors and biases in the estimated species tree. Table 1 provides a summary of loci currently being used within the GCPSR framework for fungal plant pathogens.

Methods for identifying novel DNA markers are well summarized by Thomson et al. (26) who describe three marker types: nuclear protein coding loci (NPCL), exon-primed intron crossing

Table 1
Gene loci already in use for multigene phylogenies of fungal plant pathogens

Loci	Other names	Taxonomic group	References
18S ribosomal RNA gene, nuclear small subunit ribosomal RNA	SSU	Ambrosia fungi, *Phyllosticta*, Dothideomycetes, Kingdom Fungi, *Mycosphaerella*, *Neofabraea*	(22, 59, 109–112)
28S ribosomal RNA gene, nuclear large subunit ribosomal RNA	LSU	Ambrosia fungi, *Phyllosticta*, *Ophiostoma*, Dothideomycetes, Kingdom Fungi	(59, 109–111, 113)
Internal transcribed spacer	ITS	*Cryphonectria*, *Pyricularis*, *Cylindrocladium*, *Pleospora*, *Phyllosticta*, *Ophiostoma*, Kingdom Fungi, *Mycosphaerella*, *Neofabraea*	(18–20, 22, 109, 111–114)
Elongation factor 1 alpha	EF1α, TEF1	*Sclerotinia*, *Pleospora*, *Phytophthora*, *Calonectria*, *Fusarium*, Anther smut, Kingdom Fungi, *Fusarium avenaceum*, *Fusarium poae*	(16, 19, 61, 111, 115–119)
RNA polymerase II largest subunit	RPB1, RPB220	Kingdom Fungi	(111)
RNA polymerase II subunit 2	RPB2, RPB150	Dothideomycetes, Kingdom Fungi	(110, 111)
Beta-tubulin	bTUB, benA	Ambrosia fungi, *Cryphonectria*, *Pyricularia*, *Cylindrocladium*, *Phytophthora*, *Calonectria*, *Fusarium*, Anther smut, *Neofabraea*, *Fusarium avenaceum*	(16, 18, 20, 22, 59, 61, 114, 116–118)
Calmodulin	CaM, CMD1	*Pyricularia*, *Sclerotinia*	(18, 115)
Actin		*Pyricularia*, *Mycosphaerella*	(18, 112)
Intergenic spacer region of ribosomal cistron	IGS	*Sclerotinia*, *Fusarium avenaceum*	(115, 118)
Glyceraldehyde-3-phosphate dehydrogenase	GPD, TDH	*Pleospora*	(19)
NADH dehydrogenase subunit 1	NAD1	*Phytophthora*	(116)
COX1		*Phytophthora*, *Mycosphaerella*	(112, 116)
Histone H3	HHT	*Calonectria*	(117)
Mitochondrial small subunit rDNA	mtSSU	*Fusarium*, *Fusarium poae*	(16, 119)
Ammonia ligase		*Fusarium*	(16)
Trichothecene 3-O-acetyltransferase	TRI101	*Fusarium*	(16)
Phosphate permase	PHO	*Fusarium*	(16)
Putative reductase	RED	*Fusarium*	(16)
Gamma tubulin	TUBG	Anther smut	(61)
Trichodiene synthase	Tri5	*Fusarium poae*	(119)

markers (EPIC), and anonymous nuclear markers (ANM). Each marker is described based on its appropriateness for different applications, such as phylogeny, phylogeography, population genetics, and linkage mapping. They also assess techniques for identifying novel markers, keeping in mind the resources that are actually available for identification and application of these markers. Generally, expressed sequence tags (ESTs) and genome analyses can be used for NPCL and EPIC, whereas polymerase chain reaction (PCR) techniques that amplify unknown parts of the genome, such as AFLP or microsatellites, can be used as ANM. For species level phylogenetic inference, Thomson et al. (26) suggest that ANM or EPIC make optimal markers because they offer the highest level of sequence variability. However, it is possible to identify highly variable NPCL markers when dealing with closely related or recently derived lineages. NPCLs with a high rate of single nucleotide polymorphisms (SNPs) can be identified and represent a good option for markers that would generate a single undisputable alignment across a species complex. Noncoding RNA, i.e., genes that get transcribed but not translated into proteins, is another category of markers. Nuclear large subunit rRNA (LSU, or 28S) has been used extensively for phylogenies at the species level. Ultimately, the available resources one has access to, including reference data sets and available sequencing technology, determine what methods can be used for the identification of novel markers. The publicly available plant pathogen genomes are very valuable resources for locus selection. These are cataloged in a frequently updated list at the website of the Comprehensive Phytopathogen Genome Warehouse: http://cpgr.plantbiology. msu.edu/cgi-bin/warehouse/cpgr_warehouse.cgi.

When choosing loci, one must identify candidates that are independently inherited and physically unlinked. This removes the bias of concerted selection pressure on a single genomic region and increases the chances of constructing accurate species trees (12, 27). Ideally, potential markers should be located on separate chromosomes. This can be verified if a fully annotated genome with chromosome maps is available, which allows marker's physical positions to be identified. When a genome has not been fully annotated but has been assembled into scaffolds, one should select markers that are identified on different scaffolds of large sizes, because this will decrease the chance of the markers being linked and inherited together. If a genome of the species of interest is not available, one can use genomic data from a closely related taxon under the assumption that it should have a similar gene topology to the species complex being investigated.

Another potential source of error occurs when a chosen marker is present in multiple copies within the genome; this can result in incorrect interpretation of homology during tree construction. Multiple peaks in the electropherograms or chromatograms occur when primers coamplify the multiple copy elements. When this data is subjected to further analysis, these multiple peaks may be interpreted as heterozygous SNPs instead of variations among different copies of a gene. In other words, multiple copies of a marker make it impossible to determine if variation at a nucleotide is a result of copy differences (variation between the copies of the gene in a single organism) or sample differences (having a heterozygous individual). The ability to identify heterozygous nucleotides is important when dealing with diploid or multinucleate organisms because it indicates a hybridization event or gene flow that is indicative of an interbreeding population or species. Therefore, if a double peak is detected because of gene copy differences but interpreted as a hybridization event, species numbers could be underestimated. Multiple copies of a gene have been reported in the genera *Fusarium* for beta tubulin and ITS2 (15, 28), proposed in *Phytophthora* for elongation factor 1 alpha (29), and also observed in *Pythium ultimum* for elongation factor 1 alpha (unpublished). There are three methods for detecting paralogous markers (duplicated genes), which were summarized by Thomson et al. (26) as: (1) similarity searches, (2) phylogenetic tests, and (3) characteristics of the markers themselves. Ideally, the use of paralogous markers should be avoided; however, if they must be used, the potential errors in species inferences should be noted during analysis and interpretation of resulting species tree estimates.

Phylogenetic analyses assume that incorporated markers are under stochastic and neutral selection pressures, but in reality many gene regions are under strong negative or positive selection (30). Incorporating genes that are under such selection pressures into the GCPSR can result in a competition between ancestral signal and functional or structural signals, with the latter overriding the former. Therefore, when assessing the utility of a proposed locus for the GCPSR, it is important to test for neutrality of selection. Luckily, recent computer programs have been designed to do just that (31–33). One such program, the KaKs calculator, is available at www.yale.edu/townsend/software.html. Most deep phylogenies are based on fairly conserved genes that are neutral but lack sufficient resolution to resolve a species complex. It can be difficult to identify neutral markers that provide enough variation to differentiate strains and populations, and one should be aware of the possibility of selection pressure introducing artifacts into a phylogenetic estimation.

The phylogenetic informativeness of any gene region can be quantified using the statistical tests presented by Townsend (34). These tests have been applied in the ascomycetes comparing protein genes to ribosomal genes (35). Another database tool for identifying genes in fungi with a high level of phylogenetic informativeness can be found at http://genome.jouy.inra.fr/Funybase/ (36).

Using single mitochondrial markers for species delimitation or identification has been quite common in recent years, notably the barcoding gene cytochrome *c* oxidase subunit 1 (COI). However, there has been significant controversy about the validity of mitochondrial gene phylogenies for determining speciation (37–40). These papers argue that mitochondrial genes are under different evolutionary forces than nuclear genes and therefore do not represent the evolution of the species as a whole. This argument is based on the fact that maternal inheritance of mitochondria, as demonstrated by Martin (41) in the plant pathogen *Pythium sylvaticum*, does not allow for genetic recombination. The absence of recombination, along with uniparental inheritance of the organelle, means that a hybridization event late in the speciation process could cause introgression of the mitochondrial genome, which could be maintained long after the introduced nuclear alleles are lost within the species. Mitochondrial introgression would result in conflicting species estimates extrapolated from nuclear and mitochondrial gene trees. With these limitations in mind, the use of mitochondrial genes may be more appropriate for distinguishing between well-resolved, higher level taxonomic groups. Nonetheless, we do not explicitly condemn the use of mtDNA in the GCPSR, but strongly suggest that mitochondrial loci be used in combination with multiple nuclear genes and be analyzed with the mentioned limitations in mind.

3. Choosing Strains Representative of Taxa

Although this is not a rule of the ICBN, one should avoid describing a new species based on a single strain or a single gene region (23). Using multiple genes allows for a more accurate depiction of the organism as a whole, whereas assessing multiple strains allows a more comprehensive look at the genetic make-up of the entire group in question. Sampling design of both is critical to correctly infer species boundaries, because it applies to choosing loci and individuals. Both must therefore be selected carefully to reduce potential biases (42, 43).

At the outset of such a taxonomic experiment, it is important to include the largest possible set of available strains of the taxa being analyzed and confirm species level identifications by assessing morphological characteristics. Species identification can be further verified by sequencing a commonly used DNA marker for a given genus and performing BLAST searches on GenBank (www.ncbi.nih.gov/blast) or the DNA Barcode database (www.BOLNET.ca). When dealing with a species complex, one may encounter conflicts in the identification of a strain between morphological characters and a DNA marker or between two different DNA markers; these anomalous strains are often critical for determining species limits and are important to include in a GCPSR analysis. If dealing with a relatively rare group, all available strains should be included in the analysis to provide an adequate sample size. However, when dealing with more common plant pathogens, a subset of strains should be selected that incorporates the whole spectrum of morphological variation, geographical distribution of the putative novel species, ex-type strain, strains from the center of origin, and ex-types of closely related species. Ex-types represent a living isolate that was obtained from the holotype from which a new species was described when the holotype itself is a dried specimen or a culture permanently preserved in a metabolically inactive state that is the ultimate reference for the species. The inclusion of ex-types is necessary for comparisons to be drawn to both the history of a species name and the original species circumscription (44–46). It is important to avoid overrepresenting a category in the range of characters sampled for inclusion. Avoid including multiple strains from the same site, the same host, or with identical morphology to avoid skewing your results based on sampling (47). If a new species is found, a type specimen, or acceptable element as determined by the ICBN, must be selected and preserved and normally would be both morphologically and genetically representative of the new species. A living ex-type culture should also be preserved when possible.

4. DNA Isolation, Loci Amplification, and Sequencing

The amount of DNA required depends on the number of loci to be included in the study. Enough DNA should be extracted to provide adequate template for multiple PCR amplifications for each targeted locus. For fungal plant pathogens, a recommended protocol is that of Möller, which is fairly simple and cost effective while capable of isolating high molecular weight DNA (48). There are also many commercially available DNA isolation kits that work well on most plant pathogens for PCR amplification and sequencing. However, many plant pathogens are obligate parasites, making it difficult or impossible to grow them in culture. This in turn limits the potential to obtain enough biomass for DNA extractions and often DNA extractions are of very low quality and quantity. Whole genome amplification is a technique that has been successfully applied to these kinds of organisms (49, 50). Wang et al. (50) were even successful at amplifying the genome of *Puccinia striiformis* f. sp. *tritici* from single spores. Another issue with obligate plant pathogens is the difficulty of separating pathogen DNA from the host species. One must always be aware that the amplicon of a new marker being tested could be from a contaminant or host; this can occur because of low DNA concentration of the target or by the use of using primers that anneal within contaminant DNA. This is particularly difficult to assess when amplifying a marker for which there is no data from any species related to the target.

When using PCR to amplify genes for sequencing, use of a protocol that requires minimal amounts of dNTPs and primers eliminates the need for further purification before the sequencing reaction. This reduces processing time and increases sequencing quality because PCR products cannot be lost during purification. For this method, DNA is amplified by PCR in a volume of 10 μL with primer concentrations of 0.08 μM and 0.2 mM of deoxyribonucleoside triphosphates (51).

There are often errors in sequencing runs, and low sequence quality can result in background noise being incorrectly identified as phylogenetically important characteristics such as SNPs or sequence polymorphisms. To reduce the potential effect of these errors in phylogenetic estimates, it is imperative that sequence reads have full bidirectional coverage and are of the highest quality available. Sequence quality can be assessed through PHRED values (www.phrap.com/phred/) (52).

Nevertheless, some apparent sequencing "errors" may indicate evolutionary events. In particular, a single base ambiguity, in a diploid or dikaryotic organism, may represent a historical hybridization event and represent a heterozygous allele. An abrupt shift from a good quality sequencing read to seemingly uninformative noise, perhaps appearing as if the electropherogram has been doubled and shifted slightly, may indicate an insertion or deletion (indel) in only one of the genes alleles. In this scenario, both forward and reverse strands will typically obtain a clear sequence until an indel is reached, whereupon the sequence becomes unreadable. Luckily, sequencing errors caused by allelic variation disappear if the alleles are sequenced separately. Alleles can be isolated by using plasmid vectors to clone individual PCR products (53). Sequences from the isolated alleles can then be obtained by using the cloned fragments. Individual allele sequences should also be included in the data set for phylogenetic analyses, because allelic variation may be caused by a hybridization event between two otherwise genetically distinct clades, and this independent assortment of the alleles would help to define species boundaries.

4.1. Primer Design

Sufficient care and verification of sequencing primers is necessary to select primers that are optimized for delineation. Primers should be designed on regions conserved within the taxonomic group. These regions can be identified by aligning a taxonomically representative set of gene sequences. In particular, when working with obligate parasites, it is necessary to use primers with some specificity to avoid amplification of contaminant or host DNA. There are many software programs that can aid in selection of primer sites and identify potential problems such as undesirable primer pairing, including hairpin structures, self-dimer, and primer to primer dimers (54, 55).

6. Multigene Analyses

5.1. Algorithms and Analyses

Under the ESC, all algorithms used for phylogenetic analyses and reconstructions must be based on sound evolutionary assumptions. This excludes algorithms based purely on genetic distance methods. Models that are still applicable include maximum likelihood, Bayesian and parsimony methods. There are innumerable software programs that apply these models to data sets composed of DNA sequences and they are summarized on many websites. Perhaps the most exhaustive is http://evolution.genetics.washington.edu/phylip/software.html with currently 379 described and linked phylogeny programs, along with 54 free webservers for carrying out analyses.

5.2. Gene Concatenation

Historically, the main approach to analyzing multilocus gene data was to produce a concatenated super gene, which is then analyzed as a single data set. However, this may lead to phylogenetic trees that incorrectly represent the species tree by improper inferences (56) and artificially constrain all gene genealogies to the same topology and branch lengths. When each gene is instead analyzed separately, the patterns of each locus can be observed. Because the focus of GCPSR is to produce a species tree based on the concordance of gene tree topologies, this suggests that concatenated data sets should not be analyzed. However, some gene regions lend themselves quite well to concatenation, but to assess that possibility tests for incongruence of gene signal must be carried out. For those cases where concatenation is inappropriate, it mirrors genetic events listed below as causing discordance in gene trees, and these have been described in the literature (57, 58).

There have been several statistical tests developed to test for incongruence of phylogenetic signals between gene loci. The following have been applied to systematic questions in plant pathology (15, 16, 18–20, 59–63): (1) the partition homogeneity test (PHT) also known as the incongruence length difference (ILD) test (64), (2) the nonparametric Wilcoxon signed ranks (WS-R) (65, 66), (3) the Kishino-Hasegawa test (67), and (4) the Shimodaira-Hasegawa test (SH) (68). Problems inherent to each of the tests have been described (57, 66, 68–70). Goldman et al. (68) proposed the Shimodaira-Hasegawa test as the most reliable predictor, which reduces the inference errors possible when congruence testing is used. This test is implemented in PAUP 4.0.

Producing a phylogenetic tree based on concatenated genes can be a good starting point for species level comparisons. However, the results produced must be compared to each individual gene tree, to ensure that there are not biases being produced when all genes are assessed together as a single entity. If upon visual inspection of separate gene trees incongruent nodes are recognized, we would suggest that gene region concatenation is not an option, even if an incongruence test identifies the possibility of concatenation. Concatenation could be supported by a test that had a low probability to detect incongruence, hiding essential information that could clearly demonstrate gene flow.

5.3. Rooted vs. Unrooted Analyses and Outgroup Selection

There are many methods for rooting a phylogenetic tree. Three frequently used methods are well defined in the literature: (1) the outgroup method (71), (2) midpoint rooting (72), and (3) inference of a molecular clock (73). These methods have been recently compared with large data sets from angiosperms (74, 75) and animals (73). Selection of a suitable root or outgroup for phylogenetic analysis is imperative because rooting can severely influence the topology of a tree. Because of this, it is also highly recommended that unrooted trees be produced as a step in investigations of species relationships. Unrooted trees can establish clear hypotheses about the phylogenetic relationships of the organisms included and are useful in species delimitation, although providing no evidence of the direction of evolution within the group. This will also allow the experimenter to judge if the outgroup as defined is creating a bias in the analysis.

6. Interpreting Gene Trees, Assessing Gene Tree Concordance, and Estimating Species Trees

There are five kinds of genetic events that can lead to discordance between individual gene trees. These are (1) incomplete lineage sorting, (2) horizontal gene transfer, (3) gene duplication and loss, (4) hybridization and introgression, and (5) recombination. Species limits exist at the node where the transition from concordance to conflict rests. The most frequently occurring events causing conflict are assumed to be recombination and incomplete lineage sorting, so the point where conflict arises delimits the position where recombination still occurs, pointing to an inclusive species as illustrated in Fig. 1.

The degree of concordance across the gene trees representing multiple loci can be assessed visually in simple cases. Where there is a similar overall topology of clades, it should be possible to recognize trends suggesting recombination within genes. However, this might be difficult in some situations and it does not provide statistical support for the proposed relationships or consider branch length variation among trees. We must then utilize methods that summarize potential gene genealogies so that the species history can be observed and given some measure of support. Recent advances in phylogenetics have produced more methods for measuring the congruence of independent loci into a signal phylogenetic inference, even if each locus represents a unique evolutionary history.

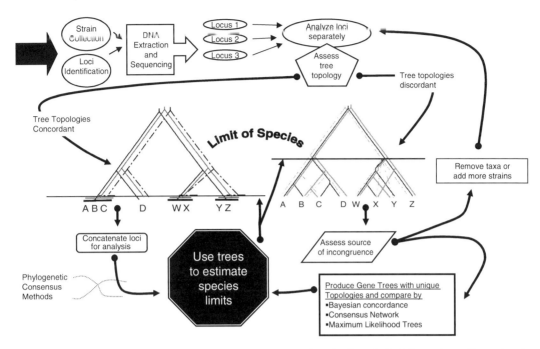

Fig. 1. Diagram outlining the workflow for identifying species using the Genealogical Concordance Concept under Phylogenetic Species Recognition. Gene genealogy scenarios show how the transition from concordance among branches to incongruity among branches can be used to determine species limit. Modified from Johnson and Soltis (108) and Taylor et al. (6).

These methods have been well summarized (42, 70, 76–79), and each has its limits. Some are challenging to implement, some are limited in the number of species they can assess, and others cannot reconstruct an overall species tree within which the gene trees are embedded. There are algorithms implemented in computer programs that can estimate species trees from either a collection of gene trees (79–81) or from a set of sequence alignments for a multilocus data set (57, 82). These are well summarized and discussed in a recent book edited by Knowles and Kubatko (83). We will discuss here methods of analysis translated into software packages that are accessible to users not heavily versed in phylogenetic methods.

6.1. Consensus Network Analysis: Reticulate Relationships

It is difficult to visualize the distribution of multiple gene genealogies so that one can easily recognize the common history among and between gene lineages. One approach is through consensus network analysis, as implemented in the SplitsTree software package (84). If data provided to SplitsTree produce wholly congruent relationships, then the SplitsGraph output will be visually equivalent to a dichotomous tree. But where incongruencies exist, the program will instead produce a tree-like network that can illuminate different and conflicting topologies (85). Data input can be in the format of aligned sequences, with distance measures or tree files in a number of formats. Liu et al. (86) utilized SplitsTree to examine evidence for genetic recombination in the scr74 family of phytotoxin-like effector genes of *Phytophthora infestans*.

6.2. Bayesian Concordance Analysis: Concordance Trees

A number of Bayesian approaches to inference have been developed in the context of the multispecies coalescent theory. This theory assumes that each gene tree represents the relationships between orthologous genes from a sample of individuals of multiple species and that no horizontal gene transfer has occurred between individuals of different species. Under this assumption, several Bayesian approaches to species tree inference have been developed (87).

- BATWING (Bayesian analysis of trees with internal node generation) is a program designed for within-species data, written in C, and available for download from: www.mas.ncl. ac.uk/~nijw/ (88). It was developed to estimate species trees from single gene trees with the inclusion of data estimating times of speciation, population sizes, and growth rates.

- MCMCcoal (89) and its recent update BP&P (90) are both available for download at http://abacus.gene.ucl.ac.uk/software.html. These algorithms are both implemented in C. These packages generate posterior probabilities of species assignments considering uncertainties caused by unknown gene trees and coalescent processes in ancestral populations. A previously estimated species tree must be provided for these computations to occur.

- BUCKy (91) is available for download from www.stat.wisc. edu/~ane/bucky/index.html. Using Bayesian concordance analysis, this program estimates species trees by detecting groups of genes that share the same tree and combining them to gain more resolution at terminal branches on the tree (92). In this model, no assumptions are made about the genetic events contributing to discordance among the gene trees. The underlying notion of a primary concordance tree is well described by Baum (93).

- BEST (Bayesian estimation of species trees) estimates the topology of a species tree based on a Bayesian analysis. It is also capable of estimating divergence times and ancestral population sizes from a set of gene trees, using an importance sampling method (57, 82, 87, 94, 95). The governing model of BEST assumes that any discrepancies between gene lineages or trees is exclusively a result of lineage sorting and the traditional view of gene evolution under the limitations of free recombination between genes, with the absence of recombination within genes. This makes it inappropriate to apply BEST for reconstruction of species trees where there is the chance of horizontal gene transfers or gene duplications and/or deletions (94). It is available for download at: www.stat.osu.edu/~dkp/BEST/introduction/ and works within the phylogenetics package MrBayes (96).

- COAL (97) is a software package that computes probabilities of gene tree topologies by enumerating coalescent histories when provided with species trees and branch lengths (98, 99).

- The DeepCoalescenceSp procedure (100) estimates the species tree by minimizing the discrepancy between gene trees and the proposed tree. Reconstruction proceeds with one assumption that the only genetic process occurring is lineage sorting. The reconstruction minimizes the depths of gene tree divergences. It has been suggested that this process does not allow for the evaluation of the statistical properties of their species estimators (101).

- The AUGIST (accommodating uncertainty in genealogies while inferring species trees) software package was designed for inferring species trees while accommodating uncertainty in gene genealogies. It is available for use with the Mesquite software package and is designed for analysis of data sets in which the underlying gene tree topologies may be unknown. It provides sampling procedures that incorporate uncertainty in gene tree reconstruction and also provides confidence measures for clades on the inferred species trees. Gene tree distributions are generated for each locus showing frequencies at which a clade occurs in the distribution of trees, then a probability of that clade is provided (102). This software package is available for download at www.lycaenid.org/augist/.

6.3. Likelihood Approaches

Currently, the only computationally active maximum likelihood approach to species tree estimation is the STEM package. It estimates the maximum likelihood (ML) species tree from a sample of gene trees under the assumption that disagreements between the observed gene trees and the species tree occur because of coalescent processes and not other genetic events. Although STEM works under the coalescent model, its use of maximum likelihood and its speed relative to other coalescent concordance methods make it different (80). This method is well represented in a simulation study by McCormack, testing its power at resolving challenging evolutionary histories (103).

6.4. Concordance Techniques Applied to Different Groups

Most multilocus genotyping data sets for fungal plant pathogens published in the last 10 years have proceeded only as far as the homogeneity test described above under gene concatenation. In most cases, all gene loci were found to have homogeneous topologies and large concatenated data sets were created, with heavy sampling done on the combined loci to create species trees. When trying to look for concordance in the context of species recognition, there is a need for powerful analysis techniques that will detect any departure from concordance, as described above. The tests for homogeneity, on the other hand, are mainly used to decide whether or not to concatenate genes for phylogenetic analyses. The power of homogeneity tests, i.e., the probability of detecting a departure from homogeneity at a given magnitude, is probably too low to be used in GCPSR. In phylogenetic analyses based on concatenated data sets, some slight topology differences that might represent gene flow are unlikely to affect final conclusions and/or will be reflected in decreased branch supports. It is these slight differences that can become crucial when applying the GCPSR. Therefore, in the context of hypothesis testing for species, concatenation of loci might be appropriate to define some of the deeper nodes with more certainty, but should still be combined with other techniques that handle the phylogenetic signal of each individual locus more appropriately.

The recent proliferation of statistical models to assist with generating species trees from multiple discordant gene trees suggests that this is a burgeoning field that will quickly provide many options for estimating the evolutionary history of a group of organisms and delimiting their species boundaries. Most of these analytical techniques are computationally intense and data sets should be limited to a few species. As more data become available, more analytical options combined with more computing power will need to be developed and communicated to the systematists. That so many analytical approaches consider nontraditional methods of reproduction and gene transfer is promising and suggests that true phylogenies will soon be able to be available for all groups of fungal plant pathogens, regardless of their reproductive lifestyle.

6.5. Phylogenomic Approaches

The emerging field of phylogenomics offers unique opportunities for resolution of evolutionary relationships in plant pathogenic groups. Advancements in this field are being driven by the continuing decrease in sequencing costs, new sequencing strategies, and improvement of bioinformatics approaches for analyzing and managing large amounts of data. To date, these techniques have been applied to the higher taxonomic levels of the Kingdom Fungi (104–107). In all of these approaches, homologous regions are identified from genome sequence data and then analyzed as phylogenetic data through either a concatenated gene alignment or using Super Tree technology. We will not go into detail about these approaches here, but suggest that these powerful tools will be applied to species recognition in the future. As more genomes become available and bioinformatics expertise is shared, it may become commonplace that full genomes are the basis on which a strain that is a putative new species will be compared to well-established species.

7. Conclusion

Figure 1 illustrates the work flow for applying the GCPSR technique to assess species limits in fungal plant pathogens. We hope that we have sufficiently emphasized the necessity for a complete consideration of all steps that would lead a researcher to accurately define species limits. The importance of strain sampling and performing the proper analysis to determine the evolutionary path of chosen loci cannot be stressed enough. It is also important that potential pitfalls of phylogenetic reconstruction methods are understood and that proper assumptions and methods of analysis are followed. With the flood of molecular data for phylogenetics or DNA barcoding, it has become very common for plant pathologists using sequencing to identify a culture, only to find that their strain differs by a few base pairs from the reference sequences in GenBank. This should not be an immediate trigger to describe a new species. Similarly, a single strain can easily differ morphologically from the holotype simply because the phenotypic plasticity of the species was not assessed comprehensively in the original description. One must demonstrate that this strain is not simply a variant of the already described species. This can only be accomplished with the sequencing of multiple genes for multiple strains, representing a broad coverage to determine with the appropriate analysis techniques if there are new species to be described.

References

1. Mishler BD, Theriot EC (2000) The Phylogenetic Species Concept (sensu Mishler and Theriot): Monophyly, Apomorphy, and Phylogenetic Species Concepts. In: Wheeler QD, Meier R (eds) Species concepts and phylogenetic theory: A debate. Columbia University Press. New York

2. Kohn LM (2005) Mechanisms of fungal speciation. Annu Rev Phytopathol 43:279–308

3. Richards TA, et al (2006) Evolution of filamentous plant pathogens: gene exchange across eukaryotic kingdoms. Curr Biol 16:1857–1864

4. Barr DJS (1992) Evolution and kingdoms of organisms from the perspective of a mycologist. Mycologia 84:1–11

5. Wiley EO (1978) The evolutionary species concept reconsidered. Syst Biol 27:17–26

6. Taylor JW, et al (2000) Phylogenetic species recognition and species concepts in fungi. Fungal Genet Biol 31:21–32

7. Baum D (1992) Phylogenetic species concepts. Trends Ecol Evol 7:1–2

8. Nixon KC, Wheeler QD (1990) An amplification of the phylogenetic species concept. Cladistics 6:211–223

9. Baum DA, Donoghue MJ (1995) Choosing among alternative "Phylogenetic" Species Concepts. Syst Bot 20:560–573

10. Wheeler QD, Meier R (eds) (2000) Species Concepts and Phylogenetic Theory. New York: Columbia University Press

11. Baum DA, Shaw KL (1995) Genealogical perspectives on the species problem. In: Hock PC, Stevenson AG (eds) Experimental and molecular approaches to plant biosystematics. Missouri Botanical Garden. St. Louis

12. Avise JC, Ball RM (1990) Principles of genealogical concordance in species concepts and biological taxonomy. In: Futuyma D, Antonovics J (eds) Oxford surveys in evolutionary biology. Oxford Univ. Press. Oxford, U.K

13. Taylor JW, Fisher MC (2003) Fungal multilocus sequence typing - It's not just for bacteria. Curr Opin Microbiol 6:351–356

14. Moralejo E, et al (2008) *Pythium recalcitrans* sp. nov. revealed by multigene phylogenetic analysis. Mycologia 100:310–319

15. O'Donnell K (2000) Molecular phylogeny of the *Nectria haematococca-Fusarium solani* species complex. Mycologia 92:919–938

16. O'Donnell K, et al (2000) Gene genealogies reveal global phylogeographic structure and reproductive isolation among lineages of *Fusarium graminearum*, the fungus causing wheat scab. Proc Natl Acad Sci USA 97: 7905–7910

17. Steenkamp ET, et al (2002) Cryptic speciation in *Fusarium subglutinans*. Mycologia 94: 1032–1043

18. Hirata K, et al (2007) Speciation in *Pyricularia* inferred from multilocus phylogenetic analysis. Mycol Res 111:799–808

19. Inderbitzin P, et al (2009) *Pleospora* species with *Stemphylium* anamorphs: A four locus phylogeny resolves new lineages yet does not distinguish among species in the *Pleospora herbarum* clade. Mycologia 101:329–339

20. Kang JC, et al (2001) Species concepts in the *Cylindrocladium floridanum* and *Cy. spathiphylli* Complexes (Hypocreaceae) based on multi-allelic sequence data, sexual compatibility and morphology. Syst Appl Microbiol 24:206–217

21. Villa NO, et al (2006) Phylogenetic relationships of *Pythium* and *Phytophthora* species based on ITS rDNA, cytochrome oxidase II and β-tubulin gene sequences. Mycologia 98:410–422

22. De Jong SN, et al (2001) Phylogenetic relationships among *Neofabraea* species causing tree cankers and bull's-eye rot of apple based on DNA sequencing of ITS nuclear rDNA, mitochondrial rDNA, and the β-tubulin gene. Mycol Res 105:658–669

23. Seifert KA, Rossman AY (2010) How to describe a new fungal species. IMA Fungus 1:109–116

24. Seifert KA (2009) Progress towards DNA barcoding of fungi. Mol Ecol Resour 9:83–89

25. Hebert PD, et al (2003) Biological identifications through DNA barcodes. Proceedings of the Royal Society of London Series B: Biological Sciences 270:313–321

26. Thomson RC, et al (2010) Genome-enabled development of DNA markers for ecology, evolution and conservation. Mol Ecol 19: 2184–2195

27. Kuo CH, Avise JC (2008) Does organismal pedigree impact the magnitude of topological congruence among gene trees for unlinked loci? Genetica 132:219–225

28. O'Donnell K, Cigelnik E (1997) Two divergent intragenomic rDNA ITS2 types within a monophyletic lineage of the fungus *Fusarium* are nonorthologous. Mol Phylogen Evol 7:103–116

29. Blair JE, et al (2008) A multi-locus phylogeny for *Phytophthora* utilizing markers derived from complete genome sequences. Fungal Genet Biol 45:266–277

30. Massey SE, et al (2008) Characterizing positive and negative selection and their phylogenetic effects. Gene 418:22–26

31. Nozawa M, et al (2009) Reliabilities of identifying positive selection by the branch-site and the site-prediction methods. Proc Natl Acad Sci USA 106:6700–6705

32. Kosakovsky Pond SL, Frost SDW (2005) Not so different after all: A comparison of methods for detecting amino acid sites under selection. Mol Biol Evol 22:1208–1222

33. Zhang Z, et al (2006) KaKs_Calculator: Calculating Ka and Ks through model selection and model averaging. Genomics Proteomics Bioinformatics 4:259–263

34. Townsend JP (2007) Profiling phylogenetic informativeness. Syst Biol 56:222–231

35. Schoch CL, et al (2009) The ascomycota tree of life: A phylum-wide phylogeny clarifies the origin and evolution of fundamental reproductive and ecological traits. Syst Biol 58:224–239

36. Marthey S, et al (2008) FUNYBASE: A FUNgal phYlogenomic dataBASE. BMC Bioinformatics 9

37. Ballard JWO, Whitlock MC (2004) The incomplete natural history of mitochondria. Mol Ecol 13:729–744

38. Hurst GDD, Jiggins FM (2005) Problems with mitochondrial DNA as a marker in population, phylogeographic and phylogenetic studies: The effects of inherited symbionts. Proc R Soc Lond, Ser B: Biol Sci 272:1525–1534

39. Rubinoff D, et al (2006) A genomic perspective on the shortcomings of mitochondrial DNA for "barcoding" identification. J Hered 97:581–594

40. Galtier N, et al (2009) Mitochondrial DNA as a marker of molecular diversity: A reappraisal. Mol Ecol 18:4541–4550

41. Martin FN (1989) Maternal inheritance of mitochondrial DNA in sexual crosses of *Pythium sylvaticum*. Curr Genet 16:373–374

42. Knowles LL, Carstens BC (2007) Delimiting species without monophyletic gene trees. Syst Biol 56:887–895

43. Knowles L (2010) Sampling strategies for species tree estimation. In: Knowles L, Kubatko LS (eds) Estimating species trees: Practical and theoretical aspects. John Wiley and Sons. Hoboken, N.J

44. Winston JE (1999) Describing Species: Practical Taxonomic Procedure for Biologists. Columbia University Press. New York

45. Rossman AY, Palm-Hernández ME (2008) Systematics of plant pathogenic fungi: Why it matters. Plant Dis 92:1376–1386

46. Prendini L (2001) Species or supraspecific taxa as terminals in cladistic analysis? Groundplans versus exemplars revisited. Syst Biol 50: 290–300

47. Dereeper A, et al (2010) BLAST-EXPLORER helps you building datasets for phylogenetic analysis. BMC Evol Biol 10:Art. No.: 8

48. Moller EM, et al (1992) A simple and efficient protocol for isolation of high molecular weight DNA from filamentous fungi, fruit bodies, and infected plant tissues. Nucleic Acids Res 20:6115–6116

49. Fernández-Ortuño D, et al (2007) Multiple displacement amplification, a powerful tool for molecular genetic analysis of powdery mildew fungi. Curr Genet 51:209–219

50. Wang Y, et al (2009) Whole genome amplification of the rust *Puccinia striiformis* f. sp. *tritici* from single spores. J Microbiol Methods 77:229–234

51. Allain-Boulé N, et al (2004) Identification of *Pythium* species associated with cavity-spot lesions on carrots in eastern Quebec. Can J Plant Pathol 26:365–370

52. Ewing B, et al (1998) Base-calling of automated sequencer traces using phred. I. Accuracy assessment. Genome Res 8:175–185

53. Goss EM, et al (2009) Ancient isolation and independent evolution of the three clonal lineages of the exotic sudden oak death pathogen *Phytophthora ramorum*. Mol Ecol 18:1161–1174

54. Biñas M (2000) Designing PCR primers on the Web. BioTechniques 29:988–990

55. Abd-Elsalam KA (2003) Bioinformatic tools and guideline for PCR primer design. Afr J Biotechnol 2:91–100

56. Kubatko LS, Degnan JH (2007) Inconsistency of phylogenetic estimates from concatenated data under coalescence. Syst Biol 56:17–24

57. Edwards SV, et al (2007) High-resolution species trees without concatenation. Proc Natl Acad Sci USA 104:5936–5941

58. Seo TK (2008) Calculating bootstrap probabilities of phylogeny using multilocus sequence data. Mol Biol Evol 25:960–971

59. Massoumi Alamouti S, et al (2009) Multigene phylogeny of filamentous ambrosia fungi associated with ambrosia and bark beetles. Mycol Res 113:822–835

60. Geiser DM, et al (1998) Cryptic speciation and recombination in the aflatoxin-producing fungus *Aspergillus flavus*. Proc Natl Acad Sci USA 95:388–393

61. Le Gac M, et al (2007) Phylogenetic evidence of host-specific cryptic species in the anther smut fungus. Evolution 61:15–26

62. Staats M, et al (2005) Molecular phylogeny of the plant pathogenic genus *Botrytis* and the evolution of host specificity. Mol Biol Evol 22:333–346

63. O'Donnell K, et al (2004) Genealogical concordance between the mating type locus and seven other nuclear genes supports formal recognition of nine phylogenetically distinct species within the *Fusarium graminearum* clade. Fungal Genet Biol 41:600–623

64. Farris JS, et al (1994) Testing significance of incongruence. Cladistics 10:315–319

65. Soltis DE, P.S. Soltis and J.J. Doyle (1998) Molecular Systematics of Plants II: DNA Sequencing. Kluwer Academic. Boston, Mass

66. Mason-Gamer RJ, Kellogg EA (1996) Testing for phylogenetic conflict among molecular data sets in the tribe triticeae (gramineae). Syst Biol 45:524–545

67. Kishino H, Hasegawa M (1989) Evaluation of the maximum likelihood estimate of the evolutionary tree topologies from DNA sequence data, and the branching order in hominoide. J Mol Evol 29:170–179

68. Goldman N, et al (2000) Likelihood-based tests of topologies in phylogenetics. Syst Biol 49:652–670

69. Hipp AL, et al (2004) Congruence versus phylogenetic accuracy: Revisiting the incongruence length difference test. Syst Biol 53:81–89

70. Edwards SV (2009) Is a new and general theory of molecular systematics emerging? Evolution 63:1–19

71. Maddison WP, et al (1984) Outgroup analysis and parsimony. Syst Zool 33:83–103

72. Swofford DL, et al (1996) Phylogenetic inference. In: Hillis DM, et al (eds) Molecular Systematics, 2nd edn. Sinauer Associates, Inc.,. Sunderland, Massachusetts

73. Huelsenbeck JP, et al (2002) Inferring the root of a phylogenetic tree. Syst Biol 51: 32–43

74. Renner SS, et al (2008) Rooting and dating maples (*Acer*) with an uncorrelated-rates molecular clock: Implications for North American/Asian disjunctions. Syst Biol 57: 795–808

75. Boykin LM, et al (2010) Comparison of methods for rooting phylogenetic trees: A case study using Orcuttieae (Poaceae: Chloridoideae). Mol Phylogen Evol 54: 687–700

76. Brito PH, Edwards SV (2009) Multilocus phylogeography and phylogenetics using sequence-based markers. Genetica 135: 439–455

77. Degnan JH, et al (2009) Properties of consensus methods for inferring species trees from gene trees. Syst Biol 58:35–54

78. Degnan JH, Rosenberg NA (2006) Discordance of species trees with their most likely gene trees. PLoS Genet 2

79. Liu L, et al (2009) Coalescent methods for estimating phylogenetic trees. Mol Phylogen Evol 53:320–328

80. Kubatko LS, et al (2009) STEM: Species tree estimation using maximum likelihood for gene trees under coalescence. Bioinformatics 25:971–973

81. Mossel E, Roch S (2010) Incomplete lineage sorting: Consistent phylogeny estimation from multiple loci. IEEE/ACM Trans Comput Biol Bioinform 7:166–171

82. Liu L, Pearl DK (2007) Species trees from gene trees: reconstructing Bayesian posterior distributions of a species phylogeny using estimated gene tree distributions. Syst Biol 56:504–514

83. Knowles LL, Kubatko LS (2010) Estimating Species Trees: Practical and Theoretical Aspects. Wiley-Blackwell

84. Moulton V (ed) (2003) SplitsTree: A network-based tool for exploring evolutionary relationships in molecular data. (The phylogenetic handbook : A practical approach to DNA and protein phylogeny). Cambridge, UK; New York Cambridge University Press

85. Huson DH (1998) SplitsTree: analyzing and visualizing evolutionary data. Bioinformatics 14:68–73

86. Liu Z, et al (2005) Patterns of diversifying selection in the phytotoxin-like scr74 gene family of *Phytophthora infestans*. Mol Biol Evol 22:659–672

87. Heled J, Drummond AJ (2010) Bayesian inference of species trees from multilocus data. Mol Biol Evol 27:570–580

88. Wilson IJ, et al (2003) Inferences from DNA data: Population histories, evolutionary processes and forensic match probabilities. J Roy Stat Soc Ser A (Stat Soc) 166:155–188

89. Rannala B, Yang Z (2003) Bayes estimation of species divergence times and ancestral population sizes using DNA sequences from multiple loci. Genetics 164:1645–1656

90. Yang Z, Rannala B (2010) Bayesian species delimitation using multilocus sequence data. Proc Natl Acad Sci USA 107:9264–9269

91. Ané C, et al (2007) Bayesian estimation of concordance among gene trees. Mol Biol Evol 24:412–426

92. Ané C (2010) Reconstructing concordance trees and testing the coalescent model from genome-wide data sets. In: Knowles L, Kubatko LS (eds) Estimating species trees: Practical and theoretical aspects. Wiley-Blackwell. Hoboken, N.J.

93. Baum DA (2007) Concordance trees, concordance factors, and the exploration of reticulate genealogy. Taxon 56:417–426

94. Liu L (2008) BEST: Bayesian estimation of species trees under the coalescent model. Bioinformatics 24:2542–2543

95. Castillo-Ramirez S, et al (2010) Bayesian estimation of species trees:A practical guide to optimal sampling and analysis. In: Knowles L, Kubatko LS (eds) Estimating species trees: Practical and theoretical aspects. Wiley Blackwell. Hoboken, N.J

96. Ronquist F, Huelsenbeck JP (2003) MrBayes 3: Bayesian phylogenetic inference under mixed models. Bioinformatics 19: 1572–1574

97. Degnan JH, Salter LA (2005) Gene tree distributions under the coalescent process. Evolution 59:24–37

98. Degnan JH, Rosenberg NA (2009) Gene tree discordance, phylogenetic inference and the multispecies coalescent. Trends Ecol Evol 24:332–340

99. Degnan JH (2010) Probabilities of gene tree topologies with intraspecific sampling given a species tree. In: Knowles L, Kubatko LS (eds) Estimating species trees: Practical and theoretical aspects. Wiley Blackwell. Hoboken, N.J

100. Maddison W, Knowles L (2006) Inferring phylogeny despite incomplete lineage sorting. Syst Biol 55:21–30

101. Liu L, et al (2009) Maximum tree: A consistent estimator of the species tree. J Math Biol 60:95–106

102. Oliver JC (2008) AUGIST: Inferring species trees while accommodating gene tree uncertainty. Bioinformatics 24:2932–2933

103. McCormack JE, et al (2009) Maximum likelihood estimates of species trees: How accuracy of phylogenetic inference depends upon the divergence history and sampling design. Syst Biol 58:501–508

104. Robbertse B, et al (2006) A phylogenomic analysis of the Ascomycota. Fungal Genet Biol 43:715–725

105. Fitzpatrick DA, et al (2006) A fungal phylogeny based on 42 complete genomes derived from supertree and combined gene analysis. BMC Evol Biol 6

106. Wang H, et al (2009) A fungal phylogeny based on 82 complete genomes using the composition vector method. BMC Evol Biol 9:Article number 195

107. Dutilh BE, et al (2007) Assessment of phylogenomic and orthology approaches for phylogenetic inference. Bioinformatics 23:815–824

108. Johnson LA, Soltis DE (1998) Assessing congruence:empirical examples from molecular data. In: Soltis DE, et al (eds) Molecular Systematics of Plants II DNA Sequencing. Kluwer Academic Press. Boston

109. Motohashi K, et al (2009) Phylogenetic analyses of Japanese species of *Phyllosticta* sensu stricto. Mycoscience 50:291–302

110. Schoch CL, et al (2006) A multigene phylogeny of the Dothideomycetes using four nuclear loci. Mycologia 98:1041–1052

111. James TY, et al (2006) Reconstructing the early evolution of Fungi using a six-gene phylogeny. Nature 443:818–822

112. Arzanlou M, et al (2008) Multiple gene genealogies and phenotypic characters differentiate several novel species of *Mycosphaerella* and related anamorphs on banana. Persoonia 20:19–37

113. Roets F, et al (2006) Multi-gene phylogeny for *Ophiostoma* spp. reveals two new species from Protea infructescences. Stud Mycol 55:199–212

114. Gryzenhout M, et al (2006) New taxonomic concepts for the important forest pathogen *Cryphonectria parasitica* and related fungi. FEMS Microbiol Lett 258:161–172

115. Carbone I, et al (1999) Patterns of descent in clonal lineages and their multilocus fingerprints are resolved with combined gene genealogies. Evolution 53:11–21

116. Kroon LPNM, et al (2004) Phylogenetic analysis of *Phytophthora* species based on mitochondrial and nuclear DNA sequences. Fungal Genet Biol 41:766–782

117. Lombard L, et al (2010) Multigene phylogeny and mating tests reveal three cryptic species related to *Calonectria pauciramosa*. Stud Mycol 66:15–30

118. Nalim FA, et al (2009) Multilocus phylogenetic diversity of *Fusarium avenaceum* pathogenic on lisianthus. Phytopathology 99:462–468

119. Stenglein SA, et al (2010) Phylogenetic relationships of *Fusarium poae* based on EF-1α and mtSSU sequences. Fungal Biol 114:96–106

Chapter 35

MAP Kinase Phosphorylation and cAMP Assessment in Fungi

Rahim Mehrabi, Sarrah Ben M'Barek, Abbas Saidi, Masoud Abrinbana, Pierre J.G.M. de Wit, and Gert H.J. Kema

Abstract

The cyclic AMP (cAMP) signaling and mitogen-activated protein (MAP) kinase pathways are the most important signal transduction pathways in eukaryotes. In many plant pathogenic fungi they play pivotal roles in virulence and development. Identification and understanding the role of signal transduction pathways in regulation of cellular responses require robust biochemical techniques. Determination of both the phosphorylation status of MAPKs and the intracellular levels of cAMP is required to unravel the function of these pathways during adaptation of fungi to environmental stress conditions or when particular fungal genes are disrupted or silenced. Here we describe protocols to determine the phosphorylation status of three different MAPKs including Fus3, Slt2 and Hog1 as well as a protocol to measure the intracellular levels of cAMP levels. These protocols can be adapted for a wide range of fungi.

Key words: cAMP measurement, Mitogen-activated protein kinase, Protein kinase A, Phosphorylation, Fus3, Slt2, Hog1

1. Introduction

Eukaryotic organisms use signal transduction pathways to perceive and respond to various extracellular signals to modulate proper developmental and physiological responses. The cyclic AMP (cAMP) signaling and mitogen-activated protein (MAP) kinase pathways are among the best studied signal transduction pathways in eukaryotes (1). Kinases are proteins with phosphotransferase activity (PKAs) and are able to covalently transfer a phosphate group (PO_4^{3-}) from a nucleoside triphosphate (usually ATP) to one of the three amino acids (serine, threonine, or tyrosine) that have a free hydroxyl group (–OH). MAPKs belong to the serine/threonine protein kinase superfamily (EC 2.7.11.1) and possess a

Melvin D. Bolton and Bart P.H.J. Thomma (eds.), *Plant Fungal Pathogens: Methods and Protocols*,
Methods in Molecular Biology, vol. 835, DOI 10.1007/978-1-61779-501-5_35, © Springer Science+Business Media, LLC 2012

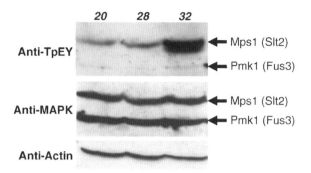

Fig. 1. Detection of the activation of Mps1 by elevated temperature. Proteins were isolated from hyphae of *Magnaporthe oryzae* strain Guy11 treated at 20, 28, and 32°C for 15 min. The *top panel* shows that Mps1 (homolog of Slt2) is phosphorylated upon heat treatment. The phosphorylation level of Pmk1 (homolog of Fus3) is relatively low and unchanged under all tested conditions. The *middle panel* shows the amount of total proteins of Mps1 and Pmk1 present in these samples. The *lower panel* shows that the total amount of proteins used for each samples were similar.

dual phosphorylation site of TxY (where x is any amino acid). They are the key components of major signal transduction pathways and are highly conserved across the fungal kingdom, particularly in their catalytic domains. A MAPK cascade normally consists of a MEK kinase (MEKK), a MAPK kinase (MEK) and a MAPK (2). The sequential phosphorylation of the MEKK, MEK and MAPK often leads to the phosphorylation of transcription factors. Phosphorylation usually results in functional changes and activation of the target transcription factors, which in turn regulate the expression of a subset of genes in response to extracellular signals. Many extracellular and intracellular signals modulate transcription of specific genes through phosphorylation or dephosphorylation of MAPKs in response to environmental stimuli.

The important roles of MAPK pathways in virulence and development have been addressed in many plant pathogenic fungi (1–3). In the budding yeast *Saccharomyces cerevisiae*, there are five distinct MAPK pathways that are involved in mating responses (Fus3 MAPK), filamentous growth (Kss1 MAPK), cell wall integrity (Slt2 MAPK), high osmotic growth stress responses (Hog1 MAPK) and ascospore formation (Smk1 MAPK) (4, 5). Filamentous fungi typically possess only three MAPK cascades that are orthologous to the yeast Fus3, Hog1 and Slt2 MAPK pathways (1; see Fig. 1).

The Fus3 ortholog is the most studied MAPK in plant pathogenic fungi. In appressorium forming pathogens including *Magnaporthe oryzae, Colletotrichum lagenarium, Pyrenophora teres,* and *Cochliobolus heterostrophus,* this MAPK pathway is important for appressorium formation and virulence (1–3). In other fungal pathogens, such as *Mycosphaerella graminicola, Fusarium oxysporum, Botrytis cinerea, Ustilago maydis, Claviceps purpurea, F. graminearum, Verticillium dahliae, Bipolaris oryzae, Stagonospora nodorum, Cryphonectria parasitica,* and *Sclerotinia sclerotiorum,* the Fus3 homologs are involved in various cellular responses as well as plant infection processes (1–3).

In fungal pathogens, the Slt2 MAPK homologs are often found to regulate the cell wall integrity and virulence. This includes *M. oryzae* Mps1, *C. lagenarium* Maf1, *C. purpurea* Cpmk2, *F. graminearum* Mgv1, *M. graminicola* MgSlt2, and *B. cinerea* Bmp3 (1–3). The Hog1 MAPK homologs in fungal pathogens are generally involved in osmoregulation and in some cases in virulence, probably due to differences in infection mechanisms of these pathogens. While the *M. graminicola* MgHog1, *C. parasitica* Cpmk1 and *B. cinerea* BcSak1 are involved in virulence, the *M. oryzae* Osm1, *C. lagenarium* Osc1, and *B. oryzae* Srm1 are dispensable for plant infection (1, 2).

The cAMP-dependent PKA pathway also is very important in eukaryotic organisms. Several proteins including the adenylyl cyclase and small GTPases are involved in the generation of cAMP in response to environmental signals (6, 7). PKA is a key protein in the cAMP/PKA pathway that is regulated by cAMP molecules. PKA consists of two catalytic and two regulatory subunits. Upon generation, cAMP binds to the regulatory subunits of PKA leading to the release of active catalytic subunits, which in turn phosphorylate serine and threonine residues on various target proteins (8).

Identification and understanding of the role of signal transduction pathways in the regulation of various cellular responses to different environmental stimuli requires the application of robust biochemical assays Here we describe protocols to determine the phosphorylation status of three MAPKs orthologous to yeast Fus3, Slt2, and Hog1 and to measure the intracellular cAMP level in plant pathogenic fungi. These protocols can be used for a wide range of fungal pathogens.

2. Materials

2.1. Equipment

1. Bead beater (e.g., Biospec Mini bead beater, Fulltech, UF-60D11 or FastPrep FP120).

2. Protein electrophoresis equipment (e.g., Protean II from BioRad® and power supply).

3. Blotting apparatus: A semidry blotting apparatus (e.g., BioRad Trans-Blot® SD semidry transfer or Hoefer SemiDry Transfer unit).

4. Phosphor imager: visualize chemiluminescence with a phosphor imager such as the Storm 840 (GE Healthcare Life Sciences).

5. Centrifuge (various cooling bench centrifuges can be used).

2.2. Reagents

1. Protein extraction buffer: 50 mM Tris–HCl pH 7.5, 100 mM NaCl, 50 mM NaF, 5 mM EDTA, 1 mM EGTA, 1% Triton X-100, and 10% glycerol. Add 10 μL of protease inhibitor cocktail (Sigma P8215) into 1 mL extraction buffer and phenyl-methylsulfonyl fluoride (PMSF) to the final concentration of 2 mM immediately before use.

2. Resolving gel buffer: 1.5 M Tris–HCl pH 8.0, 0.4% w/v SDS.

3. Stacking gel buffer: 0.5 M Tris–HCl pH 6.8, 0.4% w/v SDS.

4. Acrylamide: 30% Acrylamide/Bis solution (BioRad, Cat# 161-0158).

5. Running buffer (5×): 72 g/L glycine, 15 g/L Tris base, 5 g/L SDS.

6. TEMED (*N, N, N′, N′*-Tetramethylethylenediamine).

7. APS (ammonium persulfate).

8. Antibodies: anti-p42/44 MAPK and anti-phospho-p42/44 MAPK (Cell Signaling Technology, Boston, MA); monoclonal anti-actin antibody (Sigma Aldrich, St Louis, MO); anti-phosphate-p38 MAPK kinase; (y-40) rabbit polyclonal antibodies (Santa Cruz Biotechnology, Santa Cruz, CA); anti-rabbit-IgG antibody conjugated horseradish peroxidase (HRP) (Santa Cruz Biotechnology, Inc., Santa Cruz, CA or Amersham Biosciences).

9. Nitrocellulose membranes (e.g., Protran® nitrocellulose membranes, 0.45 μm pore size).

10. SuperSignal West Pico Chemiluminescent substrate (Thermo Fisher Scientific Inc.).

11. Laemmli sample buffer.

12. TBS (10×): 1.54 M NaCl, 0.1 M Tris–HCl pH 7.5.

13. TBS-T: TBS + 0.1% Tween 20 (w/v).

14. Chemiluminescence (ECL): Use a commercial kit such as Amersham ECL Plus™ Western Blotting Detection Reagents (RPN2132).

15. Coomassie Plus protein assay reagent kit (Pierce Biotechnology, Inc.).

16. Transfer buffer: 3.03 g Tris base, 14.4 g Glycine, and 10% methanol per L.

17. Ponceau reagent: 0.25% Ponceau S in 40% methanol and 15% acetic acid.

18. Tris-EDTA buffer: dissolve in 25 mL of distilled water to give an approximately 0.05 M solution at pH 7.5 containing 4 mM EDTA.

19. Binding protein purified from bovine muscle, freeze dried, dissolve in 15 mL of distilled water by gently swirling. Invert the vials occasionally but do not shake vigorously.

20. (8-3H) Adenosine 3′,5′-cyclic phosphate tracer 180 pmol containing approximately 185 kBq, 5 µCi, freeze-dried: remove the cap from the vial carefully to avoid loss of freeze-dried material. Add 10 mL of distilled water, replace the cap, and dissolve by swirling and inversion.

21. Adenosine 3′,5′-cyclic phosphate standard 1,600 pmol, freeze-dried: remove the cap carefully to avoid loss of freeze-dried material. Add 5 mL of distilled water, replace the cap, and dissolve by swirling and inversion.

22. Charcoal adsorbent: add 20 mL of ice-cold distilled water, stand the container in an ice-bath, and stir continuously using a magnetic stirrer.

3. Methods

3.1. MAP Kinase Phosphorylation Assay

3.1.1. Electrophoretic Separation of Proteins

1. Harvest fungal biomass such as conidia and vegetative hyphae from fresh cultures by centrifugation or filtration.

2. Transfer 100–150 mg of the fungal biomass into 2-mL screw-cap tubes, 1/3 filled with clean or acid washed glass beads (0.5 mm).

3. Immediately fill out the tube with protein extraction buffer containing PSMF and protease inhibitor cocktail.

4. Homogenize the fungal biomass with a bead beater. Adjust the bead beater to 4,000–5,000 rpm and run for 40 s. Repeat four times and keep the tube on ice between each pulse to avoid overheating.

5. Centrifuge the homogenized mycelia for 15 min at 4°C at >16,000 ×g on a bench-table centrifuge. Take supernatant and centrifuge again 15 min at the same speed.

6. Transfer supernatant to new tube and keep at –80°C for further analyses.

7. Prepare the resolving gel by mixing 2.5 mL of resolving gel buffer, 4.2 mL of 30% acrylamide mixture, and 3.4 mL water in a 15-mL tube.

8. Add 35 µL fresh 10% APS and 6 µL TEMED; swirl gently and immediately pour from the top corner within a mini gel cassette (approximately 7 cm × 10 cm × 1.5 mm). Fill up to ¾ of the cassette height (about 3.5 mL).

9. Gently overlay with water to remove air bubbles and keep it horizontally.

10. Wait 15 min, pour off water, and dry out the remaining water with the help of filter paper.

11. Prepare the stacking gel by mixing 0.75 mL of resolving gel buffer, 0.5 mL of 30% acrylamide, and 3.75 mL water in a 15-mL tube.

12. Add 23 μL fresh 10% AMPS and 7.5 μL TEMED, swirl gently, and pipette into the gel cassette from the top corner immediately.

13. Insert a 10-well gel comb immediately and gently without introducing air bubbles.

14. Add loading buffer to proteins provided in step 6, close the lid, and denature by putting it in boiling water for 5 min. Cool sample immediately on ice for 5 min.

15. Transfer the gel cassette into the electrophoresis tank and fill out with 1× running buffer. Pull out the comb gently and load 15–20 μL (approximately 30 μg) of denatured proteins into the gel slots. Load the protein marker in the first and the last lanes.

16. Perform electrophoresis at 100 V for about 1 h or till the dye front (from the bromophenol blue dye in the samples) has reached the bottom of the gel.

17. Stop electrophoresis and open the gel cassette. The gel will stick to one of the glass plates. Rinse the gel with water and transfer carefully to a container with western blot transfer buffer.

3.1.2. Transfer Proteins to Nitrocellulose Membrane

1. Cut 3 mm Whatman filter papers to the size of the gels, wet in transfer buffer, and place three of these papers onto the semidry blotter (see Note 1). Place the gel on top of the filters and then a prewetted nitrocellulose filter on top of the gel subsequently (see Note 2). Then put three additional wetted Whatman 3M filters on top of the pile. Close the apparatus and run for 30 min at 200–500 mA or adjust the apparatus on constant voltage (between 10 and 15 V).

2. After running disassemble the apparatus and remove nitrocellulose filter.

3. Stain membrane for 3 min with Ponceau reagent as a removable stain, and gently wash it with water or TBS-T, and check the transfer quality. If the proteins were blotted efficiently, they can be seen as pale pink bands.

4. Destain with regular SDS-PAGE gel destain solution (7.5% methanol, 10% acetic acid) or TBS-T.

5. Block the membrane with 3% BSA or 2–5% dry milk in TBS-T for 1 h with gentle shaking.

6. Wash the membrane with 15 mL of TBS, TBS-T, and TBS each time for 5 min.

3.1.3. Primary and Secondary Antibody Incubations

1. The following antibodies are used as the primary antibodies. For Hog1 phosphorylation a rabbit anti-phosphate-p38 MAPK kinase, for Fus3 and Slt2 phosphorylation an anti-phosphate-p44/42 MAPK kinase antibody (Cell Signaling Technology, Boston, MA) at a 1:1,000 to 1:5,000 dilution is used (see Note 3). For total Hog1 an anti-Hog1 and for detection of total Fus3 (y-40) rabbit polyclonal antibodies (Santa Cruz Biotechnology, Santa Cruz, CA) is used. For the detection of total Slt2 anti-Slt2 antibody (Santa Cruz Biotechnology, Inc., Santa Cruz, CA) is used. For the detection of total Slt2 and Fus3 an anti-p42/44 MAPK (Cell Signaling Technology, Boston, MA) can be used as well at a 1:1,000 to 1:5,000 dilution (see Note 4). For actin detection use a monoclonal anti-actin antibody (Sigma, St Louis, MO) at a 1:2,000 dilution.

2. Dilute all antibodies at the abovementioned dilution in TBS-T with 2% milk and incubate O/N at 4°C with gentle shaking.

3. Wash off the unbounded first antibody from the membrane in 15 mL TBS-T three times and each time for 10 min.

4. Incubate the membrane for 1 h at RT in diluted secondary antibody (anti-rabbit-IgG antibody conjugated HRP) in TBS-T with 2% milk. For MAPK phosphorylation dilute at 1:2,000. For MAPK detection dilute at 1:10,000. For actin detection dilute at 1:100,000.

5. Wash the membrane twice with TBS-T and twice with TBS each one for 10 min.

3.1.4. Signal Detection

1. Mix equal parts (3 mL) of the Stable Peroxide solution and the Luminol/Enhancer solution.

2. Incubate blot in SuperSignal solution for 5 min.

3. Remove blot from SuperSignal solution and place it in a plastic membrane protector. Use an absorbent tissue to remove excess liquid and to carefully press out any bubbles from between the blot and surface of the membrane protector.

4. Place the protected membrane in a film cassette with the protein side facing up. Turn off all lights except those appropriate for film exposure (e.g., a red safelight).

5. Make sure excess of substrate is removed from the membrane and the membrane protector.

6. Carefully place a piece of film on top of the membrane. A recommended first exposure time is 60 s. Increase exposure time if necessary to achieve optimal results.

7. If using a storage phosphor imaging device (e.g., Bio-Rad's Molecular Imager® System) or a CCD Camera (e.g., Alpha-Innotech Corporation's ChemiImager™ System), longer exposure times may be necessary.

8. Develop film using appropriate developing solution and fixative.

3.2. Intracellular cAMP Measurements

3.2.1. Sample Preparation

1. Grow the fungus in desired medium and collect the spores or mycelia by centrifugation or by filtration through sterile Miracloth. Wash the fungal biomass with sterile water to remove residual medium.

2. Include three biological replicates for each strain.

3. Flash freeze the fungal biomass in liquid nitrogen and lyophilize the fungal biomass over night.

4. For each sample use three 2-mL-screw cap tubes. Two tubes will be considered as replicate of the tested sample. One tube will be used for the measurement of protein concentration (protein normalization tube).

5. Transfer 100 mg of lyophilized fungal biomass in each tube. Add three glass beads (2 mm diameter) to each tube and grind to a fine powder using a bead beater. Perform bead beating at 4,000–5,000 rpm and run for 40 s (see Note 5).

6. For cAMP measurement resuspend powdered mycelia in 700 μL of ice-cold 6% trichloroacetic acid (TCA).

7. Precipitate proteins by centrifugation at $16,000 \times g$ for 2 min at 4°C, recover the supernatant, and discard the pellet.

8. Wash the supernatant: take 380 μL of the supernatant and add 1,600 μL water saturated diethyl ether and centrifuge at $16,000 \times g$ for 2 min at 4°C. The upper ether layer should be discarded after each wash. Wash the supernatant four times with water saturated diethyl ether to remove residual of TCA.

9. Stand the Eppendorf tube on ice. Open the cap of the Eppendorf tube in a hood to evaporate ether for a few minutes.

10. Alternatively, wash the supernatant four times with an equal volume of chloroform at $16,000 \times g$ for 2 min at 4°C.

11. Lyophilize the remaining aqueous extract over night.

12. For protein normalization add 700 μL of water to the powdered mycelia, vigorously vortex it, and centrifuge at $16,000 \times g$ for 2 min at 4°C. Use 100 μL of the supernatant for protein quantification using a Coomassie Plus protein assay kit with bovine serum albumin as a reference.

3.2.2. The cAMP Measurement

1. Use cAMP (^{3}H) Assay protocol for cAMP measurement (Amersham Life Science Inc., Arlington Heights, IL) (see Note 6).

2. Dissolve the lyophilized sample in 500 μL phosphate buffer (10 mM, pH 7.4).

3. Before cAMP quantification, measure the pH and if necessary adjust to pH of 6–8 with 50% $KHCO_3$.

4. Make a dilution series from the adenosine 3′,5′-cyclic phosphate standard 1,600 pmol in order to obtain five levels of standard cAMP: 0.5 mL of the buffer (reagent 1) is added to

Table 1
Schematic representation of a cAMP assay including all control standards and unknown samples

Tube no	Reagent 1 buffer	Standards	Unknowns	Reagent 3 (^{3}H) cAMP	Reagent 2 binding protein	
1, 2	150[a]	–	–	50	–	Charcoal blank
3, 4	50	–	–	50	100	Zero dose
5, 6	–	50	–	50	100	1 pmol standard
7, 8	–	50	–	50	100	2 pmol standard
9, 10	–	50	–	50	100	4 pmol standard
11, 12	–	50	–	50	100	8 pmol standard
13, 14	–	50	–	50	100	16 pmol standard
15, etc	–	–	50	50	100	Unknowns

[a]All volumes are in microlitres

each tube and to the first tube 0.5 mL of cAMP standard (reagent 4) is added. Mix thoroughly. Take 0.5 mL of this dilution to the next tube and mix again. Repeat the procedure successively with each remaining tube. At the end, 50 μL from each solution will give 168, 4, 2, and 1 pmol per assay tube. Use a fresh pipette tip for each successive operation.

5. Label 14 Eppendorf tubes and additional tubes for unknowns including duplicates as shown in Table 1 and put them on ice. It is important to keep the order of reagent addition maintained along the assay.

   1. Determine the blank counts per minute by adding 150 μL of Tris-EDTA buffer (reagent 1) into the tubes 1 and 2.

   2. Determine the binding in the absence of unlabeled cAMP by adding 50 μL of Tris-EDTA into the assay tubes 3 and 4.

   3. Add 50 μL of each dilution into each successive pair of assay tubes (tubes 5–14), starting with the lowest level of standard cAMP.

   4. Add 50 μL of each unknown, in duplicate, into the additional assay tubes as appropriate.

   5. To every tube, pipette 50 μL of the labeled cAMP, (reagent 3).

   6. Add 100 μL of the binding protein (reagent 2) to assay tubes 3–14 and to the unknown tubes.

   7. Vortex all tubes briefly for 5 s.

   8. Keep the tubes on ice or refrigerator at 2–8°C for ~2 h (equilibrium is reached within 30 min; however, to minimize variation, a reaction time of 2 h is used).

9. Add 20 mL of ice cold distilled water to the charcoal reagent at least 15 min before the end of the incubation time. Keep the bottle on ice and stir continuously during use.

10. Add 100 μL of the charcoal suspension to all tubes.

11. Vortex briefly. Do not add charcoal to more tubes than can be centrifuged in one batch.

12. Centrifuge the tubes to sediment the charcoal not less than 1 min and no more than 6 min after the addition of charcoal to the last tube in a batch (a good precipitate is obtained after 1.5–2 min at $12,000 \times g$) (see Note 7).

13. After centrifugation, take 200 μL from each tube, without disturbing the pellet and place in scintillation vials for counting. Do not leave the tubes with undisturbed charcoal pellet more than 10 min on ice.

14. Samples are added directly to a suitable water-compatible liquid scintillant mixture such as PCS which is manufactured by GE Healthcare Bio-Sciences Corp. (USA) or Optiphase-MP. A counting time of four min is recommended for samples in this assay.

15. In order to check counting efficiencies, determine the percentage bound at zero dose. Total counts can be determined by counting two 50 μL aliquots of tritiated cAMP (reagent 3) in scintillant containing 1 mL water. The actual total activity is then obtained by multiplying the counts for 50 μL (corrected for background counts) by two-thirds.

16. Determine the blank counts per minute (cpm) by averaging the cpm for tubes 1 and 2.

17. Determine the cpm bound in the absence of unlabeled cAMP (C_0) by averaging the cpm for tubes 3 and 4 and subtract the blank cpm.

18. Average the cpm for each pair of duplicates in tubes 5–14 and for the unknowns. Subtract the blank cpm from each result to give the cpm bound in the presence of standard or unknown unlabeled cAMP (C_x).

19. Determine C_0/C_x for each level of cAMP and the unknowns.

20. Draw a linear graph by plotting C_0/C_x against pmoles of inactive cAMP/ tube (see Fig. 2, Table 2). A straight line should be obtained with an intercept of 1.0 on the ordinate (see Note 8).

21. From the standard curve and from C_0/C_x value, determine the number of pmoles of inactive cAMP.

22. After quantification, normalize the cAMP values to the protein level for each strain (pmol cAMP/mg total protein).

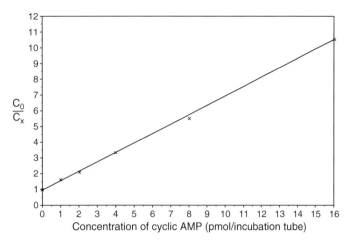

Fig. 2. A typical cAMP standard curve.

Table 2
Typical assay data[a]

Tube no	cpm[b]	Average cpm	Average cpm-blank	C_0/C_x	Standard cAMP pmol/tube
1	91				
2	89	90	–	–	–
3	6,561				
4	6,635	6,598	6,508	1.00	0.00
5	4,088				
6	3,978	4,033	3,943	1.65	1.00
7	3,167				
8	3,216	3,192	3,102	2.10	2.0
9	2,017				
10	2,022	2,020	1,930	3.37	4.0
11	1,274				
12	1,271	1,273	1,183	5.5	8.0
13	712				
14	698	705	615	10.6	16.0

[a]The above data, the standard curves shown in Fig. 2 are representative only. The users must prepare standard curve from data obtained in their own laboratory
[b]Corrected for instruments blank of 166 cpm

4. Notes

1. Following electrophoresis, the proteins must be transferred from the electrophoresis gel to a membrane. Electrophoretic transfer is most commonly used because of its efficiency. We use a semidry blotter, which is quicker and easier than fully submerged blotting methods.

2. It is important to remove the air bubbles potentially present between the filter and nitrocellulose membrane. Use a rolling pen to remove possible air bubbles.

3. The dilution should be experimentally checked but we recommend diluting these antibodies at 1:1,000.

4. We recommend to perform the first experiment at 1:2,000 dilution and optimize it based on the result.

5. We usually repeat beat-beating three times each time 5 min interval on ice. Repeat beat-beating if necessary to get a fine powder of the fungal biomass.

6. The assay is based on the competition between unlabeled cAMP and a fixed quantity of the tritium-labeled compound for binding to a protein which has a high specificity and affinity for cAMP. The amount of labeled protein-cAMP complex formed is inversely related to the amount of unlabeled cAMP present in the assay sample. The kit is suitable for the assay of cAMP in the range of 0.2–16 pmol/incubation tube.

7. During the assay, keep the charcoal reagent in an ice-bath and do not leave the tubes standing too long before centrifugation, which may increase the amount of radioactivity bound to the charcoal.

8. Construct a standard curve every time the kit is used (2 and 8 pmol are recommended).

Acknowledgments

We gratefully thank Prof. Jin-Rong Xu for critical reading of the manuscript. The cAMP assay was performed in GHJ Kema lab (Wageningen University) and phosphorylation assay in J-R Xu lab (Purdue University).

References

1. Mehrabi, R., Zhao, X., Yangseon, K., and Xu, J.-R. (2009) The cAMP Signaling and MAP Kinase Pathways in Plant Pathogenic Fungi, In: *Hloger Deising, ed. The Mycota XXII, Plant Relationships, volume V, 2nd edition, pp 157–172.*

2. Xu, J. R. (2000) MAP kinases in fungal pathogens, *Fungal Genetics and Biology 31*, 137–152.

3. Zhao, X. H., Mehrabi, R., and Xu, J. R. (2007) Mitogen-activated protein kinase pathways and fungal pathogenesis, *Eukaryotic Cell 6*, 1701–1714.

4. Posas, F., Takekawa, M., and Saito, H. (1998) Signal transduction by MAP kinase cascades in budding yeast, *Current Opinion in Microbiology 1*, 175–182.

5. Schwartz, M. A., and Madhani, H. D. (2004) Principles of map kinase signaling specificity in *Saccharomyces cerevisiae*, *Annual Review of Genetics 38*, 725–748.

6. Bores-Walmsley, M. I., and Walmsley, A. R. (2000) cAMP signalling in pathogenic fungi: control of dimorphic switching and pathogenicity, *Trends in Microbiology 8*, 133–141.

7. Lee, N., D'Souza, C. A., and Kronstad, J. W. (2003) Of smuts, blasts, mildews, and blights: cAMP signaling in phytopathogenic fungi, *Annual Review of Phytopathology 41*, 399–427.

8. Tamaki, H. (2007) Glucose-stimulated cAMP-protein kinase A pathway in yeast *Saccharomyces cerevisiae*, *Journal of Bioscience and Bioengineering 104*, 245–250.

Chapter 36

A One-Step Affinity-Purification Protocol to Purify NB-LRR Immune Receptors from Plants That Mediate Resistance to Fungal Pathogens

Wladimir I.L. Tameling

Abstract

Nucleotide-binding, leucine-rich repeat (NB-LRR) immune receptors from plants confer resistance to fungal pathogens and many other pathogenic organisms. Their low expression makes it challenging to purify these receptors from plants in sufficient quantities to be able to identify interacting proteins by mass spectrometry. Here we describe a protocol to affinity-purify recombinant NB-LRR immune receptors, fused to the streptavidin-binding peptide tag.

Key words: Affinity-purification, NB-LRR, Streptavidin-binding peptide tag, Resistance, Immune receptor

1. Introduction

Plant disease resistance (R) proteins mediate a highly effective resistance to fungal pathogens and other plant-pathogenic organisms. The nucleotide binding, leucine-rich repeat (NB-LRR) proteins form the largest group within the R protein family. They recognize virulence-proteins derived from the pathogen (so-called effectors) within the plant cell and therefore function as plant immune receptors. The activation of these receptors triggers an immune signaling cascade that activates the defense response. In most cases, defense activation eventually culminates in programmed cell death at the site of infection (1). Since inappropriate activation of immune receptors is deleterious to the plant, their activity must be tightly regulated (2–4). It is therefore also not surprising that NB-LRR immune receptors are expressed at low levels. Even when expression is driven by the strong cauliflower mosaic virus (CaMV)

Melvin D. Bolton and Bart P.H.J. Thomma (eds.), *Plant Fungal Pathogens: Methods and Protocols*,
Methods in Molecular Biology, vol. 835, DOI 10.1007/978-1-61779-501-5_36, © Springer Science+Business Media, LLC 2012

35S promoter in stably or transiently transformed plant cells, these proteins generally do not accumulate to high levels. Purification of immune receptor complexes from plants is essential to learn more about the molecular mechanisms behind the activation of the immune signaling cascade that leads to resistance to pathogens. This knowledge could identify new leads for establishing resistance of plants to pathogens. However, because of the earlier mentioned low levels of accumulation of these resistance proteins, this is a challenging task. A few examples of successful purification of NB-LRR immune receptor complexes from plants have been reported (5–8). In two of these examples, the streptavidin-binding peptide (SBP) tag, in combination with a streptavidin matrix, was used for affinity-purification of the immune receptor from *Nicotiana benthamiana* or tomato (*Solanum lycopersicum*) (7, 8). This tag has a very high affinity for the streptavidin matrix (the dissociation constant of the SBP tag is about 2.5 nM) (9) and bound SBP-tagged proteins can easily be eluted with biotin. An additional advantage of this approach is that the matrix is relatively inexpensive compared to those containing immobilized antibodies. Here a one-step affinity-purification protocol is described for purification of NB-LRR immune receptors from plants, which can easily be scaled up for the identification of interacting proteins by mass spectrometry. Note that this protocol has been optimized for affinity-purification from *N. benthamiana* leaf extracts, which can show browning during the course of purification due to polyphenol oxidases that react with phenolic compounds. The use of insoluble polyvinyl-polypyrrolidone (PVPP) in the extraction buffer and the passage of the cleared homogenate through a Sephadex G-25 desalting column, are important factors in this protocol to lose these phenolic compounds. This issue also applies for other *Solanaceous* plants (e.g., tomato and tobacco). This protocol has been used successfully to affinity-purify the transiently expressed NB-LRR immune receptor Rx, mediating resistance to *Potato virus X*, from *N. benthamiana* after agroinfiltration. A Ran GTPase-activating protein (RanGAP2) was identified as interacting protein by mass spectrometry analysis (7).

2. Materials

1. 1 M dithiothreitol (DTT) stock: Dissolve DTT in H_2O and store aliquots at –20°C.

2. Extraction buffer: 25 mM Tris–HCl (pH 7.5), 150 mM NaCl, 1 mM EDTA, 10% glycerol, 5 mM DTT, 1× protease inhibitor cocktail (Complete Protease Inhibitor Cocktail Tablets, Roche). Add DTT from the 1 M stock and protease inhibitor cocktail just before protein extraction. Optional: include 0.15% NP-40.

3. Washing buffer: 25 mM Tris–HCl (pH 7.5), 150 mM NaCl, 1 mM EDTA, 10% glycerol, 5 mM DTT, and 0.15% NP-40. Add DTT just before use.

4. Sephadex equilibration buffer: 25 mM Tris–HCl (pH 7.5), 150 mM NaCl, 1 mM EDTA, 10% glycerol, and 5 mM DTT. Add DTT just before use.

5. Elution buffer: 25 mM Tris–HCl (pH 7.5), 150 mM NaCl, 1 mM EDTA, 10% glycerol, 5 mM DTT, 0.15% NP-40, and 4 mM D-Biotin. Add DTT and D-Biotin just before use. Stir or mix tube with elution buffer for 2 h head-over-head to dissolve D-Biotin, at room temperature.

6. 15-mL tubes.

7. 5-mL syringes.

8. Glass wool.

9. Polyvinyl-polypyrrolidone (PVPP).

10. Streptavidin matrix (GE Healthcare).

3. Methods

3.1. Preparation of Sephadex G-25 Columns

1. Put a little bit of glass wool on the bottom of a 5-mL syringe and place it in a 15-mL tube. A hole should be pierced through the side at the upper part of the tube to allow an unrestrained flow through the column when the column has a tight fit (see Fig. 1).

2. Prepare Sephadex G-25 slurry by pouring 8–9-mL Sephadex equilibration buffer into a beaker (volume for one column). Add Sephadex G-25 resin and allow swelling of the resin while swirling. Add Sephadex G-25 till only a thin layer (approximately 2 mm) of buffer sits on top of the settled and swollen Sephadex G-25. Leave at room temperature for 5 min for complete equilibration and swelling.

3. Mix the Sephadex G-25 slurry vigorously by swirling and directly pour the slurry in the 5-mL syringe till the top (see Note 1).

4. Store the column at 4°C for at least 30 min to allow sufficient cooling.

3.2. Equilibration of the Streptavidin Matrix

1. After resuspension, pipette the required amount of Streptavidin matrix in a reaction tube with a tip having a large opening or cutoff the top of the tip.

2. Spin 1 min at $500 \times g$ in a microcentrifuge and discard the supernatant.

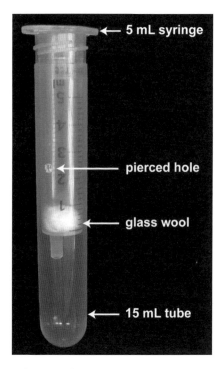

Fig. 1. Preparation of a Sephadex G-25 column. Pierce a hole in the side of the tube and then put in the 5-mL syringe with glass wool at the bottom. Pour the Sephadex G-25 slurry in the syringe till the top.

3. Wash the matrix 2–3 times with 1-mL extraction buffer.

4. Prepare a 75% slurry of the Streptavidin matrix in extraction buffer (75% estimated bed or matrix volume and 25% extraction buffer).

3.3. Protein Extraction

1. Harvest leaves from *N. benthamiana* in which the SBP-tagged NB-LRR of interest has been transiently expressed, for example through *Agrobacterium tumefaciens* transient transformation (7, 10), and grind tissue with a mortar and pestle in liquid nitrogen till a fine powder is obtained. Store powder in 50-mL tubes in liquid nitrogen.

2. Add 2% (w/v) insoluble PVPP to a 10- or 15-mL tube and add 2.5 mL of extraction buffer.

3. Add 1 g of the powdered tissue to the tube and mix vigorously by shaking. Incubate tube on ice and vortex for 10 s.

4. Spin the homogenate for 10 min at $2,000 \times g$ (4°C) in a swing-out rotor.

5. Transfer the supernatant to 2-mL tubes and spin 10 min at $16,000 \times g$ (4°C) in a microcentrifuge.

6. Meanwhile, spin the Sephadex G-25 column 2 min at $720 \times g$ (4°C) in a swing-out rotor.

7. Place the settled column in a new 15 mL tube with a hole pierced through the upper side of the tube (to allow unrestrained flow-through the column).

8. Carefully load 1.2–1.3 mL of the supernatant of the homogenate on top of the column.

9. Spin the Sephadex G-25 column 2 min at $720 \times g$ (4°C) in a swing-out rotor and collect the flow-through. Take a 50 μL sample as "input" for immunoblot analysis.

3.4. Affinity-Purification with Streptavidin Matrix

1. Add 30 μL of the 75% Streptavidin slurry to 1 mL of the cleared extract, which is the flow-through of the Sephadex G-25 column, in a 1.5-mL tube. Mix the 75% slurry vigorously before pipetting.

2. Incubate the tube for 2–3 h (4°C), with head-over-head rotation.

3. Spin the tube 15 s at $16,000 \times g$ (4°C) in a microcentrifuge to pellet the Streptavidin matrix.

4. Carefully discard the supernatant by pipetting (optional: take a 50-μL sample for immunoblot analysis, to study how much of the target protein has been captured).

5. Apply 1-mL wash buffer and mix by inverting the tube till the Streptavidin matrix has been fully resuspended.

6. Repeat this washing step five times. Carefully pipette off as much as wash buffer as possible after the last washing step.

7. Add 45 μL (2× the bed volume of the Streptavidin matrix) of elution buffer and incubate 5 min at room temperature or for a longer time on ice. Vortex very gently to resuspend the Streptavidin matrix and repeat this a couple of times during the elution.

8. Spin the tube for 15 s at $16,000 \times g$ (4°C) in a microcentrifuge and collect the eluate in a new 1.5-mL tube (see Notes 2 and 3).

9. Repeat steps 7 and 8 once and pool the elutes for further analysis.

10. Add 2× protein sample buffer to the matrix in a volume that is equal to the total amount of elution buffer used (90 μL in this case) and incubate for 5 min at 99°C. This sample can be analyzed by immunoblotting to study the elution efficiency.

4. Notes

1. Note that precasted Sephadex G-25 columns are commercially available.

2. If the results of the Streptavidin pull-down will only be analyzed by immunoblot analysis, precipitated proteins can also be

eluted by incubating the Streptavidin matrix for 5 min at 99°C in 2× protein sample buffer. For analysis on Coomassie Brilliant Blue- or silver-stained protein gels, elution with D-biotin is recommended, because this results in a more specific elution, with a lower background.

3. When scaled up (e.g., 5–10 times), incubate with the Streptavidin matrix in a 10- or 15-mL tube with a conical bottom. For washing, spin the matrix for 1 min at $500–1,000 \times g$. After the last washing step, transfer the Streptavidin matrix to a 1.5-mL reaction tube after having resuspended it in 1 mL of wash buffer. Spin the tube for 1 min at $500–1,000 \times g$ and carefully pipette off the supernatant. Then proceed from step 7 by adding two bed volumes of elution buffer.

Acknowledgments

I thank Matthieu Joosten for critical reading of the manuscript. I am grateful to the EU-funded Integrated Project Bioexploit, the Netherlands Organization for Scientific Research (NWO; VENI grant 863.08.018 to WILT) and the Centre for BioSystems Genomics (CBSG), which is part of the Netherlands Genomics Initiative and NWO, for support.

References

1. Jones, J.D.G., and Dangl, J.L. (2006). The plant immune system. *Nature* 444, 323–329.

2. Takken, F.L.W., Albrecht, M., and Tameling, W.I.L. (2006). Resistance proteins: molecular switches of plant defence. *Curr Opin Plant Biol* 9, 383–390.

3. Zhang, Y., Goritschnig, S., Dong, X., and Li, X. (2003). A gain-of-function mutation in a plant disease resistance gene leads to constitutive activation of downstream signal transduction pathways in *suppressor of npr1-1, constitutive 1*. *Plant Cell* 15, 2636–2646.

4. Shirano, Y., Kachroo, P., Shah, J., and Klessig, D.F. (2002). A gain-of-function mutation in an Arabidopsis Toll Interleukin1 receptor-nucleotide binding site-leucine-rich repeat type R gene triggers defense responses and results in enhanced disease resistance. *Plant Cell* 14, 3149–3162.

5. Qi, Y., and Katagiri, F. (2009). Purification of low-abundance Arabidopsis plasma-membrane protein complexes and identification of candidate components. *Plant J* 57, 932–944.

6. Sacco, M.A., Mansoor, S., and Moffett, P. (2007). A RanGAP protein physically interacts with the NB-LRR protein Rx, and is required for Rx-mediated viral resistance. *Plant J* 52, 82–93.

7. Tameling, W.I.L., and Baulcombe, D.C. (2007). Physical association of the NB-LRR resistance protein Rx with a Ran GTPase–activating protein is required for extreme resistance to *Potato virus X*. *Plant Cell* 19, 1682–1694.

8. Guttierez-Pulgar, J.R., Balmuth, A.L., Ntoukakis, V., Mucyn, T.S., Gimenez-Ibanez, S., Jones, A.M., and Rathjen, J.P. (2010). Prf immune complexes of tomato are oligomeric and contain multiple Pto-like kinases which diversify effector recognition. *Plant J* 61, 507–518.

9. Keefe, A.D., Wilson, D.S., Seelig, B., and Szostak, J.W. (2001). One-step purification of recombinant proteins using a nanomolar-affinity streptavidin-binding peptide, the SBP-Tag. *Protein Expr Purif* 23, 440–446.

10. Van der Hoorn, R.A.L., Laurent, F., Roth, R., and De Wit, P.J.G.M. (2000). Agroinfiltration is a versatile tool that facilitates comparative analyses of *Avr9/Cf-9*-induced and *Avr4/Cf-4*-induced necrosis. *Mol Plant-Microbe Interact* 13, 439–446.

Chapter 37

Karyotyping Methods for Fungi

Rahim Mehrabi, Masatoki Taga, Mostafa Aghaee, Pierre J.G.M. de Wit, and Gert H.J. Kema

Abstract

Pulsed field gel electrophoresis enables separation of fungal chromosomes up to several megabases and is a worthwhile tool for fungal karyotyping. The germ tube burst method is a technique to separate fungal chromosomes of any size for chromosome number determination as well as in situ hybridization. Here we provide detailed protocols for both complementary methods that have many applications in fungal biology including chromosome size and chromosome number polymorphisms, and in situ localization of genes on chromosomes.

Key words: Electrophoretic karyotyping, Pulsed field gel electrophoresis, Germ tube burst method, Chromosome polymorphisms

1. Introduction

Separation of DNA molecules by common gel electrophoresis is performed by running DNA through a solid matrix like agarose under a constant electric field. However, large DNA molecules (>50 Kb) are beyond the resolution of conventional gel electrophoresis techniques. The concept of pulsed field gel electrophoresis (PFGE) to separate large DNA molecules up to 10 Mb was first described by Schwartz et al. (1). This technique is based on a direct current electric field that periodically alters direction and/or intensity. Due to the hexagonal geometry of the electrode array (CHEF: contour clamped homogeneous electrical field) the PFGE apparatus generates a 120° reorientation field as well as periodical changes in the strength and the direction of the electric field, allowing large DNA molecules to migrate through the agarose matrix by a zigzag movement. Migration of small DNA molecules in the gel matrix

Melvin D. Bolton and Bart P.H.J. Thomma (eds.), *Plant Fungal Pathogens: Methods and Protocols*,
Methods in Molecular Biology, vol. 835, DOI 10.1007/978-1-61779-501-5_37, © Springer Science+Business Media, LLC 2012

requires a relatively short pulse time in the direction of the electric field, whereas relatively long pulse times enable the migration of large DNA molecules. The running parameters of the PFGE apparatus can be adjusted over a long range, which provides a universal method for separation of DNA molecules ranging from kilobases (Kb) to megabases (Mb). PFGE has been used efficiently to determine chromosome length (CLPs) and chromosome number polymorphisms (CNPs) among individual fungal strain from natural populations and progeny from a genetic cross. By using chromosome size standards such as *Saccharomyces cerevisiae*, *Saccharomyces pombe*, and *Hansenula wingei*, PFGE is fairly accurate in the measurement of chromosome size and detection of small size chromosomes. It can also be used for PCR detection of genes on excised chromosome bands after completion of PFGE or by Southern analysis on the complete PFGE gel and chromosome painting. However, separation of large size chromosomes (>7 Mb) by this technique is difficult. In addition, PFGE is prone to underestimate chromosome numbers due to co-migration of similar sized chromosomes in the gel. Morphological features of chromosomes cannot be obtained by PFGE. Therefore, a cytological method, the germ tube burst method (GTBM), has been employed in combination with PFGE for more accurate karyotyping of fungal genomes.

GTBM was first introduced to enhance the spread of mitotic chromosomes of *Botryotinia fuckeliana* (anamorph: *Botrytis cinerea*) (2). This method was further modified and developed for various other fungal species during the past decade (3–9). Basically, conidia are suspended in complete medium and placed on a glass slide for germination. Germ tubes adhering to a glass slide are made to burst by treatment with methanol–acetic acid solution to release metaphase chromosomes for further analyses. The GTBM has been successfully applied to karyotyping of several fungi including *Alternaria* spp. (3, 4), *Fusarium graminearum* (8) *Mycosphaerella graminicola* (5), *Cochliobolus* spp. (6), *Nectria haematococca* (7), and *Cryphonectria parasitica* (9). These studies indicated that the GTBM is very useful to visualize and enumerate fungal mitotic metaphase chromosomes independent of their size. In some cases the GTBM provides information on chromosome morphology.

Therefore, PFGE and GTBM are complementary karyotyping methods, which were used for molecular and cytological karyotyping of several fungal species that has enhanced the knowledge on structural genomics in fungal biology, including the identification of horizontal chromosome transfer among related fungal species like *Fusarium oxysporum* and several *Alternaria* species (3–13).

In an era of ongoing developments in genome sequence technologies, PFGE and GTBM provide complementary techniques for structural genomic analyses of fungi, which stimulated us to provide detailed protocols for both methods.

2. Materials

2.1. Equipment

1. CHEF (Contour-clamped homogeneous electric field)-type apparatus including all accessories.

2. Centrifuge. Various bench centrifuges can be used. Do not use angle fixed centrifuge.

3. Miracloth with typical pore size: 22–25 μm (Calbiochem).

4. Filter funnel: Cut miracloth into approximately 15 × 15 cm pieces. Place the miracloth in a medium size funnel (diameter approximately 7 cm). Wrap the funnel completely in aluminum foil and autoclave it.

5. High-speed blender base (Fungi Perfecti®) and 250 mL autoclavable container (Fungi Perfecti®).

6. Epifluorescence microscope (e.g., Olympus BH2/BHS-RFC or Nikon Eclipse E600/Y-FL).

7. Standard flat glass microscope slides (approximately 1 × 3 in. (25 × 75 mm). Use frosted-end slides and label them using ordinary pencil.

8. Cover slips 0.13–0.17 mm thick suitable for oil immersion.

9. Slide box and glass slide jar.

10. Disposable plug molds (BioRad; Cat# #170-3713).

2.2. Reagents

1. Cell wall-degrading enzymes. Sigma lysing enzymes known as Glucanex® (Sigma, Cat#: L1412) (final concentration in protoplasting buffer 10 mg/mL). Driselase® (Sigma, Cat#: D8037) (final concentration in protoplasting buffer 20 mg/mL). Chitinase (Sigma, Cat#: C8241) (final concentration in protoplasting buffer 1 mg/mL) (see Note 1).

2. Low melting agarose.

3. Lysing solution: 500 mM EDTA, 10 mM Tris (pH 8.0), 1% (w/v) N-lauroylsarcosinate, 1 mg/mL proteinase K.

4. *Saccharomyces cerevisiae* chromosomal DNA size marker, strain YNN295 (Bio-Rad, Cat#: 170-3605) appropriate as a marker for medium and small size chromosomes (0.2–2 Mb).

5. *Hansenula wingei* chromosomal DNA size marker strain YB-4662-VIA (Bio-Rad, Cat#: 170-3667) appropriate marker for medium size chromosomes (1–3 Mb).

6. *Schizosaccharomyces pombe* chromosomal DNA size marker strain 972h (Bio-Rad, Cat#: 170-3633) appropriate marker for large size chromosomes (3–6 Mb).

7. Protoplasting buffer: 1 M sorbitol or 1.2 M $MgSO_4$.

8. STC (1 M Sorbitol, 50 mM Tris (pH 8), 50 mM $CaCl_2$).

9. Running buffer: 0.5× TBE buffer (45 mM Tris, 45 mM boric acid, 1 mM EDTA, pH 8.0).

10. Thiabendazole. Dissolve 50 mg/mL in dimethylsulfoxide (DMSO) to prepare a 1,000× stock solution.

11. Poly-L-Lysine solution (Sigma). Prepare working solution as 1 mg/mL Poly-L-Lysine dissolved in water. Store for up to 2 months at 4°C.

12. Vectashield antifade mounting solution (Vector Laboratories).

13. DAPI (4′,6-diamidino-2-phenylindole). To make the 1,000× stock solution, dissolve 1 mg/mL DAPI in antifade mounting solution.

14. Propidium iodide (PI). To make the 1,000× stock solution, dissolve 1 mg/mL PI in antifade mounting solution.

3. Methods

3.1. PFGE

3.1.1. Conidium-Based Fungal Biomass Preparation

1. For some fungi that produce abundant spores in liquid or agar media, conidial cell wall digestion could be applied for protoplast preparation. This method is basically suitable for yeast or yeast-like fungi such as *Mycosphaerella graminicola*.

2. Grow fungus in appropriate liquid culture to produce conidia. Alternatively, grow the fungus on a solid medium and wash off conidia from agar plates with sterile water to prepare conidial suspension.

3. Pass the conidial suspension or liquid culture through sterile miracloth (filter funnel) to remove hyphal debris.

4. Transfer the conidial suspension to 50 mL disposable screw-cap tubes (e.g., Falcon tube).

5. Collect conidia by centrifugation at $3,000–3,500 \times g$ for 5 min.

6. Decant the supernatant and wash the pellet with sterile water. Repeat washing step with 1 M sorbitol. Use the fungal pellet for protoplast preparation and follow the protoplasting procedure (see Subheading 3.1.4).

7. Conidial digestion for some fungi like *Fusarium graminearum* does not generate enough protoplasts and thus we recommend following the germ tube-based protoplast preparation.

3.1.2. Germ Tube-Based Fungal Biomass Preparation

1. Prepare conidia as mentioned in the previous section.

2. Inoculate 10^8 conidia into 100 mL of rich liquid medium in a 250 mL Erlenmeyer flask. Incubate at optimal temperature for several hrs under continuous shaking at 120–170 rpm.

3. Check conidial germination regularly by microscopy until >95% of the conidia have germinated.

4. Centrifuge the culture at 3,000–3,500 ×*g* for 5–10 min to collect the germinated conidia. If centrifugation does not precipitate the fungal biomass, filter the culture through sterile miracloth. Germinated conidia should stay on the miracloth.

5. Wash the germinated conidia by gradually adding 50 mL of 1 M sorbitol into filter funnel.

6. Collect the germinated conidia using a sterile scalpel and transfer them to a 50 mL tube and follow the protoplasting procedure (see Subheading 3.1.4). Add protoplasting buffer (20 mL solution/100 mL initial culture).

3.1.3. Mycelium-Based Fungal Biomass Preparation

1. Fungi that do not produce abundant conidia, conidium- or germtube-based protoplast preparation techniques do not provide sufficient high quality protoplasts. In those cases, follow mycelium-based protoplast preparation.

2. Grow the fungus on nutritious agar plates (like PDA or V8 agar) for several days at appropriate condition.

3. Slice one square inch of agar culture and transfer it to blender container containing 50-mL rich liquid medium supplemented with 50 µg/mL ampicillin.

4. Blend the culture for 10–30 s, transfer the homogenized mycelia to a 250-mL flask and incubate under continuous shaking for at least 24 h (prolong if necessary) until the culture gets blurry.

5. Blend the culture again for 10–30 s and pour the mixture back into a sterile 250-mL flask. Add additional 50-mL nutritious liquid medium containing 50 µg/mL ampicillin. Shake for 24 h at appropriate temperature (prolong incubation time if necessary).

6. Collect mycelial mat by filtering through sterile miracloth. Wash the mycelial mat by gradually adding 50 mL of 1 M sorbitol into the filter funnel.

7. Transfer the mycelial mat to a 50-mL tube and follow the protoplasting procedure (see Subheading 3.1.4).

3.1.4. Protoplasting

1. Resuspend the fungal biomass in 20 mL of 1 M sorbitol containing lysing enzymes. A combination of glucanex® or driselase with chitinase or other cell wall-degrading enzymes could be used for effective cell wall digestion. Check the literature to identify which enzyme combination is most efficient or experimentally determine the optimal enzyme combination and concentration for the studied fungus.

2. Digest fungal cell walls by incubating for 30 min or longer at 30°C under gentle shaking at 60 rpm (see Note 2).

3. Monitor protoplasting by light microscope after 30 min and prolong if necessary.

4. Pass the digested fungal biomass through filter funnel and collect the flow-through (containing protoplasts) in 50-mL screw-cap tubes.

5. Concentrate protoplasts by centrifugation at $3,000 \times g$ for 10 min and resuspend in STC at a concentration of 5×10^8 protoplast/mL.

6. Make 1% low melting agarose solution prepared in STC and heat the melted agarose to 50°C in a water bath.

7. Prewarm the protoplast suspension to 40°C and mix gently but thoroughly with an equal volume of 1% low melting agarose. Keep the protoplast/agarose mixture at 50°C while mixing by gentle pipetting.

8. Transfer the mixture to plug molds using sterile pipettes. Place the molds at 4°C for 10–15 min and allow the agarose to solidify.

9. Take out the plugs and incubate in lysing solution for 24 h at 50°C. Each 1 cm² plug requires at least 3 mL of lysing solution.

10. Wash the plugs in 30 mL of 50 mM EDTA (pH 8.0) for 30 min at room temperature with gentle agitation. Repeat washing steps four times.

11. Store the plugs in 50 mM EDTA (pH 8.0) at 4°C until use. The plugs are stable at least for 3 months.

3.1.5. Pulsed Field Gel Electrophoresis

1. Assemble the gel tray and adjust the height of the comb to 2 mm above the surface of the platform. Placing a thin plastic ruler between the comb and the gel platform makes a good height gauge.

2. Prepare approximately 120 mL of 0.8% SeaKemH gold agarose (see Note 3) in 0.5× TBE buffer. Heat to boiling in microwave, cool in water bath at 55°C, and pour it into the casting stand for a thickness of approximately 6–7 mm. Allow the gel to solidify for 30 min at room temperature.

3. Carefully remove the comb. Sample plugs can be added to the wells while the gel remains in the casting stand.

4. Cut the sample plugs to the size of wells using a razor blade or spatula and gently place into the wells.

5. Samples should be less than 90% of the height of the wells. Place agarose plugs onto the front walls of the sample wells using a spatula and gently press them to the bottom of the wells. Press the plugs firmly against the front walls of the wells to remove possible air bubbles.

6. Fill each sample well with low melting agarose at an agarose concentration equal to that of the gel, and allow the agarose to solidify at room temperature for 10–15 min.

7. Disassemble the gel cast and gently move the gel into the electrophoresis tank. The electrophoresis tank should be washed 2–3 times with sterile water and once with running buffer preferably for several hours while buffer circulation is on.

8. Add running buffer to the electrophoresis tank and make sure that the gel is covered by about 2 mm of buffer. Fix the gel position at each corner using the four angular knobs provided.

9. Adjust the buffer flow, if necessary, by using the flow adjustment knob on the variable speed pump. Typically, the dial is set at 70, for about 0.75 L/min.

10. Enter run parameters. Small chromosomes can be separated in 1% SeaKemH gold agarose under 200 V at 11°C with a 60–120 s switching interval for 24 h.

11. Separate large chromosomes in 0.8% SeaKemH gold agarose (see Note 4) using the running conditions, 50 V with a ramped 3,600–1,800 s switching interval for 115 h, 50 V with a ramped 1,800–1,300 s for 24 h; 60 V with a ramped 1,300–800 s interval for 30 h and 80 V with a ramped 800–600 s interval for 27 h.

12. After electrophoresis stain the gel with ethidium bromide (0.5–1 mg/mL) for 30 min and destain in water for 1 h with gentle agitation in darkness.

13. Use *Saccharomyces cerevisiae* YPH80, *Hansenula wingei* YB-4662- VIA and *Schizosaccharomyces pombe* 972 h size standards for small, medium, and large chromosomes, respectively. Use two different size standards in the first and last lanes of the gel (Fig. 1).

3.2. Germ Tube Burst Method

3.2.1. Sample Preparation

1. Grow fungus under conditions that are appropriate for conidial production. Harvest conidia form agar plates or liquid culture and pass the conidial suspension through sterile miracloth (filter funnel).

2. Transfer the conidial suspension in 50 mL disposable screw-cap tubes and collect fungal conidia by centrifugation at $3,000$–$3,500 \times g$ for 5 min.

3. Decant the supernatant and wash the pellet twice with sterile water.

4. Resuspend the conidia in nutritious liquid medium (e.g., PDB, potato dextrose broth) and adjust the conidia concentration at 2–3×10^5 conidia/mL.

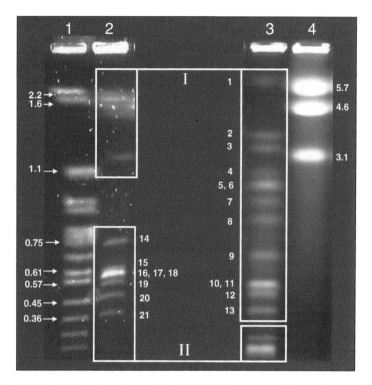

Fig. 1. PFGE of *Mycosphaerella graminicola* IPO323 chromosomes. The numbers on the *left* and *right side* of the figure represent the chromosome sizes of the standards in Mb, *Saccharomyces cerevisiae* (1), *Schizosaccharomyces pombe* (4). Lanes 2 and 3: separation of the small sized and the large sized chromosomes, respectively. Chromosome bands are numbered based on their descending order of size. Chromosome bands can be identified in Boxes I and II that contain large and small chromosomes, respectively. Bands with a higher intensity represent co-migrating chromosomes.

5. Coat the microscope slides with poly-L-lysine by dipping them in the poly-L-lysine solution. Leave for 5–10 min with gentle agitation. Let the slides to dry under flow cabinet (see Note 5).

6. On the top of the slides make a rectangular frame with the rubber cement glue (see Note 6).

7. Pipette 150–200 µL of conidial suspension on the top of slides and inside the rectangular frame.

8. Put the slides in a humid chamber and incubate at appropriate temperature. Keep the humid chamber in darkness that might stimulate conidial germination in some cases.

9. Monitor the germination status after 6 h using light microscopy and prolong incubation if necessary.

10. Pipette out the liquid from the surface of the slide. Do not allow the slides to dry.

11. Gently pipette 200 µL of the same liquid medium (e.g., PDB) containing the thiabandozole (TBZ) solution at a final concentration of 50 µg/mL to arrest mitosis at metaphase. Additionally other chemical treatments such as DNA synthesis inhibitor hydroxyurea at a final concentration of 50 mM can be used to synchronize mitosis to increase methaphase frequency.

12. Incubate for additional 2–3 h at 22°C.

13. Deplete the TBZ solution by gently pipetting.

14. Remove the rubber cement glue with a fine forceps.

15. Slowly dip the slide on sterile MQ for a while to wash off TBZ solution.

16. Wipe off excess of water with filter paper leaving the germlings wet.

17. Gently immerse the slide in fixation solution (methanol: glacial acetic acid 17:3, or 9:1 (v/v)) for 15–30 min. Fixation solution should be prepared freshly (see Note 7).

18. Dry the slide by passing it quickly a few times over smooth flame. Prevent overdrying or overheating. Put the slide in flow cabinet for several minutes to dry completely (see Note 8).

3.2.2. Staining and Observation (in Darkness)

1. Transfer 20–25 µL of 1 µg/mL DAPI solution onto the slide without touching the slide surface.

2. Alternatively stain the chromosomes with both 1 µg/mL DAPI and 1 µg/mL PI (DAPI/PI staining) (see Note 9).

3. Gently put a cover slip on the slide and keep it in darkness for 10–15 min.

4. Fix the cover slip with common nail polisher.

5. Observe chromosomes by UV-excitation using an epifluorescence microscope equipped with a 100× oil immersion objective lens.

6. In order to identify best positions, use the movement knob to monitor the sample from top-left corner along the X-axis by reciprocating movement.

7. Record the photographs on 400 ASA/ISO negative color film or an Olympus digital camera (C5050-Z or DP-70) attached to the microscope with appropriate adapters.

8. Develop negative color film and digitalize the images using a film scanner (e.g., CoolScan IV, Nikon).

9. Process the images with Photoshop version 7 (Adobe) (Fig. 2).

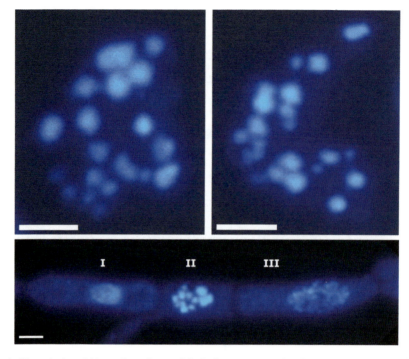

Fig. 2. Cytological karyotyping of *Mycosphaerella graminicola* chromosomes using the germ tube burst method. *Top panel* shows metaphase chromosomes released from germ tubes. *Bottom panel* shows *M. graminicola* nonreleased chromosomes in a single hypha at resting (I), metaphase (II), and prophase (III) stages. *Bars* represent 2 μm.

4. Notes

1. Dissolve enzyme mixture in protoplasting buffer and stir in Erlenmeyer flask for 30 min. Filter sterilize in a 0.45 μm filter.

2. Protoplasts are sensitive to physical shearing by vortexing or pipetting, osmotic shock, and detergents.

3. The agarose concentration affects the size range of DNA molecules separated, and the sharpness, or tightness, of the bands. Agarose concentrations of 0.8% are useful in separating DNA molecules up to 3 Mb in size. Increase the agarose concentrations in the range of 1.2–1.5% to improve band tightness. Gel concentrations below 0.8% (0.5–0.7%) are useful for separation of extremely high molecular weight DNA, greater than 3 Mb. Usually low agarose concentrations generate fuzzy bands with low sharpness and tightness.

4. Do not use high voltage for separation of large chromosomes. Good resolutions can be achieved at low voltage ranging from 50 to 80 V and longer switch time. For better band resolution adjust switch time ramp.

5. Poly-L-Lysine is a positively charged chemical and thus provides higher adhesion of chromosomal DNA or any negatively charged cells to the slide. This reduces the chances of cells or DNA loss during processing. The increased adhesion level permits easier manipulating of samples.

6. Fungal conidia will be pipetted in this rectangular frame. This prevents leaking of the fungal suspension during incubation. The rubber cement can be easily removed after conidial germination.

7. The ratio of methanol to acetic acid solution in the fixative is crucial the preparation of good quality specimens. A higher methanol proportion tends to increase the frequency of burst germlings, but damages chromosomes to a greater extend than with less methanol.

8. Slides can be maintained at room temperature for several days or weeks. The stained slides can be kept in the refrigerator for several days.

9. DAPI binds preferably to AT bases, while PI binds to DNA with almost no sequence preference. When DAPI/PI staining is used, AT-rich regions will show more intensely than with DAPI alone.

References

1. Schwartz, D. C., Saffran, W., Welsh, J., Haas, R., Goldenberg, M., and Cantor, C. R. (1983) New techniques for purifying large DNAs and studying their properties and packaging, *Cold Spring Harb Symp Quant Biol* **47 Pt 1**, 189–195.

2. Shirane, N., Masuko, M., and Hayashi, Y. (1989) Light microscopic observation of nuclei and mitotic chromosomes of *Botrytis* species, *Phytopathology* **79**, 728–730.

3. Taga, M., and Murata, M. (1994) Visualization of mitotic chromosomes in filamentous fungi by fluorescence staining and fluorescence in-situ hybridization, *Chromosoma* **103**, 408–413.

4. Akamatsu, H., Taga, M., Kodama, M., Johnson, R., Otani, H., and Kohmoto, K. (1999) Molecular karyotypes for *Alternaria* plant pathogens known to produce host-specific toxins, *Current Genetics* **35**, 647–656.

5. Mehrabi, R., Taga, M., and Kema, G. H. J. (2007) Electrophoretic and cytological karyotyping of the foliar wheat pathogen *Mycosphaerella graminicola* reveals many chromosomes with a large size range, *Mycologia* **99**, 868–876.

6. Tsuchiya, D., and Taga, M. (2001) Cytological karyotyping of three *Cochliobolus* spp. by the germ tube burst method, *Phytopathology* **91**, 354–360.

7. Taga, M., Murata, M., and Saito, H. (1998) Comparison of different karyotyping methods in filamentous ascomycetes - a case study of *Nectria haematococca*, *Mycological Research* **102**, 1355–1364.

8. Gale, L. R., Bryant, J. D., Calvo, S., Giese, H., Katan, T., O'Donnell, K., Suga, H., Taga, M., Usgaard, T. R., Ward, T. J., and Kistler, H. C. (2005) Chromosome complement of the fungal plant pathogen *Fusarium graminearum* based on genetic and physical mapping and cytological observations, *Genetics* **171**, 985–1001.

9. Eusebio-Cope, A., Suzuki, N., Sadeghi-Garmaroodi, H., and Taga, M. (2009) Cytological and electrophoretic karyotyping of the chestnut blight fungus *Cryphonectria parasitica*, *Fungal Genetics and Biology* **46**, 342–351.

10. Ma, L. J., van der Does, H. C., Borkovich, K. A., Coleman, J. J., Daboussi, M. J., Di Pietro, A., Dufresne, M., Freitag, M., Grabherr, M., Henrissat, B., Houterman, P. M., Kang, S., Shim, W. B., Woloshuk, C., Xie, X. H., Xu, J. R., Antoniw, J., Baker, S. E., Bluhm, B. H., Breakspear, A., Brown, D. W., Butchko, R. A. E., Chapman, S., Coulson, R., Coutinho, P.

M., Danchin, E. G. J., Diener, A., Gale, L. R., Gardiner, D. M., Goff, S., Hammond-Kosack, K. E., Hilburn, K., Hua-Van, A., Jonkers, W., Kazan, K., Kodira, C. D., Koehrsen, M., Kumar, L., Lee, Y. H., Li, L. D., Manners, J. M., Miranda-Saavedra, D., Mukherjee, M., Park, G., Park, J., Park, S. Y., Proctor, R. H., Regev, A., Ruiz-Roldan, M. C., Sain, D., Sakthikumar, S., Sykes, S., Schwartz, D. C., Turgeon, B. G., Wapinski, I., Yoder, O., Young, S., Zeng, Q. D., Zhou, S. G., Galagan, J., Cuomo, C. A., Kistler, H. C., and Rep, M. (2010) Comparative genomics reveals mobile pathogenicity chromosomes in *Fusarium*, *Nature* **464**, 367–373.

11. Suga, H., Ikeda, S., Taga, M., Kageyama, K., and Hyakumachi, M. (2002) Electrophoretic karyotyping and gene mapping of seven formae speciales in *Fusarium solani*, *Current Genetics* **41**, 254–260.

12. Zhong, S. B., and Steffenson, B. J. (2007) Molecular karyotyping and chromosome length polymorphism in *Cochliobolus sativus*, *Mycological Research* **111**, 78–86.

13. Taga, M., Murata, M., and VanEtten, H. D. (1999) Visualization of a conditionally dispensable chromosome in the filamentous ascomycete *Nectria haematococca* by fluorescence in situ hybridization, *Fungal Genetics and Biology* **26**, 169–177.

Chapter 38

Isolation of Apoplastic Fluid from Leaf Tissue by the Vacuum Infiltration-Centrifugation Technique

Matthieu H.A.J. Joosten

Abstract

Upon infection of plants by pathogens, at least at the early stages of infection, the interaction between the two organisms occurs in the apoplast. To study the molecular basis of host susceptibility vs. resistance on the one hand, and pathogen virulence vs. avirulence on the other, the identification of extracellular compounds such as pathogen effectors that determine the outcome of the interaction is essential. Here, I describe the vacuum infiltration-centrifugation technique, which is an extremely simple and straightforward method to explore one of the most important battlefields of a plant–pathogen interaction; the apoplast.

Key words: Vacuum infiltration-centrifugation technique, Apoplast, Apoplastic fluid, Intercellular washing fluid, Plant–pathogen interaction, Extracellular space, Communication, Effector, Avirulence, Resistance

1. Introduction

The apoplast of the plant consists of intercellular spaces and cell walls and is the first environment that plant pathogens encounter upon their ingress (1). Both for fungal and bacterial pathogens, of which the latter remain in the apoplast throughout their life cycle, and for oomycete pathogens such as *Phytophthora infestans*, the initial communication that takes place in this area can be decisive for the outcome of the interaction (1–3). For example, when immediate recognition of components of the pathogen by the plant occurs, defence will be mounted leading to plant resistance and pathogen avirulence. However, when the microbe is successful in avoiding or suppressing recognition, then susceptibility of the plant is the outcome and the pathogen is virulent (4).

Melvin D. Bolton and Bart P.H.J. Thomma (eds.), *Plant Fungal Pathogens: Methods and Protocols*,
Methods in Molecular Biology, vol. 835, DOI 10.1007/978-1-61779-501-5_38, © Springer Science+Business Media, LLC 2012

To identify plant and pathogen components that play a role on the extracellular battleground, the isolation of apoplastic fluid (AF), also referred to as intercellular washing fluid (IWF), can be essential. This approach allows the identification and detailed analysis of apoplastic components without the presence of massive amounts of interfering cellular components. For example, for studies on the interaction between the pathogenic fungus *Cladosporium fulvum* and tomato, the isolation of AF has been instrumental. *C. fulvum* is a biotrophic fungus that penetrates tomato leaf tissue through the stomata and remains fully apoplastic during its complete life cycle. Therefore, it was anticipated that all essential communication between the fungus and tomato occurs in the apoplast. Indeed, in a classic paper De Wit and Spikman (5) showed that *C. fulvum*-derived "race and cultivar-specific elicitors of necrosis" (now referred to as "effectors") are present in AF isolated from susceptible tomato genotypes colonised by *C. fulvum*. A small peptide secreted by *C. fulvum* and specifically inducing necrosis (also referred to as the hypersensitive response (HR)) in tomato carrying the *Cf-9* gene for resistance to *C. fulvum* was purified from the AF. The peptide was referred to as Avr9 and the encoding gene (*Avr9*) was the first fungal avirulence gene ever cloned (6). Subsequently, it was shown that the *Cf-9* gene encodes a receptor-like protein (RLP) that mediates recognition of Avr9 by tomato. Interestingly, Cf-9 appeared to be a transmembrane immune receptor that co-localises with Avr9, as it mainly consists of extracellular leucine-rich repeats (7). Soon other *Avrs* were cloned following the same approach, such as *Avr4* (8) and *Avr4E* (9). In addition, it was found that Rcr3, an extracellular cysteine protease of tomato, is the virulence target of the secreted Avr2 effector protein of *C. fulvum*, which proved to be a cysteine protease inhibitor. Rcr3 was found to be guarded by the Cf-2 resistance protein, which is also an RLP, resulting in triggering of resistance by Cf-2 upon manipulation of Rcr3 by Avr2 (10).

The analysis of AF also revealed that in the apoplast of resistant tomato there is a swift accumulation of various classical pathogenesis-related (PR) proteins secreted by the host upon inoculation with an avirulent strain of *C. fulvum*. In susceptible tomato, these PR proteins also accumulate, but this happens after the intercellular spaces have already been fully colonised by the fungus (11, 12). In the apoplast of colonised leaflets, a depletion of sucrose and a massive accumulation of mannitol occurs. The levels of glucose and fructose also increased, suggesting that the fungus hydrolyses sucrose and eventually converts glucose and fructose into mannitol. The latter carbohydrate cannot be metabolised by tomato, whereby the equilibrium of nutrient resources is shifted to the side of the fungus (13).

The examples given above apply to the interaction between the fully extracellular fungus *C. fulvum* and its host, tomato. However, for interactions involving pathogens with quite different lifestyles, the detailed analysis of AF obtained from infected leaflets has also proven to be very informative. For example, for the oomycete pathogen *P. infestans*, which feeds on the plant through haustoria that have an intimate interaction with the host cells, similar observations were made by focussing on events that take place in the apoplast. Also here, secreted pathogen effectors target extracellular host proteases (3), and upon recognition of the pathogen by resistant potato genotypes there is a substantial increase in the amount of extracellular PR proteins (2). These examples illustrate that the isolation of AF, which is a very simple and elegant technique, will be beneficial to studies on the molecular mechanisms underlying plant–pathogen interactions. As reviewed by Lohaus et al. (14), various methods have been developed to obtain AF. However, the vacuum infiltration-centrifugation technique is by far most widely used and is described here.

2. Materials

1. Perspex centrifuge tubes consist of two parts (see Fig. 1). The upper part has a perforated bottom and loosely fits on the lower part. In the lower part, the apoplastic fluid is collected upon centrifugation (see Note 1).

2. Vacuum pump.

3. Centrifuge with swing-out rotor.

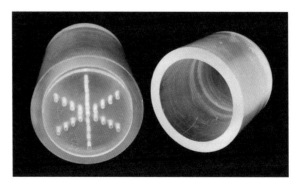

Fig. 1. Centrifuge tube used for apoplastic fluid isolation. The tube shown has an internal diameter of 4.2 cm and an external diameter of 5.3 cm. The upper part (*left*) has a perforated bottom and has an inner depth of 6.5 cm. It loosely fits on the lower part (*right*, inner depth 4.3 cm), in which the apoplastic fluid is collected upon centrifugation.

3. Methods

3.1. Infiltration of Leaves

1. Pick leaves, or leaflets and stack them in a beaker of sufficient volume. For example, for 20 tomato (*Solanum lycopersicum*) leaflets a beaker with a volume of 400 mL is used. Although this will increase contamination of the AF by cellular compounds (see below), big leaves, like those from tobacco (*Nicotiana tabacum*) can be cut into strips of about 10 cm wide or folded carefully to fit the tube after infiltration. From Arabidopsis (*Arabidopsis thaliana*) entire rosettes can be used to infiltrate.

2. To prevent the leaves from floating, place a weight (for example flat stones or lead rings) on them. Fill the beaker with deionised water, or another solution to be used for infiltration (see Fig. 2, Note 2).

3. Place the beaker in a desiccator jar and apply a vacuum employing a vacuum pump. A final pressure of about 60 mbar (hPa), which is about 46 Torr (mmHg) is sufficient.

4. Slowly release the vacuum by opening the vent on the desiccator jar. Ideally, it takes about 10 min to fully release the vacuum. During vacuum release the leaves are slowly infiltrated with the surrounding solution, a process which is clearly visible as the

Fig. 2. Tomato leaflets ready to be infiltrated. Stones have been placed on the leaves to prevent them from floating and water has been added.

Fig. 3. Tomato leaflets after infiltration. Note that the leaf tissue becomes dark as a result of the infiltration, which results in water-soaking. The *arrow* indicates a small area of tissue that has not been fully infiltrated.

tissue becomes water-soaked and dark in colour (see Fig. 3). This procedure normally results in 90–100% infiltration of the leaf area.

3.2. Isolation of AF by Centrifugation

1. Remove the weight, pour off the excess of water and remove most of the water between the leaves by gentle shaking the stack of infiltrated leaves.

2. Carefully remove the remaining water from the leaf surface by drying the leaves between a few layers of tissue paper. When large amounts of leaves have to be processed a salad-spinner can be used at low speed for this purpose.

3. Stack and roll up the infiltrated leaves and place them in the upper part of the centrifuge tube (see Fig. 4, Note 3). Place this on the lower part of the centrifuge tube and centrifuge at 3,000 $\times g$ for 10 min at 4°C (5) (see Note 4).

4. During centrifugation, the AF is extruded from the leaves and is collected in the lower part of the centrifuge tube (see Fig. 5). After centrifugation, immediately recover the AF with a pipette and place on ice. Aliquot and store at –20°C until further use. This procedure typically yields 0.5 mL of AF per gram of tomato leaflet tissue (fresh weight) (see Notes 5 and 6).

Fig. 4. Infiltrated tomato leaflets, dried, stacked and rolled up and subsequently inserted into the upper part of the centrifuge tube.

Fig. 5. Tube after centrifugation, with extruded leaves in the upper part and AF collected in the lower part. Note that the leaflets are pressed together due to the centrifugal force.

4. Notes

1. Instead of the custom-made centrifuge tubes shown in Fig. 1, spin columns with filter in the bottom, or disposable 25 or 50 mL syringes (without plunger) can be used to isolate small volumes of AF from a few leaflets. These are placed in a centrifuge tube, after which the AF is collected in the tube upon centrifugation.

2. In addition to deionised water, particular solutions with varying osmolality and/or pH can also be used for vacuum infiltration (14). In combination with different centrifugation forces this affects the concentration of, for example, ions and metabolites in the AF.

3. Infiltrated long and narrow leaves, such as those from the monocotyledonous cereals, can be rolled in Miracloth before placing them in the centrifuge tube. This allows faster handling of the leaves and will result in less cellular contamination of the AF.

4. Centrifugation at high centrifugation forces, such as 3,000 xg, ensures complete recovery of the AF and therefore provides high yields of strictly extracellular compounds. However, it also results in slight contamination of the AF with cytoplasmic and/or vacuolar components, which in most cases is not a problem, as only enrichment for apoplastic components is required. If cellular contamination needs to be avoided, a centrifugation force not exceeding 1,000 xg should be used. Cellular contamination of the AF can be determined by measuring the activity of malate dehydrogenase (MDH) (14), hexose phosphate isomerase (HPI) (14) or α-mannosidase (15).

5. When deionised water or a solution with only a low amount of additives is used, the AF can directly be analysed for its contents. For example, its protein composition can be determined by sodium dodecyl sulphate-polyacrylamide gel electrophoresis (SDS-PAGE) followed by staining with Coomassie Brilliant Blue. For this, aliquots of the AF of 25–100 μL are freeze-dried and the residue is dissolved in loading buffer. After boiling for 5 min and centrifugation in a microcentrifuge at 16,000 xg for 10 min, the supernatant can be loaded on gel.

6. AF isolation from leaf tissue heavily infected with the pathogen of choice can result in AF containing large amounts of phenolic defence compounds of the plant and pathogen-derived metabolites which interfere with analyses like SDS-PAGE. In such cases the AF can be cleaned up by passing it through a (pre-casted) column containing Sephadex G-10. Alternatively, proteins can be precipitated from the AF by adding acetone to a final concentration of 50–90% and the pellet is then analysed by SDS-PAGE.

Acknowledgments

Wladimir Tameling is thanked for his help with making the pictures and for critical reading of the manuscript. I thank the Centre for BioSystems Genomics (CBSG), which is part of the Netherlands Genomics Initiative and NWO, for support.

References

1. Hammerschmidt R (2010) The dynamic apoplast. Physiol Mol Plant Pathol 74:199–200.

2. Kombrink E, Schröder M, Hahlbrock K (1988) Several "pathogenesis-related" proteins in potato are 1,3-β-glucanases and chitinases. Proc Natl Acad Sci USA 85:782–786.

3. Kaschani F, Shabab M, Bozkurt T, Shindo T, Schornack S, Gu C, Ilyas M, Kamoun S, Van der Hoorn RAL (2010) An effector-targeted protease contributes to defense against *Phytophthora infestans* and is under diversifying selection in natural hosts. Plant Physiol 154: 1794–1804.

4. Thomma BPHJ, Nürnberger T, Joosten MHAJ (2011) Of PAMPs and effectors: the blurred PTI-ETI dichotomy. Plant Cell 23:4–15.

5. De Wit PJGM, Spikman G (1982) Evidence for the occurrence of race and cultivar-specific elicitors of necrosis in intercellular fluids of compatible interactions of *Cladosporium fulvum* and tomato. Physiol Plant Pathol 21:1–11.

6. Van Kan JAL, Van den Ackerveken GFJM, De Wit PJGM (1991) Cloning and characterisation of cDNA of avirulence gene *avr*9 of the fungal tomato pathogen *Cladosporium fulvum*, causal agent of tomato leaf mold. Mol Plant-Micobe Int 4:52–59.

7. Jones DA, Thomas CM, Hammond-Kosack KE, Balint-Kurti PJ, Jones JDG (1994) Isolation of the tomato *Cf-9* gene for resistance to *Cladosporium fulvum* by transposon tagging. Science 266:789–793.

8. Joosten MHAJ, Cozijnsen TJ, De Wit PJGM (1994) Host resistance to a fungal tomato pathogen lost by a single base-pair change in an avirulence gene. Nature 367:384–387.

9. Westerink N, Brandwagt BF, De Wit PJGM, Joosten MHAJ (2004) Cladosporium fulvum circumvents the second functional resistance gene homologue at the Cf-4 locus (Hcr9–4E) by secretion of a stable avr4E isoform. Mol Microbiol 54:533–545.

10. Rooney HCE, Van 't Klooster JW, Van der Hoorn RAL, Joosten MHAJ, Jones JDG, De Wit PJGM (2005) Cladosporium Avr2 inhibits tomato Rcr3 protease required for Cf-2-dependent disease resistance. Science 308: 1783–1786.

11. Joosten MHAJ, De Wit PJGM (1989) Identification of several pathogenesis-related proteins in tomato leaves inoculated with *Cladosporium fulvum* (syn. *Fulvia fulva*) as 1,3-β-glucanases and chitinases. Plant Physiol 89:945–951.

12. Joosten MHAJ, Bergmans CJB, Meulenhoff EJS, Cornelissen BJC, De Wit PJGM (1990) Purification and serological characterization of three basic 15-kilodalton pathogenesis-related proteins from tomato. Plant Physiol 94: 585–591.

13. Joosten MHAJ, Hendrickx LJM, De Wit PJGM (1990) Carbohydrate composition of apoplastic fluids isolated from tomato leaves inoculated with virulent or avirulent races of *Cladosporium fulvum* (syn. *Fulvia fulva*). Neth J Plant Pathol 96:103–112.

14. Lohaus G, Pennewiss K, Sattelmacher B, Hussmann M, Muehling, KH (2001) Is the infiltration-centrifugation technique appropriate for the isolation of apoplastic fluid? A critical evaluation with different plant species. Physiol Plant 111:457–465.

15. Delannoy M, Alves G, Vertommen D, Ma J, Boutry M, Navarre C (2008) Identification of peptidases in *Nicotiana tabacum* leaf intercellular fluid. Proteomics 8:2285–2298.

Chapter 39

Gene Cloning Using Degenerate Primers and Genome Walking

Javier A. Delgado, Steven Meinhardt, Samuel G. Markell, and Rubella S. Goswami

Abstract

Gene cloning is the first step of targeted gene replacement for functional studies, discovery of gene alleles, and gene expression among other applications. In this chapter, we will describe a cloning technique suitable for fungal species where the genomic information and sequences available are limited. This strategy involves obtaining protein sequences of the gene of interest from various organisms to identify at least two conserved regions. Degenerate primers are designed from these two conserved regions and the resulting PCR products are sequenced. The sequence of the PCR products can be analyzed using suitable databases to determine their similarity to the gene/protein of interest. In cases where the entire gene cannot be cloned directly using these primers, this initial nucleotide sequence can be used as a template for further primer design and genome walking in both directions for either the cloning of a longer fragment or even the cloning of the complete gene. Here, we describe the partial cloning of a reducing polyketide synthase gene from the fungal plant pathogen *Ascochyta rabiei* using this strategy.

Key words: Cloning, Degenerate primers, Polyketide synthases, DNA walking

1. Introduction

The availability of complete genome sequences for many model fungi has helped to alleviate many of the technical difficulties associated with gene cloning. However, many economically important plant pathogens still do not have a whole genome sequence available. The approach presented in this chapter is especially important for fungal genomes that have not been sequenced or those species which lack comprehensive expressed sequence tag (EST) databases. The process begins by identifying at least two conserved amino acid regions from the protein family of interest and designing

Melvin D. Bolton and Bart P.H.J. Thomma (eds.), *Plant Fungal Pathogens: Methods and Protocols*,
Methods in Molecular Biology, vol. 835, DOI 10.1007/978-1-61779-501-5_39, © Springer Science+Business Media, LLC 2012

degenerate primers aligned to these regions. The products from the PCR reaction using these primers are sequenced and gene-specific primers are designed that can then be used to extend the size of the cloned fragment using a DNA walking strategy. This strategy allows the user to clone and sequence the flanking regions of any given DNA fragment by two consecutive PCRs.

In this chapter, we present an approach that was used for cloning reducing polyketide synthases (PKSs) from *Ascochyta rabiei*, a fungal pathogen of chickpea, for which no genome sequence is currently available and published genetic studies are very limited. This pathogen is known to produce a phytotoxin solanapyrone (1), the biosynthesis of which involves a reducing type PKS. Therefore, cloning of a reducing PKS from *A. rabiei* provides a good example of gene cloning from a nonsequenced fungal species. The experimental steps involved in cloning a reducing PKS are illustrated, including selection of conserved regions, design of degenerate primers, PCR amplification of a region of the gene of interest, and further extension of the sequence using DNA walking.

2. Materials

2.1. Cultures and Media

1. Potato dextrose agar (PDA): 24 g potato dextrose broth powder and 15 g Bacto Agar are added to 1 L distilled water (dH$_2$O) and autoclaved.

2. Potato dextrose broth (PDB): 24 g potato dextrose broth powder is added to 1 L distilled water (dH$_2$O) and autoclaved.

3. Luria Bertini/Ampicillin (LB/Amp) broth: 10 g tryptone, 5 g yeast extract, 10 g NaCl are added to 1 L dH$_2$O. The pH is adjusted to 7.0 with 5 N NaOH and autoclaved. The medium is allowed to cool to 55°C before adding 1 mL of a filter-sterilized 50 mg/mL solution of ampicillin. Aliquots (4 mL) of this LB/Amp broth are transferred to 15 mL-tubes.

4. LB/Amp agar: 10 g tryptone, 5 g yeast extract, and 10 g NaCl are added to 1 L dH$_2$O. The pH is adjusted to 7.0 with 5 N NaOH. Bacto agar (15 g) is added before autoclaving. Ampicillin is added as described above.

5. LB/Amp/X-gal/IPTG plates: 40 μL of 2% X-gal (5-bromo-4-chloro-3-indolyl-beta-D-galacto-pyranoside) solution prepared by dissolving X-gal in DMSO and 40 μL of 2.5% IPTG (iso-propyl-β-D-thio-galactoside) solution prepared by dissolving IPTG in dH$_2$O. Both X-gal and IPTG are spread onto each LB/Amp plate with a sterile spreading rod. The plates are then kept inverted at 37°C for at least 2 h to ensure absorption of the chemicals.

2.2. Genomic DNA Extraction from Ascochyta rabiei Cultures

1. Genomic DNA purification kit or procedure of preference.

2. Commercially available DNase-free RNase A solution (may be included in the DNA purification kit).

3. Nuclease-free 1.5-mL microcentrifuge tubes.

4. Water bath at 65°C.

5. Water bath at 37°C.

6. Molecular grade isopropanol at room temperature.

7. 70% molecular grade ethanol at room temperature.

8. For fungal cell disruption, we use nuclease-free lysing matrix A tubes with Fastprep® 24 instrument. Also, other implements and supplies such as a mortar and pestle, bead-beater, etc. can be used for disrupting fungal tissue.

2.3. Endpoint PCR and Genome Walking

1. Commercially available PCR master mix.

2. DNA staining dye.

3. 10× Tris/Borate/EDTA (TBE): Add 108 g of Tris base, 55 g of boric acid, 40 mL of 0.5 M EDTA pH 8.0 and bring the volume to 1 L with dH_2O Autoclave for 20 min.

4. Agarose.

5. Commercially available agarose gel purification kit.

6. Commercially available TA cloning kit and competent *E. coli* cells.

7. M13 universal primers, 10 μM.

8. PCR tubes.

9. GenomeWalker™ Universal kit from Clontech.

3. Methods

3.1. Genomic DNA Extraction from Ascochyta rabiei Cultures

Here, we describe the DNA extraction procedure that is routinely used in our lab. This procedure is carried out using the Wizard® Genomic DNA Purification Kit from Promega and cell disruption is performed using a Fast-Prep® instrument and supplies. However, fungal DNA extraction can be conducted using any other suitable protocol and cell disruption can also be achieved using a sterile pestle and mortar or another method. Therefore, this protocol is a modification of the instructions provided by the manufacturer.

1. Cultures of *Ascochyta rabiei* are grown on PDA at 20–22°C for 7 days with a 12 h light regimen.

2. Agar plugs (0.5 cm diameter) containing *Ascochyta* mycelia are cut with a sterile cork borer from the growing edge of each PDA plate. The agar plugs are placed in 15-mL tubes containing

4 mL of PDB. The inoculated tubes are then incubated in a slightly slanted position at 25°C for 2–3 days with orbital shaking at 200 rpm. Subsequently, the cultures are frozen for 24 h and freeze-dried overnight.

3. Freeze-dried fungal tissue (10 mg) is placed in a lysing matrix A tube containing 700 μL of Nuclei lysis solution. Cell disruption is performed by setting the Fast-Prep® instrument at a speed of 6.0 for 40 s. The lysed tissue is then incubated at 65°C for 15 min.

4. After initial incubation, 3 μL of RNase A solution (included in the commercial kit) is added to the cell lysate and mixed by inversion. The mixture is incubated for 15 min in a water bath set at 37°C. The lysate is cooled to room temperature for 5 min. Protein Precipitation solution (200 μL) is added and the lysate is mixed vigorously using a vortex set at maximum speed for 20 s. Undesirable components, such as proteins, are separated by centrifugation at 16,000 ×g for 5 min.

5. The supernatant containing the DNA is carefully removed from the tube and transferred to a clean 1.5-mL tube containing 600 μL of isopropanol at room temperature. The solution is gently mixed by inverting the tube 8–10 times. The DNA is separated by centrifugation at 16,000 ×g for 1 min at room temperature (see Note 1).

6. The supernatant is then carefully discarded and 600 μL of 70% ethanol at room temperature is added. The tube is gently inverted (8–10 times) to wash the DNA. The DNA is collected by centrifugation at 13,000–16,000 ×g for 1 min at room temperature. The ethanol is carefully aspirated using a suitable pipette and tip. The DNA pellet is dried in an incubator set at 37°C for 20 min.

7. The DNA is rehydrated by adding 100 μL of DNA Rehydration solution and incubating the solution overnight at 4°C. The DNA can then be quantified with a Nanodrop ND-1000 spectrophotometer at 260 nm using a 2-μL aliquot or any other mode of quantification. DNA dilutions can be prepared (25 ng/μL) and stored at –20°C. DNA purity can be assessed using the A260/A280 ratio that indicates the levels of protein contamination. A range between 1.8 and 2.0 indicates that DNA purity is acceptable and suitable for cloning and enzymatic digestion.

3.2. Initial Cloning Using Degenerate Primers

Initially, designing degenerate primers involves searching for well-characterized proteins encoded by the gene of interest. An extensive literature search for protein accession numbers of well-characterized homologs of the gene of interest (in our case the reducing polyketide synthases) from various fungi is required

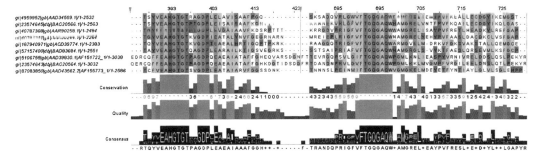

Fig. 1. Representation of a fragment of a multiple alignment of nine reducing PKSs. Two conserved regions were identified among these PKSs (EAHGTGT and FTGQGAQW). This figure was edited with the multiple alignment editor using Jalview (http://www.jalview.org/).

before cloning (2, 3). The selected protein sequences can be retrieved from corresponding databases such as Genbank (http://www.ncbi.nlm.nih.gov/). Multiple alignments of the protein sequences can be performed using ClustalX2 (http://www.clustal.org/) in order to identify conserved regions suitable for designing degenerate primers (see Fig. 1) and primers can be designed manually. In the study presented here, two conserved regions were identified in fungal PKSs. The first conserved region is located between amino acid positions 391 and 397 and the second between 691 and 698. Both conserved regions were located within the beta-ketoacyl synthase domain of this protein family, which is a domain essential for functioning of these proteins. Degenerate primers were manually designed using the genetic code. The forward degenerate primer (PKS391F) was designed from the amino acid sequence EAHGTG that presented a degeneracy of 1,024 and reverse degenerate primer (PKS693R) was designed from the amino acid sequence GQGAQW with a degeneracy of 256 (see Note 2).

PKS391F (18 bp): 5′-GAR GCN CAY GGN CAN GGN-3′

PKS693R (18 bp): 5′-CCA YTG NGC NCC YTG NCC-3′

1. The initial cloning is conducted using PCR with degenerate primers. The reaction is carried out in a 30 µL volume using the TopTaq DNA polymerase kit (Qiagen) as follows: 1× of TopTaq PCR buffer (pH 8.7), 2 mM of $MgCl_2$, 300 µM of each dNTP, 300 nM of each degenerate primer, 0.75 U of TopTaq DNA polymerase, and 1 ng/µL of fungal DNA.

2. The PCR amplification program used for the primers mentioned above is as follows: initial denaturation at 94°C for 3 min; 35 cycles of denaturation at 94°C for 1 min, primer alignment at 60°C for 1 min and extension at 72°C for 3 min; and a final extension at 72°C for 10 min.

3. PCR products are analyzed by electrophoresis in a 1% agarose gel precast with a suitable DNA staining dye in 1× TBE at

80 mV for 2 h and visualized over a transilluminator at a wavelength of 302 nm.

4. PCR products between 500 and 1,500 bp are individually excised from the agarose gel with a sterile scalpel and transferred to a nuclease-free 1.5-mL tube. The PCR products are individually cleaned with a commercially available agarose gel purification kit.

5. The recovered PCR products are cloned using the TA cloning procedures using suitable vectors and competent *E.coli* cells. In our lab, we routinely use the TOPO TA cloning kit with One Shot Mach1™ phage-resistance chemically competent *E. coli* kit (Invitrogen) for sequencing. The ligation is conducted with 4 µL of fresh, clean PCR product for 30 min at room temperature. A 3 µL aliquot of this ligation mix is added to a vial containing 50 µL of competent cells and is incubated on ice for 30 min.

6. Aliquots of the transformed competent cell solution are spread onto prewarmed (at 37°C) LB/Amp/X-gal plates and incubated overnight at 37°C. After overnight incubation, the plates may be kept at 4°C for 2–4 h for further development of the blue colored cells. Insertion of the DNA product in the vector carried in the transformed white colonies is confirmed by PCR using M13 primers (see Notes 3 and 4).

7. The PCR reaction for confirmation of the transformants is prepared as described in step 1 of this section. Bacterial cells are used as reaction template (see Note 5). The PCR amplification is conducted as follows: initial denaturation at 94°C for 10 min; 25°C for 3 min; second denaturation at 94°C for 40 s; 30 cycles of denaturation at 94°C for 20 s, primer alignment at 60°C for 25 s and extension at 72°C for 30 s; and a final extension at 72°C for 10 min.

8. Positive colonies are picked and grown overnight in 3.5 mL of LB/Amp broth at 37°C (see Note 6). Plasmid extraction is then carried out using 3 mL of the culture with a commercially available kit. Plasmid concentration can be determined with a Nanodrop ND-1000 spectrophotometer at 260 nm using a 2-µL aliquot and adjusted before sequencing according to the recommendations of the sequencing service provider.

9. Sequence similarity of the PCR products to PKSs can be assessed using megaBLAST and BLASTX against the NCBI nucleotide (nr/nt) and nonredundant protein (nr) databases, respectively. In the example shown in this chapter, an 842 bp PCR product was found to be highly similar to the KS domain of reducing fungal PKSs and was chosen for the extension of the gene sequence using genome walking.

3.3. Extension of Nucleotide Sequence by Genome Walking

During the presented study, we conducted the extension of the PCR product using a commercially available kit, GenomeWalker™ Universal kit from Clontech. This method allows extension of DNA sequences to obtain larger portions of the gene of interest or the complete gene of interest, depending on gene size. Gene-specific primers that are used are designed based on small initial DNA sequence coding for a region of the gene. This protocol involves creation of four libraries from the organism of interest by digesting its genomic DNA with four restriction enzymes. The digested DNA is then ligated to an adaptor of known sequence. The adaptor contains two sites for adaptor-specific primers (provided by the manufacturer) that when coupled with two gene-specific primers (designed by the user) from the known initial DNA sequence enables amplification of a longer amplicon that contains a portion of the initial DNA sequence and a portion of newly amplified DNA.

Recommendations for the preparation of genomic DNA libraries, design of gene-specific primers, and PCR amplifications can be found in the manufacturer's manual. Here, we present a summarized protocol and recommendations that we followed during the cloning of PKS genes from *Ascochyta rabiei*.

1. A good genomic DNA library requires genomic DNA of good quality. Therefore, the quality of the DNA must be checked by assessing both the size and purity of the fungal genomic DNA. The size of a band seen during electrophoresis of good genomic DNA should be no smaller than 50 kb with minimal smearing. This is determined by electrophoresis of 1 μL of genomic DNA in a 0.6% agarose gel. We routinely conduct this DNA electrophoresis at 80 mV for 2 h and it is visualized over a transilluminator at a wavelength of 302 nm. The size can be estimated by comparison to a suitable DNA ladder. For this particular library, the genomic DNA purity can be assessed by electrophoresis of 5 μL of *Dra*I-digested fungal genomic DNA and undigested DNA as described above. The digestion is conducted as described in the manufacturer's instructions. The digested DNA should produce a smear showing DNA fragments of different sizes. These DNA fragments should not produce any distinctive, well-defined bands. The undigested DNA should be observed as described above, no smaller than 50 kb with minimal smearing.

2. After the assessment of DNA quality, the fungal genomic DNA is ready to be digested for preparation of the four genomic libraries. The DNA is digested in four nuclease-free 1.5-mL tubes with four restriction enzymes (*Dra*I, *Eco*RV, *Stu*I, and *Pvu*II). Upon completion of digestion, the enzymes are inactivated by phenol and the DNA is precipitated with ice cold ethanol, sodium acetate, and glycogen according to the manufacturer's

instructions. The DNA is then rehydrated with TE buffer before ligation. All four blunt-end digested products are ligated to the provided adaptor creating four digested fungal genomic DNA libraries, namely *Dra*I, *Eco*RV, *Stu*I, and *Pvu*II-library.

3. The criteria for primer design using the GenomeWalker™ kit are as follows: the primers must be 26–30 bp in length, the GC content must be between 40 and 60%, and the Tm should be as close to 67°C as possible. The DNA polymerase used with this kit is able to perform at 67°C allowing the alignment and extension step to happen simultaneously. This kit uses a long-distance PCR enzyme from the Advantage® 2 PCR kit also from Clontech. During our work, two sets of primers were designed for genome walking, with one set to extend the nucleotide sequence toward the N-terminal of the PKS proteins (KSN1 and KSN2 primers) and another to extend it toward the C-terminal (KSC1 and KSC2 primers). All primers were 28 bp in length and had a Tm between 59 and 60°C. Primers KSN1, KSC1, and KSC2 had a GC content of 46% while primer KSN2 had 57% GC content. Primer KSN1 aligned to the initial DNA fragment from nucleotide (nt) 58 to 85, primer KSN2 from 61 to 34 nt, primer KSC1 from 513 to 540 nt, and primer KSC2 from 622 to 649 nt. Two separate primary PCRs were carried out according to the manufacturer's instructions using the corresponding gene-specific primers and adaptor-specific primers (AP). The primers KSN1 and AP1 were used for extension toward the 5′-end of the gene; and primers KSC1 and AP1 were used for extension toward the 3′-end of the gene. A 1 μL aliquot of the diluted (1:50) primary PCR product was then used for the secondary or nested PCR with primers KSN2/AP2 and KSC2/AP2, respectively. PCR amplifications were performed according to the manufacturer's recommendations.

KSN1: 5′-GCTTATTTCTGCCTTCTCCGAAGACTCT-3′

KSN2: 5′-CTCTATGGAGAGCCCCTACCTCGATAGG-3′

KSC1: 5′-GAACGTAGGAAGATGAAAGACGAGGTTG-3′

KSC2: 5′-TCTACCTCGAACAACGACCAGAGATCTT-3′

4. PCR products can be analyzed by electrophoresis in a 1% agarose gel and selected PCR products can be excised from the agarose gel, cleaned using suitable gel extraction protocols, cloned, and sent for sequencing as previously described.

5. The sequenced products can be aligned to the original fragment using pairwise global alignment in BioEdit to create a consensus sequence.

Our multiple alignments created a consensus sequence of 1,083 bp. From this sequence, new primers (KSN3, KSN4, KSC3, and KSC4) were designed for further extension of the nucleotide

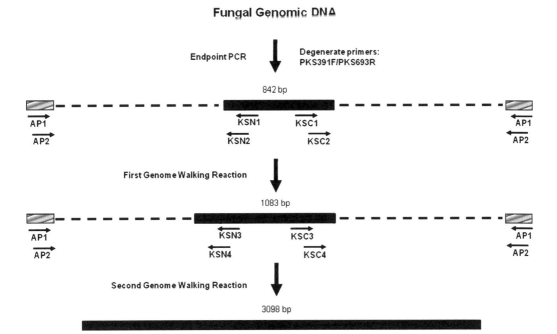

Fig. 2. Work flow of the cloning of reducing PKS. The cloning of this particular gene was initiated using degenerate primers and the DNA fragment was extended by two subsequent genome walking reactions (adapted from Clontech's GenomeWalker™ Universal kit manual).

sequence, in the same fashion as previously described. With the approach described in this chapter, we obtained a final fragment of 3,098 bp (see Fig. 2) which encoded 780 amino acids. Conserved Domain Search (CDS) analysis showed that the amino acid sequence contained a PKS ketoacyl synthase domain from amino acid 221 to 484, and a PKS acetyltransferase domain from amino acid 562 to 778.

4. Notes

1. Some of the supernatants may remain in the original tube containing the protein pellet. It is best to leave this residual liquid in the tube to avoid contaminating the DNA solution with the precipitated protein.

2. Two conserved regions within a group of proteins have to be identified by multiple alignments. These regions should not be very far from each other and should preferably be between 5 and 7 amino acids long. The degree of degeneracy of the region can be calculated to assess the number of different possible PCR products that they might yield. The degree of degeneracy is calculated by multiplying the number of codons that code

for a certain amino acid. For example, glutamic acid (E) is coded by two codons and alanine is coded by four codons, and so on.

Amino acid code

Amino acid	One-letter code	Codons
Alanine	A	GCT, GCC, GCA, GCG
Cysteine	C	TGT, TGC
Aspartic acid	D	GAT, GAC
Glutamic acid	E	GAA, GAG
Phenylalanine	F	TTT, TTC
Glycine	G	GGT, GGC, GGA, GGG
Histidine	H	CAT, CAC
Isoleucine	I	ATT, ATC, ATA
Lysine	K	AAA, AAG
Leucine	L	TTA, TTG, CTT, CTC, CTA, CTG
Methionine	M	ATG
Aspargine	N	AAT, AAC
Proline	P	CCT, CCC, CCA, CCG
Glutamine	Q	CAA, CAG
Arginine	R	CGT, CGC, CGA, CGG, AGA, AGG
Serine	S	TCT, TCC, TCA, TCG, AGT, AGC
Threonine	T	ACT, ACC, ACA, ACG
Valine	V	GTT, GTC, GTA, GTG
Tryptophan	W	TGG
Tyrosine	Y	TAT, TAC
Stop	*	TAA, TAG, TGA

In the *A. rabiei* PKS cloning described in this chapter, the degeneracy of the first conserved region (EAHGTGT) calculated using this method was found to be 4,096 and the degeneracy of the second region (FTGQGAQW) was 2,048. The degeneracy of the first region was improved by using only the first six amino acids (EAHGTG) for primer design. Thus, the degeneracy of the region used for designing the forward primer, EAHGTG, was 1,024 (2*4*2*4*4*4). In the case of the reverse primer, only the last six amino acids from the second conserved region were used for primer design and degeneracy of the region used for designing the reverse primer (GQGAQW) was 256 (4*2*4*4*2*1). We have found that good degenerate primers should be 20–24 bp long with degeneracy values below 2,000. In this case, both forward and reverse primers were 24 bp in size.

We used the degeneracy code found in the Integrated DNA Technologies international product catalog (http://cdn.idtdna.com/Catalog/Catalog_2008_International.pdf). Letter R includes adenine and guanine bases, letter Y includes thymine and cytosine bases, and letter N includes all four bases. The bases of the forward primer were deduced from the genetic code as follows:

Conserved region	Amino acid	Codons	Degenerate codon
EAHGTG	E	GAA, GAG	GAR
	A	GCT, GCC, GCA, GCG	GCN
	H	CAT, CAC	CAY
	G	GGT, GGC, GGA, GGG	GGN
	T	ACT, ACC, ACA, ACG	ACN
	G	GGT, GGC, GGA, GGG	GGN
GQGAQW	G	GGT, GGC, GGA, GGG	GGN
	Q	CAA, CAG	CAR
	G	GGT, GGC, GGA, GGG	GGN
	A	GCT, GCC, GCA, GCG	GCN
	Q	CAA, CAG	CAR
	W	TGG	TGG

Therefore, the sequence of the forward primer was 5′-GAR GCN CAY GGN ACN GGN-3′. The nucleotide sequence of the second conserved region was 5′-GGN CAR GGN GCN CAR TGG-3′ and the sequence of the reverse primer was deduced using BioEdit to generate a reverse complement of the sequence. Thus, the sequence of the reverse primer was 5′-CCA YTG NGC NCC YTG NCC-3′.

3. It is recommended to spread at least two plates with different volumes of the transformed competent cell solution because some plates may show only a few colonies while others may show too many. The chances of finding a positive clone in plates that have very few colonies are less. Also, picking up a single positive clone without contamination of the neighboring colonies is very difficult when there are too many colonies in a plate. We have found that using 50 and 100 μL aliquots helps to overcome these situations.

4. The use of M13 primers with pCR4-TOPO vector ligations will produce amplicons that are 165 bp longer than the original amplicons.

5. Bacterial cells can be picked from selected white colonies using a pipette with a sterile long reach 10-µL pipette tip. For a good PCR amplification, it is important not to use too many cells. Therefore, barely touch the bacterial colony when retrieving the bacterial cells. You should not see the colonies on the tip. Submerge the tip in the PCR mix that has already been dispensed in each PCR tube. Also, we have found that pipetting up and down helps the dislodging of the bacterial cells from the tip surface.

6. Positive colonies can be picked with a sterile long reach 10-µL pipette tip as described in Note 5. The tip can then be propelled into the tube, which can be incubated overnight in an upright position. It is important that no more than 50% of the tip is submerged in the liquid medium. We have found that no more than 4 mL of LB/Amp broth should be used with these tips in order to prevent contamination. Smaller volumes of medium should be dispensed in culture tubes if shorter tips are being used.

References

1. Mizushina, Yoshiyuki, et al (2002) A plant phytoxin, solanapyrone A, is an inhibitor of DNA polymerase β and λ. J Biol Chem 277: 630–638

2. Cox, Russell, et al (2004) Rapid cloning and expression of a fungal polyketide synthase gene involved in squalestatin biosynthesis. Chem Commun 20:2260–2261

3. Fujii, Isao, et al (2005) An iterative type I polyketide synthase PKSN catalyzes synthesis of the decaketide alternapyrone with region-specific octa-methylation. Chem Biol 12:1301–1309

Chapter 40

Construction of Hairpin RNA-Expressing Vectors for RNA-Mediated Gene Silencing in Fungi

Shaobin Zhong, Yueqiang Leng, and Melvin D. Bolton

Abstract

RNA-mediated gene silencing is one of the major tools for functional genomics in fungi and can be achieved by transformation with constructs that express hairpin (hp) RNA with sequences homologous to the target gene(s). To make an hpRNA expression construct, a portion of the target gene can be amplified by PCR and cloned into a vector as an inverted repeat. The generic gene-silencing vectors such as the pSilent1 and pSGate1 have been developed and are available for RNA-mediated gene-silencing studies. In this protocol, we describe construction of hpRNA-expressing constructs using both pSilent1 and pSGate1. With pSilent1, the PCR products of the target gene are inserted into the vector by conventional cloning (i.e., restriction enzyme digestion and ligation). For pSGate1, the PCR products of the target gene are inserted into the vector through the Gateway-directed recombination system. In this chapter, we describe the construction of RNAi vectors for RNA-mediated gene silencing using both pSilent1 and pSGate1.

Key words: RNAi, Gene silencing, Gateway cloning system, Hairpin RNA, pSilent1, pSGate1

1. Introduction

RNA-mediated gene silencing or RNA interference (1) has been found to be a common gene-silencing phenomenon in eukaryotic organisms. In this process, double-stranded RNA (dsRNA), which can be induced by viral RNA, hairpin RNA (hpRNA), or virus-induced gene-silencing encoded RNA (2), is degraded into small interfering RNAs (siRNAs). These siRNAs are incorporated into the RNA-induced silencing complex to target and degrade mRNA of genes with complementary sequence (2–5). RNAi has been demonstrated to silence genes in a number of fungi (for review see ref. (6)). Examples include *Cryptococcus neoformans* (5), *Magnaporthe oryzae* (7–9), *Venturia inaequalis* (10), *Histoplasma*

Melvin D. Bolton and Bart P.H.J. Thomma (eds.), *Plant Fungal Pathogens: Methods and Protocols*,
Methods in Molecular Biology, vol. 835, DOI 10.1007/978-1-61779-501-5_40, © Springer Science+Business Media, LLC 2012

capsulatum (11), *Schizophyllum commune* (12), *Coprinopsis cinerea* (13, 14), *Mortierella alpina* (15), *Dictyostelium discoideum* (16), *Aspergillus* and *Fusarium* species (4), *Bipolaris oryzae* (17), *Cladosporium fulvum* (18–20), and *Ophiostoma novo-ulmi* (21). Most of the studies mentioned above used RNAi vectors containing inverted repeats of the target gene or its partial sequence separated by a spacer for transformation. Construction of these RNAi vectors often involves several cloning steps and thus is laborious and time-consuming. Nakayashiki et al. (9) developed the pHANNIBAL-like silencing vector, pSilent-1 (see Fig. 1), for RNA silencing studies in filamentous fungi. To make an RNAi construct expressing the hpRNA of the target gene with

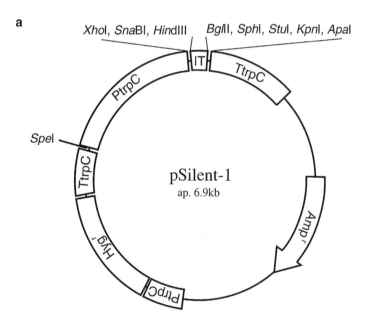

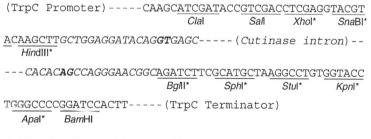

Fig. 1. (**a**) A schematic map of the vector pSilent1. Unique restriction sites are indicated. *AmpR* ampicillin-resistant gene; *HphR* hygromycin-resistant gene; *Intron* intron 2 of the cutinase (CUT) gene from *Magnaporthe oryzae*; *PtrpC Aspergillus nidulans trpC* promoter; *TtrpC A. nidulans trpC* terminator. (**b**) Restriction sites are *underlined*. *Asterisks* indicate unique sites in the silencing vector. *Italic* nucleotides indicate sequence from the CUT gene and *bold letters* represent 5 and 3 splice sites (9).

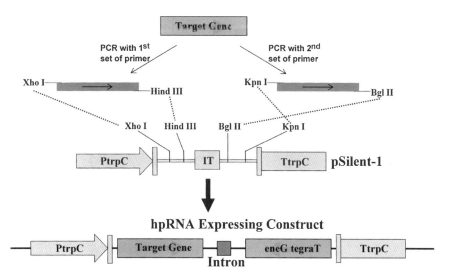

Fig. 2. A flowchart showing the construction of a hairpin (hp) RNA-expressing RNAi vector using pSilent1. A DNA fragment from a target gene was amplified by PCR with primers containing appropriate restriction sites. The PCR product was cloned into each side of the intron in pSilent1 sequentially, resulting in an RNAi construct expressing the hpRNA of the target gene.

pSilent-1, a DNA fragment from a target gene is amplified by PCR with primers containing appropriate restriction sites and cloned sequentially into each side of the intron spacer in the vector (see Fig. 2). The applicability of pSilent-1 was demonstrated in several phytopathogenic ascomycete fungi, including *M. oryzae*, *Colletotrichum lagenarium*, and *B. oryzae* (9, 17). However, pSilent-1 is a restriction enzyme-based cloning vector and therefore may not be suitable for the construction of large numbers of RNAi vectors for high-throughput gene-silencing studies. Recently, a pHELLSGATE-like RNAi vector (pTroya) based on Invitrogen's Gateway technology has been developed and shown to be useful in *C. gloeosporioides* (22). A similar type of RNAi vector (pSGate1) (see Fig. 3) based on the Gateway system has also been developed and used for gene silencing in the spot blotch fungal pathogen of barley and wheat *Cochliobolus sativus* (23). To make an RNAi construct with pSGate1, a portion of the target gene is amplified by PCR with primers containing the *att*B1 and *att*B2 sites, and cloned into an entry plasmid (pDONR221) to create an entry clone by BP reaction (see Fig. 4). The insert in the entry clone is then recombined into both sides of the intron spacer in pSGate1 in opposite orientation with the LR reaction, resulting in an RNAi construct expressing the hpRNA of the target gene (see Fig. 4). In this protocol, we describe the construction of RNAi vectors for RNA-mediated gene silencing using both pSilent1 and pSGate1.

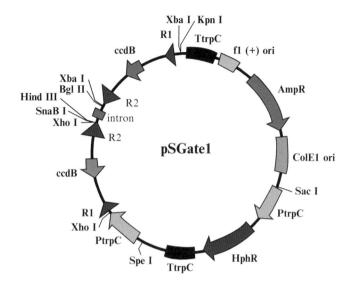

Fig. 3. A schematic map of the Gateway cloning system-based RNAi vector, pSGate1. The *ccdB* cassettes were derived from pHellsgate-12 (24) and cloned into the silencing vector pSilent1 (9). Unique restriction sites are indicated. *AmpR* ampicillin-resistant gene; *HphR* hygromycin-resistant gene; *Intron* intron 2 of the cutinase (CUT) gene from *Magnaporthe oryzae*; *PtrpC Aspergilus nidulans trpC* promoter; *TtrpC A. nidulans trpC* terminator.

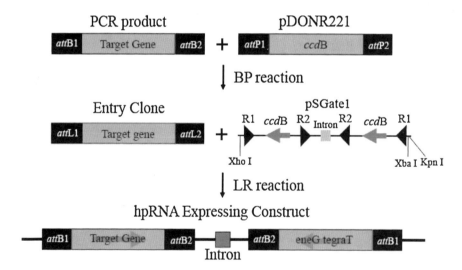

Fig. 4. Flowchart showing the construction of a hairpin (hp) RNA-expressing RNAi vector using pSGate1. A DNA fragment from a target gene is amplified by PCR with primers containing the *att*B1 and *att*B2 sites. The PCR product is cloned into the plasmid pDONR221 to create an entry clone by BP reaction. The insert in the entry clone is then replaced by the *ccdB* genes of pSGate1 by LR reaction, resulting in an RNAi construct expressing the hpRNA of the target gene.

2. Materials

2.1. Equipment and Consumables

1. Bench top microcentrifuge.

2. Environmental incubator shaker.

3. pH meter.

4. Thermal cycler.

5. Electrophoresis equipment.

6. Water bath.

7. Petri plates.

8. Cell culture tubes.

9. Microcentrifuge tubes.

10. PCR tubes.

2.2. Media and Reagents (see Note 1)

1. LB medium: 10 g tryptone, 5 g yeast extract, 10 g NaCl, 20 g of agar (for plates only) to 900 mL deionized H_2O. Adjust pH to 7.0 with NaOH solution. Add deionized H_2O to a final volume of 1 L. Autoclave. Cool to 55–60°C and add 1 mL of ampicillin (100 mg/mL; filter-sterilized with 0.2-µm membrane filter) or kanamycin (50 mg/mL; filter-sterilized with 0.2-µm membrane filter) as necessary.

2. SOC medium: 2% (w/v) tryptone, 0.5% (w/v) yeast extract, 10 mM NaCl, 2.5 mM KCl, 10 mM $MgCl_2$, 10 mM $MgSO_4$, 20 mM glucose. Autoclave.

3. Electrophoresis-grade agarose.

4. *Xho*I and *Hind*III restriction enzymes, including 10× reaction buffers as provided by supplier.

5. T4 ligase, including 10× ligation buffer as provided by supplier.

6. PCR-grade Taq polymerase, including 10× PCR amplification buffer as provided by supplier.

7. PCR-grade dNTPs.

8. Chemically competent *Escherichia coli* cells.

9. Proteinase K: supplied with Gateway® BP or LR kits.

10. TE (10 mM Tris–HCl pH 8.0, 1 mM EDTA).

11. 30%PEG/$MgCl_2$: supplied with Gateway® BP or LR kits.

12. DNA purification Kit.

13. Plasmid miniprep Kit.

14. Gene-specific primers with appropriate restriction sites at the 5′ end (for construction of RNAi vectors with pSilent1) (see Note 2).

15. pSilent1 (for construction of RNAi vectors with pSilent1; see Note 3).

16. Gene-specific primers with *att*B1 and *att*B2 sequences at the 5′ end (for construction of RNAi vectors with pSGate1; see Note 4).

17. pSGate1 (for construction of RNAi vectors with pSGate1; see Note 5).

18. pDONR221 (Invitrogen, Carlsbad, CA).

19. Gateway® BP Clonase® II enzyme mix (Invitrogen).

20. Gateway® LR Clonase® II enzyme mix (Invitrogen).

3. Methods

3.1. Construction of RNAi Vectors with pSilent1

1. Design the first set of primers containing appropriate restriction sites added to the 5′ end of each primer that amplify the sense sequence of the target gene (see Note 2). For example purposes, we have chosen the forward primer to contain the XhoI restriction site and the reverse primer to contain HindIII restriction site.

2. Amplify the targeted gene using the primer pair designed above in a standard PCR reaction following Taq polymerase supplier's recommendations. The thermal cycling conditions are as follows: initial denaturation (95°C, 2 min), followed by 35 cycles of denaturation (94°C, 30 s), annealing (58°C, 30 s; see Note 6), and extension (72°C, 1 min), and then one final cycle of extension (72°C, 10 min).

3. Digest the PCR product with XhoI and HindIII following restriction enzyme supplier's digestion recommendations and buffers. Digestion is performed in a 50 μL reaction which contains 40 μL of PCR product, 5 μL of 10× restriction enzyme buffer, 0.5 μL 100× bovine serum albumin (BSA), 5 units of each enzyme, and water up to 50 μL. The reaction is incubated at 37°C overnight (see Note 7).

4. Purify the PCR product using a commercial PCR purification kit following the manufacturers' recommendations.

5. Linearize pSilent-1 by HindIII and XhoI double-digestion following restriction enzyme supplier's digestion recommendations and buffers.

6. Purify the linearized pSilent-1 vector using a commercial purification kit following supplier's recommendations.

7. Ligate the XhoI/HindIII-digested PCR product in the XhoI/HindIII-digested pSilent1. Ligation is performed in a 10 μL reaction which contains 50 ng of purified PCR product, 100 ng of pSilent-1, 1 μL of 10× ligation buffer, 1 μL of T4 Ligase, and water up to 10 μL. Incubate the reaction at room temperature for 1–2 h (see Note 8).

8. Add 2–5 μL of the ligation reaction to 50 μL of competent E. coli cells and incubate on ice for 30 min. Heat-shock cells by incubating at 42°C for 60 s without shaking. Remove the vial from 42°C water bath and place on ice for 2 min.

9. Add 250 μL of SOC medium and incubate at 37°C for 1 h with shaking at 150 rpm.

10. Spread an aliquot of the transformation reaction onto LB agar plates with 100 μg/mL ampicillin (see Note 9). Incubate plates at 37°C for 24 h.

11. Pick single colonies and inoculate culture tubes containing 2 mL of LB containing 100 μg/mL ampicillin and incubate at 37°C with shaking at 250 rpm overnight (around 14–16 h).

12. Isolate the plasmid DNA using a commercial plasmid purification kit following the manufacturer's recommendations.

13. Verify the integration of the PCR product into pSilent1 by double digestion of the plasmid with *Hind*III and *Xho*I following restriction enzyme supplier's digestion recommendations and buffers. Alternatively, use PCR to confirm integration.

14. Design the second set of primers containing appropriate restriction sites to amplify the sense sequence of the target gene (see Note 2). For example purposes, we have chosen the forward primer to contain the *Kpn*I restriction site and the reverse primer to contain *Bgl*II restriction site.

15. Repeat steps 2–13 above but utilize *Kpn*I and *Bgl*II for the design, integration, and confirmation of the anti-sense copy of the gene to be silenced into pSilent1.

16. Verify the hpRNA expression construct by double digestion (*Xho*I/*Hind*III or *Bgl*II/*Kpn*I) and visualize pattern in standard gel electrophoresis (see Note 10).

17. The verified silencing construct is linearized by *Spe*I and can be used for RNAi via PEG-mediated transformation (see Note 11).

3.2. Construction of RNAi Vectors with pSGate1

1. Design a primer pair with the appropriate attB sequences added at the 5′ end of each primer (see Note 4).

2. Amplify the targeted gene using the primer pair designed above in a standard PCR reaction following Taq polymerase supplier's recommendations. The thermal cycling conditions are as follows: initial denaturation (95°C, 2 min), followed by 35 cycles of denaturation (94°C, 30 s), annealing (58°C, 30 s; see Note 6), and extension (72°C, 1 min), and then one final cycle of extension (72°C, 10 min).

3. Add 75 μL of TE and 50 μL of 30% PEG/MgCl$_2$ solution to 25 μL of the PCR reaction to purify the PCR product. Mix well and centrifuge for 15 min at 14,000 rpm in a microcentrifuge.

4. Remove the supernatant carefully and re-suspend the pellet in 10 μL TE.

5. Quantify the PCR product. The concentration of the purified PCR product should be more than 10 ng/μL (see Note 12).

6. Mix 1–7 μL (15–150 ng) of the purified PCR products with 1 μL of pDONR221, and add TE to the final volume of 8 μL. Keep the reaction on ice and add 2 μL BP Clonase II enzyme mix (see Note 13). Mix well and incubate at 25°C for 2 h.

7. Add 1 μL Proteinase K solution to the sample and incubate at 37°C for 10 min to terminate the reaction.

8. Add 3–5 μL of BP reaction to 50 μL of competent *E. coli* cells and incubate on ice for 30 min. Heat-shock cells by incubating at 42°C for 60 s without shaking. Remove the vial from 42°C water bath and place on ice for 2 min.

9. Add 250 μL of SOC medium and incubate at 37°C for 1 h with shaking at 150 rpm.

10. Spread an aliquot of the transformation reaction onto LB agar plates containing 50 μg/mL kanamycin. Incubate plates at 37°C for 24 h (see Note 9).

11. Pick single colonies and inoculate culture tubes containing 2 mL of LB with 50 μg/mL kanamycin and incubate at 37°C with shaking at 250 rpm overnight (around 14–16 h).

12. Isolate the plasmid DNA using a commercial plasmid purification kit following the manufacturer's recommendations.

13. Carry out PCR using the isolated plasmid DNA as template and the primers designed in step 1 to verify whether the plasmid contains the DNA insert from the target gene. The verified plasmid is then used as entry clone for the LR reaction.

14. Perform LR reaction by mixing 1 μL entry clone (150 ng/μL), 1 μL destination clone (pSGate1, 100 ng/μL), and 6 μL TE. Keep the reaction on ice and add 2 μL LR clonase enzyme mix (see Note 14). Mix the reaction well and incubate at 25°C for 2 h.

15. Add 1 μL Proteinase K solution to the sample and incubate at 37°C for 10 min to terminate the reaction.

16. Add 3–5 μL of the BP reaction to 50 μL of competent *E. coli* cells and incubate on ice for 30 min. Heat-shock cells by incubating at 42°C for 60 s without shaking. Remove the vial from 42°C water bath and place on ice for 2 min.

17. Add 250 μL of SOC medium and incubate at 30°C (see Note 15) for 1 h with shaking at 150 rpm. Spread 100 μL of transformation reaction onto LB agar plates containing 50 μg/mL ampicillin. Incubate plates at 30°C (see Note 15) for 24 h.

18. Pick 6–10 single colonies and inoculate them individually to culture tubes containing LB medium with 50 μg/mL ampicillin. Incubate the cultures at 30°C (see Note 15) with shaking at 250 rpm overnight (around 14–16 h).

19. Isolate the plasmid DNA using a commercial plasmid purification kit following the manufacturer's recommendations.

20. Verify the hpRNA-expressing construct by single restriction enzyme (*Xho*I) and double restriction enzyme (*Bgl*II and *Kpn*I) digestions (see Note 16).

21. The verified silencing construct is linearized by *Spe*I and can be used for RNAi via PEG-mediated transformation (see Note 11).

4. Notes

1. All media used for culturing bacteria should be sterilized by autoclaving at 121°C for 20 min.

2. Two sets of primers must be designed to amplify the target gene. The first primer pair amplifies a fragment that is cloned into the *Hind*III-*Sna*BI-*Xho*I polylinker of pSilent1 and the second primer pair amplifies the same fragment, which is cloned into the *Bgl*II-*Sph*I-*Stu*I-*Kpn*I-*Apa*I polylinker of pSilent1 (see Fig. 1). The region of the target gene to be amplified should not contain the same restriction enzyme sites as those used in primers. Gene-specific primers should be 18–25 bp with melting temperature (T_M) ranging from 55 to 60°C. The size of fragment to be amplified from the target gene should range from 300 to 600 bp to achieve effective gene silencing.

3. pSilent-1 is available from the Fungal Genetics Stock Center (FGSC#634).

4. The forward primer is 5′-GGGGACAAGTTTGTACAAAAAA GCAGGCT-*gene-specific primer*-3′ and the reverse primer is 5′-GGGGACCACTTTGTACAAGAAAGCTGGGT-*gene-specific primer*-3′. Gene-specific sequences should be 18–25 bp with annealing temperature ranging from 55 to 60°C. The size of fragment to be amplified from the target gene should range from 300 to 600 bp to achieve effective gene silencing.

5. pSGate1 is available from the Fungal Genetics Stock Center.

6. Annealing temperature depends on T_M of the primers used.

7. Digestion of PCR productions may take as little as 2 h. However, if a large amount of DNA is digested, an overnight digestion is recommended.

8. For more efficient ligation, the ligation reaction can be incubated at 16°C overnight. The following formula shows how to calculate the amount of insert and vector used in the ligation reaction: amount of insert/100 ng vector = 100 × size of insert (Kb) × 3/size of vector (Kb).

9. When plating *E. coli* cells on LB agar plates, plate with two different volumes (50 and 100 μL) to ensure that at least one plate has well-separated colonies.

10. Double-digestion by *Hind*III and *Xho*I or *Bgl*II and *Kpn*I should generate a fragment with the same size as the PCR product amplified from the target gene. Sequencing could be used to further confirm the insert in the vector construct.

11. The hpRNA-expressing construct cassette can also be released with appropriate restriction enzymes and cloned into an appropriate T-DNA vector for *Agrobacterium*-mediated transformation.

12. After purification, 10 μL of TE is usually added to the purified PCR products and 3 μL of the DNA solution is used for BP reaction. If the concentration of the purified PCR product is too low, a larger volume (50 μL or more) of the PCR reaction can be purified to achieve a higher concentration.

13. Thaw the BP clonase enzyme mix on ice before use and store at –20 or –80°C immediately after use.

14. Thaw the LR clonase enzyme mix on ice before use and store at –20 or –80°C immediately after use.

15. Incubation and culture of bacteria containing the hrRNA-expressing plasmid construct at 30°C is highly recommended to optimize bacterial growth and stabilize the construct.

16. Digestion by single restriction enzyme (*Xho*I) and double restriction enzymes (*Bgl*II and *Kpn*I) should generate the same size of fragments. Sequencing could be used to further confirm the vector. Approximately 50% of the colonies are likely to contain the right construct.

Acknowledgments

The authors thank Dr. H. Nakayashiki (Kobe University, Japan) for generously providing pSilent-1.

References

1. Coppin E, Debuchy R, Arnaise S, Picard M (1997) Mating types and sexual development in filamentous ascomycetes. Microbiol Mol Biol Rev 61:411–428

2. Waterhouse PM, Helliwell CA (2003) Exploring plant genomes by RNA-induced gene silencing. Nat Rev Genet 4:29–38

3. Nakayashiki H, Nguyen QB (2008) RNA interference: roles in fungal biology. Curr Opin Microbiol 11:494–502

4. McDonald T, Brown D, Keller NP, Hammond TM (2005) RNA silencing of mycotoxin production in *Aspergillus* and *Fusarium* species. Mol Plant-Microbe Interact 18:539–545

5. Liu H, Cottrell TR, Pierini LM et al (2002) RNA interference in the pathogenic fungus *Cryptococcus neoformans*. Genetics 160:463–470

6. Kück U, Hoff B (2010) New tools for the genetic manipulation of filamentous fungi. Appl Microbiol Biotechnol 86:51–62

7. Jeon J, Park SY, Chi MH et al (2007) Genome-wide functional analysis of pathogenicity genes in the rice blast fungus. Nat Genet 39: 561–565

8. Kadotani N, Nakayashiki H, Tosa Y, Mayama S (2003) RNA silencing in the phytopathogenic fungus *Magnaporthe oryzae*. Mol Plant-Microbe Interact 16:769–776

9. Nakayashiki H, Hanada S, Quoc NB et al (2005) RNA silencing as a tool for exploring gene function in ascomycete fungi. Fungal Genet Biol 42:275–283

10. Fitzgerald A, van Kan JAL, Plummer KM (2004) Simultaneous silencing of multiple genes in the apple scab fungus, *Venturia inaequalis*, by expression of RNA with chimeric inverted repeats. Fungal Genet Biol 41: 963–971

11. Rappleye CA, Engle JT, Goldman WE (2004) RNA interference in *Histoplasma capsulatum* demonstrates a role for α-(1, 3)-glucan in virulence. Mol Microbiol 53:153–165

12. De Jong JF, Deelstra HJ, Wosten HAB, Lugones LG (2006) RNA-mediated gene silencing in monokaryons and dikaryons of *Schizophyllum commune*. Appl Environ Microbiol 72: 1267–1269

13. Namekawa SH, Iwabata K, Sugawara H et al (2005) Knockdown of *LIM15/DMC1* in the mushroom *Coprinus cinereus* by double-stranded RNA-mediated gene silencing. Microbiology 151:3669–3678

14. Walti MA, Villalba C, Buser RM et al (2006) Targeted gene silencing in the model mushroom *Coprinopsis cinerea* (*Coprinus cinereus*) by expression of homologous hairpin RNAs. Eukaryot Cell 5:732–744

15. Takeno S, Sakuradani E, Tomi A et al (2005) Improvement of the fatty acid composition of an oil-producing filamentous fungus, *Mortierella alpina* 1S-4, through RNA interference with

Δ12-desaturase gene expression Appl Environ Microbiol 71:5124–5128

16. Martens H, Novotny J, Oberstrass J et al (2002) RNAi in *Dictyostelium*: the role of RNA-directed RNA polymerases and double-stranded RNase. Mol Biol Cell 13:445–453

17. Moriwaki A, Ueno M, Arase S, Kihara J (2007) RNA mediated gene silencing in the phytopathogenic fungus *Bipolaris oryzae*. FEMS Microbiol Lett 269:85–89

18. van Esse HP, Bolton MD, Stergiopoulos I et al (2007) The Chitin-Binding *Cladosporium fulvum* effector protein Avr4 is a virulence factor. Mol Plant-Microbe Interact 20:1092–1101

19. Bolton MD, van Esse HP, Vossen JH et al (2008) The novel *Cladosporium fulvum* lysin motif effector Ecp6 is a virulence factor with orthologues in other fungal species. Mol Microbiol 69:119–136

20. van Esse HP, van't Klooster JW, Bolton MD et al (2008) The *Cladosporium fulvum* virulence protein Avr2 inhibits host proteases required for basal defense. Plant Cell 20:1948–1963

21. Carneiro JS, de la Bastide PY, Chabot M et al (2010) Suppression of polygalacturonase gene expression in the phytopathogenic fungus *Ophiostoma novo-ulmi* by RNA interference. Fungal Genet Biol 47:399–405

22. Shafran H, Miyara I, Eshed R et al (2008) Development of new tools for studying gene function in fungi based on the Gateway system. Fungal Genet Biol 45:1147–1154

23. Leng Y, Wu C, Liu Z et al (2011) RNA-mediated gene silencing in the cereal fungal pathogen *Cochliobolus sativus*. Mol Plant Pathol 12:289–298

24. Wesley SV, Helliwell CA, Smith NA et al (2001) Construct design for efficient, effective and high throughput gene silencing in plants. Plant J 27:581–590

Chapter 41

An Unbiased Method for the Quantitation of Disease Phenotypes Using a Custom-Built Macro Plugin for the Program ImageJ

Ahmed Abd-El-Haliem

Abstract

Accurate evaluation of disease phenotypes is considered a key step to study plant–microbe interactions, as the rate of host colonization by the pathogenic microbe directly reflects whether the defense response of the plant is compromised. Although several techniques were developed to quantitate the amount of infection, only a few of them are inherently suitable for large disease screens. Here, I describe an unbiased method to quantitate disease phenotypes which manifest themselves by visible symptoms contrasting with the remaining unaffected parts of the host tissue. The method utilizes a macro plugin written for the image processing program "ImageJ" to calculate two values which determine the disease index for a specific treatment. In case the disease symptoms are not clear, a transgenic pathogenic fungus expressing the GUS gene is suitable for high-throughput disease screens, since staining for GUS activity facilitates an easy detection of the blue-stained pathogen. I illustrate the versatility of this method by analyzing a data set from a functional silencing screening experiment in resistant tomato that was inoculated with a GUS-expressing strain of the fungus *Cladosporium fulvum*. The method calculates a disease index for each silenced plant and thereby provides a basis for the unbiased identification of candidate host genes required for full resistance to this fungus.

Key words: Disease phenotype, GUS, Quantitation, Imaging, Software, ImageJ, Disease index, Phenotype Quant, Clado GUS Quant, *Cladosporium fulvum*, Tomato

1. Introduction

A vital step in studies on plant–microbe interactions is to be able to accurately perform an unbiased evaluation of the severity of a disease phenotype. This is for example required to judge the effect of a specific genotype, mutation, gene knockdown, or a particular treatment on the level of resistance of a host plant to a pathogenic

Melvin D. Bolton and Bart P.H.J. Thomma (eds.), *Plant Fungal Pathogens: Methods and Protocols*,
Methods in Molecular Biology, vol. 835, DOI 10.1007/978-1-61779-501-5_41, © Springer Science+Business Media, LLC 2012

microbe. Several methods exist to achieve this goal. Perhaps the most prominent current approach is based on the quantitation of pathogen-specific transcripts using real-time PCR. When carried out in a control/treatment setup, the relative difference in the amount of colonization by the microbe can be estimated. Although these methods can be applied to a large set of different genera of microbes that are hosted by plants, there are limitations to this approach, which are mainly due to the required optimization to detect a specific pathogen, sampling issues and general laboriousness of the procedure.

Depending on the nature of the microbe, other standard detection techniques exist to determine the rate of microbial colonization of the host tissue. For example, counting the amount of colonies growing from an extract prepared from infected plant material is very common to quantitate bacterial colonization, while propagating microbes from infected stem sections is used to detect and quantitate pathogens which reside in the xylem tissue.

In principle, the method described below can be used for the quantitation of any disease phenotype that causes visible symptoms that contrast with the remaining unaffected parts of the host tissue. When this is not the case, techniques are available to reveal the presence of microbial structures colonizing plant tissues. These methods offer a direct visualization of microbes by utilizing chemical stains which specifically bind to microbial cell wall components. Subsequent steps of fixation and destaining are often required to achieve acceptable contrast allowing detection of the microbe in the treated sample (*1*). Although these direct visualization methods are well established, they are suboptimal for high-throughput screens because they are laborious and require large amounts of hazardous chemicals.

A recent modification of the latter method makes use of biological markers (also known as reporter gene systems) to non-invasively detect wide arrays of pathogens such as viruses, bacteria, and fungi. The marker can be either florescent, such as the green fluorescent protein (GFP), or encompass a visually detectable enzyme-substrate activity such as the beta-glucuronidase (GUS) enzyme. The latter produces an insoluble blue stain from a provided substrate as activity readout. GFP is often used as a marker for microscopic detection and localization of pathogens. In some cases, detection of the fluorescence is possible by the naked eye when using UV as a light source as was for example used for detecting several transgenic viruses (*2, 3*). Visual detection is possible only when the transgenic pathogen produces significant amounts of the fluorescent protein. Accordingly, the fluorescence signal produced by transgenic fungi upon partial colonization of plant tissues showing near-full resistance (a plant of which the resistance has been partially compromised due to, for example, gene knockdown or a mutation) requires microscopic detection. This limitation makes it difficult to use GFP as a quantitative reporter for fungal colonization in high-throughput resistance screens. In contrast, GUS activity can be readily used to detect and quantitate the presence of GUS-transgenic fungi in high-throughput experimental settings. An example is the use of the GUS marker to detect a transgenic strain of *Cladosporium fulvum* in inoculated leaflets of its only host, tomato (*Solanum lycopersicum*) (*4*). This strategy is frequently used to study the function of host genes related to resistance, as it is expected that for example knockdown of genes required for full resistance of tomato to the pathogen will result in increased colonization (*5, 6*). Although detection of the colonizing fungus is easy, as in this case it only requires harvesting infected leaves from treated plants and subjecting them to the GUS assay, the actual unbiased determination of the infection levels and comparison between all the leaflets from one treatment and the controls remains difficult. As a consequence, small differences in infection levels, although biologically reproducible, are difficult to record by only using the GUS reporter system (*6*). These drawbacks of course also hold for the evaluation of natural symptoms or colonization by pathogens visualized in other ways.

Here, I describe an unbiased method to automatically quantitate a disease phenotype. The versatility of the method is illustrated by quantitating the amount of blue stain produced by the GUS reporter. The obtained values are a measure for the rate of infection by a transgenic *C. fulvum-GUS* strain growing in leaves of (partially) resistant tomato plants. These plants were first inoculated with recombinant tobacco rattle virus (TRV), to specifically knock down host target genes by virus-induced gene silencing (VIGS), in order to study their contribution to full resistance. The method makes use of a custom-built color recognition macro plugin for the program ImageJ (*7*). The plugin is termed Phenotype Quant and the variant which was used here is referred to as Clado GUS Quant (CGQ), according to the settings of the image segmentation filters, HSB, which allows specific recognition of the blue stain (see Figs. 1–3). The CGQ macro plugin can run in two modes: single image (CGQ single image) or batch mode (CGQ batch mode). "CGQ single image" is used to analyze a single image and shows the progress of the analysis on screen in real time. It generates the output values in addition to a selection mask (see Figs. 1 and 2) which can be overlaid on the image to recognize the regions which were selected to produce these output values. "CGQ batch mode" carries out the same functions as "CGQ single image" except that it can process a stack of images, runs faster, and does not show the complete progress of the analysis on screen. In both cases, the analyzed images contain tomato leaflets of which some show blue spots indicative of GUS activity and hence colonization by the pathogen. Running "CGQ batch mode" from the plugin menu of ImageJ results in the execution of an automatic identification

Fig. 1. Image analysis using the macro plugin "Clado GUS Quant single image." An image showing leaflets from tomato plants expressing the *Cf-4* resistance gene was opened in ImageJ. Patches of *blue stain* are indicative for leaf colonization by *C. fulvum-*GUS. Although the leaflets originated from an originally resistant plant, Cf-4-mediated resistance to *C. fulvum* was compromised by silencing the *Cf-4* gene using TRV-Cf-4. The location of the installed macro plugin is indicated in the opened "Plugins" menu list.

and quantitation of the area (in pixels) occupied either by the blue stain or by the total leaf surface of one or more images in a given folder. The program generates two output values (see Figs. 2 and 3) for the blue and total leaf areas in each image. The ratio of these two values represents the disease index which can be used to compare infection levels between different treatments. CGQ macro plugins use the existing "Threshold Colour" plugin (*8*) to filter for a specific target in the image within a preset value in the HSB color space filters. This quantitation method was applied to a data set from a functional silencing screening experiment in resistant tomato and was used for the identification of candidate host genes putatively playing a role in resistance to *C. fulvum* (see Fig. 4). Notably, the same plugin can be simply modified to automate the quantitation of other types of spots or patches with different colors, or to identify and quantitate other imaged materials or shapes. This modification only requires changing the values of the used HSB filters to select a new type of target within a specific image (see Note 1).

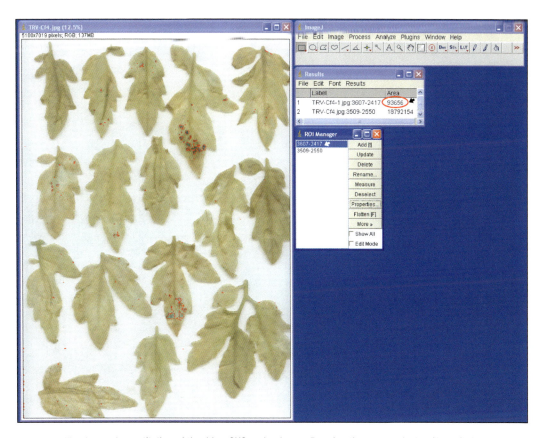

Fig. 2. Identification and quantitation of the *blue*, GUS-stained area. Running the macro plugin "Clado GUS Quant single image" efficiently identifies the blue patches (confined by the *red line*), which are randomly distributed on the tomato leaflets. The total area of the *blue patches* in the image is indicated by the *x* and *y* coordinates (*white arrow*) and calculated in pixels (*black arrow*).

2. Materials

1. Plant organs with visible symptoms. As an example, destained leaflets showing blue spots after performing a GUS assay were used (see Fig. 1).

2. Flatbed photo scanner (for example an "Epson Perfection V500 photo" scanner with LED illumination) (see Note 2).

3. The free image processing software ImageJ (version 1.4 or later), with an installed "Threshold Colour" plugin. ImageJ can be downloaded from: http://imagej.nih.gov/ij/.

4. The macro plugin "Phenotype Quant," including the file "Threshold_Colour.jar" can be downloaded from: http://www.php.wur.nl/UK/downloads (see Fig. 1).

5. Microsoft Excel software program.

Fig. 3. Identification and quantitation of the total leaflet area. Running the macro plugin "Clado GUS Quant single image" identifies the leaflet area (surrounded by *red contour*) after having quantified the *blue patches*. Similar as shown in Fig. 2, the total area of the leaflets in the image is indicated by the *x* and *y* coordinates (*white arrow*) and calculated in pixels (*black arrow*).

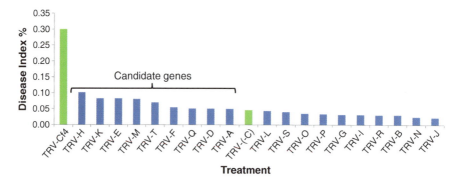

Fig. 4. Identification of host gene candidates putatively required for full resistance to *C. fulvum*. The quantitation method using the macro plugin "Clado GUS Quant batch mode" was applied to a data set of silenced tomato plants and led to the identification of a number of host genes of which silencing compromised full resistance to *C. fulvum*. The different TRV constructs used to silence either the Cf-4 (TRV-Cf-4), nonplant gene (TRV-(-C)) or host genes (TRV-one letter code) are indicated on the *horizontal axis*. The *vertical axis* represents the disease index, which is determined by dividing the pixel values of the total blue area by the total leaf area, for each treatment.

3. Methods

3.1. Plant Material

To help illustrate the utility of this protocol, seedlings of tomato expressing the *Cf-4* resistance gene, which confers resistance to *C. fulvum* strains expressing the effector protein Avr4, were agro-infiltrated to deliver either TRV-Cf-4 to silence the *Cf-4* gene (positive control) or TRV-(-C) to target a nonplant gene (negative control). Importantly, other seedlings were treated with additional TRV-constructs to specifically target several plant genes, as their function in Cf-4/Avr4-triggered resistance is under investigation. Subsequently, silenced plants were inoculated with a transgenic *C. fulvum* strain expressing Avr4 and GUS. After the infection was established, all the inoculated leaves were harvested, subjected to the GUS assay, and finally destained to remove natural leaf pigments. This allows increasing the visual contrast between the blue stain resulting from GUS activity and the rest of the leaf surface (*6*) (see Fig. 1, Note 3).

3.2. Imaging

1. Combine all destained leaflets from each TRV treatment to successively produce one or more images, using a regular flat-bed scanner. Remove the leaflets one by one from the 70% ethanol (where they were temporally stored after destaining) contacting their surfaces with tissue paper to briefly dry them and finally place them on the glass plate of the photo scanner. It is preferable to have the abaxial leaf surface oriented upward. This should reduce light reflection and thereby increase color contrast.

2. Use a white background which does not reflect light.

3. Generate images at a resolution of 600 dots per inch (dpi), 24-bit colors, and in *.JPG as an image compression format. It is preferable to have all leaflets, on the same image, separated from each other by a small distance (see Note 4).

4. After generating an image, rename it so that the name indicates the TRV treatment and the number of the image within that treatment. This name appears in front of the calculated output values for this image in the "Results" window after running the macro plugin.

5. Save all images from the same experiment in one single folder. This folder is accessed by the analysis software in such a way that the pictures in that folder are analyzed automatically one by one when the batch mode (CGQ batch mode) has been selected.

3.3. Image Analysis

1. Install a recent version of ImageJ.

2. Unzip the file named "Phenotype Quant" to generate a folder with the same name.

3. Transfer the folder "Phenotype Quant" to the folder called "plugins" in the installation directory of ImageJ (see Notes 5 and 6).

4. Run ImageJ, go to the menu: Plugins > Phenotype Quant (see Note 7), and run "Clado GUS Quant batch mode." The software will ask you to select the folder where the images to be analyzed are located, this will start the analysis of the images in that folder.

5. After analysis of all the images in the selected folder, go to the results window and then to the menu: File > Save as, and choose a location where you can save the results in Microsoft *.xls format.

6. Open the *.xls file and simply divide the combined total values of the blue area (appearing as the first value in the "Results" window) by the combined total value of the leaf area for each treatment. This generates the disease index for that treatment. Comparison between the disease index of the different treatments and the controls defines the treatments which show a putative effect on resistance (see Fig. 4).

4. Notes

1. The plugin can be calibrated to modify the detection accuracy of the blue GUS stain or to detect another stain, such as chlorotic spots, or different shapes, such as brown necrotic lesions. This will require to open a sample image in ImageJ and run the "Threshold Colour" plugin (by going to menu: plugin > Phenotype Quant > Threshold Colour) and to change the values of the HSB filters until the target object in the image is correctly selected. These values can be entered into the respective fields in the file "Blue_.ijm" in the "Phenotype Quant" folder. You need to use a text editor (for example: Microsoft WordPad) to open this file and edit the values before saving the changes that were made. The default values of the HSB filters (in the file Blue_.ijm) for the recognition of the blue GUS stain are Hue = 70, 255, Saturation = 19, 255, and Brightness = 0, 255. The values of the "Saturation" filter can be adjusted to ~60, 219 in case of suboptimal detection of the blue GUS stain due to the presence of too many brown spots or a faint blue background color on the leaflets. It is essential to analyze images of the controls and the treatments using the same values of the HSB filters otherwise the results of the analysis, in which the treatments are compared with the controls, are not reliable.

2. The use of a photo scanner with LED light technology reduces the warm-up time and therefore allows faster generation of

images. A feature which is useful when plant material from large experiments needs to be handled.

3. After destaining, the leaflets should not contain too much brownish spots as these can slightly interfere with the specific recognition of the blue stain and thus the obtained final disease index. These spots are usually the result of exposing the plants to excessive light, high temperature, or high humidity during the course of the experiment.

4. Increasing the image resolution to values higher than the indicated values or saving the images in the *.TIF compression format may increase the detection sensitivity of the blue stain (within the employed values of the HSB filters). However, these changes will require more memory and longer processing times.

5. If the "Threshold Colour" plugin was preinstalled, which you can check by going to the menu: Plugins > Segmentation, then you need to delete the "Threshold_Colour.jar" from the folder "Phenotype Quant" which you have made previously, otherwise the software will not run properly.

6. Run ImageJ and go to the menu: Edit > Options > Memory & Threads, and increase the memory size available for ImageJ to less than 70% of the available PC-RAM with a minimum of 1,400 MB and up to 1,500 MB. This will allow faster image processing and prevents ImageJ from generating an "Out of memory" message.

7. The "Phenotype Quant" macro plugin contains the following tools:

   (a) Clado GUS Quant batch mode: Calculates both blue and total leaf area separately for a stack of images.

   (b) Clado GUS Quant single image: Calculates both blue and total leaf area separately for a single image and indicates the identified blue or leaf areas for visual inspection (see Figs. 1–3).

   (c) Blue: Calculates only the total blue area.

   (d) Leaf: Calculates only the total leaf area.

   (e) Threshold Colour: A plugin for ImageJ (8) which is used to threshold a colored RGB images in the HSB, RGB, CIE Lab, and YUV spaces.

Acknowledgments

I wish to thank Matthieu Joosten for critically reading the manuscript. This work was supported by a Mosaic grant from the Dutch Organization for Scientific Research (NWO), grant number 017.003.046.

References

1. Narayanasamy P (2001) Plant pathogen detection and disease diagnosis Books in soils, plants, and the environment, Ed 2nd. M. Dekker, New York, 10:0824705912, 518 p.

2. Berg RH, Beachy RN (2008) Fluorescent protein applications in plants. Methods Cell Biol 85:153–177.

3. Leffel SM, Mabon SA, Stewart CN, Jr. (1997) Applications of green fluorescent protein in plants. Biotechniques 23:912–918.

4. Gabriëls SHEJ, Vossen JH, Ekengren SK, van Ooijen G, Abd-El-Haliem AM, van den Berg GCM, Rainey DY, Martin GB, Takken FLW, de Wit PJGM, Joosten MHAJ (2007) An NB-LRR protein required for HR signalling mediated by both extra- and intracellular resistance proteins. Plant J 50:14–28.

5. Stulemeijer IJE, Stratmann JW, Joosten MHAJ (2007) Tomato mitogen-activated protein kinases LeMPK1, LeMPK2, and LeMPK3 are activated during the Cf-4/Avr4-induced hypersensitive response and have distinct phosphorylation specificities. Plant Physiol 144: 1481–1494.

6. Vossen JH, Abd-El-Haliem A, Fradin EF, Van Den Berg GCM, Ekengren SK, Meijer HJG, Seifi A, Bai Y, Ten Have A, Munnik T, Thomma BPHJ, Joosten MHAJ (2010) Identification of tomato phosphatidylinositol-specific phospholipase-C (PI-PLC) family members and the role of PLC4 and PLC6 in HR and disease resistance. Plant J 62:224–239.

7. Wayne Rasband (2010) http://imagej.nih.gov/ij/.

8. Landini G, a modification of Bob Dougherty's BandPass2 filter. http://www.dentistry.bham.ac.uk/landinig/software/software.html.

INDEX

Melvin D. Bolton and Bart P.H.J. Thomma (eds.), *Plant Fungal Pathogens: Methods and Protocols*,
Methods in Molecular Biology, vol. 835, DOI 10.1007/978-1-61779-501-5, © Springer Science+Business Media, LLC 2012